T0297762

CAMBRIDGE LIBRARY COLLECTION

Books of enduring scholarly value

Earth Sciences

In the nineteenth century, geology emerged as a distinct academic discipline. It pointed the way towards the theory of evolution, as scientists including Gideon Mantell, Adam Sedgwick, Charles Lyell and Roderick Murchison began to use the evidence of minerals, rock formations and fossils to demonstrate that the earth was older by millions of years than the conventional, Bible-based wisdom had supposed. They argued convincingly that the climate, flora and fauna of the distant past could be deduced from geological evidence. Volcanic activity, the formation of mountains, and the action of glaciers and rivers, tides and ocean currents also became better understood. This series includes landmark publications by pioneers of the modern earth sciences, who advanced the scientific understanding of our planet and the processes by which it is constantly re-shaped.

Fossil Plants

A.C. Seward (1863–1941) was an eminent English geologist and botanist who pioneered the study of palaeobotany. After graduating from St John's College, Cambridge, in 1886 Seward was appointed a University Lecturer in Botany in 1890. In 1898 he was elected a Fellow of the Royal Society, and was appointed Professor of Botany in 1906. These volumes, published to great acclaim between 1898 and 1919, provide a detailed discussion and study of an emerging science. In the early nineteenth century, research and critical literature concerning palaeobotany was scattered across disciplines. In these volumes Seward synthesised and revised this research and also included a substantial amount of new material. Furnished with concise descriptions of fossil plants, detailed figures and extensive bibliographies these volumes became the standard reference for palaeobotany well into the twentieth century. Volume 2, first published in 1910, contains systematic descriptions of fossil ferns.

Cambridge University Press has long been a pioneer in the reissuing of out-of-print titles from its own backlist, producing digital reprints of books that are still sought after by scholars and students but could not be reprinted economically using traditional technology. The Cambridge Library Collection extends this activity to a wider range of books which are still of importance to researchers and professionals, either for the source material they contain, or as landmarks in the history of their academic discipline.

Drawing from the world-renowned collections in the Cambridge University Library, and guided by the advice of experts in each subject area, Cambridge University Press is using state-of-the-art scanning machines in its own Printing House to capture the content of each book selected for inclusion. The files are processed to give a consistently clear, crisp image, and the books finished to the high quality standard for which the Press is recognised around the world. The latest print-on-demand technology ensures that the books will remain available indefinitely, and that orders for single or multiple copies can quickly be supplied.

The Cambridge Library Collection will bring back to life books of enduring scholarly value (including out-of-copyright works originally issued by other publishers) across a wide range of disciplines in the humanities and social sciences and in science and technology.

Fossil Plants

*A Text-Book for Students of
Botany and Geology*

VOLUME 2

A.C. SEWARD

CAMBRIDGE
UNIVERSITY PRESS

CAMBRIDGE UNIVERSITY PRESS

Cambridge, New York, Melbourne, Madrid, Cape Town, Singapore,
São Paolo, Delhi, Dubai, Tokyo, Mexico City

Published in the United States of America by Cambridge University Press, New York

www.cambridge.org
Information on this title: www.cambridge.org/9781108015967

© in this compilation Cambridge University Press 2010

This edition first published 1910
This digitally printed version 2010

ISBN 978-1-108-01596-7 Paperback

CAMBRIDGE BIOLOGICAL SERIES

GENERAL EDITOR:—ARTHUR E. SHIPLEY, M.A., F.R.S.
FELLOW AND TUTOR OF CHRIST'S COLLEGE, CAMBRIDGE

FOSSIL PLANTS

CAMBRIDGE UNIVERSITY PRESS

London: FETTER LANE, E.C.

C. F. CLAY, Manager

Edinburgh: 100, PRINCES STREET
London: H. K. LEWIS, 136, GOWER STREET, W.C.
Berlin: A. ASHER AND CO.
Leipzig: F. A. BROCKHAUS
New York: G. P. PUTNAM'S SONS
Bombay and Calcutta: MACMILLAN AND CO., Ltd.

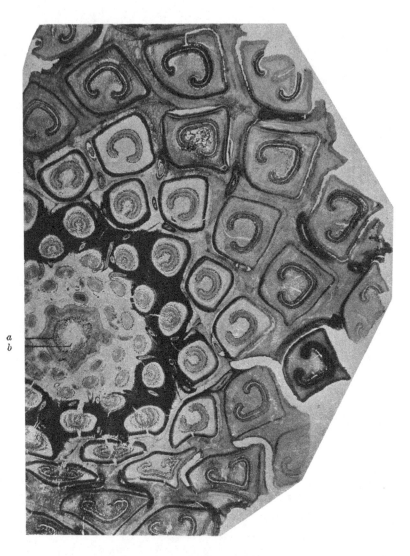

a
b

Part of a transverse section of a Permian Osmundaceous Fern stem, *Thamnopteris Schlechtendalii* (Eichwald). *a*, outer xylem; *b*, inner xylem. For description, see page 329. (After Kidston and Gwynne-Vaughan. Very slightly reduced.)

FOSSIL PLANTS

A TEXT-BOOK FOR STUDENTS
OF BOTANY AND GEOLOGY

BY

A. C. SEWARD, M.A., F.R.S.

PROFESSOR OF BOTANY IN THE UNIVERSITY; FELLOW OF ST JOHN'S
COLLEGE AND HONORARY FELLOW OF EMMANUEL COLLEGE, CAMBRIDGE

WITH 265 ILLUSTRATIONS

VOL. II

CAMBRIDGE:
AT THE UNIVERSITY PRESS
1910

Cambridge:

PRINTED BY JOHN CLAY, M.A.
AT THE UNIVERSITY PRESS.

PREFACE

I REGRET that pressure of other work has prevented the completion of this Volume within a reasonable time since the publication of Volume I. Had Volume II been written ten years ago, the discoveries made in the course of the last decade would have given an out-of-date character to much of the subject-matter. It is more especially in regard to the Ferns and the extinct members of the Gymnosperms that our outlook has been materially altered by recent contributions to Palaeobotany. It is, however, some satisfaction to be able to add that recent progress has been relatively slight in that part of the subject dealt with in the first volume.

The original intention was to complete the whole work in two volumes. Soon after the second volume was begun, it became evident that the remaining divisions of the plant-kingdom could not be included within the compass of a single volume. I decided, therefore, to take the consequences of having embarked on too ambitious a plan of treatment, and to preserve uniformity of proportion by reserving the seed-bearing plants for a third volume. The third volume will include the Pteridosperms, other than those briefly described in the final chapter of the present volume, and other classes of Gymnosperms. I propose also to devote such space as is available within the limits of a text-book to the neglected subject of the geographical distribution of plants at different

stages in the history of the earth. It is my intention to
complete Volume III with as little delay as possible. As I
have written elsewhere, the past history of the Flowering
plants needs special treatment, and anything more than a mere
compilation can be adequately attempted only after considerable research and with the assistance of botanists possessing
a special knowledge of different families of Angiosperms. The
need of a critical examination of available data in regard to
the geological history of this dominant group will not be lost
sight of.

I am well aware that while certain genera have received
an undue share of attention in the present volume, others
have been ignored or treated with scant consideration. For
this inconsistency I have no excuse to offer, beyond the statement that the subject is a large one, and selection is necessary
even though the work consists of three volumes.

The publication in 1909 of a collection of excellent photographs of Palaeozoic Plants, with brief descriptive notes, by
Mr Newell Arber, as one of a series of popular "Nature Books,"
bears striking testimony to the remarkable spread of interest
in the study of the vegetation of the past, which is one of the
outstanding features in the recent history of botanical science.

In the list of illustrations I have mentioned the source of
all figures which have been previously published. I would,
however, supplement the statement of fact with an expression
of thanks to corporate bodies and to individuals who have
allowed me to make use of blocks, drawings, or photographs.

I wish to thank my colleague, Mr A. G. Tansley, for placing
at my disposal several blocks originally published in the pages
of the *New Phytologist*. To Professor Bertrand of Lille and to
his son Dr Paul Bertrand I am indebted for several prints and

descriptive notes of specimens in their possession. My friends
Dr Nathorst of Stockholm and Dr Zeiller of Paris have generously responded to my requests for information on various
points. I wish especially to thank Dr Kidston for several
excellent prints of specimens in his collection and for the loan
of sections. I have profited by more than one examination of
his splendid collection at Stirling. Professor Weiss has generously allowed me to borrow sections from the Manchester
University collections, more especially several which have been
reproduced in the chapter devoted to the genus *Lepidodendron*.
To Professor F. W. Oliver my thanks are due for the loan
of sections from the collection under his charge at University
College. I have pleasure also in thanking Dr Scott, not only
for lending me sections of a Lepidodendron and for allowing
me to use some drawings of *Miadesmia* originally made by
Mrs Scott for reproduction in his invaluable book, *Studies in
Fossil Botany*, but for kindly undertaking the laborious task
of reading the proofs of this volume. It would be unfair to
express my gratitude to Dr Scott for many helpful suggestions
and criticisms, without explicitly stating that thanks to a
friend for reading proofs must not be interpreted as an attempt
to claim his support for all statements or views expressed.
The General Editor of the Series, Mr A. E. Shipley, has also
kindly read the proofs. I am under obligations also for assistance of various kinds to Prof. Thomas of Auckland, New
Zealand, to Mr Boodle of Kew, to Mr D. M. S. Watson of
Manchester, to Mr T. G. Hill of University College, and to
Mr Gordon of Emmanuel College, Cambridge. I am indebted
to the kind offices of Miss M. C. Knowles for the photograph
of the specimen of *Archaeopteris hibernica* in the Irish National
Museum, Dublin, reproduced on page 561.

Many of the illustrations are reproduced from drawings by my wife: those made from the actual specimens are distinguished by the addition of the initials M. S. I am grateful to her also for some improvements in the letter-press. For the drawings made from sections and for some of the outline sketches I am responsible. I have availed myself freely of the facilities afforded by Professor McKenny Hughes in the Sedgwick Museum of Geology for the examination of specimens under the charge of Mr Newell Arber, the University Demonstrator in Palaeobotany. It is a pleasure to add that, as on former occasions, I am indebted to the vigilance of the Readers of the University Press for the detection of several errors which escaped my notice in the revision of the proofs.

A. C. SEWARD.

BOTANY SCHOOL, CAMBRIDGE.
March 12, 1910.

TABLE OF CONTENTS

CHAPTER XII

SPENOPHYLLALES (continued from Volume I.). Pp. 1—16.

CHAPTER XIII

PSILOTALES. Pp. 17—29.

CHAPTER XIV

LYCOPODIALES. Pp. 30—91.

CHAPTER XV

ARBORESCENT LYCOPODIALES. Pp. 92—195.

CHAPTER XVI

SIGILLARIA. Pp. 196—226.

CHAPTER XVII

STIGMARIA. Pp. 227—247.

CHAPTER XVIII

BOTHRODENDREAE. Pp. 248—270.

CHAPTER XIX

SEED-BEARING PLANTS CLOSELY ALLIED TO MEMBERS OF THE LYCOPODIALES. Pp. 271—279.

CHAPTER XX

FILICALES. Pp. 280—323.

CHAPTER XXVII

GENERA OF PTERIDOSPERMS, FERNS, AND *PLANTAE INCERTAE SEDIS*. Pp. 484—580.

LIST OF ILLUSTRATIONS

Several of the illustrations are printed from blocks for which I am indebted to learned societies or to individuals. The sources from which clichés were obtained are mentioned within square brackets.

ERRATA IN VOL. I

Page 16, line 4. For "The North American Tulip tree" read The Tulip
    tree of North America and China.
„ 66, line 2 from the bottom. For "Browera" read Berowra.
„ 127, line 3 and 4 from bottom. For *Achyla* and *Palaeachyla* read
    Achlya and *Palaeachlya.*
„ 145, lines 4 and 5. For "Upper Greensand" read Lower Eocene.
„ 162, line 3 from bottom. For "*Corallina barbata*" read *Cymopolia
    barbata.*
„ 170, line 20. For "sporangiaphore" read sporangiophore.
„ 185, line 2. The genera *Udotea* and *Halimeda,* members of the
    Siphoneae, are incorrectly included under the Corallinaceae.
„ 191, line 11 from bottom. Omit *Chondrus crispus,* which is one of
    the Florideae and not a Brown Alga.
„ 202, line 13. For "*Halmeda*" read *Halimeda.*
„ 250, line 11. For "three" read the.
„ 381, line 10. For "*Calamopytus*" read *Calamopitys.*

CHAPTER XII[1].

SPHENOPHYLLALES (*concluded*).

Sphenophyllum.

THE account of the Sphenophyllales given in the first volume[2] of this work must be extended and somewhat modified in the light of recent work on the fertile shoots of *Sphenophyllum.*

Sphenophyllostachys Dawsoni (Will.) was described as consisting of an axis bearing superposed whorls of bracts connate at the base in the form of a shallow funnel-shaped collar giving off from the upper surface and close to the axis of the cone two concentric series of sporangiophores. Occasionally there are three series, as represented in fig. 112. In another type of strobilus, *Sphenophyllostachys Römeri*[3] each sporangiophore terminates in two pendulous sporangia (fig. 113, A; see also fig. 107, C, vol. I.). It has already been pointed out that the common occurrence of detached strobili necessitates their description under distinct specific names; it is only by a rare accident that we can assign fossil cones to their vegetative shoots. There are, however, reasons for believing that *Sphenophyllostachys Dawsoni* is the strobilus of the plant originally described by Sternberg[4] from impressions of foliage-shoots as *Rotularia cuneifolia.* Another difficulty presented by petrified material is that of determining, with certainty, whether two imperfect specimens, differing from one another in features which do not appear to

---

[1] The full titles of books and papers referred to in footnotes distinguished by the addition of A after the date are given in the Bibliography at the end of Volume I.

[2] Chap. XI.   [3] *ibid.* p. 405.

[4] Sternberg (23) A. p. 33, Pl. XXVI. figs. 4 *a*, 4 *b*.

be of sufficient importance to warrant specific separation, are forms of one species or portions of specifically distinct cones. It has been pointed out by Scott[1] that the strobilus known as *Sphenophyllostachys Dawsoni* probably includes two distinct species, one being the cone of *Sphenophyllum cuneifolium* Sternb., and the other the cone of *S. myriophyllum* Crép[2]. The stem of *S. myriophyllum* agrees anatomically with the type known as *Sphenophyllum plurifoliatum* Will. and Scott[3].

FIG. 112. Sketch of a radial longitudinal section of *Sphenophyllostachys*. There are usually two concentric series of sporangia on the sporophylls, not three as shown in the figure. The upper figure (after Zeiller) shows the linear bracts in surface-view.

In addition to the two types of cone already mentioned, *Sphenophyllostachys Dawsoni* and *S. Römeri*, others have been described by Kidston from carbonised impressions. One of these is the fertile branch of *Sphenophyllum majus*[4]. The basal portions of the bracts of each whorl form a narrow collar round the axis of the cone; the free portion of each bract consists of a lamina divided into two equal bifid lobes bearing

[1] Scott (05) p. 34.
[2] Zeiller (88) A. Pl. LXII. figs. 2—4.
[3] Vol. I., p. 397.
[4] Kidston (01) p. 128, fig. 25; (02) p. 361, fig. 13.

on its upper surface one group, or possibly two groups, of four
sessile sporangia between the narrow coherent bases of the
laminae and the sinus between the terminal lobes (fig. 113, C).
Another characteristic feature is the greater length of the
internodes; this renders the cone less compact and less sharply
differentiated from the vegetative shoots than those of other
species. A specimen in Dr Kidston's collection illustrates the

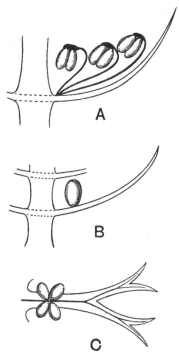

FIG. 113.   A. *Sphenophyllostachys Römeri.*   (Solms-Laubach.)
            B. *Sphenophyllum trichomatosum* Stur.
            C. *Sphenophyllum majus.* Bronn. (A—C. After Kidston.)

peculiar character of the fertile portion of this species; it consists of an axis bearing a succession of lax sporophylls succeeded
above and below by whorls of sterile leaves. In this species,
therefore, we cannot speak of a compact strobilus at the end of
a shoot of limited growth, but of axes in which sterile and
fertile leaves are borne alternately[1], a condition recalling the

---

[1] Bower (08) p. 404, fig. 221.

alternation of foliage leaves and sporophylls in *Tmesipteris* and in *Lycopodium Selago.*

Another form of cone, also from the Middle Coal Measures, is referred by Kidston to *Sphenophyllum trichomatosum* Stur[1] (fig. 113, B): this is characterised by the more horizontal position of the bracts, which "do not appear to be so much or so suddenly bent upwards in their distal portion as in some other species of *Sphenophyllum,*" and by sessile sporangia borne singly on the upper face of each bract.

A more recent addition to our knowledge of the fertile shoots of *Sphenophyllum* is due to Scott who has described a new type of cone under the name *Sphenophyllum fertile*[2]. The

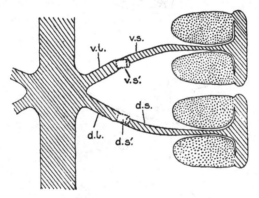

FIG. 114. *Sphenophyllostachys fertilis* (Scott). (After Scott.) Diagram of a node in longitudinal section, showing one sporophyll and the base of the opposite one. *v.l.* ventral lobe of sporophyll; *v.s.* one of the segments into which it divides; *v.s'.* stump of another segment; *d.l.* dorsal lobe; *d.s., d.s'.* segments of dorsal lobe.

petrified specimen on which the species was founded was discovered by Mr James Lomax in the Lower Coal Measures of Lancashire; it represents a portion of a cone 6 cm. long and approximately 12 mm. broad. The axis contains a single vascular cylinder agreeing in essentials with the type of stem structure known as *Sphenophyllum plurifoliatum.* The nodal regions, which exhibit the slight swelling characteristic of the genus, bear several (probably twelve) appendages connate at

[1] Kidston (91) p. 59, Pl. I. ; (01) p. 123, fig. 22.
[2] Scott (05).

the base and forming a narrow flange encircling the axis.
Each bract, the base of which forms part of the narrow collar
surrounding the axis, consists of two lobes, ventral and dorsal,
divided palmately into several (sometimes four) segments or
sporangiophores (fig. 115). Each sporangiophore terminates
distally in an oblong or oval lamina bearing two sporangia on
its adaxial face (fig. 114). The space between the axis and the
periphery of the cone is thus occupied by crowded peltate
laminae, each with its pair of sporangia. A single vascular
bundle supplies each sporangiophore and bifurcates in the

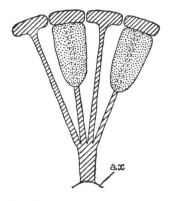

Fig. 115. *Sphenophyllostachys fertilis* (Scott). (After Scott.) Diagram of a
single sporophyll as it would appear in a transverse section of the cone;
showing one lobe (dorsal or ventral). *ax*, part of axis to which the
sporophylls are attached.

distal lamina into two branches which extend to the bases of
the sporangia. The sporangia agree in structure with those of
other species of *Sphenophyllum*: the spores are of one size and
elliptical, characterised by the presence of several sharp ridges
or flanges encircling the spore-wall in the direction of the major-
axis. *Sphenophyllostachys fertilis* differs from all previously
recorded types in the absence of sterile bracts. The appendages
of the cone-axis are all fertile, a striking contrast to the
differentiation into protective and sporangia-bearing bracts which
constitutes a constant feature in the cones of *Sphenophyllum*
and *Calamites*. It is possible, as Scott suggests, that the
absence of sterile segments is the result of modification of the

more usual type of strobilus; instead of the dorsal and ventral
lobes of the bracts sharing between them the duties of protection and spore-production, the whole of each bract is constructed
on the plan of the maximum spore-output, the laminar terminations of the sporangiophores serving the purpose of protection.
The cone may be described as more specialised than the normal
type of strobilus for reproductive purposes[1].

It has been stated, on evidence which is unsatisfactory, that
Sphenophyllum possesses two kinds of spores. While regarding
the genus as homosporous on the evidence before us, it is

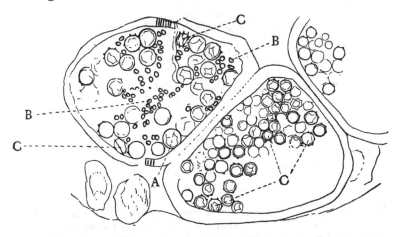

Fig. 116.   *Sphenophyllostachys Dawsoni.* (After Thoday.)   A. Larger spores;
B, abortive spores; C, mature spores showing the characteristic spines.

interesting to find that cases occur in which the spores in the
same sporangium exhibit a marked difference in size. Attention
has been called by Williamson and Scott[2] to variation in the
dimensions of spores: a more pronounced difference in size has
been recorded by Mr Thoday[3] who gives 120μ as the maximum
and 90μ as the minimum diameter of the spores in a cone of
Sphenophyllostachys Dawsoni. The presence of several abortive
spores in the sporangium (fig. 116) containing the larger
spores favours the view that this difference in size may be the
first step towards the development of heterospory.

¹ See also Browne, Lady Isabel (09) p. 4.
² Williamson and Scott (94) A. p. 911.   ³ Thoday (06).

It is clear that the types of strobilus designated *Spheno-phyllostachys* (figs. 112—114) present a divergence of characters too great to be comprised under one genus; but in the absence of fuller information, we cannot do otherwise than follow the only logical custom of grouping them together as examples of strobili borne by plants which, in the present state of our knowledge, are most conveniently referred to the genus *Spheno-phyllum*.

Cheirostrobus.

This generic name was applied by Dr Scott[1] to a calcified cone obtained by Mr James Bennie in 1883 from the Lower Carboniferous plant-beds of Pettycur near Burntisland on the Firth of 'Forth. *Cheirostrobus* is distinguished from *Spheno-phyllostachys* by its greater breadth (3·5 cm.); externally it agrees more closely with the fertile shoots of *Lepidodendron* than with those of *Sphenophyllum*. A single vascular cylinder having the form of a fluted Doric column (fig. 117, B, *x*) occupies the axis of the cone: it consists for the most part of reticulate tracheae which tend to assume a short or isodiametric form in the central region; the smaller protoxylem tracheids with the spiral form of pitting constitute the sharp and prominent ridges at the periphery of the xylem-cylinder. In the outer part of the cylinder the metaxylem[2] consists exclusively of tracheae, but towards the centre of the axis these are associated with numerous parenchymatous cells.

The xylem is therefore centripetal in origin as in *Spheno-phyllum* and in nearly all recent and fossil members of the Lycopodiales. In the type-specimen of *Cheirostrobus* the vascular cylinder of the cone consists entirely of primary xylem, but secondary xylem has been found in a more recently discovered specimen[3]. Secondary xylem occurs also in the peduncle of the cone. No appreciable remains of phloem have

---

[1] Scott (97) A. ; see also Scott (00) p. 106.

[2] The term metaxylem may be conveniently applied to the primary xylem other than protoxylem; the latter is usually but by no means invariably characterised by spiral thickening bands.

[3] Scott (05) p. 21 (footnote).

been found. The cortex consists of slightly elongated rather thick-walled tissue containing secretory sacs. Crowded superposed whorls of bracts (or sporophylls), usually twelve in each whorl, are borne on the axis and each sporophyll receives a single vascular bundle from one of the vertical ridges of the xylem column (fig. 117, A, *lt*). The members of each whorl are connate at the base: from this narrow collar each sporophyll

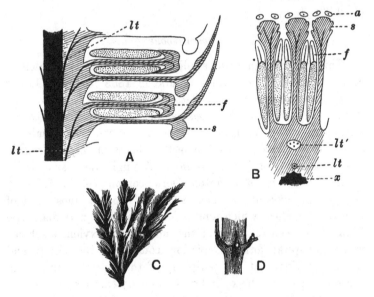

Fig. 117. A, B. *Cheirostrobus pettycurensis* Scott. (After Scott.)
C, D. *Pseudobornia ursina* Nath. (After Nathorst.)

A. Diagrammatic radial longitudinal section of part of the cone-axis and two sporophylls. *lt*, bundle passing out to sporophyll; *f*, fertile segment of sporophyll showing two sporangia; *s*, sterile (lower) segment.
B. Part of transverse section. *x*, stele; *lt*, *lt'*, bundles on their way to sporophylls; *a*, tips of sterile segments of lower sporophylls.
C. Palmately branched leaf (½ natural size).
D. Node of stem showing leaf-bases.

branches into an upper or dorsal and a lower or ventral limb (fig. 117, A, *f* and *s*). Each limb divides palmately at a short distance from its origin into three slender segments, which extend in a horizontal direction and terminate in large laminar expansions (fig. 117, B, *s*) to afford a protective covering to the surface of the cone. The upper set of three segments, constituting sporangiophores (fig. 117, A, B, f) or fertile divisions of the sporophyll, expand distally into comparatively bulky laminae; each of these bears on its adaxial face four diagonally placed outgrowths which form the short pedicels of very long and narrow sporangia. The three lower segments—the sterile divisions of the sporophylls—(fig. 117, A, B, s) are similar to the upper set except in their greater length and in the kite-shaped form of their distal laminae which are provided with lateral lobes. The single vascular strand which supplies each sporophyll is represented at lt in fig. 117, B; at lt' the strand has divided into four, the three upper bundles in the figure supply the sterile segments and the single lower bundle ultimately divides into three which supply the fertile segments. A pair of blunt processes (fig. A, s) extend downwards over the ends of the underlying fertile lamina and two slender prolongations extend upwards through several internodes.

An economical arrangement of the long and narrow sporangia and of the sporophyll-segments between the axis and the periphery of the cone is rendered possible by the interlocking of the sterile and fertile segments by means of a groove in the upper face of the latter for the accommodation of the former. The sporangia are characterised by their unusually long and narrow form: the length of a sporangium may reach 1 centimetre. In the structure of the wall the sporangia of *Cheirostrobus* agree closely with those of *Calamostachys*[1] and *Sphenophyllostachys*. The spores are of one size only. The vascular cylinder of the peduncle, originally described by Williamson[2] as the peduncle of a large *Lepidostrobus* (the cone of *Lepidodendron*), is characterised by the presence of a short radially disposed zone of secondary tracheids, a feature, as Scott points out, which may extend into the axis of the cone. It is noteworthy that the protoxylem elements are not always external, but occasionally occur internal to one or two of the outermost metaxylem tracheae: the usual exarch[3] structure of

[1] Vol. I. p. 354, fig. 95, C.
[2] Williamson (72) Pl. XLIV. p. 297, figs. 29, 30.
[3] 'Exarch' denotes that the protoxylem is on the outside of the primary xylem; 'endarch' that it is on the inner edge or in a central position; 'mesarch' that it is internal, either near the inner or the outer edge of the metaxylem.

the central cylinder is not therefore absolutely constant, but may be replaced by a mesarch arrangement.

The presence of a few sterile leaves on the peduncle below the fertile portion of the cone, which agree in their lobed laminae with the sporophylls, is the only fact which we possess as to the form of the vegetative characters of the genus.

The above description is sufficient to indicate the extraordinary complexity and high degree of specialisation of *Cheirostrobus*. The sporophylls, with their trilobed segments, and the crowded sporangia of exceptional length attached only by a narrow base constitute striking peculiarities of the genus.

It is unfortunate that we are still without any satisfactory evidence as to the nature of the plant the cones of which have been made the type of a new genus and a new family. *Cheirostrobus* affords an interesting example of a type of reproductive shoot constructed on a plan *sui generis*, and may be classed with some other extinct genera as instances of the production in the course of evolution of architectural schemes which appear to have been ill adapted for competition with equally efficient though much simpler types. But the discovery of these isolated forms of restricted geological range among the relics of the Palaeozoic vegetation frequently supplies a key to phylogenetic problems. *Cheirostrobus* by its complex combination of features characteristic of the Equisetales, the Lycopodiales and the genus *Sphenophyllum* throws a welcome light on the inter-relationships of groups which represent divergent series. The combination of morphological features in this generalised type led the author of the genus to describe it as a descendant of an old stock which existed prior to the divergence of the Equisetales and Lycopodiales.

The discovery of this new type of strobilus naturally led to a search among Lower Carboniferous plants for vegetative shoots exhibiting characters conformable with the whorled and branched leaves of *Cheirostrobus*. In *Sphenophyllum* we have a genus obviously comparable with *Cheirostrobus* as regards the form and disposition of the leaves, but the differences between the cones and the striking similarity of the vascular cylinder of the latter to that of *Lepidodendron* demonstrate conclusively that we must look elsewhere for the vegetative members of the plant which produced cones of the *Cheirostrobus* type.

In 1902 Professor Nathorst[1] instituted the generic name *Pseudobornia* for plants of which imperfect examples had previously been referred by Heer[2] to *Calamites* under the name *C. radiatus.* Heer's plants were obtained from Upper Devonian rocks of Bear Island in the Arctic seas and additional specimens were brought from the same locality by the Swedish Polar Expedition of 1898. *Pseudobornia* possesses jointed stems (fig. 117, D) bearing whorled and shortly stalked leaves, often four in number, at each node. The leaves are palmately branched with fine serrated edges (fig. 117, C). Certain specimens, which are no doubt correctly described by Nathorst as cones, are characterised by a thick axis bearing whorled leaves with sporangia on their lower surfaces, but the material is not sufficiently well preserved to render possible a recognition of structural details. It has been suggested by Scott that *Pseudobornia* may possibly be referable to the Sphenophyllales and that the stem of *Cheirostrobus* "may have had something in common with" Nathorst's genus[3]. The beds in which the stems occur are of Upper Devonian age, while *Cheirostrobus* was found in Lower Carboniferous rocks: this difference in age is not, however, a serious objection to the validity of the comparison. We cannot do more than express the view that *Pseudobornia*, so far as can be ascertained without an examination of petrified material or of more perfect impressions of strobili, exhibits vegetative features not inconsistent with the morphological characters of the fertile shoots known as *Cheirostrobus.*

The institution of a special group-name for the reception of *Sphenophyllum* is justified by the sum of its morphological features, which do not sufficiently conform to those of any existing group of Pteridophytes to warrant its inclusion in a system of classification based on recent genera. In the case of *Cheirostrobus* we are limited to the characters of the cone and

---

[1] Nathorst (02) p. 24.   [2] Heer (71) p. 32, Pls. I—VI.
[3] Scott (07) p. 155.

its peduncle. The suggestion that the Devonian fossils known
as *Pseudobornia* may represent the foliage shoots of a plant
closely related to *Cheirostrobus* has still to be proved correct.
Although we may find justification in the highly complex and
peculiar structure of *Cheirostrobus* for the recognition of the
genus as a type of still another group of Pteridophytes, it would
be unwise to take this step without additional knowledge.

The undoubted similarity between *Cheirostrobus* and
Sphenophyllum coupled with striking points of difference favours
the inclusion of the two genera in distinct families placed, for
the present at least, in the group Sphenophyllales.

Group **SPHENOPHYLLALES**.

Sphenophylleae: genus *Sphenophyllum*.

Cheirostrobeae: genus *Cheirostrobus*.

It has recently been proposed to include the family
Psilotaceae, comprising the two recent genera *Psilotum* and
Tmesipteris, as another subdivision of the Sphenophyllales.
This proposal had been made by Professor Thomas[1] primarily
on the ground that the sporophylls of *Tmesipteris* and *Psilotum*
appear to afford the closest parallel among existing plants to
the peculiar form of sporophyll characteristic of the Sphenophyllales. The morphological interpretation of the sporophylls
of both *Sphenophyllum* and *Cheirostrobus* has been the source
of considerable discussion[2]. If we regard each sporophyll as a
leaf with two lobes, one fertile and one sterile, except in the
case of *Sphenophyllostachys fertilis* in which both are fertile, an
obvious comparison may be made with the fern *Ophioglossum*;
but the difference between a single fern frond, consisting of a
comparatively large sterile lamina bearing a fertile branch
composed of a long axis with two rows of sporangia embedded
in its tissues, and the whorled sporophylls of *Sphenophyllum* is
considerable.

A brief reference may be made to the principal reasons which
have led to the suggestion that the Psilotaceae should be included

---

[1] Thomas, A. P. W. (02) p. 350.   [2] Bower (04) p. 227; (08) p. 424.

in the Sphenophyllales. The shoots of *Tmesipteris* bear simple
foliage leaves spirally disposed on a slender axis, and in
association with these occur sporophylls consisting of a short
axis bearing a pair of small lobes and a bilocular synangium[1]
(fig. 120, B). The synangium is seated on a very short stalk
given off from its sporophyll at the base of the pair of laminae:
the synangium with its short stalk may be spoken of as the
sporangiophore. In most cases the synangium appears to be
sessile on the sporophyll, but occasionally the much reduced stalk
is prolonged and forms an obvious feature. Dr Scott[2] suggested
that the Tmesipteris synangium with its axis may correspond
to the ventral lobe (or sporangiophore) of *Sphenophyllum*. In the
latter genus the whorled sporophylls consist in most species of
a dorsal and a ventral lobe, the latter serving as a sporangiophore bearing one or more sporangia; in *Tmesipteris* the
sporophylls are spirally disposed and each consists of a bilobed
sterile portion bearing a septate sporangium or bilocular
synangium on a very short ventral lobe. Professor Bower[3], in
his account of the development and structure of the sporophylls
of *Tmesipteris*, drew attention to the comparatively frequent
occurrence of abnormal sporophylls and spoke of the plant as
unstable. More recently Professor Thomas[4] of Auckland has
carefully examined living plants, with the result that variations
of different kinds are proved to be exceedingly common. He
finds that sporophylls occur which exhibit repeated dichotomy
of the axis (fig. 120, D, F) and thus each may bear four instead
of two leaf-lobes and three synangia, one at the first fork and
one at each of the forks of the second order[5].

Other abnormalities occur in which the synangium is raised
on a distinct stalk instead of being more or less sessile at
the point from which the leaf-lobes diverge. A third form of
departure from the normal is that in which there is no synangium
on the bilobed sporophyll, its place being taken by a leaf-lobe.
The deduction from the occurrence of these abnormalities is
that the synangium of *Tmesipteris* represents a ventral leaf-

[1] See p. 19.   [2] Scott (00) p. 499.
[3] Bower (94) p. 545.   [4] Thomas (02). See also Sykes (08).
[5] Sykes (08).

lobe, as Scott suggested. Professor Thomas draws attention to
the resemblance between *Tmesipteris* sporophylls and the foliage-
leaves of *Sphenophyllum*, which are either simple with dichotomously branched veins or the lamina is deeply divided into
two or more segments. In some types of *Sphenophyllostachys*
the bracts are simple (*S. Dawsoni*), but in others (*Sphenophyllum
majus*, fig. 113, C) they are forked like the foliage-leaves
and bear a close resemblance to the abormal sporophylls of
Tmesipteris. Moreover, in *Sphenophyllostachys Römeri* (fig.
113, A) each ventral lobe of a sporophyll bears two sporangia,
a condition almost identical with that represented by the
occasional occurrence of a synangium on a comparatively
long stalk in *Tmesipteris*. Similarly the more elaborate
sporophylls of *Cheirostrobus* may be compared with the branched
sporophylls of *Tmesipteris* (fig. 120). This agreement between
the sporophylls of the Palaeozoic and recent genera acquires
additional importance from the very close resemblance between
the exarch stele of *Sphenophyllum* and that of the genus
Psilotum, which conforms to the Palaeozoic type not only in
the centripetal character of the primary xylem and in its exarch
structure, but also in the occasional occurrence of secondary
xylem[1], and in the stellate form of its transverse section. The
occasional mesarch structure of the stele of *Cheirostrobus* finds
a parallel in the mesarch xylem groups in the stem of *Tmesipteris*.
It is thus on the strength of these resemblances that Thomas
and Bower would remove the Psilotaceae from the group
Lycopodiales and unite them with *Sphenophyllum* and *Cheirostrobus* in the Sphenophyllales. While admitting the validity
of the comparison briefly referred to above, I prefer to retain
the Psilotaceae as a division of the Pteridophyta including only
Psilotum and *Tmesipteris*.

In his recent book on *The Origin of Land Flora*, Prof. Bower
raises objection to the use of the term ventral lobe in speaking of
the sporangium-bearing stalk or sporangiophore borne on the
sporophyll of *Sphenophyllum*. He points out that the use of
this term implies the derivation of the sporangiophore by
metamorphosis of part of a vegetative leaf, an opinion untenable

---

[1] Boodle (04) ; see *postea* p. 21.

in the absence of proof. The designation sporangiophore is no doubt preferable to that of ventral lobe as it carries with it no admission of particular morphological value; as a further concession to a non-committal attitude we may provisionally at least regard a sporangiophore as an organ *sui generis* "and not the result of modification of any other part[1]."

The view put forward by Prof. Lignier[2] that the Sphenophyllales are descendants of primitive ferns is not convincing, and his comparison of *Sphenophyllum* with *Archaeopteris* lacks force in view of our ignorance as to the nature of the reproductive organs of the latter genus. That the Sphenophyllales are connected with the Equisetales and with the Psilotales by important morphological features is clear; but the comparison between the sporophylls of the extinct genera with those of the existing genus *Tmesipteris*, though helpful and possibly based on true homology, cannot be considered as settling the morphological value of the sporangiophores of *Sphenophyllum* and *Cheirostrobus*.

I do not propose to discuss at length the different views in regard to the morphological nature of the sporangiophore of *Sphenophyllum*. The comparison, which we owe in the first instance to Scott, with the synangium of the Psilotales with its short stalk, though not accepted by Lignier as a comparison based on true homology, is one which appeals to many botanists and is probably the best so far suggested. The further question, whether these sporangiophores are to be called foliar or axial structures is one which has been answered by several authors, but it is improbable that we shall soon arrive at a decision likely to be accepted as final. Discussions of this kind tend to assume an exaggerated importance and frequently carry with them the implication that every appendage of the nature of a sporangiophore can be labelled either shoot or leaf. We treat the question from an academic standpoint and run a risk of ignoring the fact that the conception of stem and leaf is based on morphological characteristics, which have been evolved as the result of gradual differentiation of parts of one originally

---

[1] Bower (08) p. 426.   [2] Lignier (03); (08).

homogeneous whole.   There is much that is attractive in the
view recently propounded by Mr Tansley that a leaf is not an
appendicular organ differing *ab initio* from the axis on which it
is borne, but that it is in phylogenetic origin a "branch-system
of a primitive undifferentiated sporangium-bearing thallus[1]."
Admitting the probability that this view is correct, our faith
in the importance of discussions on the morphological nature of
sporangiophores is shaken, and we realise the possibility that
our zeal for formality and classification may lead to results
inconsistent with an evolutionary standpoint[2].

[1] Tansley (08) p. 26, who refers to similar views held by Potonié and by
Hallier.
[2] On the morphology of Sporangiophores, see also Benson (08[2]) and
Scott, D. H. (09) p. 623.

CHAPTER XIII.

PSILOTALES.

THE two recent genera *Psilotum* and *Tmesipteris* are usually spoken of as members of the family Psilotaceae which is included as one of the subdivisions of the Lycopodiales. It is probable, as Scott[1] first suggested, that these two plants are more nearly allied than are any other existing types to the Palaeozoic genus *Sphenophyllum*. We may give expression to the undoubted resemblances between *Tmesipteris* and *Psilotum* and the Sphenophyllales by including the recent genera as members of that group, originally founded on the extinct genus *Sphenophyllum*; this is the course adopted by Thomas[2] and by Bower[3]: or we may emphasise the fact that these two recent genera differ in certain important respects from *Lycopodium* and *Selaginella* by removing them to a separate group, the Psilotales. The latter course is preferred on the ground that the inclusion of *Psilotum* and *Tmesipteris* in a group founded on an extinct and necessarily imperfectly known type, is based on insufficient evidence and carries with it an assumption of closer relationship than has been satisfactorily established.

The genus *Tmesipteris* (fig. 120, A) is represented by a single species *T. tannensis* Bertr.[4] which usually occurs as an epiphyte on the stems of tree-ferns in Australia, New Zealand, and Polynesia. *Psilotum*, with two species *P. triquetrum*

---

[1] Scott (00). [2] Thomas (02). [3] Bower (08) p. 398.
[4] Dangeard (91) and Bertrand, C. E.(81) recognise other species of *Tmesipteris*, but it is doubtful how far such differences as exist are worthy of specific recognition.

Sw. (fig. 118) and *P. complanatum* Sw., flourishes in moist tropical regions of both hemispheres, growing either on soil rich in organic substances or as an epiphyte. Both genera are considered to be more or less saprophytic.

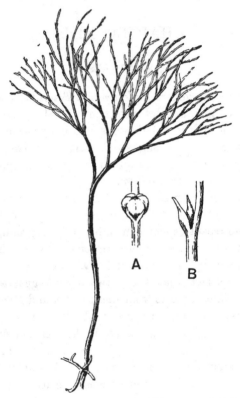

FIG. 118. *Psilotum triquetrum* ($\frac{1}{2}$ natural size).
A. Synangium.
B. Sporophyll after removal of the synangium. (M.S.)

Psilotum. The common tropical species *P. triquetrum* (fig. 118) is characterised by an underground rhizome which forms a confused mass of dark brown branches covered with filamentous hairs as substitutes for roots and gives off erect repeatedly forked aerial shoots. In *P. complanatum*[1] the habit

[1] Baker (87) A. p. 30.

is similar to that of the more abundant and better-known species, but the pendulous shoots are characterised by their broader and flatter form. In both species the function of carbon-assimilation is performed by the outer cortex of the green branches, as the small size of the widely-separated foliage leaves renders them practically useless as assimilating organs.

The sporophylls consist of a short axis terminating in two small divergent forks and bearing on its adaxial surface a trilocular or in rare cases a bilocular synangium (fig. 118, A and B). The walls of the loculi are composed of several layers of cells and dehiscence takes place along three lines radiating from the centre of the synangium. Professor Thomas[1] has recorded "fairly numerous instances in *Psilotum* of a second dichotomy of one branch of the first fork, or, less frequently, of both branches": instead of one synangium subtended by the two slender leaflets of the forked sporophyll-axis, there may be two synangia and three leaf-lobes or three synangia and four leaf-lobes. The occurrence of both these abnormalites in *Psilotum* and *Tmesipteris* shows a decided tendency in the Psilotales to a repeated dichotomy of the sporophylls[2].

A single stele[3] with a fluted surface occupies the axis of an aerial shoot (fig. 119, A); the axial region is occupied by a core of elongated mechanical elements (s), which may occasionally extend to the periphery of the xylem and break the continuity of the band of scalariform tracheae (fig. 119, A, a). The tracheae form the arms of an irregularly stellate stele and each arm is terminated by protoxylem elements (fig. 119, B, px). The rays of the xylem cylinder, which may be as many as six or eight in the upper part of the aerial shoots, become reduced in number as the rhizome is approached, assuming a diarch structure near the junction. In the rhizome the xylem forms an approximately triangular group of tracheae without any core of mechanical elements. Three to four layers of paren-

---

[1] Thomas (02) p. 349.

[2] Another form of abnormality in the sporophylls of *Psilotum* has recently been described by Miss Sykes. Sykes (08[2]).

[3] Bertrand, C. E. (81) ; Ford (04).

chyma succeeded externally by an ill-defined phloem (fig. 119, A,
p) surround the xylem and a fairly distinct endodermis (fig. 119,
A and B, *e*) encloses the whole.   To Mr Boodle[1] is due the

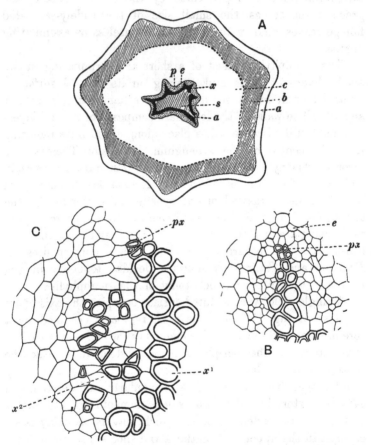

FIG. 119. A. Diagram of transverse section of aerial shoot of *Psilotum
            triquetrum. a—c*, cortex ; *p*, phloem ; *e*, endodermis; *s*, stereome;
            x, xylem ; *a*, gap in xylem.
         B. Enlarged view of one of the angles of the xylem shown in A.
            px, protoxylem.
         C. Part of transverse section of an approximately triangular
            rhizome stele showing a portion of the metaxylem x^1 ; *px*, proto-
            xylem elements ; x^2, secondary xylem.

[1] Boodle (04).

XIII]   TMESIPTERIS   21

interesting discovery that in some parts of the rhizome the
parenchymatous zone surrounding the scalariform tracheae
may become the seat of meristematic activity which results
in the production of secondary tracheae often characterised
by a sinuous longitudinal course. There is no definite
cambium, but the radially disposed tracheae and the adjacent
parenchymatous elements clearly demonstrate the secondary
nature of the tissue immediately external to the group of
primary xylem. Fig. 119, C, drawn from a section kindly
supplied by Mr Boodle, shows the secondary xylem elements at
x^2 associated with radially disposed thin-walled cells abutting
on the primary xylem, x^1. It is probable that this added tissue
may be a remnant of a more extensive secondary thickening
characteristic of the ancestors of the recent species. In their
manner of occurrence and sinuous course these secondary
tracheids bear a resemblance to the secondary xylem of
Lepidodendron fuliginosum[1]. The stele of the aerial shoot bears
a fairly close resemblance to the vascular axis of *Cheirostrobus*,
and its three-rayed form in the lower portions of the green
branches recalls that of the *Sphenophyllum* stele, except that
the axial xylem elements of the Palaeozoic genus are usually
represented in *Psilotum* by mechanical tissue. The cortex
consists of three regions (fig. 119, A), an outer zone of chlorophyllous tissue (*a*) rich in intercellular spaces succeeded by a
band of mechanical tissue (*b*) which gradually passes into an
inner region of larger and thinner-walled cells (*c*).

The genus *Tmesipteris*[2] agrees with *Psilotum* in general habit
and in its epiphytic and probably in some degree saprophytic
mode of life. Its brown rootless rhizome, which grows among
the roots of tree-ferns or rarely in the ground, gives off
pendulous or erect shoots reaching a length of two feet and
bearing lanceolate mucronate leaves 2—3 cm. long (fig. 120, A)
attached by decurrent leaf-bases. The sporophylls, replacing
the upper leaves or occurring in more or less well-defined zones
alternating with the foliage leaves, consist of a short axis
terminating in a pair of lanceolate lobes and bearing on its

[1] See p. 150.   [2] Bertrand (81) ; Jennings and Hall (91).

adaxial surface an elongated bilocular synangium attached to a
very short stalk (fig. 120, B). Reference has already been

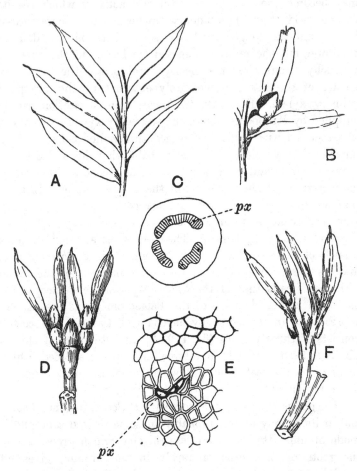

FIG. 120. *Tmesipteris*. A. Foliage leaves.
            B. Sporophyll and bilocular synangium.
            C. Diagram of transverse section of stele. *px*, protoxylem.
            D, F. Abnormal sporophylls. (From drawings made by Prof.
                    Thomas and generously placed at my disposal. A.C.S.)
            E. Portion of C enlarged.

made to the divergent opinions as to the morphological nature
of the sporophylls or sporangiophores, but recent investigations

distinctly favour the view that a sporophyll is best interpreted as a stalked leaf with two sterile laminae and an almost sessile, or in some cases a more obviously stalked, synangium; the whole sporophyll is characterised by the possession of a ventral and a dorsal lobe[1]. The drawings reproduced in fig. 120, D and F, illustrate some of the frequent variations described by Thomas in plants which he observed in the New Zealand forests. The sporophyll shown in fig. 120, D and F, has branched twice and bears three synangia.

The aerial branches of *Tmesipteris* possess a central cylinder of separate xylem groups in which the protoxylem occupies an internal position (fig. 120, C and E, px) enclosing an axial parenchymatous region. The cells of a few layers of the inner cortex immediately outside the endodermis are rendered conspicuous by a dark brown deposit. The cortex as a whole is composed of uniform parenchymatous tissue. In the lower part of the aerial shoots and in the rhizome the xylem forms a solid strand without protoxylem elements and conforms more clearly to that of *Psilotum*.

In this short account of the anatomy of *Tmesipteris* no mention is made of the effect produced on the stele by the departure of leaf-traces and of vascular stands to supply branches. Miss Sykes[2] in a recently published paper on the genus has shown that the exit of a leaf-trace does not break the continuity of the xylem of the stele, while the exit of a sporophyll-trace is marked by an obvious gap. Evidence is adduced in support of the conclusion that this difference, which at first sight appears to be one of morphological import-ance, is in reality merely a question of degree and " is due to the earlier preparation for the formation of 'sporophyll' than leaf-traces." Miss Sykes gives her adherence to the view that the "sporophylls" of *Tmesipteris* are branches and not leaves, but despite the arguments advanced this interpretation seems to me less probable than that which recognises the sporophyll as a foliar organ. Prof. Lignier[3] has pointed out that if Miss Sykes's conclusion as to the axial nature of the sporophyll in *Tmesipteris* is accepted, it diminishes the force of the com-

[1] Sykes (08).   [2] *ibid.* (08).   [3] Lignier (08).

parison between the sporophylls of that genus and *Sphenophyllum* as those of the latter can hardly be regarded as other than foliar organs.

Both members of the Psilotales may, as Boodle has suggested, be regarded as descendants of a common parent in which the aerial stems possessed a fluted or stellate cylinder of mesarch xylem. There can be no doubt as to the significance of the morphological resemblances between the Psilotales and the genera *Sphenophyllum* and *Cheirostrobus*, but the position of *Tmesipteris* and *Psilotum* in the plant-kingdom may probably be best expressed by adopting the group-name Psilotales rather than by transferring the recent genera to the Sphenophyllales. One of the most striking differences between the Psilotales and the genus *Lycopodium* is in the form of the sporophylls and sporangia; in *Lycopodium* a single sporophyll bears a unilocular sporangium, but in the Psilotales the sporophyll may be described as a bilobed structure homologous with a foliage-leaf, bearing a sporangiophore which consists of a short stalk terminating in a bilocular or trilocular synanguim; the short stalk receives a special branch from the vascular bundle of the sterile portion of the sporophyll[1].

Fossils described by authors as being closely allied to Psilotum.

A search through palaeobotanical literature reveals the existence of a very small number of specimens which have been identified as representatives of the Psilotales. An inspection of the material or published drawings leads one to the conclusion that practically no information of a satisfactory kind is available in regard to the past history of the two southern genera *Psilotum* and *Tmesipteris*, which are regarded by some botanists as relics of an ancient branch[2] of pteridophytes.

In 1842 Münster[3] instituted the genus *Psilotites* for a small impression of a slender branched axis from Jurassic rocks near Mannheim in Germany which he named *Psilotites filiformis*;

---

[1] Bower (94); (08).   [2] Bertrand (81) p. 254.
[3] Münster (42) p. 108, Pl. xiii. fig. 11; Pl. xv. fig. 20.

Schimper[1] spoke of the specimens as too doubtful for determination, an opinion with which every botanist would cordially agree. Goldenberg's species *Psilotites lithanthracis*[2] from the Saarbrücken coal-field is founded on impressions of axes: some of these are dichotomously branched and bear small oval projections, which may be rudimentary leaves or possibly leaf-scars. More recently Kidston[3] described specimens of branched axes from the Lanarkshire coal-field bearing a row of lateral thorn-like projections under the title *Psilotites unilateralis*; but these fragments, as Dr Kidston himself admits, are of no botanical value.

In a paper on fossil Salvinias, Hollick[4] mentions *Salvinia reticulata*, originally described by Heer and by Ettingshausen and *S. Alleni* Lesq.[5] a Tertiary species, and calls attention to their very close resemblance in form, nervation, and apex to the leaves of the genus *Tmesipteris*: he refers both species to that genus. The drawings reproduced by Hollick represent leaves with a midrib and numerous anastomosing lateral veins, whereas in *Tmesipteris* the lamina of the leaf has a midrib without lateral branches. An enlarged drawing of the outlines of the epidermal cells would correspond closely with the small reticulations in the fossil leaves and it may be that there has been some confusion between veins and cell-outlines. In any case there would seem to be no reason for the use of the recent generic name[6].

Among other fossils assigned to the Psilotales we have Marion's genus *Gomphostrobus* from the Permian of France and Germany[7]. Marion placed this plant in the Coniferales on the strength of its resemblance to *Walchia* and *Araucaria*, but Potonié[8] is inclined to recognise in the leaves and monospermic sporophylls characters suggestive of Lycopodiaceous affinity.

---

[1] Schimper (70) A. p. 75. [2] Goldenberg (55) p. 13, Pl. II. fig 7.
[3] Kidston (86²). [4] Hollick (94) p. 255, figs. 12, 13.
[5] Lesquereux (78) Pl. v. fig. 11.

[6] Since this was written I have had an opportunity of seeing a leaf labelled *Tmesipteris* from the Tertiary plant-beds of Florissant in a collection recently acquired by the British Museum : the specimen bears no resemblance to a leaf of the recent genus.

[7] Marion (90). [8] Potonié (93) A. p. 197, Pls. xxvii., xxviii., xxxiii.

The latter author in 1891[1], in ignorance of Marion's proposal to adopt the name *Gomphostrobus*, instituted a genus *Psilotiphyllum* for the sporophylls of a species originally described by Geinitz[2] as *Sigillariostrobus bifidus*, but he subsequently adopted Marion's designation and with some hesitation included the French and German specimens in the Psilotales. As stated elsewhere[3], Potonié's arguments in favour of his view hardly carry conviction, and it is probably more in accordance with truth to deal with *Gomphostrobus* in the chapter devoted to the Coniferales.

Psilophyton.

The generic title *Psilophyton*, instituted by the late Sir William Dawson[4], has become familiar to geologists as that of a Pre-Carboniferous plant characteristic of Devonian and Silurian rocks in Canada, the United States of America, and Europe. From the botanist's point of view the name stands for miscellaneous remains of plants of different types and in many cases unworthy of record. The genus was founded on impressions of branched axes from the Devonian strata of New Brunswick resembling the rachis and portions of lateral pinnae of ferns or the forked slender twigs of a Lycopod. The type-species *Psilophyton princeps* Daws. as represented on somewhat slender evidence in Dawson's restoration, which accompanies the original description of the genus and has since been copied by several authors, is characterised by the possession of a horizontal rhizome bearing numerous rootlets and giving off dichotomously branched aerial shoots with spinous appendages, compared with rudimentary leaves, and terminating in slender branchlets bearing pendulous oval " spore-cases " from their tips. Some of the branchlets exhibit a fern-like vernation. The plant is spoken of by Dawson as apparently a generalised type[5], resembling in habit and in its rudimentary leaves the recent genus *Psilotum* and presenting points of contact with ferns.

---

[1] Potonié (91); (93) A. p. 197.
[2] Geinitz (73) p. 700, Pl. III. figs. 5—7.
[3] Seward and Gowan (00) p. 137 ; Seward and Ford (06) p. 374.
[4] Dawson (59) A. p. 478, fig. 1.   [5] *ibid.* (71) A. p. 38.

Specimens were found in an imperfectly petrified state showing a central cylinder of scalariform tracheae surrounded by a broad cortical zone of parenchyma and fibrous tissue.

Among other species described by the author of the genus we need only mention *Psilophyton robustius*, characterised by vegetative shoots and "spore-cases" similar to those of the type-species; but, as Solms-Laubach[1] has pointed out, the petrified sections referred by Dawson to *P. robustius* are of an entirely different anatomical type from that of *P. princeps*[2].

British fossils from the Old Red Sandstone from the north of Scotland, Orkney and Caithness, originally figured by Hugh Miller and compared by him with algae but more especially with recent Lycopods, were subsequently placed by Carruthers[3] in the genus *Psilophyton* as *P. Dechianum*, the specific designation being chosen on the ground that the Scotch specimens are specifically identical with fossils described by Goeppert[4] as *Haliserites Dechianus*.

Various opinions have been expressed in regard to the nature of the Devonian species *Haliserites Dechianus* Goepp. with which Carruthers[5] identified Miller's Old Red Sandstone plant: reference may be made to a paper by White[6] containing figures of dichotomously branched impressions described as species of *Thamnocladus* which he includes among the algae.

In describing some Belgian impressions of Devonian age as *Lepidodendron gaspianum* Daws. Crépin[7] states that Carruthers has come to regard the specimens named by him *Psilophyton Dechianum* as branches of a Lepidodendron; he also quotes Carruthers as having expressed the opinion that the name *Psilophyton* had been employed by Dawson for two kinds of fossils, some being twigs of *Lepidodendron* while others, identified by Dawson as the reproductive branches of species of *Psilophyton*, represent the spore-cases of ferns comparable with Stur's genus *Rhodea*[8]. One of the examples figured by Carruthers[9] as *P. Dechianum* from Thurso (preserved in the British Museum, no. 52636), measuring 34 cm. in length and

[1] Solms-Laubach (95) A.   [2] Dawson (71) A. Cf. Pl. xi. figs. 131, 134, etc.
[3] Carruthers (73).   [4] Goeppert (52) A.   [5] Carruthers (73).
[6] White (02).   [7] Crépin (75).   [8] Stur (75) A. p. 33.
[9] Carruthers (73).

8 mm. broad, bears a close resemblance to a fern rhizome covered with ramental scales such as that of a species of *Davallia*. Other Belgian specimens described by Gilkinet[1] as *Lepidodendron burnotense*, like Crépin's species, are no doubt generically identical with some of the Scotch and Canadian fossils placed in the genus *Psilophyton*, though Penhallow[2] considers that the species *Lycopodites Milleri* is more correctly referred to *Lycopodites* than to *Psilophyton*.

A more recent paper on the Geology of the Perry basin in South-eastern Maine by Smith and White[3] contains a critical summary of the literature on *Psilophyton* and drawings of specimens. The latter afford good examples of Pre-Carboniferous plant fragments, such as are often met with in various parts of the world, which conform in habit to the New Brunswick specimens made by Dawson the type of his genus.

An examination of material in the Montreal Museum and of Hugh Miller's specimens in the Edinburgh collection leads me to share the opinion of Count Solms-Laubach that the name *Psilophyton* has been applied to plants which should not be included under one generic title. As Kidston[4] pointed out, the Canadian species *Psilophyton robustius* is not generically distinct from British and Belgian specimens referred to *Lepidodendron*; it may possibly be identical with the Bohemian plants on which Stur founded his genus *Hostinella*[5]. The Devonian plants described by Stur have since been examined by Jahn[6] who regards them as vascular plants, and not as algae to which Stur referred them; he mentions two species of *Psilophyton* but gives no figures.

The "spore-cases" of Dawson may be found to be the micro-sporangia or perhaps the small seeds of some pteridosperm; the forked axes with a smooth surface and others figured by Miller and by Dawson, with the surface covered with scales suggesting the ramenta of a fern, may be the rachises or rhizomes of filicinean plants. Other specimens may be Lepidodendron twigs, as for example the petrified fragments figured by

---

[1] Gilkinet (75) figs. 2—5.   [2] Penhallow (92) p. 8.
[3] Smith and White (05) p. 58, Pls. v. vi.   [4] Kidston (86²) p. 232.
[5] Stur (81) Pls. iii. iv.   [6] Jahn (03) p. 77.

Dawson as *Psilophyton princeps*; while the stem identified as
P. robustius is most probably that of a Gymnosperm. It is
doubtful whether a useful purpose is served by retaining the
genus *Psilophyton*. It was in the first instance instituted on
the assumption, which cannot be upheld, that the abundant
material in the New Brunswick beds bore a sufficiently close
resemblance to the rhizome and aerial branches of *Psilotum*.
Psilophyton has served as a name for miscellaneous plant
fragments, many of which are indeterminable. Dr White
concludes his account of the genus with the following words[1]:

"The examination of such so-called Psilophyton material as
I have seen shows the existence in America of two or more
groups, represented by several fairly well-marked species which
possess stratigraphical value, and which should be carefully
diagnosed and illustrated. It is probable also that additional
material throwing light on the structure and relationships of
these very remarkable early types of land-plants will be
discovered at some locality. The inspection of the material in
hand emphasises the need, as was pointed out by Solms-
Laubach, for the revision of the material referred by various
authors to *Psilophyton*, together with a thorough re-examination
and re-publication of the types."

Until a thorough re-examination has been made of the
Canadian material, with a view to determine whether there
exist substantial reasons for the retention of Dawson's genus,
it is undesirable to continue to make use of this name for Pre-
Carboniferous fossils which are too incomplete to be assigned
with certainty to a definite group of plants. Dr White draws
attention to the similarity of some of the Perry basin specimens
to Nathorst's genus *Cephalotheca*[2] from Devonian rocks of Bear
Island in the Arctic regions, a comparison which might be
extended to other genera and which serves to illustrate the
possibility that many of the specimens labelled *Psilophyton* may
eventually be recognised as examples of well defined generic
types belonging to more than one group of plants.

---

[1] Smith and White (05) p. 63.
[2] Nathorst (02) p. 15, Pl. i. figs. 18—35.

CHAPTER XIV.

LYCOPODIALES.

THE recent members of the Lycopodiales are considered apart from the extinct genera in order that our examination of the latter may be facilitated by a knowledge of the salient characteristics of the surviving types of this important section of the Pteridophyta. A general acquaintance with the extinct as well as with the recent genera will enable us to appreciate the contrasts between the living and the fossil forms and to realise the prominent position occupied by this group in the Palaeozoic period, a position in striking contrast to the part played by the diminutive survivors in the vegetation of the present day. In the account of the recent genera special attention is drawn to such features as afford a clue to the interpretation of the fossils, and the point of view adopted, which at times may appear to lead to an excessive attention to details, is necessarily somewhat different from that represented in botanical text-books[1].

A. HOMOSPOREAE.

Lycopodiaceae: genera *Phylloglossum, Lycopodium.*

B. HETEROSPOREAE.

Selaginellaceae: genus *Selaginella.*
Isoetaceae: genus *Isoetes.*

The existing plants included in the Lycopodiales are in nearly all cases perennial herbaceous pteridophytes, exhibiting

[1] For a general account of recent Lycopodiales see Pritzel (02) ; Campbell (05); Bower (08).

in their life-histories a well marked alternation of generations. The sporophyte (asexual generation) is characterised by the relatively small size of the leaves except in the genus *Isoetes* (fig. 132) and in the Australian and New Zealand genus *Phylloglossum*. The stems are usually erect or trailing, pendulous in epiphytic species or small and tuberous in *Isoetes* and *Phylloglossum*. The repeated forking of the shoots (monopodial and dichotomous branching) is a prominent feature of the group. The vascular tissue of the stem usually assumes the form of a single axial strand (stele) (fig. 125), but the shoots of some species of *Selaginella* often contain two or more distinct steles (fig. 131). The group as a whole is characterised by the centripetal development of the xylem composed almost entirely of scalariform tracheids: secondary xylem and phloem of a peculiar type occur in *Isoetes*, and the production of secondary xylem elements in a very slight degree has been noticed in one species of *Selaginella* (*S. spinosa*)[1]. The roots are constructed on a simple plan, having in most cases only one strand of spiral protoxylem elements (monarch structure). In *Lycopodium*, in which stem and root anatomy are more nearly of the same type than in the majority of plants, several protoxylem strands may be present. The sporangia are axillary or, more frequently, borne on the upper surface of sporophylls, which are either identical with or more or less distinct from the foliage leaves; in the latter case the sporophylls often occur in the form of a well defined strobilus (cone) at the tips of branches.

The gametophyte (sexual generation) is represented by prothalli which, in the homosporous genera, may live underground as saprophytes, or the upper portion may develop chlorophyll and project above the surface of the ground as an irregularly lobed green structure (*e.g. Lycopodium cernuum*)[2]. In the heterosporous forms the prothalli are much reduced and do not lead an independent existence outside the spore by the membrane of which they are always more or less enclosed. The sexual organs are represented by antheridia and archegonia;

[1] Bruchmann (97).
[2] Treub (84—90); see also Lang (99) and Bruchmann (98).

the male cells are provided with two cilia except in *Isoetes* which has multiciliate antherozoids like those of the ferns. The existing Lycopods, though widely distributed, never grow in sufficiently dense masses to the exclusion of other plants to form a conspicuous feature in the vegetation of a country. The inconspicuous rôle which they play among the plant-associations of the present era affords a striking contrast to the abundance of the arborescent species in the Palaeozoic forests of the northern hemisphere.

Lycopodiaceae. *Lycopodium*, represented by nearly 100 species, forms a constituent of most floras: epiphytic species predominate in tropical regions, while others flourish on the mountains and moorlands of Britain and in other extra-tropical countries. For the most part *Lycopodium* exhibits a preference for a moist climate and appears to be well adapted to habitats where the amount of sunlight is relatively small and the conditions of life unfavourable for dense vegetation. Mountains and islands constantly recur as situations from which species have been recorded. Some species are essentially swamp-plants, *e.g. Lycopodium inundatum*, a British species, and *L. cruentum* from the marshes of Sierra Nevada. A variety of the American species, *L. alopecuroides* (var. *aquaticum*) affords an instance of a submerged form, which has been collected from an altitude of 12—14,000 ft. on the Andes and Himalayas. It is noteworthy that a considerable variety of habitats is represented within the limits of the genus and that many species are sufficiently hardy to exist in circumstances which would be intolerable to the majority of flowering plants[1].

The British species frequently spoken of as Club Mosses, include *Lycopodium Selago*, *L. annotinium*, *L. clavatum*, *L. alpinum*, and *L. inundatum*.

Selaginellaceae. The species of *Selaginella*, over 300 in number, are widely spread in tropical and subtropical forests, growing on the ground with trailing, suberect or erect stems climbing over taller and stouter plants or as pendulous epiphytes on forest trees.

---

[1] See Baker (87) A.

Selaginella lepidophylla, a tropical American type, popularly known as the Resurrection plant, and often erroneously spoken of as the Rose of Jericho[1], possesses the power of rolling up its shoots during periods of drought and furnishes an example of a species adapted to conditions in marked contrast to those which are most favourable to the majority of species.

The only British species is *Selaginella spinosa* named by Linnaeus *Lycopodium selaginoides* and occasionally referred to as *Selaginella spinulosa* A. Br. (not to be confounded with a Javan species *S. spinulosa* Spring[2]).

Isoetaceae. *Isoetes* (fig. 132), of which Mr Baker in his *Handbook of the Fern-Allies* enumerates 49 species, is a type apart, differing in habit as in certain other characters from the other members of the Lycopodiales. Some botanists[3] prefer to include the genus among the Filicales, but the balance of evidence, including resemblances between *Isoetes* and extinct Lycopodiaceous plants, would seem to favour its retention as an aberrant genus of the group Lycopodiales. Some species are permanently submerged, others occur in situations intermittently covered with water, and a few grow in damp soil. *Isoetes lacustris* is found in mountain tarns and lakes of Britain and elsewhere in Central and Northern Europe and North America. *Isoetes hystrix*[4], a land-form occurs in Guernsey, North-East France, Spain and Asia Minor.

Lycopodiaceae.

The monotypic genus *Phylloglossum*, represented by *P. Drummondii* of Australia and New Zealand, though interesting from the point of view of its probable claim to be considered the most primitive type of existing Lycopodiaceous plants, need not be dealt with in detail. A complete individual, which does not exceed 4 or 5 cm. in length, consists of a very small tubercle or protocorm bearing a rosette of slender subulate leaves and prolonged distally as a simple naked axis which overtops the foliage leaves and terminates in a compact cluster of

[1] The Rose of Jericho is *Anastatica Hierochuntina* L. a Cruciferous plant.
[2] Baker (87) A. p. 34.   [3] Vines (88).   [4] Scott and Hill (00).

small scale-like sporophylls, each subtending a single sporangium[1].

Lycopodium.   It would be out of place in a volume devoted mainly to fossil plants to attempt a comprehensive account of the general morphology of recent species, and indeed our knowledge of the anatomical characters of the genus is still somewhat meagre.   For purposes of comparison with extinct types, it is essential that some of the more important morphological features of existing species should be briefly considered. The additions made to our knowledge of the gameophyte[2] of European and tropical species during the last two decades have revealed a striking diversity in habit.

In several species, grouped round the widely distributed type *Lycopodium Selago* Linn., the comparatively short, erect or suberect, shoots form fairly compact tufts ; the ordinary foliage-leaves function as sporophylls, and the sporangia are not localised on special portions of shoots.   From this type, we pass to others in which the fertile leaves tend to be confined to the tips of branches, but hardly differ in form from the sterile.   A further degree of specialisation is exhibited by species with well-defined cones composed of leaves (or bracts), the primary function of which is to bear sporangia and to afford a protective covering to the strobilus[3].

Lycopodium rufescens Hook.   An Andian species with stout dichotomously branched erect stems bears on the younger shoots crowded leaves with their thick and broadly triangular laminae pointing upwards, but on the older and thick shoots the laminae are strongly reflexed (fig. 121, A).   The lower part of the specimen represented in fig. 121, A, shows tangentially elongated scars and persistent leaf-bases or cushions left on the stem after the removal of the free portions of the leathery leaves, a surface-feature which also characterises the Palaeozoic genus *Lepidodendron*.   The reflexed leaves and persistent leaf-cushions are clearly seen in the piece of old stem of *Lycopodium dichotomum* Jacq., a tropical American species

---

[1] For *Phylloglossum*, see Bertrand (82) ; Bower (94), (08) ; Campbell (05).
[2] Treub (84—90) ; Bruchmann (98) ; Lang (99).
[3] Sykes (08³).

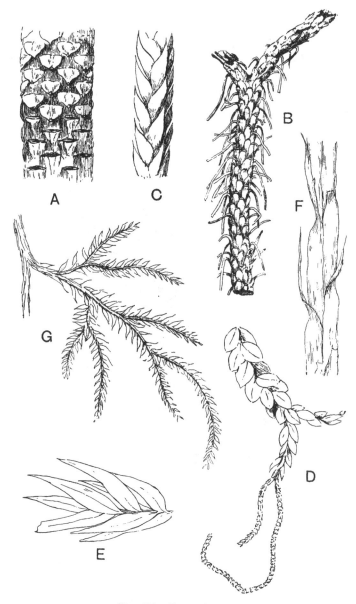

Fig. 121. *Lycopodium*.

A. *Lycopodium rufescens*.   B. *L. dichotomum*.
C. *L. tetragonum*.   D. *L. nummularifolium*.
E. *L. Dalhousianum*.   F. *L. casuarinoides*.
G. *L. volubile*.

From specimens in the Cambridge Herbarium and Botanic Garden. M.S.)

reproduced in fig. 121, B. Such species as *L. erythraeum*
Spring, and others with stiff lanceolate leaves exhibit a striking
resemblance to the more slender shoots of some recent conifers,
more especially *Araucaria excelsa, A. Balansae, Cryptomeria,
Dacrydium* and other genera.

FIG. 122. *Lycopodium squarrosum*. The branches of the larger shoot terminate
in cones. (From a plant in the Cambridge Botanic Garden. Reduced.)

In *Lycopodium tetragonum* Hook., (fig. 121, C), a species
from the Alpine region of the Andes, the long, pendulous and
repeatedly forked branches bear four rows of fleshy ovate leaves
and simulate the vegetative characters of certain conifers.

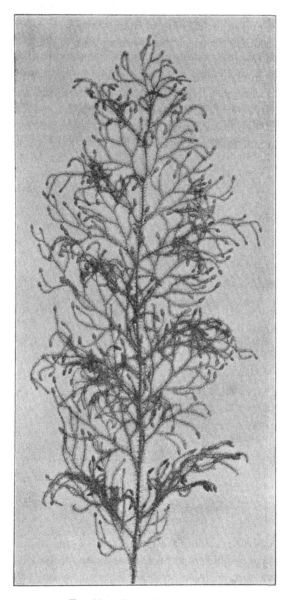

FIG. 123. *Lycopodium cernuum.*
(From a specimen in the Cambridge Herbarium. ½ nat. size.)

L. squarrosum Forst. (fig. 122) a tropical species from India, Polynesia, and other regions, is characterised by its stout stems reaching a diameter of 2·5 cm., bearing long pendulous branches with large terminal cones composed of sporophylls differing but slightly from the foliage leaves. The plant represented in the photograph serves as a good illustration of the practical identity in habit between Palaeozoic and recent genera.

L. Dalhousianum Spring, from the mountains of the Malay Peninsula and Borneo, has larger leaves of finer texture with a

Fig. 124. *Lycopodium obscurum.*

distinct midrib reaching a length of 2—3 cm. (fig. 121, E). Another type is illustrated by *L. nummularifolium* Blume, also a Malayan species, in which the leaves are shorter, broadly oblong or suborbicular, and the branches terminate in narrow and often very long strobili (sometimes reaching a length of

30 cm.) with small bracts in striking contrast to the foliage leaves (fig. 121, D). A similar form of long and slender strobilus occurs in *L. Phlegmaria* Linn., a common tropical Lycopod: the frequent forking of the strobili noticed in this and other species is a character not unknown among fossil cones (*Lepidostrobi*).

L. cernuum Linn. (fig. 123), another widely spread tropical type, offers an even closer resemblance than *L. squarrosum* to the fossil Lepidodendra. The stiff erect stem, reaching in some cases a length of several feet, bears numerous repeatedly forked branches, with crowded linear leaves, terminating in short cylindrical cones with broadly ovate sporophylls. A similar habit characterises the North American species *L. obscurum* Linn. (fig. 124) bearing cones several centimetres in length.

L. casuarinoides Spring (fig. 121, F) an eastern tropical species, is worthy of notice as exhibiting a peculiar form of leaf consisting of a very small lamina, 3 mm. in length, borne on the top of a long decurrent base, which forms a narrow type of leaf-cushion, bearing some resemblance to the long and rib-like cushions of certain species of *Sigillaria,* and recalling the habit of slender fossil twigs referred to the Coniferae under such names as *Widdringtonites, Cyparissidium, Sphenolepidium.*

L. volubile Forst. (fig. 121, G) a New Zealand species, in habit and leaf-form bears a close resemblance to the Jurassic *Lycopodites falcatus* Lind. and Hutt. (fig. 137): it is also a representative of a few species of *Lycopodium* which agree with the majority of species of *Selaginella* in having two kinds of sterile leaves, comparatively long falcate leaves forming two lateral rows and smaller appressed leaves on the upper surface of the branches.

These examples suffice to illustrate the general appearance presented by the vegetative shoots of recent species of which the foliage leaves vary considerably—from the small scale-leaves of *Lycopodium tetragonum,* to the very slender linear subulate leaves of such a species as *L. verticillatum* Linn. or the long and broader lamina of *L. Dalhousianum* (fig. 121, E). It is obvious that fragments of the various types preserved as fossils might well be mistaken either for some of the larger mosses or

for twigs of conifers. As Dr Bommer[1] has pointed out in his
interesting paper on " Les causes d'erreur dans l'étude des
empreintes végétales" some dicotyledonous plants may also
simulate the habit of Lycopods: he cites *Phyllachne clavigera*
Hook (Candolleaceae), *Tafalla graveolens* Wedd (Compositae)
and *Lavoisiera lycopodioides* Gard. (Melastomaceae). Another
point illustrated by fig. 121 is the close agreement in habit and
in the form of the leaves and leaf-cushions between the recent
plants and the Palaeozoic Lepidodendreae.

In his masterly essay "On the vegetation of the Carboniferous Period, as compared with that of the present day" Sir
Joseph Hooker called attention to the variation in the shape
and arrangement of the leaves in the same species of *Lycopodium*. The three woodcuts which he publishes of *Lycopodium
densum*, a New Zealand species, afford striking examples of the
diversity in habit and leaf-form and justify his warning "that
if the species of *Lepidodendron* were as prone to vary in the
foliage as are those of *Lycopodium*, our available means for
distinguishing them are wholly insufficient[2]."

As we have already noticed, there is a considerable
diversity among recent species, both as regards habitat and
habit; in the anatomy of the stem also corresponding variations
occur within the limits of a well-defined generic type of stele.
In species with creeping stems, such as *L. clavatum*[3], the
stele exhibits an arrangement of vascular tissue characteristic
of the plagiotropic forms. The xylem consists of more or less
horizontal plates of scalariform tracheae, each surrounded by
small-celled parenchyma, alternating with bands or groups of
somewhat ill-defined phloem. The protoxylem and protophloem elements occupy an external position (exarch), pointing
to a centripetal development of the metaxylem. This centripetal or root-like character of the primary xylem is an important
feature in recent as in fossil Lycopods. The close agreement
between the roots and stems of recent species in the disposition
of the vascular elements also denotes a simpler type of anatomy

[1] Bommer (03) Pl. ix. figs. 140, 141.
[2] Hooker (48) p. 423, figs. 12—14.
[3] Jones (05).

than occurs in the majority of vascular plants in which stem
and root have more pronounced structural peculiarities. A
pericycle, 2—6 cells in breadth, encloses the xylem and phloem

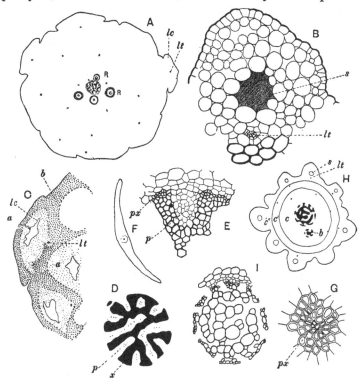

FIG. 125. A. *Lycopodium dichotomum.* Transverse section of stem: *lc*, leaf-
cushion; *lt*, leaf-trace; R, roots.

        B. *L. cernuum*, portion of cortex of fig. H, enlarged.

        C. *L. saururus.* Cortex: *lt*, leaf-trace; *a*, thin-walled tissue;
b, thick-walled tissue; *lc*, lacuna.

        D. *L. saururus.* Stele: *x*, xylem; *p*, phloem.

        E. Portion of fig. D, enlarged: *px*, protoxylem; *p*, phloem.

        F. Transverse section of leaf of *Lycopodium.*

        G. Vascular bundle of leaf: *px*, protoxylem.

        H. *L. cernuum*: *b*, branch of stele; *c—c''*, cortex; *s*, space in
cortex; *lt*, leaf-trace.

        I. Stele of fig. H, enlarged (phloem omitted).

bands and this is succeeded by an endodermis, 2—3 cells broad,
with vaguely defined limits. In *L. clavatum*, as in *L. alpinum*,
another British species, the broad cortex is differentiated into

three fairly distinct regions; abutting on the endodermis is a
zone several layers broad of thick-walled cells constituting an
inner cortex modified for protection and support; the central
region consists of larger and thinner-walled cells adapted for
water-storage and aeration; beyond this is an outer cortical
zone of firmer and thicker elements. The prominent leaf-bases
or leaf-cushions (fig. 125, A, *lc*) give to the surface of a transverse
section a characteristic appearance which presents the closest
agreement with that of the younger shoots of *Lepidodendron*.
From the peripheral protoxylem groups small strands of xylem
are given off, which follow a steeply ascending course through
the cortex to the single-veined leaves. The leaf-traces, in several
species at least, are characterised by a mesarch structure
(fig. 125, F, G), the spiral protoxylem elements occupying an
approximately central position. The mesophyll of the leaves
varies in regard to the extent of differentiation into a palisade
and spongy parenchyma; in all cases there is a single vascular
bundle occasionally accompanied by a secretory duct.

In erect stems of *Lycopodium*, as represented by *L. cernuum*
(figs. 123,125, H, I), *L. Dalhousianum, L. squarrosum* (fig. 122)
and many others, the stele presents a characteristic appearance
due to the xylem plates being broken up into detached groups
or short uniseriate bands with the interspaces occupied by
phloem islands. This type of structure bears a superficial
resemblance to that in the single stele of certain species of the
fern *Lygodium*[1], but it is distinguished by the islands of phloem
scattered through the stele. In other species the xylem tends
to assume the form of a Maltese cross (*e.g. L. serratum* Thbg.)
or it may be disposed as V-shaped and sinuous bands terminating in broad truncate ends composed of protoxylem elements.
This form of the xylem and the distribution of the phloem
groups are shown in fig. 125, D, E, drawn from a section of a
plant of *Lycopodium saururus* Lam.[2] collected by Mr A. W. Hill
at an altitude of 15,000 feet on the Andes of Peru. The
position of the protoxylem is shown fig. 125, E, *px*.

[1] Boodle (01) Pl. xix.
[2] This species is figured under the name *Lycopodium crassum* by Hooker
and Greville (31) Pl. 224. See also Brongniart (37) Pl. I. fig. 1.

While several species possess a cortex of three distinct zones (fig. 125, H, c, c′, c″), in others the extra-stelar tissue is much more homogeneous, consisting of thin-walled parenchyma or in some cases of thick-walled elements; as a general rule, however, there is a tendency towards a more compact arrangement in the inner and outer˙ portions of the cortex as contrasted with the larger and more loosely connected cells of the middle region. In certain types the middle cortex contains fairly large spaces, as in the swamp-species *L. inundatum*, which with *L. alopecuroides* exhibits another feature of some interest first described by Hegelmaier[1] If a transverse section of the stem of *L. inundatum* be examined the leaf-traces are seen to be accompanied by a circular canal containing mucilage which extends into the lamina of the leaf. In a specimen of *L. cernuum*[2] obtained at a height of 2500 ft. by Professor Stanley Gardiner in the Fiji Islands, the leaf-traces (fig. 125, B *lt*,) were found to be accompanied for part of their course by a well-marked secretory space (fig. 125, B, *s*). There is little doubt that the presence of these mucilage canals is directly connected with a certain type of habitat[3] and attention is called to them in view of a resemblance which they offer to a characteristic strand of tissue, known as the parichnos, which is associated with the leaf-traces of *Lepidodendreae* and *Sigillarieae*. In the section shown in fig. 125, H, the xylem of the stele forms more continuous bands than is often the case in *L. cernuum* which has already been described as having its xylem in small detached groups. The presence of the smaller branch-stele (fig. 125, H, *b*) affords an example of monopodial branching. The outer cortex of *L. saururus* (fig. 125, C) exhibits a somewhat unusual feature in the distribution of the thicker-walled tissue (*b*) which encloses a patch of more delicate parenchyma (*a*) with large lacunae (*lc*) in the region of the

---

[1] Hegelmaier (72). See also Hill, T.G.(06) p. 269; this author draws attention to the fact that in some species of *Lycopodium* the mucilage canals are confined to the sporophylls.

[2] Professor Yapp has drawn my attention to the very close anatomical resemblance between a specimen of *Lycopodium salakense* obtained by him from Gunong Inas in the Malay Peninsula and *L. cernuum* as represented in fig. 125, H and I.

[3] Jones (05).

leaf-bases, and presents the appearance of an irregular reticulum. This arrangement of the mechanical tissue in the outer cortex is comparable with that in stems of some species of *Sigillaria*.

In certain species of *Lycopodium* the roots[1], which arise endogenously from the axial vascular cylinder, instead of passing through the cortex of the stem by the shortest route, bend downwards and bore their way in a more or less vertical direction before emerging at or near the base of the aerial shoot. The transverse section of *L. dichotomum* represented in fig. 125, A, shows several roots (R) in the cortex; they consist of a xylem strand of circular or crescentric form accompanied by phloem and enclosed by several layers of root-cortex. The roots of *Lycopodium* do not always present so simple a structure as those of *L. dichotomum*; the xylem may have an irregularly stellate form with as many as ten protoxylem groups.

Reproductive Shoots[2]. In *Lycopodium Selago* the foliage leaves serve also as sporophylls and, as Professor Bower[3] has pointed out, the branches exhibit to some extent a zonal alternation of sterile and fertile leaves; in other species, in which foliage leaves and sporophylls are practically identical, the sporangia occur sporadically on the ordinary leaves. In species with well-defined terminal cones the lower sporophylls may bear arrested sporangia and thus form transitional stages between sterile and fertile leaves, a feature which occurs also in the male and female flowers of many recent Araucarieae[4] The sporangia[5] (fig. 126, D, F) are usually reniform and compressed in a direction parallel to the surface of the cone-scales; they are developed from the upper surface and close to the base of the fertile leaf to which they are attached by a short and thick stalk (*e.g. L. inundatum*) or by a longer and more slender pedicel (*L. Phlegmaria*, fig. 126, E). On maturity the sporangia open as two valves in the plane of compression

[1] Strasburger (73) p. 109; Brongniart (37) Pl. 8; (39) A. Pl. 32: Brongniart figures stems of *L. Phlegmaria* and other species showing roots in the cortex. See also Goldenberg (55); Bruchmann (74); Saxelby (08).

[2] Since this was written a comparative account of the sporophylls of *Lycopodium* has been published by Miss Sykes. [Sykes (08³).]

[3] Bower (94) p. 514; (08).   [4] Seward and Ford (06).

[5] Goebel (05) p. 579.

and the line of dehiscence is determined in some species at least by the occurrence of smaller cells in the wall. In

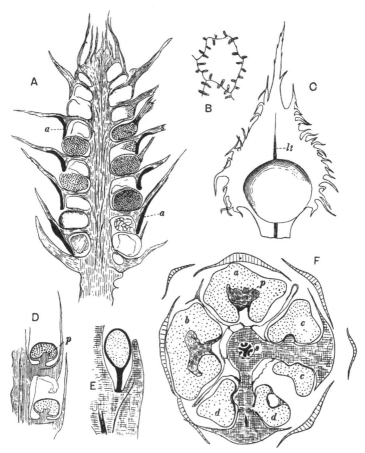

FIG. 126. A. *Lycopodium cernuum*, longitudinal section of strobilus; *a*, band of lignified cells.

B. *L. cernuum.* Cell from sporangium wall.

C. *L. cernuum.* Sporophyll and sporangium; *lt*, vascular bundle.

D. *L. clavatum.* Part of radial longitudinal section of strobilus; *p*, sterile tissue.

E. *L. Phlegmaria.* Sporophyll and stalked sporangium.

F. *L. clavatum.* Transverse section of strobilus; *p*, sterile pad.

transverse sections of cones in which the sporangia are strongly saddle-shaped, the sporophylls may appear to bear two sporangia.

This is well shown in the section of a cone of *L. clavatum* shown in fig. 126, F. The sporangia *a* and *b* are cut through in an approximately median plane showing the irregular outline of the sterile pad (*p*) of tissue in the sporogenous cavity. Those at *c* and *d* have been traversed at a lower level and the two lobes of the saddle-shaped sporangia are cut below the attachment to the sporophyll. The distal laminae of the sporophylls, cut at different levels, are seen at the periphery of the cone.

In longitudinal radial section of some cones the sporangia appear to occupy an axillary position, but in others (*e.g. L. clavatum*) they are attached to the horizontal portion of the sporophyll almost midway between the axis of the cone and the upturned distal end of the sporophyll (fig. 126, D). The wall of a sporangium frequently consists of 2—3 cell-layers and in some cases (*e.g. L. dichotomum*), it may reach a thickness of seven layers, resembling in this respect the more bulky sporangia of a certain type of Lepidodendroid cone. The sporogenous tissue is separated from the stalk of the sporangium by a mass of parenchymatous tissue which may project as a prominent pad (fig. 126, D, F, *p*) into the interior of the sporogenous cavity. This basal tissue (the subarchesporial pad of Bower[1]) has been observed in *L. clavatum* to send up irregular processes of sterile cells among the developing spores, suggesting a comparison with the trabeculae which form a characteristic feature of the sporangia of *Isoetes* and with similar sterile strands noticed by Bower[2] in *Lepidostrobus* (cone of *Lepidodendron*).

Each sporophyll is supplied by a single vascular bundle which according to published statements never sends a branch to the sporangium base. The fertile tips of the foliage shoots of *L. cernuum* (figs. 126, A—C) afford good examples of specialised cones. The surface of the cone is covered by the broadly triangular laminae of sporophylls (fig. 126, C) which in their fimbriate margins resemble the Palaeozoic cone-scales described by Dr Kidston[3] as *Lepidostrobus fimbriatus*. The distal portions of the sporophylls are prolonged downwards

[1] Bower (94).   [2] *ibid.* (94) Pl. XLVIII.
[3] Kidston (83) Pl. XXXI. figs. 2—4.

(fig. 126, A) to afford protection to the lower sporangia, their efficiency being increased by the lignified and thicker walls (A, *a*) of the cells in the lower portion of the laminar expansion. The cells of the sporangial wall are provided with strengthening bands which in surface-view (fig. 126, B) present the appearance of prominent pegs. Since the appearance of Miss Sykes's paper on the sporangium-bearing organs of the Lycopodiaceae, Dr Lang[1] has published a more complete account of the structure of the strobilus of *Lycopodium cernuum* in which he

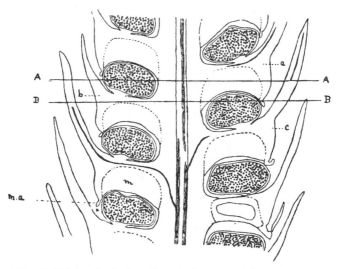

FIG. 127. Radial longitudinal section of the cone of *Lycopodium cernuum*. (After Lang.)

records certain features of special interest. The importance of these morphological characters is increased by their agreement, as shown by Lang, with those of the Palaeozoic cone *Spencerites*[2] The sporophylls of a cone (12 mm. long by 3 mm. in diameter) of *Lycopodium cernuum* show an abrupt transition from the foliage leaves, but like these they occur in alternate whorls of five. A large sporangium is attached to the upper face of each sporophyll close to the base of the obliquely vertical distal lamina

---

[1] Lang (08).   [2] See page 192, and Watson (09).

(fig.127); each sporophyll,which is supplied with a single vascular bundle, has a large mucilage-cavity (*m*) in its lower region. " The mucilaginous change " in the sub-sporangial portion of a sporophyll " extends to the surface involving the epidermis, so that this portion of the sporophyll-base may be described as consisting of a mass of mucilage bounded below by a structureless membrane[1]." Dehiscence of the sporangia occurs at the middle of the distal face (fig. 127, *x*). As seen in the radial section (fig. 127, *ma*) the outer margin of the base of the sporophyll bears a short outgrowth. The leaf-bases of each whorl hang down between the sporangia of the alternating whorl below, and

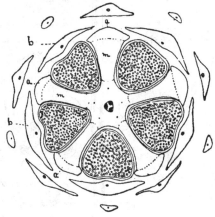

FIG. 128. Transverse section of the cone of *Lycopodium cernuum*, in its plane
AA of fig. 127. (After Lang.)

the base of each sporophyll is coherent with the margins of the two sporophylls of the next lower whorl between which it lies, the sporangia being thus closely packed and lying in a pocket " open only on the outer surface of the cone." Fig. 128 represents a transverse section through a cone in the plane AA of fig. 127 ; this traverses the sporangia and their subtending bracts (*b*) of one whorl and the dependent bases of the sporophylls of the next higher whorl in the region of the mucilage-sacs (*m*), which are bounded at the periphery by the

[1] Lang (08) p. 357.

outer tissue of the sporophylls (*a*). A transverse section in the
plane BB of fig. 127 is shown in fig. 129: the pedicels and a
part of each vascular strand are seen at *b* radiating from the
axis of the cone; one sporophyll (*sp, a*) is cut through in the

Fig. 129.   Transverse section of the cone of *Lycopodium cernuum* in the plane
BB of fig. 127.   (After Lang.)]

region of the pad of tracheal tissue that characterises the short
sporangial stalks. The upper portions of the sporangia of the
next lower whorl, which project upwards against the mucilaginous bases of the sporophylls above (cf. fig. 127, BB) are shown
at *c* and external to them, at *a*, the section has cut through
the outer persistent portions of these sporophyll bases.

As Lang points out, this highly complex structure is an
expression of the complete protection afforded to the sporangia
of a plant met with in exposed situations in the tropics; it
is also of importance from a morphological standpoint as exhibiting an agreement with the extinct type of Lycopod cone
represented by *Spencerites*.

Selaginellaceae.

Selaginella differs from *Lycopodium* in the production of
two kinds of spores, megaspores and microspores, and, in the

great majority of species, in the dimorphic character of the foliage leaves, which are usually arranged in four rows, the laminae of the upper rows being very much smaller than those of the lower (fig. 130, 1—3). The smaller leaves are shown more clearly in fig. 130, 1a. It is obvious from an examination of a

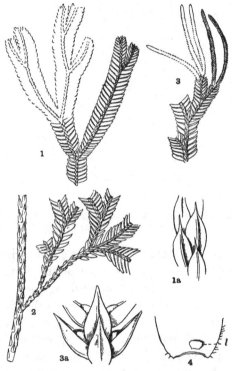

Fig. 130. *Selaginella grandis*. (1—3, nat. size.)

Selaginella shoot, such as is shown in fig. 130, that in fossil specimens it would often be almost impossible to recognise the existence of two kinds of leaves. Some species, *e.g. Selaginella spinosa*[1], the sole British representative of the genus, are homophyllous and agree in this respect with most species of *Lycopodium*. Another feature characteristic of *Selaginella*, as

[1] Bruchmann (97).

contrasted with *Lycopodium*, is the presence of a ligule in both foliage leaves and sporophylls. This is a colourless thin lamina attached by a comparatively stout foot to the base of a pit on the upper surface and close to the lower edge of the leaf (fig. 130, 4, *l*; fig. 131, E, F, *l*).

In an erect species, such as *S. grandis* Moore[1] (fig. 130 and fig. 131, G) from Borneo, the main shoots, which may attain a height of 2—3 feet, bear small and inconspicuous leaves of one kind, but the lateral and repeatedly forked shoots are heterophyllous. The passage from the homophyllous to the heterophyllous arrangement is shown in the transition from the erect to the dorsiventral habit of the lateral shoots (fig. 130, 2). The monopodially or dichotomously branched shoots produce long naked axes at the forks; these grow downwards to the ground where they develop numerous dichotomously forked branches. For certain reasons these naked aerial axes were named rhizophores and have always been styled shoots, the term root being restricted to repeatedly forked branches which the rhizophores produce in the soil. It has, however, been shown by Professor Harvey-Gibson[2] that there is no sufficient reason for drawing any morphological distinction between rhizophores and roots, the term root being applicable to both.

Our knowledge of the anatomy of *Selaginella*, thanks chiefly to the researches of Harvey-Gibson[3], is much more complete than in the case of *Lycopodium*. The stems, which may be either trailing or erect, are usually dorsiventral, and it is noteworthy that different shoots of the same plant or even the same axis in different regions may exhibit considerable variation in the structure and arrangement of the vascular tissue. In the well-known species, *Selaginella Martensii*, the stem, which is partly trailing, partly ascending, possesses a single ribbon-shaped stele composed of scalariform tracheids with two marginal protoxylems formed by the fusion of the leaf-traces of the dorsal and ventral leaves respectively. As in *Lycopodium* the metaxylem tracheae are as a rule scalariform, but reticulate xylem elements are by no means unknown. The tracheal band,

---

[1] Gard. Chron. (82).   [2] Harvey-Gibson (02).   [3] *ibid.* (94) (97) (02).

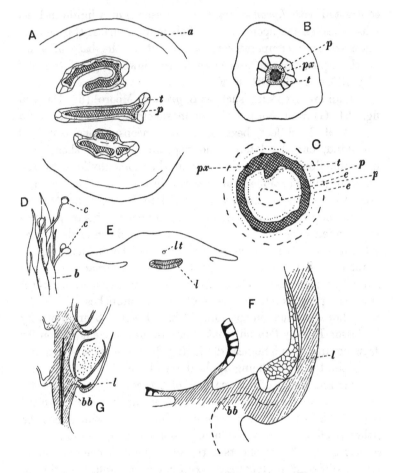

FIG. 131.   A.   *Selaginella Willdenowii.*   Transverse section of stem: *a*, outer
              cortex; *p*, phloem; *t*, trabeculae.
           B.   *S. spinosa*, stem: *px*, protoxylem.
           C.   *S. laevigata* var. *Lyallii*, section of stele: *t*, ridge of xylem
              cylinder; *e*, endodermis.
           D.   *S. rupestris*, seedlings with cotyledons (*c*) protruding beyond the
              sporophylls (*b*).
           E.   Transverse section of *Selaginella* leaf-base: *l*, ligule; *lt*, leaf-
              trace.
           F.   Portion of G. enlarged.
           G.   *S. grandis.* Longitudinal section of strobilus: *bb*, sporophyll-
              trace; *l*, ligule.
              (A, B, C, E, F, after Harvey-Gibson; D, after Miss Lyon.)

surrounded by parenchymatous elements, is enclosed by phloem
with external protophloem elements. The characteristic features
of the stele are shown in the diagrammatic drawing of a section
of another species—*S. Willdenowii*—represented in fig. 131, A.

A pericycle composed of one or two layers of chlorophyll-
containing cells encircles the whole stele which is suspended
in a lacuna by trabeculae (fig. 131, A, B, *t*) connecting the
pericycle with the inner edge of the broad cortex. The
trabeculae consist in part of endodermal cells characterised by
cuticular bands. The cortex is usually differentiated into three
fairly distinct regions. Mechanical tissue of thick-walled fibres
constitutes the outer region (*a*); the middle cortex consists of
thinner-walled parenchyma, the elements of which become
smaller and rather more compactly arranged in the inner zone.
The middle cortex is frequently characterised by the presence
of spaces and by the hyphal or trabecular structure of the
tissue, a feature which, as Bower[1] pointed out, is common to
many recent and fossil members of the Lycopodiales. In some
cases, *e.g. S. erythropus*, from tropical America, the cortex
of the creeping stem consists entirely of thick-walled cells.
Selaginella grandis (fig. 130) has "a short decumbent stem
rooted at close intervals[2]," from which thick erect aerial shoots
rise to a height of one foot or more. In the apical region
these erect axes give off repeatedly forked foliage shoots
on which the spiral phyllotaxis of the homophyllous axis
is gradually replaced by four rows of two kinds of leaves
(fig. 130, 2). The anatomy of this species agrees with that
of *S. Martensii*. The trailing or semi-erect and homophyllous
shoots of *Selaginella spinosa*[3] present a distinct type of vascular
anatomy. The upper part of the ascending stem has an axial
strand of xylem with seven peripheral groups of spiral protoxylem tracheae (fig. 131, B); in the trailing portion of the
shoot the protoxylem elements occur as one central group in
the solid rod of metaxylem through which the leaf-traces pass
on their way to the axial protoxylem. This type is important

---

[1] Bower (93).   [2] Harvey-Gibson (94) p. 152.
[3] *ibid.* (94) p. 194; Scott (96) p. 9.

as affording an exception, in the endarch structure of the xylem, to the usual exarch plan of the stelar tissues. This species is the only one in which any indication of the production of secondary xylem elements has so far been recorded. Bruchmann[1] has shown that, in the small tuberous swelling which occurs at the base of the young shoot (hypocotyl), a meristematic zone is formed round the axial vascular strand and by its activity a few secondary tracheids are added to the primary xylem. With this exception *Selaginella* appears to have lost the power of secondary thickening, the possession of which constitutes so striking a feature of the Palaeozoic Lycopods. Another type is represented by *S. inaequalifolia*, an Indian species, the shoots of which may have either a single stele or as many as five, each in its separate lacuna. The homophyllous *S. laevigata* var. *Lyallii* Spr., a Madagascan species, affords a further illustration of the variation in plan of the vascular tissues within the genus. There is a considerable difference in structure between the erect and creeping shoots ; in the former there may be as many as 12—13 steles, which gradually coalesce before the vertical axis joins the creeping rhizome to form one central and four peripheral steles. In the rhizome there is usually a distinct axial stele without protoxylem, surrounded by an ill-defined lacuna and enclosed by a cylindrical stele (solenostele)[2] usually two tracheae in width with four protoxylem strands on its outer edge. The continuity of the tubular stele is broken and, in transverse section, it assumes the form of a horseshoe close to the base of an erect shoot to which a crescentric vascular strand is given off. Harvey-Gibson[3] has figured a section of the rhizome of this type in which the axial vascular strand is represented by a slight ridge of tracheae (fig. 131, C, *t*) projecting towards the

---

[1] Bruchmann (97).

[2] The term solenostele, first used by Van Tieghem and revived by Gwynne-Vaughan, may be applied to a stem in which the vascular tissue has the form of a hollow cylinder with phloem and endodermis on each side of the xylem. As each leaf-trace is given off the continuity of the vascular tube is interrupted. See Gwynne-Vaughan (01) p. 73.

[3] Harvey-Gibson (94) Pl. xii, fig. 93.

centre of the axis of the tubular stele. The cylindrical stele consists of xylem with external and internal phloem (*p*): cuticularised endodermal cells occur at *e* and *e*.

Reference has already been made to the descending naked branches given off from the points of ramification of the foliage shoots of *Selaginella*. It has been shown by Harvey-Gibson[1] that these branches, originally designated rhizophores by Nägeli and Leitgeb, as well as the dichotomously branched roots which they produce below the level of the ground, possess a single vascular strand of monarch type. It is interesting to find that in some species the aerial portion of the rhizophore has a xylem strand with a central protoxylem, an instance of endarch structure like that in certain portions of the shoot-system of *S. spinosa*. The root-anatomy of *Selaginella* and the dichotomous habit of branching afford points of agreement with the subterranean organs of *Lepido-dendron* and *Sigillaria*.

Leaves. The leaves of *Selaginella*[2] usually consist of a reticulum of loosely arranged cells, but in some cases part of the mesophyll assumes the palisade form. The single vascular bundle consists of a few small annular or spiral tracheae and at the apex of the lamina the protoxylem elements are accompanied by several short reticulated pitted elements. Both foliage leaves and sporophylls are characterised by the possession of a ligule, a structure which may present the appearance of a somewhat rectangular plate (fig. 130, 4, *l*, and fig. 131, E—G, *l*) or assume a fan-shaped form with a lobed or papillate margin. The base, composed of large cells, is sunk in the tissue of the leaf close to its insertion on the stem (fig. 131, E, *l*) and enclosed by a well-marked parenchymatous sheath. The sheath is separated from the vascular bundle of the leaf by one or more layers of cells, and in some species these become transformed into short tracheids. The ligule is regarded by Harvey-Gibson[3] as a specialised ramentum which serves the temporary function of keeping moist the growing-point and young leaves.

Cones. The terminal portions of the branches of *Selaginella* usually bear smaller leaves of uniform size which function as

---

[1] Harvey-Gibson (02).   [2] *ibid.* (97).   [3] *ibid.* (96).

sporophylls, but in this genus the fertile shoots do not generally
form such distinct cones as in many species of *Lycopodium*.
In *S. grandis* (figs. 130, 3; 131, G) the long and narrow strobili
consist of a slender axis bearing imbricate sporophylls in four
rows: each sporophyll subtends a sporangium situated between
the ligule and the axis of the shoot. The sporangium may be
developed from the axis of the cone or, as in *Lycopodium*, from
the cells of the sporophyll[1]. In some species the lower sporophylls bear only megasporangia, each normally containing four
megaspores, the microsporangia being confined to the upper
part of the cone. This distribution of the two kinds of sporangia
is, however, by no means constant[2]: in some cases, *e.g. S.
rupestris*, cones may bear megasporangia only, and in the cone
of *S. grandis*, of which a small piece is represented in fig. 131, G,
all the sporangia were found to contain microspores.

The occurrence of two kinds of spores in *Selaginella*
constitutes a feature of special importance from the point of
view of the relationship between the Phanerogams, in which
heterospory is a constant character, and the heterosporous
Pteridophytes. One of the most striking distinctions between
the Phanerogams and the rest of the vegetable kingdom lies in
the production of seeds. Recent work has, however, shown that
seed-production can no longer be regarded as a distinguishing
feature of the Gymnosperms and Angiosperms. Palaeozoic
plants which combined filicinean and cycadean features resembled the existing Phanerogams in the possession of highly
specialised seeds. This discovery adds point to the comparison
of the true seed with structures concerned with reproduction in
seedless plants, which in the course of evolution gave rise to the
more efficient arrangement for the nursing, protection, and
ultimate dispersal of the embryo. In the megaspore of
Selaginella we have, as Hofmeister was the first to recognise
in 1851, a structure homologous with the embryo-sac of the
Phanerogam. The embryo-sac consists of a large cell produced
in a mass of parenchymatous tissue known as the nucellus
which is almost completely enclosed by one or more integuments. Fertilisation of the egg-cell within the embryo-sac

[1] Bower (08) p. 315.   [2] Hieronymus (02).

takes place as a rule while the female reproductive organ is still attached to the parent-plant and separation does not occur until the ovule has become the seed.

In a few cases, notably in certain plants characteristic of Mangrove swamps, continuity between the seed and its parent is retained until after germination. The megasporangium of *Selaginella* dehisces[1] along a line marked out by the occurrence of smaller cells over the crest of the wall. It has been customary to describe the megaspores as being fertilised after ejection from the sporangia. This earlier separation from the parent and the absence of any protective covering external to the spore-wall constitute two distinguishing features between seeds and megaspores. In *Selaginella apus*, a Californian species, Miss Lyon has shown that fertilisation of the egg-cell usually takes place while the megaspore is still in the strobilus. On examining withered decayed strobili of this species which had been partially covered with the soil for some months after fertilisation of the megaspores, several young plants were found with cotyledons and roots projecting through the crevices of the megasporangia[2]. From this, adds Miss Lyon, "it seems safe to assume that an embryo may have two periods of growth separated by one of quiescence quite comparable to those of seed plants with marked xerophilous features."

In another Western American species *S. rupestris* described by the same writer the cotyledons of young plants were found protruding from the imbricate sporophylls of a withered cone (fig. 131, D). This species is interesting also from the occasional occurrence of one instead of four megasporangia in a sporangium; a condition which affords another connecting link between the heterosporous Pteridophytes, on the one hand, and the seed-bearing Phanerogams in which the occurrence of a single embryo-sac (megaspore) in each ovule is the rule. The cones of *Selaginella rupestris* retain connexion with the plant through the winter and fertilisation occurs in the following spring. After the embryo has been formed the megasporangium "becomes sunken in a shallow pit formed by the cushion-like outgrowth of the sporophyll around the pedicel." It is

---

[1] Goebel (05) p. 581.   [2] Lyon (01) p. 135.

suggested that this outgrowth may be comparable with the integument which grows up from the sporophyll in the fossil genus *Lepidocarpon*[1] and almost completely encloses the sporangium. In the drawings given by Miss Lyon no features are recognisable which afford a parallel to the integument of *Lepidocarpon*. I have, however, endeavoured to show, by a brief reference to this author's interesting account of the two Californian species, that the physiological and morphological resemblances between the megasporangia of *Selaginella* and the integumented ovules of the seed-bearing plants are sufficiently close to enable us to recognise possible lines of advance towards the development of the true seed.

Professor Campbell[2] records an additional example of a *Selaginella*—probably *S. Bigelovii*—from the dry region of Southern California in which the spores become completely dried up after the embryo has attained some size, remaining in that state until the more favourable conditions succeeding the dry season induce renewed activity.

Isoetaceae.

The genus *Isoetes* is peculiar among Pteridophytes both in habit and in anatomical features. In its short and relatively thick tuberous stem, terminating in a crowded rosette of subulate leaves like those of *Juncus* and bearing numerous adventitious roots, *Isoetes* presents an appearance similar to that of many monocotyledonous plants. The habit of the genus is well represented by such species as *Isoetes lacustris* and *I. echinospora*[3] (fig. 132) both of which grow in freshwater lakes in Britain and in other north European countries. The latter species bears leaves reaching a length of 18 cm. The resemblance in habit between this isolated member of the Pteridophytes and certain Flowering plants, although in itself of no morphological significance, is consistent with the view expressed by Campbell that *Isoetes* may be directly related to the Monocotyledons[4].

---

[1] See p. 271.   [2] Campbell (05) p. 522.
[3] Motelay and Vendryès (82).   [4] Campbell (05) p. 561.

There is as a rule little or no difference between the foliage
leaves and sporophylls; in *I. lacustris* the latter are rather

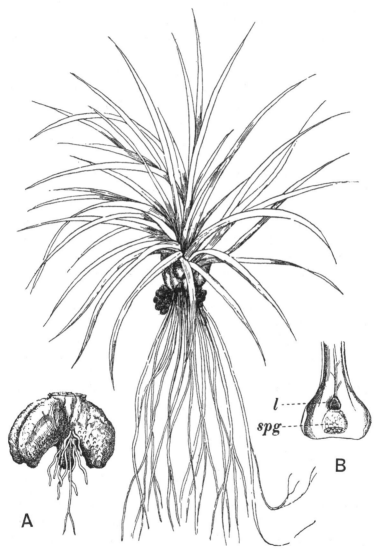

FIG. 132. *Isoetes echinospora* (After Motelay and Vendryès).
    A. Stem of *I. lacustris*.
    B. Base of sporophyll: *l*, ligule; *spg*, sporangium partially
        covered by velum.

larger and in the terrestrial species *I. hystrix*[1] the sterile leaves are represented by the expanded basal portions only, which persist like the leaf-bases of *Lepidodendron* as dark brown scales to form a protective investment to the older part of the stem. The innermost leaves are usually sterile; next to these are sporophylls bearing megasporangia, and on the outside are the older sporophylls with microsporangia. The long and slender portion of the leaf becomes suddenly expanded close to its attachment to the stem into a broad base of crescentic section which bears a fairly conspicuous ligule (figs. 132, B, *l*, 133, E, *l*) inserted by a foot or glossopodium in a pit near the upper part of the concave inner face. The ligule is usually larger than that of *Selaginella*, though of the same type. The free awl-like lamina contains four large canals bridged across at intervals by transverse diaphragms, and in the axial region a single vascular bundle of collateral structure. Other vascular elements, in the form of numerous short tracheids occur below the base of the transversely elongated ligule.

Stomata are found on the leaves of *I. hystrix, I. Boryana*[2], and in other species which are not permanently submerged. Both microsporangia and megasporangia are characterised by their large size and by the presence of trabeculae or strands of sterile tissue (fig. 133, E, H, *t*) completely bridging across the sporangial cavity or extending as irregular ingrowths among the spore-producing tissue. Similar sterile bands, though less abundant and smaller, are occasionally met with in the still larger sporangia of *Lepidostrobus*; these may be regarded as a further development of the prominent pad of cells which projects into the sporangial cavity in recent species of *Lycopodium* (fig. 126, D, *p*). The sporangia are attached by a very short stalk to the base of a large depression in the leaf-base below the ligule, from the pit of which they are separated by a ridge of tissue known as the saddle, and from this ridge a veil of tissue (the velum) extends as a roof over the sporangial chamber (fig. 133, E, *v*). In most species there is a large gap between the lower edge of the velum and that of the sporangial

---

[1] Scott and Hill (00). [2] Motelay and Vendryès (82) Pls. XVI, XVII.

pit, but in *I. hystrix* this protective membrane is separated
from the base of the leaf by a narrow opening, the resemblance
of which to the micropyle of an ovule suggested to one of the
older botanists the employment of the same term[1]. Mr T. G.
Hill[2] has called attention to the presence of mucilage canals in
the base of the sporophylls of *I. hystrix*, which he compares with
the strands of tissue known as the parichnos accompanying the
leaf-traces of *Lepidodendron* and *Sigillaria* in the outer cortex
of the stem. The transverse section shown in fig. 133, H and I,
shows two of these mucilage canals in an early stage of
development; a strand of parenchymatous elements distinguished by their partially disorganised condition and more
deeply stained membranes (fig. 133, I) runs through the
spandrels of the sporophyll tissue close to the upper surface.
There is a close resemblance between the structure of these
partially formed mucilage-canals and the tissue which has been
called the secretory zone in Lepidodendron stems. Fig. 133, H,
also shows a large microsporangium with prominent trabeculae
(*t*) lying below the velum. A longitudinal section (fig. 133, E)
through a sporophyll-base presents an appearance comparable
with that of an Araucarian cone-scale with its integumented
ovule and micropyle. The megaspores are characterised by
ridges, spines, and other surface-ornamentation[3]. Though usually
unbranched, the perennial stem of *Isoetes* (fig. 132) has in rare
cases been found to exhibit dichotomous branching, a feature, as
Solms-Laubach[4] points out, consistent with a Lycopodiaceous
affinity. The apex is situated at the base of a funnel-shaped
depression. The stem is always grooved; in some species two
and in others three deep furrows extend from the base up the sides
of the short and thick axis towards the leaves: from the sides of
these furrows numerous slender roots are given off in acropetal
succession. A stele of peculiar structure occupies the centre of
the stem; cylindrical in the upper part (fig. 133, A), it assumes
a narrow elliptical or, in species in which there are three furrows,
a triangular form in the lower portion of the tuberous stem.

The stem of *I. lacustris* represented in fig. 132, A, from which

[1] Braun (63).   [2] Hill, T. G. (04) (06).
[3] For figures, see Motelay and Vendryès (82); Bennie and Kidston (88)
Pl. vi.   [4] Solms-Laubach (02).

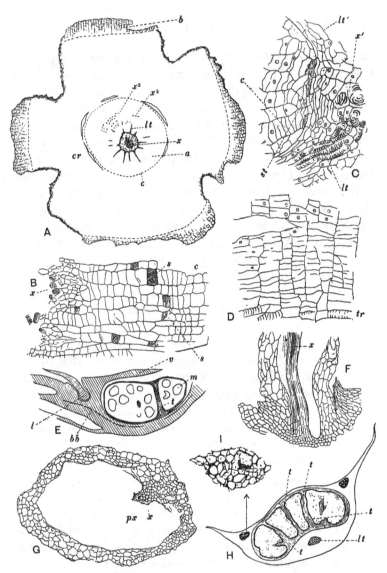

Fig. 133. *Isoetes lacustris.*

A. Transverse section of stem: *cr*, cortex; *x, x^2* xylem; *c*, cambium; *a*, thin-walled tissue; *lt*, leaf-traces; *b*, dead tissue.

B, C, D. Portions of A enlarged.

E. Longitudinal radial section of sporophyll-base: *v*, velum; *l*, ligule; *bb*, vascular bundle; *m*, megaspores; *t*, sterile tissue.

F. Longitudinal section through the base of a root.

G. Transverse section of root.

H. Transverse section of sporophyll, showing sporangium with trabeculae, *t*; leaf-trace, (*lt*), and two groups of secretory cells.

I. A group of secretory cells enlarged.

the laminae of the leaves have been removed from the summit affords an example of a species with two furrows. The drawing shows the widely gaping sides of the broad furrow with circular root-scars and a few simple and dichotomously branched roots. A short thick column of parenchymatous tissue projects from a slightly eccentric position on the base of the stem.

The primary vascular cylinder[1] consists of numerous spiral, annular or reticulate tracheids (fig. 133, A, x), which are either isodiametric or longer in a horizontal than in a vertical direction, associated with parenchyma. Lower in the stem crushed and disorganised xylem elements are scattered through a still living trabecular network of parenchymatous tissue. From the axial cylinder numerous leaf-traces (fig. 133, A, lt) radiate outwards, at first in a horizontal direction and then gradually ascending towards the leaves. The vascular cylinder is of the type known as cauline ; that is, some of the xylem is distinct in origin from that which consists solely of the lower ends of leaf-traces. As in *Lycopodium* the development of the metaxylem is centripetal.

Von Mohl[2], and a few years later Hofmeister[3], were the first botanists to give a satisfactory account of the anatomy of *Isoetes*, but it is only recently[4] that fresh light has been thrown upon the structural features of the genus the interest of which is enhanced by the many points of resemblance between the recent type and the Palaeozoic Lepidodendreae. A striking anatomical feature is the power of the stem to produce secondary vascular and non-vascular tissue ; the genus is also charac- terised by the early appearance of secondary meristematic activity which renders it practically impossible to draw any distinct line between primary and secondary growth. A cylinder of thin-walled tissue (fig. 133, A, a) surrounds the primary central cylinder and in this a cambial zone, c, is recognised even close to the stem-apex ; this zone of dividing cells is separated from the xylem by a few layers of rectangular cells to which the term prismatic zone has been applied.

---

[1] See von Mohl (40); Farmer (90).   [2] Von Mohl (40).
[3] Hofmeister (62).   [4] Farmer (90) ; Scott and Hill (00).

The early appearance of the cambial activity on the edge of the vascular cylinder is shown in fig. 133, C, which represents part of a transverse section of a young stem. A leaf-trace, *lt*, is in connexion with the primary xylem, *x'*, which consists of short tracheids, often represented only by their spiral or reticulately thickened bands of lignified wall, and scattered parenchyma. Some of the radially elongated cells on the sides of the leaf-trace are seen to be in continuity on the outer edge of the stele, at *st*, with flattened elements, some of which are sieve-tubes. The position of a second leaf-trace is shown at *lt'*. External to the sieve-tubes the tissue consists of radially arranged series of rectangular cells, some of which have already assumed the function of a cambium (*c*). The tissue produced by the cambium on its inner edge consists of a varying amount of secondary xylem composed of very short spiral tracheids; a few of these may be lignified (fig. 133, A, x^2) while others remain thin.

Phloem elements, recognisable by the presence of a thickened reticulum enclosing small sieve-areas (fig 133, B, *s*) are fairly abundant, and for the rest this intracambial region is composed of thin-walled parenchyma. In longitudinal section these tissues present an appearance almost identical with that observed in a transverse section. Fig. 133, B represents a longitudinal section, through the intracambial zone and the edge of the stele, of a younger stem than that shown in fig. 133, A. Most of the radially disposed cells internal to the meristematic region are parenchymatous without any distinctive features; a few scattered sieve-tubes (*s*) are recognised by their elliptical sieve-areas and an occasional tracheid can be detected. The cambium cuts off externally a succession of segments which constitute additional cortical tissue (fig. 133, A, *cr*) of homogeneous structure, composed of parenchymatous cells containing starch and rich in intercellular spaces. As the stem grows in thickness the secondary cortex reaches a considerable breadth and the superficial layers are from time to time exfoliated as strips of dead and crushed tissue (fig. 133, A, *b*). The diagrammatic sketch reproduced in fig. 133, A, serves to illustrate the arrangement and relative size of the tissue-regions

XIV] ISOETES

in an *Isoetes* stem. In the centre occur numerous spirally or reticulate tracheae scattered in parenchymatous tissue which has been considerably stretched and torn in the peripheral region of the stele; the radiating lines mark the position of the leaf-traces (*lt*) in the more horizontal part of their course. The zone between the cambium (*c*) and the edge of the central cylinder consists of radially disposed secondary tissue of short, and for the most part unlignified, elements including sieve-tubes and parenchyma; the secondary xylem elements consist largely of thin-walled rectangular cells with delicate spiral bands, but discontinuous rows of lignified tracheae (x^2) occur in certain regions of the intracambial zone. The rest of the stem consists of secondary cortex (*cr*) with patches of dead tissue (*b*) still adhering to the irregularly furrowed surface. The structure of the cambium and its products is shown in the detailed drawing reproduced in fig. 133, D. Many of the elements cut off on the inner side of the cambium exhibit the characters of tracheids: most of these are unlignified, but others have thicker and lignified walls (*tr*).

I. hystrix appears to be exceptional in retaining its leaf-bases, which form a complete protective investment and prevent the exfoliation of dead cortex. Each leaf-trace consists of a few spiral tracheids accompanied by narrow phloem elements directly continuous with the secondary phloem of the intracambial zone. Dr Scott and Mr Hill have pointed out that a normal cambium is occasionally present in the stem of *I. hystrix* during the early stages of growth; this gives rise to xylem internally. The few phloem elements observed external to the cambium may be regarded as primary phloem, a tissue not usually represented in an Isoetes stem[1]. The occasional occurrence of this normal cambium, may, as Scott and Hill suggest, be a survival from a former condition in which the secondary thickening followed a less peculiar course. The lower leaf-traces become more or less obliterated as the result of the constant increase in thickness of the broad zone of secondary tissues through which they pass.

The adventitious roots are developed acropetally and

---

[1] Miss Stokey (09), in a paper which appeared since this account was written, criticises the conclusions of Scott and Hill (00).

arranged in parallel series on each side of the median line of the two or three furrows. The three arms of the triangular stele of *I. hystrix* and the two narrow ends of the long axis of the stele of *I. lacustris,* which in transverse section has the form of a flattened ellipse, are built up of successive root-bases. A root of *Isoetes* (fig. 133, G) possesses one vascular bundle, *x,* with a single strand of protoxylem, *px,* thus agreeing in its monarch structure with the root-bundle in *Selaginella* and many species of *Lycopodium.* The cortical region of the root consists of a few layers of outer cortex succeeded by a large space, formed by the breaking down of the inner cortical tissue, into which the vascular bundle projects (fig. 133, F). The peculiarity of the roots in having a hollow cortex and an eccentric vascular bundle was noticed by Von Mohl[1]. In the monarch bundles, as in the fistular cortex and dichotomous branching, the roots of *Isoetes* present a striking resemblance to the slender rootlets of the Palaeozoic *Stigmaria* (see Page 246). The longitudinal section through the base of a root of *Isoetes lacustris* shown in fig. 133, F, affords a further illustration of certain features common to the fossil and recent types.

FOSSIL LYCOPODIALES.

Isoetaceae

The geological history of this division of the Pteridophyta is exceedingly meagre, a fact all the more regrettable as it is by no means improbable that in the surviving genus *Isoetes* we have an isolated type possibly of considerable antiquity and closely akin to such extinct genera as *Pleuromeia* and *Sigillaria.* If Saporta's Lower Cretaceous species *Isoetes Choffati*[2], or more appropriately *Isoetites Choffati,* is correctly determined, it is the oldest fossil member of the family and indeed the most satisfactory among the more than doubtful species described as extinct forms of *Isoetes.*

[1] Von Mohl (40).
[2] Saporta (94) p. 134, Pls. xxiv. xxv. xxvii.

Isoetites.

The generic name *Isoetites* was first used by Münster[1] in the description of a specimen, from the Jurassic lithographic slates of Solenhofen in Bavaria, which he named *Isoetites crociformis.* The specific name was chosen to express a resemblance of the tuberous appearance of the lower part of the imperfectly preserved and indeterminable fossil to a Crocus corm.

Impressions of Isoetes-like leaves from the Inferior Oolite of Yorkshire figured by Phillips[2] and afterwards by Lindley[3] as *Solenites Murrayana* were compared by the latter author with *Isoetes* and *Pilularia,* but these leaves are now generally assigned to Heer's gymnospermous genus *Czekanowskia.* An examination of the structure of the epidermal cells of these Jurassic impressions convinced me that they resemble recent coniferous needles more closely than the leaves of any Pteridophyte. The genus *Czekanowskia*[4] is recognised by several authors as a probable member of the Ginkgoales.

Isoetites Choffati. Saporta.

The late Marquis of Saporta founded this species on two sets of impressions from the Urgonian (Lower Cretaceous) of Portugal which, though not found in actual organic connexion, may possibly be portions of the same plant. Small relatively broad tuberous bodies reaching a breadth of 1 cm. are compared with the short and broad stem of *Isoetes,* which they resemble in bearing numerous appendages radiating from the surface like the roots of the recent species; on the exposed face of the stem occur scattered circular scars representing the position of roots which were detached before fossilisation. Other impressions are identified as the basal portions of sporophylls bearing sporangia : these suggest the expanded base of the fertile leaves of *Isoetes* with vertically elongated sporangia, some of which have a smooth surface while in others traces of internal structure are exposed; the interior consists of an irregular network with depressions containing carbonised remains of spores.

[1] Münster (42) p. 107, Pl. IV. fig. 4.   [2] Phillips (29) A Pl. x. fig. 12.
[3] Lindley and Hutton A (34) Pl. CXXI.   [4] Nathorst (06); Seward (00) p. 278.

While recognising a general resemblance to the sporophylls
of *Isoetes*, certain differences are obvious: there is no ligule in
the fossil leaves nor are there any distinct traces of vascular
strands such as occur in the leaves of recent species. The
form of the sporangium, more elongated than in the majority
of recent forms, is compared by Saporta with that in a south
European species *Isoetes setacea* Spr.

Such evidence as we have lends support to the inclusion of
these Portuguese fossils in the genus *Isoetites*, but apart from
the fact that we have no proof of any connexion between the
stems and supposed sporophylls, the resemblance of the latter
to those of *Isoetes* is, perhaps, hardly sufficient to satisfy all
reasonable scepticism.

The generic name *Isoetopsis* was used by Saporta as more
appropriate than *Isoetes* for some Eocene fossils from Aix-en-
Provence which are too doubtful to rank as trustworthy evidence
of the existence of the recent genus. The species, *Isoetopsis
subaphylla*[1] is founded on impressions of small scales, 4 mm.
long, bearing circular bodies which are compared with sporangia
or spores.

Other records of fossils referred to *Isoetes* need not be
described as they have no claim to be regarded as contributions
towards the past history of the genus. Heer's Miocene species
Isoetites Scheuzeri and *I. Braunii* Unger[2] from Switzerland are
based on unsatisfactory material and are of no importance.

Pleuromeia.

The generic name *Pleuromeia*, was suggested by Corda[3] for
a fossil from the Bunter Sandstone, the original description of
which was based by Münster[4] on a specimen discovered in a
split stone from the tower of Magdeburg Cathedral.

The majority of the specimens have been obtained from the
neighbourhood of Bernburg, but a few examples are recorded
from Commern and other German localities: all are now included
under the name *Pleuromeia Sternbergi*. Germar, who published

---

[1] Saporta (88) p. 28, Pl. ii. pp. 16—20.   [2] Heer (76) A.
[3] Corda, in Germar (52).   [4] Münster (42) A.

one of the earlier accounts of the species, states that Corda dissented from Münster's choice of the name *Sigillaria* and proposed the new generic title *Pleuromeia*. One of the best descriptions of the genus we owe to Solms-Laubach[1] whose paper contains references to earlier writers. Illustrations have been published by Münster, Germar[2], Bischof[3], Solms-Laubach and Potonié[4].

Pleuromeia Sternbergi. (Münster.)
Fig. 134.

1842. *Sigillaria Sternbergii*, Münster.
1854. *Sagenaria Bischofii*, Goeppert[5].
1885. *Sigillaria oculina*, Blanckenhorn.
1904. *Pleuromeia oculina*, Potonié.

Pleuromeia Sternbergi is represented by casts of vegetative and fertile axes, but the preservation of the latter is not sufficiently good to enable us to draw any very definite conclusions as to the nature of the reproductive organs. Casts of the stems reach a length of about 1 metre and a diameter of 5—6 cm., or in some cases 10 cm.; all of them are in a more or less decorticated state, the degree of decortication being responsible for differences in the external features which led Spieker[6] to adopt more than one specific name.

Fig. 134, A, represents a sketch, made some years ago, of a specimen in the Breslau Museum which contains several examples of this species, among others those described by Germar in 1852. The cylindrical cast (38 cm. long by 12 cm. in circumference), which has been slightly squeezed towards the upper end, bears spirally arranged imperfectly preserved leaf-scars and the lower end shows the truncated base of one of the short Stigmaria-like arms characteristic of the plant. As shown clearly in a specimen originally figured by Bischof and more recently by Potonié[7], the stem-base is divided by a double dichotomy into four short and broad lobes with blunt apices and bent upwards like the arms of a grappling iron (fig. 134, D).

---

[1] Solms-Laubach (99).   [2] Germar (52).   [3] Bischof (53).
[4] Potonié (01) p. 754; (04) Lief ii.   [5] Goeppert, in Römer (54) Pl. xv. fig. 7.
[6] Spieker (53).   [7] Potonié (*loc. cit.*).

The surface of this basal region is characterised by numerous circular scars (fig. 134, D; 4 scars enlarged) in the form of

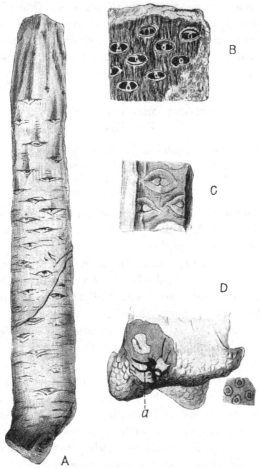

FIG. 134. *Pleuromeia Sternbergi.*
    A. Cast of stem in the Breslau Museum (⅓ nat. size). (A.C.S.)
    B. "*Sigillaria oculina*" Blanckenhorn. (After Weiss).
    C, D. Leaf-scars and base of stem: *a*, vascular tissue. (After Solms-
        Laubach.)

slightly projecting areas with a depression in the centre of each. These are undoubtedly the scars of rootlets, remains of which are occasionally seen radiating through the surrounding

rock. As seen in fig. 134, D, *a*, the fractured surface of a
basal area may reveal the existence of an axial vascular cylinder
giving off slender branches to the rootlets.

The bulbous enlargement at the base of the Brown sea-
weed *Laminaria bulbosa* Lam.[1] simulates the swollen base of
Pleuromeia; but a confusion between these two plants is hardly
likely to occur. Above the Stigmaria-like base the gradually
tapered axis, in the less decorticated specimens, bears spirally
disposed transversely elongated areas consisting of two triangular scars between which is the point of exit of a leaf-trace.
The form of the leaf-scars is best seen on the face of a mould
figured by Solms-Laubach (fig. 134, C): in this case the two
triangular areas appear as slight projections separated by a
narrow groove marking the position of the vascular bundle of
the leaf. The curved lines above and below the leaf-scar
probably mark the boundary of the leaf-base. The two
triangular scars are compared by Solms-Laubach and by
Potonié with the parichnos-scars of *Sigillaria* and *Lepidodendron* (cf. fig. 146, C.), but the large size of the Pleuromeia
scars constitutes an obvious difference though possibly not a
distinction of importance.

The occurrence of a vertical canal filled with carbonaceous
material in some of the stems throws light on the internal
structure: the canal, which is described by Solms-Laubach as
having a stellate outline in transverse section recalls the narrow
central cylinder of a Lepidodendron stem, and this comparison
is strengthened by the presence of obliquely ascending grooves
which represent leaf-traces passing through the cortex. In
specimens which have lost more of the cortical tissues the
surface is characterised by spirally disposed, discontinuous
vertical grooves representing portions of leaf-traces precisely as
they appear in similar casts of *Lepidodendron*. There is no
direct evidence of the existence of secondary wood in the stem,
but, as Potonié has pointed out, the greater transverse elongation
of the leaf-scars in the lower part of a cast (fig. 134, A) points
to the production of some secondary tissue either in the
vascular cylinder or cortex, or possibly in both regions.

[1] Barber (89) Pls. v. vi.

In some specimens of *Pleuromeia* the upper portion is clothed with crowded and imbricate sporophylls which reach a length of 2·5 cm., a maximum breadth of 2·7 cm., and a thickness of 1 mm. Each sporophyll has a thin wing-like border, and on the lower face are several parallel lines. Solms-Laubach describes the sporangium or ovule as attached to the lower surface of the sporophyll and this opinion has been confirmed by Fitting[1] who has also brought forward satisfactory evidence in favour of the sporangial nature of the reproductive organs. Fitting found numerous spores in the Bunter Sandstone near Halle; these are flattened circular bodies 0·5—0·7 mm. in diameter with a granulated surface and the three converging lines characteristic of spores produced in tetrads. The comparison made by this author between the sporophylls of *Pleuromeia*, which bore the sporangia on the lower surface instead of on the upper as in other lycopodiaceous plants, and the pollen-sacs of Conifers, is worthy of note in reference to the possible relationship between Conifers and Lycopods.

A comparison of the *Isoetes* stem represented in fig. 132, A, with the base of a *Pleuromeia* shows a striking similarity, but, as Fitting points out, the Stigmaria-like arms of the fossil contained a vascular cylinder whereas the blunt lobes of *Isoetes* consist exclusively of cortical tissue, the roots being given off from the grooves between the lobes of the tuberous stem.

The position of *Pleuromeia* must for the present be left an open question; it is, however, clear that the plant bears a close resemblance in the form of its base to the Stigmarian branches of *Lepidodendron* and *Sigillaria*. The vegetative shoot appears to be constructed on a plan similar to that of these two Palaeozoic genera, but the strobilus is of a different type. It would seem probable that *Pleuromeia* may be closely allied to *Isoetes* and to the arborescent Lycopods of Palaeozoic floras. It is not improbably a link in a chain of types which includes *Sigillaria* on the one hand and *Isoetes* on the other.

It is not improbable that a specimen from the Lower Bunter of Commern which Blanckenhorn made the type of a new species, *Sigillaria oculina* (fig. 134, B) is specifically

[1] Fitting (07).

identical with *Pleuromeia Sternbergi*. An examination of a cast of the type-specimen in the Berlin Bergakademie led me to regard the fossil with some hesitation as a true *Sigillaria*, but a more extended knowledge of *Pleuromeia* lends support to the view adopted by Potonié[1] that Blanckenhorn's plant is not generically distinct from *Pleuromeia Sternbergi*. The resemblance between *Sigillaria oculina* and some of the Palaeozoic species of *Sigillaria* emphasised by Weiss[2] has given rise to the belief that the genus *Sigillaria* persisted into the Triassic era ; it is, however, highly probable that the Bunter specimen has no claim to the generic name under which it has hitherto been known.

The Bunter Sandstone in which *Pleuromeia* is the sole representative of plant-life, at least in certain localities, is usually considered to be a desert formation. We may not be far wrong in accepting Fitting's suggestion that in this isolated species we have a relic of the sparse vegetation which was able to exist where the presence of lakes added a touch of life to the deadness of the Triassic desert.

Pleuromeia is recorded by Fliche as a rare fossil in the Middle Trias of France in the neighbourhood of Lunéville[3].

Herbaceous fossil species of Lycopodiales.

The history of our knowledge of fossil representatives of the Lycopodiales, as also of the Equisetales, affords a striking illustration of the danger of attempting to found a classification on such differences as are expressed by the terms herbaceous and arborescent in the sense in which they are usually employed. As we have seen[4], the presence of secondary wood in stems of the Palaeozoic plant now known as *Calamites* led so competent a botanist as Adolphe Brongniart to recognise a distinct generic type *Calamodendron*, which he placed in the Gymnosperms, reserving the designation *Calamites* for species in which no indication of secondary thickening had been found.

Similarly, the genus *Sigillaria* was regarded as a Gymnosperm because it was believed to be distinguished from

[1] Potonié (04) Lief ii.   [2] Weiss, C. E. (86).
[3] Fliche (03).   [4] Vol. I, p. 300.

Lepidodendron by the power of forming secondary vascular tissues; the latter genus, originally thought to be always herbaceous, was classed with the Pteridophytes. At the time when this unnatural separation was made between stems with secondary wood and those in which no secondary wood was known to exist, botanists were not aware of the occurrence of any recent Pteridophyte which shared with the higher plants the power of secondary growth in thickness provided by means of a meristematic zone. It is true that the presence or absence of a cambium does not in practice always coincide with the division into herbaceous and arborescent plants: no one would speak of a Date-Palm as a herbaceous plant despite the absence of secondary wood.

The danger which should be borne in mind, in adopting as a matter of convenience the term herbaceous as a sectional heading, is that it should not be taken to imply a complete inability of the so-called herbaceous types to make secondary additions to their conducting tissues. The specimens on which the species of *Lycopodites* and *Selaginellites*, (genera which may be designated herbaceous,) are founded are preserved as impressions and not as petrifications; we can, therefore, base definitions only on habit and on such features as are shown by fertile leaves and sporangia. We are fully justified in concluding from evidence adduced by Goldenberg more than fifty years ago and from similar evidence brought to light by more recent researches, that there existed in the Palaeozoic era lycopodiaceous species in close agreement in their herbaceous habit with the lycopods of present-day floras. It has been suggested[1] that the direct ancestors of the genera *Lycopodium* and *Selaginella* are represented by the species of *Lycopodites* and *Selaginellites* rather than by *Lepidodendron* and *Sigillaria*, the arborescent habit of which has been rendered familiar by the numerous attempts to furnish pictorial reproductions of a Palaeozoic forest. Until we are able to subject the species classed as herbaceous to microscopical examination we cannot make any positive statement as to the correctness of this view,

---

[1] Halle (07) p. 1.

but such facts as we possess lead us to regard the suggestion as resting on a sound basis.

Palaeobotanical literature abounds in records of species of *Lycopodites, Lycopodium, Selaginella* and *Selaginites,* which have been so named in the belief that their vegetative shoots bear a greater resemblance to those of recent lycopodiaceous plants than to the foliage shoots of *Lepidodendron.* Many of these records are valueless: *Lepidodendra,* twigs of *Bothrodendron*[1] species of conifers, fern rhizomes, and *Aphlebiae*[2] have masqueraded as herbaceous lycopods. It is obvious that an attempt to identify fossils presenting a general agreement in habit and leaf-form with recent species of lycopods must be attended with considerable risk of error. Recent Conifers include several species the smaller branches of which simulate the leafy shoots of certain species of *Lycopodium* and *Selaginella,* and it is not surprising to find that this similarity has been responsible for many false determinations. Among Mosses and the larger foliose Liverworts there are species which in the condition of imperfectly preserved impressions, might easily be mistaken for lycopodiaceous shoots: an equally close resemblance is apparent in the case of some flowering plants, such as New Zealand species of *Veronica, Tafalla graveolens* (a Composite), *Lavoisiera lycopodiodes* Gard.[3] (a species of Melastomaceae), all of which have the habit of Cupressineae among the conifers as well as of certain lycopodiaceous plants. It may be impossible to decide whether fossil impressions of branches, which are presumably lycopodiaceous, bear two kinds of leaves[4] like the great majority of recent species of *Selaginella. Selaginella grandis,* if seen from the under surface, would appear to have two rows of leaves only and might be confused with a small twig of such a conifer as *Dacrydium Kirkii,* a New Zealand species.

The New Zealand conifers *Dacrydium cupressinum* Soland., and *Podocarpus dacrydioides* Rich. closely simulate species of *Selaginellites* and *Lycopodites*: in the British Museum a

---

[1] Feistmantel (75) A. p. 183, Pl. xxx. pp. 1 and 2.
[2] Germar (49) Pl. xxvi; Geinitz (55) A. Pl. i. pp. 5, 6.
[3] Bommer (03) p. 29, Pl. ix, figs. 138—141.
[4] Solms-Laubach (91) A. p. 137.

specimen of the latter species bears a label describing it as *Lycopodium arboreum* (Sir Joseph Hooker and Dr Solander; 1769). The twigs of the Tasmanian conifer *Microcachrys tetragona* Hook. f. are very similar in habit to shoots of the recent *Lycopodium tetragonum* (fig. 121, C).

In the description of examples of *Lycopodites* and *Selaginellites* I have confined myself to such as appear to be above suspicion either because of the presence of spore-bearing organs or, in a few cases, because the specimens of sterile shoots are sufficiently large to show the form of branching in addition to the texture of the leaves. The two generic names *Lycopodites* and *Selaginellites* are employed for fossil species which there are substantial grounds for regarding as representatives of *Lycopodium* and *Selaginella*. The designation *Selaginellites* is adopted only for species which afford evidence of heterospory; the name *Lycopodites*, on the other hand, is used in a comprehensive sense to include all forms—whether homophyllous or heterophyllous—which are not known to be heterosporous. This restricted use of the generic name *Selaginellites* is advocated by Zeiller[1], who instituted the genus, and by Halle[2] in his recent paper on herbaceous lycopods.

Lycopodites.

The generic term *Lycopodites* was used by Brongniart in 1822[3] in describing some Tertiary examples of slender axes clothed with small scale-like leaves which he named *Lycopodites squamatus*. These are fragments of coniferous shoots. In the *Prodrome d'une histoire des végétaux fossiles*[4] Brongniart included several Palaeozoic and Jurassic species in *Lycopodites* and instituted a new genus *Selaginites*, expressing a doubt as to the wisdom of attempting to draw a generic distinction between the two sets of species. In a later work[5] he recognised only one undoubted species, *Lycopodites falcatus*. The first satisfactory account of fossils referred to *Lycopodites* is by Goldenberg[6]

[1] Zeiller (06) p. 140. [2] Halle (07).

[3] Brongniart (22) A. p. 304, Pl. VI, fig. 1. [4] Brongniart (28) A. p. 83.

[5] Brongniart (49) A. p. 40. [6] Goldenberg (55) p. 9.

who gave the following definition of the genus:—" Branches with leaves spirally disposed or in whorls. Sporangia in the axil of foliage leaves or borne in terminal strobili."

It was suggested by Lesquereux[1] that Goldenberg's definition, which was intended to apply to herbaceous species, should be extended so as to include forms with woody stems but which do not in all respects agree with *Lepidodendron.* Kidston[2] subsequently adopted Lesquereux's modification of Goldenberg's definition. We cannot draw any well-defined line between impressions of herbaceous forms and those of small arborescent species. We use the name *Lycopodites* for such plants as appear to agree in habit with recent species of *Lycopodium* and *Selaginella* and which, so far as we know, were not heterosporous : it is highly probable that some of the species so named had the power of producing secondary wood, a power possessed by some recent Pteridophytes which never attain the dimensions of arborescent plants.

It has been shown by Halle[3], who has re-examined several of Goldenberg's specimens which have been acquired by the Stockholm Palaeobotanical Museum, that some of his species of *Lycopodites* are heterosporous and therefore referable to Zeiller's genus *Selaginellites.*

In 1869 Renault described two species of supposed Palaeozoic Lycopods as *Lycopodium punctatum* and *L. Renaultii*[4], the latter name having been suggested by Brongniart to whom specimens were submitted. These species were afterwards recognised by their author as wrongly named and were transferred to the genus *Heterangium*[5], a determination which is probably correct; it is at least certain that the use of the name *Lycopodium* cannot be upheld.

We have unfortunately to rely on specimens without petrified tissues for our information in regard to the history of *Lycopodites* and *Selaginellites.* Among the older fossils referred to *Lycopodites* are specimens from Lower Carboniferous rocks at Shap in Westmoreland which Kidston originally described

[1] Lesquereux (84) A. p. 777.   [2] Kidston (86³) p. 561.
[3] Halle (07).   [4] Renault (69) p. 178, Pls. xii—xiv.
[5] Renault (96) A. p. 249.

as *Lycopodites Vanuxemi*[1], identifying them with Goeppert's *Sigillaria Vanuxemi*[2] founded on German material. In a later paper Kidston transferred the British specimens of vegetative shoots to a new genus *Archaeosigillaria*[3].

Lycopodites Stockii Kidston[4].

The plant so named was discovered in Lower Carboniferous strata of Eskdale, Dumfries, Scotland; it is represented by imperfectly preserved shoots bearing a terminal strobilus and was originally described by Kidston as apparently possessing two kinds of foliage leaves borne in whorls. The larger leaves have an ovate cordate lamina with an acuminate apex, while the smaller leaves, which are less distinct, are transversely elongated, and simulate sporangia in appearance. Dr Kidston's figure of this species has recently been reproduced by Professor Bower[5] who speaks of the supposed smaller leaves as sporangia, a view with which the author of the species agrees. It would appear that this identification is, however, based solely on external resemblance and has not been confirmed by the discovery of any spores. Assuming the sporangial nature of these structures, this Palaeozoic type represents, as Bower points out, a condition similar to that in some recent species of *Lycopodium* in which sporangia are not confined to a terminal strobilus but occur also in association with ordinary foliage leaves. The strobilus consists of crowded sporophylls which are too imperfect to afford any definite evidence as to their homosporous or heterosporous nature. As Solms-Laubach[6] points out, this type recalls *Lycopodium Phlegmaria* among recent species.

Lycopodites Reidii Penhallow.

Professor Penhallow[7] instituted this name for a specimen measuring 8 cm. long by 6 mm. in breadth, collected by Mr

[1] Kidston (86[3]).   [2] Goeppert (52) A.
[3] Kidston (01) p. 38.   [4] Kidston (84) Pl. v; (01) p. 37.
[5] Bower (08) p. 298, fig. 147.   [6] Solms-Laubach (91) A. p. 186.
[7] Penhallow (92) Pl. I. fig. 2, p. 8.

Reid from the Old Red Sandstone of Caithness, consisting of an axis bearing narrow lanceolate leaves some of which bear sporangia at the base.

Lycopodites Gutbieri Goeppert[1].

1894, *Lycopodites elongatus* Kidston[2] (not Goldenberg).

The species, figured by Geinitz as *Lycopodites Gutbieri*[3], from the Coal-Measures of Saxony is probably a true representative of the genus. The Saxon specimens are heterophyllous; the larger lanceolate and slightly falcate leaves arranged in two rows, are 4—5 mm. long while the smaller leaves are one half or one third this size; some of the dichotomously branched shoots terminate in long and narrow strobili not unlike those of Zeiller's species *Selaginellites Suissei*[4] Kidston[5] has included under this specific name some fragments collected by Hemingway from the Upper Coal-Measures of Radstock, Somersetshire, but as only one form of leaf is seen the reasons for adopting Goeppert's designation are perhaps hardly adequate.

Lycopodites ciliatus Kidston[6].

Under this name Kidston describes a small specimen, obtained by Hemingway from the Middle Coal-Measures of Barnsley in Yorkshire, consisting of a slender forked axis bearing oval-acuminate leaves approximately 5 mm. long with a finely ciliate margin. Associated with the leaves were found spores which Kidston regards as megaspores.

Lycopodites macrophyllus Goldenberg[7].

This species, originally described by Goldenberg from the Coal-Measures of Saarbrücken has been re-examined by Halle[8] who is unable to confirm Goldenberg's statement as to hetero-

[1] Goeppert (52) p. 440.   [2] Kidston (94) A. p. 254.
[3] Geinitz (55) A. p. 32, Pl. I. fig. 1.   [4] Page 88.
[5] Kidston (01) p. 36, fig. 2, B.   [6] Kidston (01) p. 37, fig. 2, A.
[7] Goldenberg (55) Pl. I. fig. 5.   [8] Halle (07) Pl. I. fig. 5.

phylly.   The shoots closely resemble *Selaginellites primaevus*[1] (Gold).

Lycopodites Zeilleri Halle[2].   Fig. 135, C.

Halle has founded this species on specimens, from the Coal-Measures of Zwickau in Saxony, characterised by dimorphic

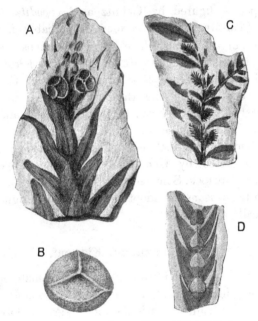

Fig. 135.   *Selaginellites and Lycopodites.*   (After Halle.)
A.   *Selaginellites primaevus* (Gold.).   × 10.
   B.   Megaspore of *Selaginellites elongatus* (Gold.).   × 50.
   C.   *Lycopodites Zeilleri* Halle.   (Nat. size.)
   D.   *Selaginellites elongatus* (Gold.).   × 2.

lanceolate leaves in four rows, the larger being 4—6 mm. long: the smaller leaves have a ciliate edge.   A comparison is made with the recent species *Selaginella arabica* Baker, *S. revoluta* Bak., and *S. armata* Bak. in which the leaves are described as ciliate. In the absence of sporangia and spores the species is placed in the genus *Lycopodites*.

[1] Page 89.   [2] Halle (07).

Lycopodites lanceolatus (Brodie). Fig. 136.

1845 *Naiadita lanceolata*, Brodie[1].
  Naiadea acuminata, Buckman[2].
1850 *Naiadea lanceolata*, Buckman[3].
  Naiadea petiolata, Buckman[4]
1900 *Naiadites acuminatus*, Wickes[5].
1901 *Naiadita lanceolata*, Sollas[6] (figures showing habit of the plant).
1904 *Lycopodites lanceolatus*, Seward[7] (figure showing habit of the plant).

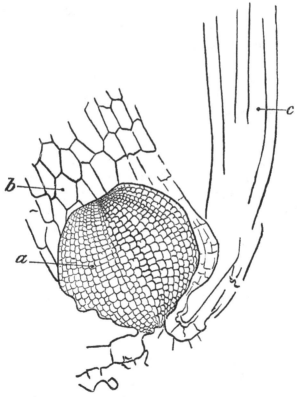

FIG. 136. *Lycopodites lanceolatus* (Brodie). (After Miss Sollas. × 40.)
 a, Sporangium wall; *b*, leaf.
 c, remains of tubular elements in stem.

[1] Brodie (45) p. 93.   [2] Buckman in Murchison (45) p. 6.
[3] Buckman (50) p. 415, fig. 2.   [4] Buckman (50) p. 415, fig. 4.
[5] Wickes (00) p. 422.   [6] Sollas (01).
[7] Seward (04) p. 14, Pl. II. figs. 2, 3.

Specimens referred to this species were originally recorded by Brodie from Rhaetic rocks in the Severn valley, the name *Naiadita* being chosen as the result of Lindley's comparison of the small and delicate leaves with those of recent species of the Monocotyledonous family Naiadaceae. The species may be described as follows:

Plant slender and moss-like in habit. The axis, which is delicate and thread-like, bears numerous linear acuminate or narrow ovate leaves reaching a length of approximately 5 mm. Under a low magnifying power the thin lamina of the leaves is seen to have a superficial layer of polygonal or rectangular cells arranged in parallel series (fig. 136 *b*). There is no trace of midrib or stomata. The sporangia are more or less spherical and short-stalked, situated at the base of the foliage leaves and containing numerous tetrads of spores. The spores have a diameter of 0·08 mm.

Buckman founded additional species on differences in the shape of the leaves but, as Miss Sollas has pointed out, such differences as he noticed may be detected on the same axis. It was stated in an earlier chapter[1] that Starkie Gardner, on insufficient evidence, proposed to place Brodie's plant among the Mosses. The discovery by Mr Wickes of new material at Pylle hill near Bristol afforded an opportunity for a re-examination of the species: this was successfully undertaken by Miss Sollas who was able to dissolve out spores from the matrix by dilute hydrochloric acid, and to recognise the remains of internal structure in the slender axes by exposing successive surfaces with the aid of a hone. It was found that sporangia occurred at the base of some of the leaves containing numerous tetrads of spores, the individual spores having a diameter of 0·08 mm., apparently twice as large as those of any recent species of *Lycopodium*. Fig. 136 shows a sporangium, *a*, at the base of a leaf, *b*. Indications of tubular elements were recognised in the stem and it is noteworthy that although the outlines of epidermal cells on the leaves are well preserved no stomata were found. The leaves of the recent American species *Lycopodium alopecuroides* Linn. var. *aquaticum* Spring[2], which lives

[1] Vol. i. p. 240.   [2] Sollas (01) p. 311.

under water, possess stomata.   It is probable that in *Lycopodites lanceolatus* the leaves had a very thin lamina and may have been similar in structure to those of recent Mosses; the plant possibly lived in very humid situations or grew submerged. Miss Sollas's investigations afford a satisfactory demonstration of the lycopodiaceous nature of this small Rhaetic species: as I have elsewhere suggested[1], the generic name *Lycopodites* should be substituted for that of *Naiadita*.   Examples of this species may be seen in the British Museum.

The Rhaetic species from Scania, *Lycopodites scanicus* Nath.[2] (*in litt.*), recently re-described by Halle and originally referred by Nathorst to *Gleichenia* affords another example of the occurrence of a small herbaceous lycopod of Rhaetic age.

Fig. 137.   *Lycopodites falcatus* L. and H.   From the Inferior Oolite of York-shire.   (Nat. size.   M.S.)

Lycopodites falcatus Lind. and Hutt.   Fig. 137.

1831   *Lycopodites falcatus*, Lindley and Hutton[3].
1838   *Muscites falcatus*, Sternberg[4]
1870   *Lycopodium falcatum*, Schimper[5].

In 1822 Young and Bird[6] figured a specimen from the

---

[1] Seward (04) p. 14.   [2] Halle (07) p. 14, Pl. III. figs. 6—12.
[3] Lindley and Hutton (31) A. Pl. LXI.   [4] Sternberg (38) A. p. 38.
[5] Schimper (70) A. p. 9.   [6] Young and Bird (22) A. Pl. II. fig. 7.

Inferior Oolite rocks of the Yorkshire coast bearing "small round crowded leaves," which was afterwards described by Lindley from additional material obtained from Cloughton near Scarborough as *Lycopodites falcatus*. The example represented in fig. 137 shows the dichotomously branched shoots bearing two rows of broadly falcate leaves. A careful examination of the type-specimen[1] revealed traces of what appeared to be smaller leaves, but there is no satisfactory proof of heterophylly. No sporangia or spores have been found. This British species has been recorded from Lower Jurassic or Rhaetic rocks of Bornholm[2] and a similar though probably not identical type, *Lycopodites Victoriae*[3], has been recognised in Jurassic strata of Australia (South Gippsland, Victoria). An Indian plant described by Oldham and Morris[4] from the Jurassic flora of the Rajmahal hills as *Araucarites* (?) *gracilis* and subsequently transferred by Feistmantel to Schimper's genus *Cheirolepis*[5] may be identical with the Yorkshire species. The Jurassic fragments described by Heer from Siberia as *Lycopodites tenerrimus*[6] may be lycopodiaceous, but they are of no botanical interest.

Other examples of Mesozoic Lycopods have been recorded, but in the absence of well-preserved shoots and sporangia they are noteworthy only as pointing to a wide distribution of *Lycopodites* in Jurassic and Cretaceous floras[7].

From Tertiary strata species of supposed herbaceous lycopods have been figured by several authors, one of the best of which is *Selaginella Berthoudi* Lesq.[8] from Tertiary beds in Colorado. This species agrees very closely in the two forms of leaf with *Selaginella grandis*, but as the specimens are sterile we have not sufficient justification for the employment of the generic name *Selaginellites*.

---

[1] No. 39314, Brit. Mus.   [2] Möller (02) Pl. vi. fig. 21.
[3] Seward (04²) p. 161, Pl. viii. figs. 2—4. The drawing is reproduced twice natural size.
[4] Oldham and Morris (63) Pls. xxxiii. xxxv.
[5] Feistmantel (77) p. 87.   [6] Heer (76) Pl. xv. figs. 1—8.
[7] Nathorst (90) A. Pl. ii. fig. 3. Saporta (94) Pls. xxiii.—xxvi. Knowlton (98) p. 136.
[8] Lesquereux (78) Pl. v. fig. 12. See also Knowlton *loc. cit.*

Selaginellites.

This generic name has been instituted by Zeiller[1] for specimens from the coal basis of Blanzy (France). It is applied to heterosporous species with the habit of *Selaginella*: Zeiller preferred the designation *Selaginellites* to *Selaginella* on the ground that the type species differs from recent forms in having more than four megaspores in each megasporangium. It is, however, convenient to extend the term to all heterosporous fossil species irrespective of the spore-output.

Selaginellites Suissei Zeiller.

This species was described in Zeiller's preliminary note[2] as *Lycopodites Suissei*, but he afterwards transferred it to the genus *Selaginellites*. In habit the plant bears a close resemblance to *Lycopodites macrophyllus* of Goldenberg; the shoots, 1—3 mm. thick, are branched in a more or less dichotomous fashion and bear tetrastichous leaves. The larger leaves reach a length of 4—6 mm. and a breadth of 2—3 mm ; the smaller leaves are described as almost invisible, closely applied to the axis, oval-lanceolate and 1—2 mm. long with a breadth of 0·5—0·75 mm. Long and narrow strobili (15 cm. by 8—10 mm.) terminate the fertile branches; these bear crowded sporophylls with a triangular lamina and finely denticulate margin. Oval sporangia were found on the lower sporophylls containing 16—24 spherical megaspores 0·6—0·65 mm. in diameter. The outer membrane of the spore is characterised by fine anastomosing ridges and thin plates radiating from the apex and forming an equatorial collarette. The microspores have a diameter of 40—60μ and the same type of outer membrane as in the megaspores. The megaspores of the recent species *Selaginella caulescens*, as figured by Bennie and Kidston[3], resemble those of the Palaeozoic type in the presence of an equatorial flange. It is interesting to find that, in spite of the occurrence of 16—24 megaspores in a single sporangium the size of the fossil spores exceeds that of the recent species.

[1] Zeiller (06) p. 141, Pls. xxxix. xli. [2] Zeiller (00) p. 1077.
[3] Bennie and Kidston (88) Pl. vi. fig. 22.

Selaginellites primaevus (Gold.).   Fig. 135, A,   fig. 138.

1855   *Lycopodites primaevus,* Goldenberg[1]
1870   *Lycopodium primaevum,* Schimper[2]
1907   *Selaginellites primaevus,* Halle[3]

FIG. 138.   *Selaginellites primaevus* (Gold.).   (After Goldenberg.)

In habit this species, first recorded by Goldenberg from the Coal-Measures of Saarbrücken, is similar to *S. Suissei* Zeill.

---

[1] Goldenberg (55) Pl. I. fig. 3.   [2] Schimper (70) A. Pl. LVII. fig. 2.
[3] Halle (07).

The drawing reproduced in fig. 138 is a copy of that of the type-specimen: another specimen, named by Goldenberg, is figured by Halle in his recently published paper. The leaves appear to be distichous: no smaller leaves have been detected, though Halle is inclined to regard the plant as heterophyllous. The sporophylls, borne in slender terminal strobili, are smaller than the foliage leaves and spirally disposed (fig. 138; smaller specimen). Halle succeeded in demonstrating that some of the sporangia contained a single tetrad of spores, each spore having a diameter of 0·4—0·5 mm. No microspores were found, but it is clear that the species was heterosporous and that it agrees with recent species in having only four spores in the megasporangium.

Selaginellites elongatus (Gold.). Fig. 135, B, D.

1855   Lycopodites elongatus, Goldenberg[1].
1870   Lycopodium elongatum, Schimper[2].

The shoots of this species resemble the recent Lycopodium complanatum; they differ from those of Selaginellites primaevus in their long and narrow branches which bear two forms of leaf. The longer leaves, arranged in opposite pairs, are slightly falcate; the smaller leaves are appressed to the axis and have a triangular cordate lamina. Another peculiarity of this species is the occurrence of sporangia in the axil of the foliage leaves, a feature characteristic of the recent Lycopodium Selago. In recent species of Selaginella the sporophylls are always in strobili. No microspores have been found nor the walls of megasporangia, but tetrads of megaspores were isolated by Halle: the spores have three radiating ridges (fig. 135, B) connected by an equatorial ridge. Halle estimates the number of spores (0·45 mm. in diameter) in a sporangium at 20 to 30. In size as in number the spores exceed those of recent species and agree more nearly with the megaspores of S. Suissei.

It would seem to be a general rule that the spores (megaspores) of the fossil herbaceous species exceeded considerably in

[1] Goldenberg (55) Pl. I. fig. 2.   [2] Schimper (70) A. p. 10.

dimensions those of recent forms and on the other hand were smaller than those of the Palaeozoic arborescent species.

There can be little doubt that some of the Mesozoic and Tertiary species included under *Lycopodites* agree more closely with the recent genus *Selaginella* than with *Lycopodium*, but this does not constitute an argument of any importance against the restricted use of the designation *Selaginellites* which we have adopted. From a botanical point of view the various records of *Lycopodites* and *Selaginellites* have but a minor importance; they are not sufficiently numerous to throw any light on questions of distribution in former periods, nor is the preservation of the material such as to enable us to compare the fossil with recent types either as regards their anatomy or, except in a few cases, their sporangia and spores. The Palaeozoic species are interesting as revealing less reduction in the number of spores produced in the megasporangia. Among existing Pteridophytes the genus *Isoetes* agrees more closely than *Selaginella*, as regards the number of megaspores in each sporangium, with such fossils as *Selaginellites Suissei* and *S. elongatus*.

It would seem that in most Palaeozoic species heterospory had not reached the same stage of development as in the recent genus *Selaginella* in which the megaspores do not exceed four in each sporangium. In *Selaginellites primaevus*, however, the heterospory appears to be precisely of the same type as in existing species.

Lycostrobus.

The generic name *Lycostrobus* has recently been instituted by Nathorst[1] for certain specimens of a lycopodiaceous strobilus, from the Rhaetic strata of Scania, which he formerly referred to the genus *Androstrobus*[2].

Lycostrobus Scotti Nathorst. Fig. 139.

The fossil described under this name is of special interest as affording an example of a Mesozoic lycopodiaceous cone comparable in habit and in size with some of the largest examples

---

[1] Nathorst (08).　　　[2] Nathorst (02[2]) p. 5, Pl. i. fig. 1.

of Palaeozoic Lepidostrobi, the cones of *Lepidodendron*. The Swedish fossil from Upper Rhaetic strata of Helsingborg (Scania) was originally designated *Androstrobus Scotti*, the generic name being adopted in view of the close resemblance of the form of the strobilus to the male flower of a Cycad. A more complete examination has shown that the bodies, which were thought to

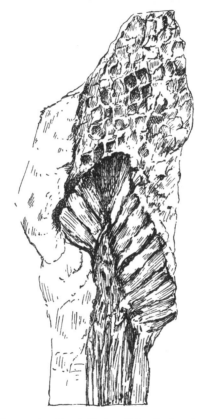

Fig. 139.   *Lycostrobus Scotti*, Nath.   (After Nathorst; ⅔ nat. size.)

be pollen-sacs—though Nathorst recognised certain differences between them and the pollen-sacs of lycopods—are the megaspores of a lycopod. Microspores have also been identified. The axis of the cone has a breadth of 2 cm. with a peduncle which may be naked or provided with a few small scales; the

sporophyll region of the axis reached a length of at least 12 cm.
The spirally disposed sporophylls terminate in a rhombic distal
end which may represent the original termination or they
may have been prolonged upwards as free laminae. Each
sporophyll bears on its upper face a single large sporangium
containing either megaspores or microspores: the megaspores,
0·55—0·60 mm. in diameter, are finely granulate and bear small
warty thorns or more slender pointed appendages. The microspores, after treatment with eau de Javelle, were found to
measure 36—44μ while others which had been treated with
ammonia reached 54μ in diameter. Nathorst describes the
microspores as occurring in spherical groups or balls, which
it is suggested may be compared with the groups of spores
separated by strands of sterile tissue (trabeculae) in the large
sporangia of *Isoetes* (cf. fig. 133, H). If this comparison is sound
it would point to a more complete septation of the sporangium
in *Lycostrobus* than in any recent species of *Isoetes*. The size
of the strobilus would seem to indicate the persistence into
the Rhaetic era of an arborescent lycopodiaceous type ; but the
appearance and manner of preservation of the axis is interpreted by Nathorst as evidence of a herbaceous rather than
a woody structure. He is disposed to regard *Isoetes* as the
most nearly allied existing genus.

The comparison made by Nathorst with *Isoetes* is based on
a resemblance between the spores of the two genera and on the
evidence, which is not decisive, of the existence of sterile
strands of tissue in the sporangia of *Lycostrobus*. This
similarity is however hardly of sufficient importance to justify
the inclusion of the Rhaetic strobilus in the Isoetaceae. In
size and in the arrangement and form of the sporophylls the
cone presents a much closer resemblance to *Lepidodendron* than
to *Isoetes*. It is probably advisable to regard this Rhaetic
type simply as a lycopodiaceous genus which we are unable,
without additional information, to assign to a particular position.

The opinion expressed by Professor Fliche[1] that the plant
described by Schimper and Mougeot as *Caulopteris tessellata*,
a supposed tree-fern stem, from Triassic rocks of Lorraine, is

[1] Fliche (03).

more probably a large lycopodiaceous stem, either a *Lepido-dendron* or a new genus, is worthy of note in reference to Nathorst's account of *Lycostrobus*.

In habit the fossil strobilus may be compared with the triassic genus *Pleuromeia*, but the position of the sporangia on the sporophylls constitutes a well-marked difference. The most important result of Nathorst's skillful treatment of this interesting fossil by chemical microscopic methods is the demonstration of the existence of a large heterosporous type of lycopodiaceous cone in a Rhaetic flora.

Poecilitostachys.

Under this generic name M. Fliche[1] has briefly described a fertile lycopodiaceous shoot from the Triassic rocks of Epinal in France: the type species *Poecilitostachys Hangi* consists of a cylindrical axis, 10 cm. × 5 mm., deprived of leaves and terminating in a rounded receptacle bearing a capitulum of bracts or fertile leaves. Detached megasporangia containing small globular bodies found in association with the capitulum are compared with the megasporangia of *Isoetes*.

[1] Fliche (09).

CHAPTER XV.

Arborescent Lycopodiales.

AMONG the best known plants in the Palaeozoic floras are the genera *Lepidodendron* and *Sigillaria*, types which are often spoken of as Giant Club-Mosses or as ancestors of existing species of *Lycopodium* and *Selaginella*. Of these genera, but more particularly of *Lepidodendron*, we possess abundant records in a condition which have made it possible to obtain fairly complete information not only in regard to habit and external features but as to the anatomical characters of both vegetative and reproductive shoots. The structure of *Lepidodendron* differs too widely from that of recent Club-Mosses (species of *Lycopodium*) to justify the statement that this prominent member of the Palaeozoic vegetation may be regarded as a direct ancestor of any living plant. There is at least no doubt that *Lepidodendron* and *Sigillaria* must be included in the Pterido-phyta. The description by Dr Scott[1] of the genus *Lepidocarpon*, founded on petrified specimens of strobili, demonstrated the existence of a type of lycopodiaceous plant in the Carboniferous period distinguished from all living representatives of the group by the possession of integumented megaspores, which may fairly be styled seeds. *Lepidocarpon* and another seed-bearing plant *Miadesmia* are described under a separate heading as lycopodiaceous types characterised by an important morpho-logical feature, which among recent plants constitutes a differentiating character between the Pteridophytes and the Phanerogams.

[1] Scott (01).

Lepidodendron.

i. *General.*

The genus *Lepidodendron* included species comparable in size with existing forest trees. A tapered trunk rose vertically to a height of 100 feet or upwards from a dichotomously branched subterranean axis of which the spreading branches, clothed with numerous rootlets, grew in a horizontal direction probably in a swampy soil or possibly under water. A description by Mr Rodway[1] of Lycopods on the border of a savannah in Guiana forming a miniature forest of Pine-like Lycopodiums might, with the omission of the qualifying adjective, be applied with equal force to a grove of Lepidodendra. The equal dichotomy of many of the branches gave to the tree a habit in striking contrast to that of our modern forest trees, but, on the other hand, in close agreement with that of such recent species of *Lycopodium* as *L. cernuum* (fig. 123), *L. obscurum* (fig. 124) and other types. Linear or oval cones terminated some of the more slender branches (fig. 188) agreeing in size and form with the cones of the Spruce Fir and other conifers or with the male flowers of species of *Araucaria, e.g. A. imbricata.* Needle-like leaves, varying considerably in length in different species, covered the surface of young shoots in crowded spirals and their decurrent bases or leaf-cushions formed an encasing cylinder continuous with the outer cortex. The fact that leaves are usually found attached only to branches of comparatively small diameter would seem to show that *Lepidodendron,* though an evergreen, did not retain its foliage even for so long a period as do some recent conifers.

By the activity of a zone of growing tissue encircling the cylinder of wood the main trunk and branches grew in thickness year by year: the general uniformity in size of the secondary conducting elements affords no indication of changing seasons. As the branches grew stouter and shed their leaves the surface of the bark resembled in some degree that of a Spruce Fir and other species of *Picea*, in which the leaf-scars form the upper limit of

[1] Rodway (95) A. p. 153.

prominent peg-like projections, which, at first contiguous and regular in contour, afterwards become less regular and separated by grooves (fig. 140) and at a later stage lose their outline as the bark is stretched to the tearing point (fig. 140, C). The leafless branches of *Lepidodendron* were covered with spirally disposed oval cushions less peg-like and larger than the decurrent leaf-bases of *Picea*, which show in the upper third of their length a clean-cut triangular area and swell out below into two prominent cheeks separated by a median groove and tapering with decreasing thickness to a pointed base, which in some forms (*e.g. Lepidodendron Veltheimianum*, fig. 185, C, D), is prolonged as a curved ridge to the summit of a lower leaf-cushion.

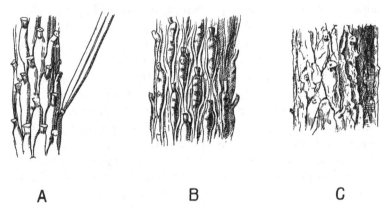

A   B   C

Fɪɢ. 140.   *Picea excelsa.* Shoots of different ages showing changes in the appearance of the leaf-cushions : a leaf attached to a cushion in fig. A. (Slightly enlarged.)

A portion of the cushion below the triangular leaf-scar often shows transverse gaping cracks or depressions (fig. 185, C) such as occur on a smaller scale on the older cushions of a Fir twig (fig. 140). Secondary thickening, as in recent trees, is not confined to the vascular cylinder but at an early stage, frequently before there are any signs of secondary wood, the outer region of the broad cortex becomes the seat of active cell-formation which results in the addition of a considerable thickness to the bark. At a later stage of increase in girth, the leaf-cushions are stretched apart and the original surface-features become obliterated by vertical cracks and by the exfoliation of the superficial tissues[1].

Some species of *Lepidodendron* produced branches characterised by spiral or vertical series of scars; these in older shoots were replaced by depressions having a diameter of several inches and comparable in appearance, as also perhaps in manner of formation, with the scars left on the stem of a Kauri Pine (*Agathis australis*)[2] on the abscission of lateral branches by a natural process. These shoots, known as *Ulodendron*, are described in a subsequent section. (Page 128.)

A fully-grown *Lepidodendron* must have been an impressive tree, probably of sombre colour, relieved by the encircling felt of green needles on the young pendulous twigs. The leaves of some species were similar to those of a fir while in others they resembled the filiform needles of the Himalayan Pine (*Pinus longifolia*). The occasional presence of delicate hyphae in the tissues of *Lepidodendron* demonstrates susceptibility to fungal pests.

Architecturally, if one may use the term, *Lepidodendron* owed its power of resistance to the bending force of the wind to its stout outer bark formed of thick-walled elements produced by the activity of a cylinder of cortical meristem (figs. 148, 172, etc.). The vascular axis, of insignificant diameter in proportion to the size of the stem (figs. 152, 153, 172, 181, A), must have played a subordinate part, from a mechanical point of view, as compared with the solid mass of wood of a Pine or an Oak.

Within the compass of a text-book it is impossible, even if it were desirable, to include an account of the majority of the species of the widely distributed Palaeozoic genus *Lepidodendron*. In spite of the great number of known species of this common member of Carboniferous floras, our knowledge of the type as a whole is deficient in many points, and such information as we possess needs systematising and extending by comparative treatment based on a re-examination of available data.

In order to appreciate the meaning of certain external

[1] A good example of an old *Lepidodendron* stem (*L. aculeatum*) is figured by Zalessky (04) Pl. i. fig. 3.

[2] Seward and Ford (06) Pl. xxiii. fig. C.

features characteristic of Lepidodendron stems it is essential to have some knowledge of the internal structure.

A dual system of terminology has been unavoidably adopted for species of *Lepidodendron*: the majority of specific names have been assigned to fossils known only in the form of casts or impressions, while petrified fragments, which unfortunately seldom show the surface-features, have received another set of names. A glance at the older palaeobotanical literature reveals the existence of several generic designations, which fuller information has shown to have been applied to lepidodendroid shoots deprived of some of their superficial tissues before fossilisation and differing considerably in appearance from the more complete branches of the same species[1]. It has in some instances been possible to correlate the two sets of specimens, casts or impressions, showing external features, and petrified fragments. We may reasonably expect that future discoveries will enable us to piece together as definite specific types specimens at present labelled with different names.

A well-preserved leaf-cushion of a *Lepidodendron*—the most obvious distinguishing feature of the genus—is rhomboidal or fusiform and vertically elongated (fig. 146, C, E; fig. 185, C, D): in exceptional cases it may reach a length of 8 cm. and a breadth of 2 cm. The cushion as a whole represents a prominent portion of the stem or branch comparable with the elevation on the twig of a Spruce Fir and the leaf-base of a *Lycopodium* (cf. fig. 121, A, lower portion) which appears in a transverse section of a branch as a rounded prominence (cf. *Lycopodium*, fig. 125, A and H). Disregarding differences in detail, a typical Lepidodendron leaf-cushion is characterised by a clearly defined smooth area often situated in the middle region (fig. 146, C, *s*). This is the leaf-scar or place of attachment of the base of the leaf which was cut off by an absciss-layer while the branch was comparatively young, as in recent forest trees and in some species of Ferns. On the leaf-scar are three smaller scars or cicatricules, the central one is circular or more or less triangular in outline, the two lateral scars being usually oval or circular. The central pit marks the position of the single vascular bundle

---

[1] See Fischer (04).

which constituted the conducting tissue connecting the leaf with the main vascular system of the stem. The two lateral scars (figs. 145, A, *p*; 146, C, *s*; 147, *p*) represent the exposed ends of two strands of tissue, the forked branches of a strand which pass from the middle cortex of the stem into the leaf; this is known as the parichnos, a name proposed by Professor Bertrand in 1891[1].

The specimen shown in fig. 141 shows the linear leaves attached to their respective cushions.

Fig. 141. *Lepidodendron Sternbergii*. From a specimen in the British Museum (No. v. 1235) from the Coal-Measures of Shropshire. (Nat. size.)

The lamina has a well-defined median keel on the lower surface and on either side a groove in which sections of petrified leaves have demonstrated the occurrence of stomata (cf. fig. 142).

ii. *Leaves and Leaf-cushions.*

All Lepidodendron leaves, so far as we know, possessed a single median vein only. In some species, as for example in *Lepidodendron longifolium* Brongn., they have the form of long

---

[1] Bertrand, C. E. (91) p. 84: derived from παρά, by the side of, and ἴχνος, trace or foot-print.

and slender acicular needles very similar to those of *Pinus
longifolium*; in *L. Sternbergii* (fig. 141) they are much broader
and shorter.   In external form as in internal structure it is often
impossible to distinguish between the leaves of *Lepidodendron*
and *Sigillaria*.   The distinguishing features enumerated by the
late M. Renault cannot be employed, with any great degree of
confidence, as diagnostic characters.   In transverse section the
lamina of a Lepidodendron leaf presents the same appearance as
that of the Sigillarian leaves represented in fig. 142.   Near the

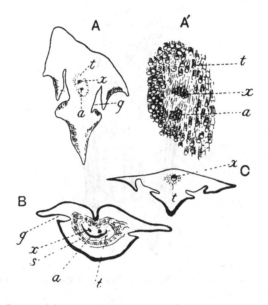

Fig. 142.   Leaves of *Sigillaria* in transverse section.
A, A'.   Section in the Manchester University Museum (Q. 631).
B, C.   Sections in Dr Kidston's Collection.

base the free part of the leaf is usually subrhomboidal in section
with short lateral wings, a ventral keel and two stomatal
grooves (fig. 142, A, B, *g*).   The form and arrangement of stomata are shown in fig. 143, A, which was drawn from a piece of a
leaf shown in surface-view in a section lent to me by Professor
Weiss.   It should, however, be pointed out that the leaf cannot
be certainly identified with *Lepidodendron* rather than with

Sigillaria, but as the leaves of these two genera are constructed on the same plan the identification is of secondary importance.

The single xylem bundle consists of primary tracheae only, at least in such laminae as have been identified as Lepidodendroid. Surrounding the xylem strand occur delicate parenchymatous cells in some cases accompanied by darker and thicker-walled elements. As in *Sigillaria,* the leaves of which are more fully described on page 210, a fairly broad sheath of wider and shorter scalariform or spiral transfusion tracheids surrounds the con-

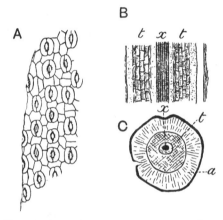

Fig. 143. A. Stomata in surface-view (*Lepidodendron?*). *a,* parenchyma; *t,* transfusion tracheae; *x,* xylem. (Manchester University Collection R. 723).
B, C. *Lepidodendron esnostense* Ren. (After Renault.)

ducting strand (figs. 142, *t*; 143, B, C, *t*). As Renault shows in the case of *Lepidodendron esnostense*[1], the small leaves of which are 1·5—2 mm. broad at the base and several centimetres long, the stomatal grooves and keel die out towards the apex when the lamina assumes a more nearly circular form (fig. 143, C).

The area of the cushion excluding the leaf-scar is spoken of by some writers as the field. Below the leaf-scar the kite-shaped cushion tapers to a gradually narrowing basal position: in *Lepidodendron Veltheimianum,* a species characteristic of

[1] Renault (96) A. Pls. xxxiii. xxxiv. p. 178. For a good section of another Lepidodendron leaf, see Scott (08) p. 160, figs. 64, 65.

Lower Carboniferous strata, it is seen to be continuous, as a ridge with sloping sides, with a lower cushion (fig. 185).

Below a leaf-scar the cushion frequently shows a pair of oval areas on which a fine pitting may be detected in well-preserved impressions, these oval scars, as seen in fig. 185, D, are practically continuous at the upper end with the parichnos scars on the leaf-scar area; this is explained by the fact that these infra-foliar scars also owe their existence to patches of lacunar, aerenchymatous tissue in close connexion with the parichnos[1].

Shortly before entering the base of the leaf-lamina the parichnos divides into two arms which diverge in the outer cortical region right and left of the vascular bundle, and passing obliquely upwards they come close to the surface of the leaf-cushion just below the leaf-scar. The diagram—fig. 144, B—shows a leaf-trace, lt, in the leaf-cushion, as seen in a diagrammatic drawing of a vertical radial section of a stem, the dotted lines, p, p', show the two parichnos arms which are represented as impinging on the surface of the leaf-cushion at p', and then bending upwards to pass into the leaf-base right and left of the vascular bundle or leaf-trace. For convenience the arms of the parichnos are represented in one plane though actually in different vertical planes.

Fig. 144, A, shows the difference between a view of the original surface of a *Lepidodendron,* as at a, where a leaf-cushion with a leaf-scar is seen, and a view of an impression representing the outer cortex, b, a short distance below the surface. The surface b, in fig. 144, A, corresponds to the face d—e in the diagrammatic longitudinal section fig. 144, B: the outline of each cushion is clearly visible and in the centre is seen the leaf-trace, lt, with its parichnos.

The surface-features, a (fig. 144, A), have been impressed on the rock, c, (fig. 144, B) in which the specimen was entombed and by the removal of the cast of the stem, that is the thickness b to e in fig. 144, B, the form of the leaf-cushion is revealed. The presence of the two infra-foliar parichnos scars at p' (fig. 144, A) is explained by the diagram, fig. 144, B, p'.

The relation of the parichnos to the oval scars below a

[1] Weiss, F. E. (07).

Lepidodendron leaf-cushion has been worked out in detail by
Weiss who shows that, at least in some species, the two arms
do not bend downwards as shown in the diagram, fig. 144, B,
but pursue a straight gradually ascending course as seen in
fig. 145, A. Just below the leaf-scar region of the cushion each
arm comes into association with a group of lacunar, aerenchymatous tissue, such as occurs in the roots of certain Mangroove

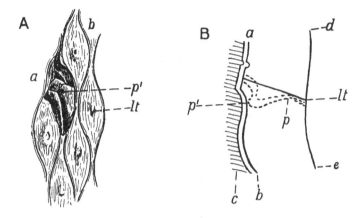

FIG. 144. *Lepidodendron Veltheimianum* Sternb.
    A. Leaf-cushion and leaf-scar seen in surface-view at *a* ; on the rest
        of the specimen a slightly lower surface is exposed. (After
        Stur.)
    B. Diagrammatic longitudinal section to explain the differences
        between its two surfaces *a* and *b* shown in fig. A.
        The shaded portion *c* represents the rock matrix, the surfaces
        ab, *ed*, mark the outer and inner edge of the outer portion of
        the bark of the Lepidodendron stem.
        lt, leaf-trace ; *p*, *p'*, parichnos.

plants, and it is this aerenchyma which is exposed on the two
oval depressions below the leaf-scar. The structure of this
aerenchyma is shown in fig. 145, B ; it consists in this species
(*L. Hickii* Wats.) of stellate cells which would constitute an
efficient aerating system. Probably, as Weiss suggests, these
patches of aerenchyma were originally covered by an epidermis
provided with stomata, and it is owing to the destruction of this
superficial layer that the two oval scars often form a prominent

feature on Lepidodendron leaf-bases[1]. The diagram reproduced
in fig. 144, B, may be taken as practically correct, as the patches
of aerenchyma described by Weiss do not differ essentially from
the parichnos tissue.

The parichnos scars are shown on the leaf-scar and cushion
in fig. 146, C. In the lower leaf-cushion shown in fig. 146, E,
the infra-foliar parichnos scars, *p*, are clearly seen, but the
preservation of the leaf-scar is not sufficiently good to show them

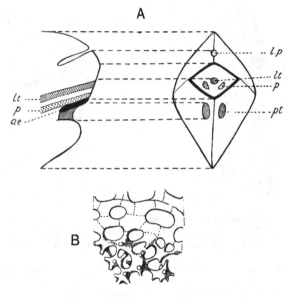

FIG. 145. A. Diagrammatic surface-view and longitudinal section of a
              Lepidodendron leaf-cushion.
           B. Aerenchyma below the leaf-scar. (After F. E. Weiss.)

on that part of the fossil. In the upper cushion (fig. 146, E) the
position of the parichnos arms is shown on the leaf-scar, but the
infra-foliar parichnos scars are hidden by two small spiral
shells. The genus *Spirorbis*, to which these shells are referred,
appears to have persisted from the Silurian epoch to the present
day. The comparatively frequent occurrence of *Spirorbis* shells
on the leaves and other parts of Palaeozoic plants, has recently

────────

[1] For a fuller account of the parichnos, see Hill, T. G. (06) and other papers
quoted by F. E. Weiss (07).

been dealt with in a paper by Barrois[1] who discusses in detail the habitats of these small animals from the point of view of the conditions under which the plants were preserved. In a note by Malaquin appended to Barrois' paper the belief is expressed that *Spirorbis* lived on pieces of Palaeozoic plants which lay under water.

The fact that with one exception all the Spirorbis shells on the specimen of *Lepidodendron*, of which two leaf-cushions are shown in fig. 146, E, occur on the large parichnos scars on the cheeks of the cushions, suggests the possibility that the escape of gases from the parichnos tissue may have rendered the position attractive to the *Spirorbis*. It can hardly be accidental that the shells occur on the parichnos strands. This fact recalls the view held by Binney[2] and accepted with favour by Darwin[3] that *Lepidodendron* and other coal-forest trees may have lived with the lower parts of the stems in sea water.

Above the leaf-scar is a fairly deep triangular or cresentic pit (fig. 146, C, *l*) known as the ligular pit from the occurrence on younger shoots of a delicate organ like the ligule of *Isoetes* (fig. 132) embedded in a depression in the upper part of the leaf-cushion. The ligule was first figured in *Lepidodendron* by Solms-Laubach[4] and described in English material by Williamson under the name of the adenoid organ[5].

In some Lepidodendron stems a second triangular depression may occur above the ligular pit, the meaning of which is not clear: this has been called the triangulum by Potonié[6]. Stur[7] suggested that it may represent the position occupied by a sporangium in Lepidodendron cones.

It is important to remember that as a branch increases in girth the leaf-cushions are capable of only a certain amount of growth: when the limit is reached they are stretched farther apart and thus the narrow groove which separates them is converted in older stems into a comparatively broad and flat channel, thus altering the surface characters.

[1] Barrois (04). See also Etheridge (80); Geikie (03) p. 1049.
[2] Binney (48).　　　　[3] Darwin (03) vol. II. pp. 217, 220.
[4] Solms-Laubach (92) Pl. II. figs. 2, 4.　　[5] Williamson (93) p. 10.
[6] Potonié (05) Lief. iii., p. 41.　　[7] Stur (75) A. Heft II. p. 277.

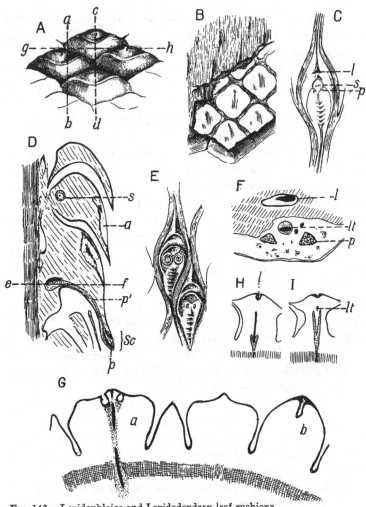

Fig. 146. Lepidophloios and Lepidodendron leaf-cushions.
 A, B, D, F, G, H, I. *Lepidophloios.* (Fig. A should be reversed.)
 C, E. *Lepidodendron aculeatum.*
 A, B. From a specimen in the Sedgwick Museum, Cambridge (leaf-
   cushion 3 cm. broad).
 C. From a specimen in the Sedgwick Museum, Cambridge (leaf-
   cushion 4 cm. long.
 D. From a section in the Cambridge Botany School Collection.
 E. From a specimen in the Bunbury Collection, Cambridge Botany
   School, showing *Spirorbis* shells (leaf-cushion 2 cm. long).
 F. From a section in the Williamson Collection, British Museum
   No. 1, 973.
 G, H, I. From sections in the Cambridge Botany School Collection.

Another feature worthy of notice in reference to the leaf-cushions of *Lepidodendron* is the occurrence in rare instances of alternate zones of larger and smaller cushions. This variation in the size of the leaf-cushions is by no means uncommon in the closely allied germs *Sigillaria*; in *Lepidodendron* it has been described by Potonié[1] in *L. volkmannianum* and more recently by Mr Leslie and myself[2] in a South African species *L. vereenigense*.

Owing to the natural exfoliation of the superficial layers of the outer bark at a certain stage in the growth of the plant, or in some instances no doubt as the result of *post-mortem* decay, which destroys the delicate cells of the meristematic zone in the outer cortex, isolated leaf-cushions and strips of the external surface are occasionally met with as carbonised impressions.

The appearance presented by a Lepidodendron stem which has been deprived of its superficial tissues may be dealt with more intelligibly after we have become familiar with the anatomical characters.

iii. *Lepidophloios.*

Before proceeding further with the genus *Lepidodendron* a short account may be intercalated of the external features of a lepidodendroid type of stem which it is customary to describe under a distinct generic title *Lepidophloios*. This name is convenient for diagnostic purposes though it seems clear that apart from the form of the leaf-cushion (fig. 146, A) we are at present unable to recognise any well-defined differences between the two forms *Lepidodendron* and *Lepidophloios*. For general purposes the name *Lepidodendron* will be used as including plants possessing leaf-cushions of the type already described as well as those with the Lepidophloios form of cushion.

The generic name *Lepidophloios* was first used by Sternberg[3] for a Carboniferous species which he had previously described as *Lepidodendron laricinum*. In 1845 Corda[4] instituted the name *Lomatophloios* for specimens possessing the same external

---

[1] Potonié (05) fig. 4.   [2] Seward and Leslie (08) Pl. x. figs. 1 and 2.
[3] Sternberg (26) A. Pl. xi. figs. 2—4 ; (02) p. 23.
[4] Corda (45) A. Pls. i.—iv.

characters as those for which Sternberg had chosen the name
Lepidophloios. The leaf-cushions of *Lepidophloios* differ from
those of the true *Lepidodendron* in their relatively greater lateral
extension (cf. fig. 146, A and C), in their imbricate arrangement
and in bearing the leaf, or leaf-scar, at the summit. In some
species referred to *Lepidophloios* the cushions are however
vertically elongated and in this respect similar to those of
Lepidodendron: an example of this type is afforded by *Lepido-*
phloios Dessorti a French species described by Zeiller[1]. In
younger branches the cushions may be directed upwards having
the leaf-scar at the top; but in the majority of specimens
the cushions are deflexed as in figs. 146, D; 160, A. The shoot
of *Lycopodium dichotomum* shown in fig. 121, B, with the leaves
in the reversed position bears a close resemblance to a branch
of *Lepidophloios*.

The photograph of *Lepidophloios scoticus* Kidst.[2] reproduced
in fig. 160, A, illustrates the dichotomous branching of the
stem and the form of the cushions with the leaf-scars pointing
downwards. In the fertile branch of the same species shown in
fig. 160, B, the leaf-scars face upwards.

In most species the cushions are simply convex without a
median keel, but in some cases a median ridge divides the
cushion into two cheeks as in the genus *Lepidodendron*. The
leaf-scar bears three small scars, the larger median scar marking
the position of the leaf-trace, while the lateral scars are formed
by the two arms of the parichnos: in some examples of deflexed
cushions, though not in all, a ligular pit occurs on the cushion a
short distance above the leaf-scar.

The drawing reproduced in fig. 146, A, showing the leaf-scar
on the upper edge of the cushion should have been reversed
with the leaf-scars pointing downwards. This figure represents
part of the surface of a specimen consisting of the outer cortex
of a stem with leaf-cushions 3 cm. broad. The thickness of
this specimen is 4 cm.: a section through the line *ab* is represented in fig. 146, D (reproduced in the correct position, with
the leaf-scars, *sc*, pointing downwards): internal to the cushions
is a band of secondary cortex (the shaded strip on the outer

---

[1] Zeiller (92) A.   [2] Kidston (93) p. 561, Pls. I. and II.

edge of the section) which was formed on the outside of the phellogen. The phellogen is a cylinder of actively dividing cells in the outer part of the cortex of the stem, often spoken of as the cork-cambium or cortical meristem, which produces a considerable amount of secondary cortical tissue on its inner face and a much smaller amount towards the stem surface. This delicate cylinder frequently forms a natural line of separation between the outer shell of bark and the rest of the stem. In the specimen before us, the thin-walled cells of the phellogen were ruptured before petrification and the outer shell of bark was thus separated as a hollow cylinder from the rest of the stem: this cylinder was then flattened, the two inner surfaces coming into contact. Fig. 146, D, represents a section of one half of the thickness of the flattened shell.

This separation of the outer cortex, and its preservation apart from the rest of the stem, is of frequent occurrence in fossil lycopodiaceous stems. The flattened outer cortical shell of a *Lepidophloios*, specifically identical with that shown in fig. 146, A and D, was erroneously described by Dr C. E. Weiss in 1881 as a large lepidodendroid cone[1].

Fig. 146, B, affords a view of the inner face of the specimen of which the outer surface is seen in fig. 146, A: the surface shown in the lower part of the drawing, on which the boundaries of the cushions are represented by a reticulum, corresponds to the inner edge of the strip of secondary cortical tissue represented by the vertically shaded band in fig. 146, D.

The shaded surface in fig. 146, B, represents a slightly deeper level in the stem which corresponds to the outer edge of the vertically shaded band of fig. 146, D: the narrow tapered ridges (fig. 146 B) represent the leaf-traces passing through the secondary cortex, and the fine vertical shading indicates the elongated elements of which this strip of secondary cortex is composed.

In the longitudinal section diagrammatically reproduced in fig. 146, D, cut along the line *ab* of fig. 146, A, the parenchymatous tissue of the stout cushions has been partially destroyed, as at *a*; at *s* is seen the section of a Stigmarian rootlet which has

[1] Seward (90).

found its way into the interior of a cushion. Each leaf-trace is accompanied by a parichnos strand as in the true *Lepidodendron*; at the base of the leaf-cushion the parichnos branches into two arms which diverge slightly right and left of the leaf-trace, finally entering the base of the leaf lamina as two lateral strands (fig. 147, *p*). At one point in fig. 146, D the section has shaved a leaf-trace represented by a black patch resting on the parichnos

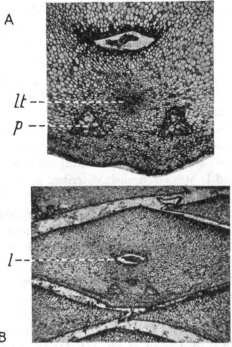

Fɪɢ. 147. *Lepidophloios* leaf-cushion in tangential section. (From a section in the Williamson Collection, British Museum, No. 1973.)

just above the line *ef*, but it passes through one of the parichnos arms *p'* which debouches on to the leaf-scar *sc* at *p*. Had the section been cut along the line *cd* of fig. 146, A the leaf-trace would have been seen in a position similar to that occupied by the parichnos *p'* in fig. 146, D.

Fig. 147, A, affords a good example of a tangential section through a *Lepidophloios* leaf-cushion, 1 cm. broad, like that represented in fig. 146, A, showing the vascular bundle *lt*, the two parichnos strands, *p*, composed of large thin-walled cells (cf. *Isoetes*, fig. 133, H, I), and the ligular pit near the upper edge of the section enclosing the shrunken remains of the ligule (fig. 147, B, *l*).

Fig. 147, B shows the form of the tangentially elongated leaf-cushions of *Lepidophloios* and their spiral disposition.

Fig. 146, F, represents a section similar to that shown in figs. 147, A and B, but in this case the leaf-trace, *lt*, and the parichnos strands, *p*, lie in a cavity formed by the destruction of some of the leaf-cushion tissue. It is worthy of notice that the parichnos cells have resisted decay more successfully than the adjacent tissue of the cushion.

The diagrammatic sketches reproduced in fig. 146, H and I, were made from a transverse section similar to one originally figured by Williamson[1]: fig. 146, H, corresponding in position to the line *gh* in fig. 146, A, passes through the ligular pit, *l*, and cuts across the parichnos in the act of branching; the leaf-trace passes outwards beyond the Y-shaped parichnos strand. In the other section, fig. 146, I, the parichnos is shown in a horizontal plane and the leaf-trace, *lt*, appears in oblique transverse section. In both sections and in fig. 146, G the shaded band at the base represents the secondary cortical tissue external to the phellogen.

The transverse section represented in fig. 146, G shows in the left-hand cushion, *a*, the exit of the two parichnos arms and the leaf-trace between them: it illustrates also the various forms assumed by lepidodendroid leaf-cushions when cut across at different levels.

iv. *The Anatomy of* Lepidodendron vasculare *Binney*[2].
Figs. 148—155, 168, A.

In the earlier literature dealing with the anatomy of *Lepidodendron* and *Sigillaria* the presence or absence of secondary vascular tissue was made the criterion of generic distinction and the distinguishing feature between the classes Pteridophytes and

---

[1] Williamson (93) Pl. IV. figs. 30—32.   [2] Binney (62).

Gymnosperms, *Lepidodendron* being relegated to the former class because it was supposed to have no power of forming secondary wood, while *Sigillaria*, characterised by a considerable development of such tissue, was classed by Brongniart and afterwards by Renault as a Gymnosperm. Binney[1] in 1865 recognised that the two types of stem pass into one another, but it was Williamson[2] who provided complete demonstration of the fallacy of the Brongniartian view.

These two undoubted Pteridophytes agree very closely in anatomical structure and both are now recognised as arborescent genera of Lycopodiaceous plants. In a paper published by Lomax and Weiss in 1905[3] a specimen is described from the Coal-Measures of Huddersfield, in which a decorticated stem with the anatomical characters of Binney's *Sigillaria vascularis* gives off a branch having the anatomical structure which it has been customary to associate with the species *Lepidodendron selaginoides*, so-called by Sternberg and founded by him on impressions showing well-preserved external characters.

In 1862 Binney[4] described petrified specimens of vegetative shoots from the Lower Coal-Measures of Lancashire under the names *Sigillaria vascularis* and *Lepidodendron vasculare*. These were afterwards recognised as different states of the same species. A few years after the publication of Binney's paper Carruthers[5] identified Binney's species *Lepidodendron vasculare* with Sternberg's *L. selaginoides*. The evidence on which this identification rests has not been stated, but many writers have retained this specific designation for the well-defined type of anatomical structure first described by Binney as *L. vasculare*. The use of the specific name *selaginoides* is, however, open to objection. The species *Lepidodendron selaginoides*, as pointed out by Kidston[6], is probably identical with the plant which Brongniart had named *L. Sternbergii* before the institution of Sternberg's species, and we are not in possession of convincing evidence as to the connection of *L. Sternbergii* (= *L. selaginoides*) with specimens possessing the anatomy of Binney's type.

[1] Binney (65); see also Binney (72).   [2] Williamson (72).
[3] Weiss, F. E. and Lomax (05).   [4] Binney (62).
[5] Carruthers (69) p. 179.   [6] Kidston (86) A. p. 151.

Binney's designation is therefore retained for the anatomical type described in the following pages[1].

The most detailed account hitherto published of the anatomy of *Lepidodendron vasculare* is that by the late M. Hovelacque[2], based on material from the Lower Coal-Measures of England.

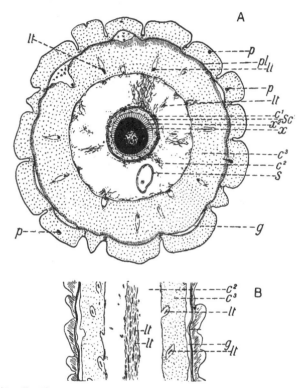

FIG. 148.   *Lepidodendron vasculare* Binney.
    A.   Transverse section. (Based on a section 2·5 cm. in diameter, in the Cambridge Botany School Collection.)
    B.   Longitudinal section. (Drawn from a section in Dr Kidston's Collection.)

The small shoot, represented somewhat diagrammatically in fig. 148, A, illustrates the anatomical features of a typical example of the species: the shoot has a diameter of 2·5 cm. and its central cylinder (x—sc) is 2·5 mm. in width.

---

[1] Seward (06) p. 372.   [2] Hovelacque (92).

Noticeable features are (i) the small size of the central cylinder (or stele) in proportion to the diameter of the branch, (ii) the production at a comparatively early stage of growth of a zone of secondary wood, x^2, which gradually assumes the form of a complete cylinder of unequal breadth, surrounding the primary xylem, x, (iii) the formation of a secondary cortical tissue by a meristematic cylinder (phellogen, pl) situated close to the leaf-cushion region of the outer cortex. On the outer edge the stele consists of narrow tracheae some of which show in longitudinal section the spiral form of thickening characteristic of most protoxylem elements: towards the centre of the stele the

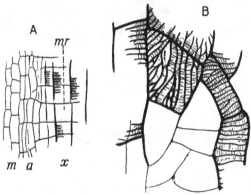

Fig. 149. *Lepidodendron vasculare*. *a*, immature tracheae; *m*, meristem; *mr*, medullary ray; *x*, xylem.
A. Longitudinal section through the edge of the secondary wood.
B. Short tracheae in the centre of the stele. (From a specimen from the Halifax Hard bed in Dr Kidston's Collection.)

diameter of the tracheae gradually increases and parenchymatous cells become associated with the elongated scalariform elements. In the central region the stele is composed of parenchymatous tissue arranged in vertical series of short cells, interspersed with short tracheae distinguished by the greater thickness of their walls and by their scalariform and reticulate thickening bands. Some of these short tracheae are shown in vertical section in fig. 149, B: the fine and broken lines connecting adjacent thickening bands probably represent the remains of the original wall. These delicate bands, which have been figured in various species

XV]   LEPIDODENDRON   113

of lepidodendroid plants[1], are worthy of notice in connexion
with the recent work of Mr Gwynne-Vaughan[2] who has shown
that in many recent ferns the scalariform bands in the xylem
elements are not connected by a thin pit-closing membrane, but
are separated from one another by open spaces. In the Lepidodendron tracheae we seem to have a stage in which the intervening membrane is in process of absorption. It is, however,
possible that the threads may be the result of contraction and
splitting of the membrane during drying or decay.

The stele of *Lepidodendron vasculare*, before the addition of
any secondary xylem, may be described as a protostele, a term
originally proposed by Professor Jeffrey[3], in which the central part
of the conducting strand of xylem elements has been converted
into rows of parenchyma and short tracheids, the latter being
better adapted to storage than to conduction. It is probable
that this type of stelar anatomy, which distinguishes *L. vasculare*
from other species, represents a comparatively primitive arrangement forming a transition between the stele of *L. esnostense*,
which consists of a solid rod of tracheids, and the stele of
L. Harcourtii (fig. 179, A) and other species in which the xylem
forms a cylinder enclosing a large parenchymatous pith.

Parenchymatous cells occur in contact with the outer edge
of the xylem-cylinder some of which are distinguished by an
irregular reticulate pitting. The tangential section represented in fig. 148, B, illustrates the appearance of a shoot of
L. vasculare in which no secondary xylem is present: the central
strand of tissue consists of the parenchyma abutting on the
xylem with several leaf-traces (*lt*) passing upwards in an almost
vertical course from the outer edge of the stele.

The secondary xylem (fig. 148, A, x^2) consists of radially
arranged scalariform tracheae with associated rows of parenchymatous cells which form medullary rays (fig. 149, *mr*).
Leaf-traces pass through the medullary rays in the secondary
xylem cylinder in a direction at right angles to the primary xylem
stele from which they are given off, but at the outer edge of the

---

[1] Solms-Laubach (92) Pl. II. fig. 6; Seward and Hill (00) Pl. IV. fig. 26.
See p. 910 of the latter paper for other references.
[2] Gwynne-Vaughan (08).   [3] Jeffrey (98). See also Tansley (08) p. 37.

S. II.   8

secondary xylem they bend suddenly upwards and for a time follow a steep and almost vertical course.

In well-preserved longitudinal sections the outermost secondary xylem tracheae are seen to be succeeded by a few narrow and vertically elongated elements (fig. 149, A, a), which represent young unlignified tracheae: these are followed by shorter parenchymatous cells (m) forming part of a meristematic zone from which the secondary xylem receives additions.

Returning to fig. 148, A; the zone of secondary wood, x^2, composed of scalariform tracheids and medullary rays, is succeeded by a few layers of parenchymatous cells and beyond this is a broader zone, sc, to which the term secretory zone has been applied[1]; this is made up of small parenchymatous cells varying in size and of larger spaces which appear to have been formed by the disorganisation of thin-walled elements. The whole zone presents a characteristic appearance due to the association of small cells, large clear spaces, and a certain amount of dark-coloured material suggestive of tissue disorganisation and secreted products. The anatomical characters of the secretory zone are shown in the photograph, fig. 168, A, sc. Several leaf-traces are seen in transverse section in the secretory zone.(black dots in fig. 148, A, sc; fig. 154, C, lt): each trace consists of a strand of narrow tracheae accompanied by a few encircling layers of small parenchymatous cells. As a trace continues its steeply ascending course through the secretory zone, it becomes associated with a strand of that tissue and assumes the form of a collateral vascular bundle, the outer part of which does not consist of typical phloem but of shorter elements derived from the secretory zone. Beyond the secretory zone we find a more homogeneous tissue composed of parenchymatous elements slightly extended tangentially (figs. 148, A, c^1; fig. 168, A, c); this is spoken of as the inner cortical region. In the great majority of sections of L. vasculare as of other species of the genus, the broader middle cortex (fig. 148, c^2) is occupied by mineral matter, introduced subsequent to decay of the tissue; or it is represented by patches of delicate tissue composed of loosely arranged parenchymatous cells varying considerably in

[1] Seward (99) p. 144.

size and shape, some being small, oval or polygonal elements while others have the form of sinuous hypha-like tubes.

In this middle cortical region may be seen leaf-traces passing outwards in an almost horizontal course (fig. 148, A, *lt*): after leaving the inner cortex the leaf-traces bend somewhat abruptly outwards to follow a more direct path through the middle and outer cortex. The ring of tissue, *s*, seen in the middle cortex of fig. 148, A, belongs to a Stigmarian rootlet.

The outer cortex (fig. 148, A and B, c^3) consists of homogeneous parenchyma which is stronger and more resistant to decay than the looser middle cortex. The leaf-traces, as shown in fig. 148, B, pass through this region in a rather steeply ascending direction: each is seen to be enclosed by a space originally occupied by a strand of middle cortical tissue which accompanies lepidodendroid leaf-traces on their under side and has already been described as the parichnos, (pp. 97, 100—103; figs. 146, 147).

The surface of the stem shown in section in fig. 148, A, is composed of broad leaf-cushions. A single leaf-trace with its parichnos passes into each cushion, but in the neighbourhood of the base of a cushion the parichnos bifurcates (cf. fig. 146, H, I) and the arms diverge slightly to the right and left finally passing beyond the cushion into the lamina of the leaf, their position being shown, as already explained, by the two small lateral scars on the leaf-scar area.

The diagrammatic sketch of a radial longitudinal section through a leaf-cushion represented in fig. 150 illustrates the relation of the leaf-trace to the leaf-cushion. The trace consists of xylem, *x*, above and a strand of the secretory zone, *st*, below; the parichnos tissue was originally present on the under side of the leaf-trace at *a*. The external surface, *bc*, marks the limit of the leaf-scar through the middle of which passes the vascular strand *lt*.

The lower gap *a* has been formed by the tearing of thin-walled cells of the phellogen, the meristematic tissue from which a considerable amount of secondary cortical tissue or phelloderm has been produced at *pd*. On the outside of the cushion, *c*, the cells are somewhat crushed and distinguished

by their darker colour from the bulk of the parenchymatous
tissue *d*.

This section also illustrates another characteristic feature of
Lepidodendron, namely the presence of a ligule and a ligular
pit: the former is represented by a carbonised patch of tissue and
the latter extends from the surface of the cushion at *b*, just
above the leaf-scar, almost to the level of the leaf-trace, *lt*.   A
comparison of this section with figs. 146 and 147 will make
clear the relation of the several parts of the cushion and leaf-scar.

The gaps *gg*, seen in fig. 148, A and B, mark the position of

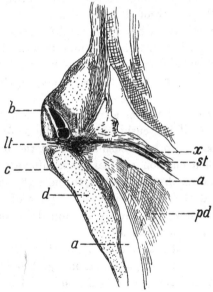

Fig. 150. *Lepidodendron vasculare*. Leaf-cushions in longitudinal section.
(From a specimen in Dr Kidston's Collection.)

the delicate meristematic zone or phellogen which arises close
to the bases of the leaf-cushions; the phellogen has already produced a few rows of radially disposed elements, represented by
short radial lines in the drawing, which constitute secondary
cortical tissue.

In older shoots the amount of the secondary cortical
tissue developed on the inner side of the phellogen is considerable (cf. figs. 152, 153).

The structure of the cortex of a shoot in which secondary growth, both in the stele and in the outer cortex, has progressed further than in the specimen shown in fig. 148 is represented in fig. 151.

The section (fig. 151, A) measures 7 × 3·8 cm. in diameter; the primary xylem is surrounded by a fairly broad cylinder of secondary wood (fig. 151, E, x and x^2). The almost smooth surface of the primary wood (fig. 151, E, x) is succeeded by the secondary xylem, x^2, characterised at its inner edge by the tapered ends of the radial rows of scalariform tracheids between

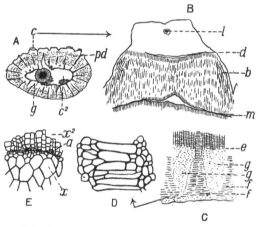

Fig. 151. *Lepidodendron vasculare.* An older stem than that shown in fig. 148. (From a section in the Manchester Museum. No. 351.)

which occur several delicate parenchymatous cells (fig. 151, E, a). The occurrence of such isodiametric elements, often exhibiting a delicate spiral thickening band, is a characteristic feature of the boundary between primary and secondary wood in lepidodendroid stems. The secondary wood is penetrated by numerous medullary rays and in some of them are seen strands of narrow spirally thickened tracheae—the leaf-traces—which are in organic continuity with the exarch protoxylem of the primary wood. The leaf-traces are oval and mesarch. The space, c^2, (fig. 151, A) originally occupied by the delicate middle cortex, is succeeded by a shell of outer cortex composed chiefly of

secondary tissue (phelloderm, *pd*) passing towards the inner
boundary of this region into the primary outer cortex *g* (fig. 151,
A and C).   The radially disposed elements which make up the
bulk of the phelloderm are associated with concentric rows of
secretory strands, represented by tangentially arranged dots in
fig. 151, A: on the outer edge of the phelloderm a few patches of
primary cortex are still preserved, as at *c*, fig. A.   One of these
is shown on a larger scale in fig. B; at *m* the phelloderm is
interrupted by a gap beyond which the cells have thinner walls
and show signs of recent division; this is probably the position
of the phellogen.   The tissue *b*, fig. 151, B, consists of secondary
cortex succeeded beyond *d* by the parenchymatous tissue of the
leaf-cushion, in which the remains of a ligule, *l*, are seen in the
ligular pit.   This section corresponds in position to a line
drawn across fig. 150 at the level of *b*.   In this specimen we
have two kinds of secondary cortical tissue: that formed external
to the phellogen, from *m* to *d* in fig. 151, B, is less in amount
than that produced internal to the phellogen.   We cannot make
any satisfactory statement as to the nature of this secondary
tissue, whether or not any of it agreed in composition with the
cork which is usually formed external to the phellogen in recent
plants.   As the stem of a *Lepidodendron* grew in girth the leaf-
cushions became separated by intervening depressions composed
of the secondary cortex formed external to the phellogen, but at
a later stage the cushions were thrown off, leaving the outer
edge of the phelloderm as the superficial tissue.   This exposed
tissue became fissured as growth and consequent stretching
continued, producing the appearance seen on the surface of
the still older stem represented in fig. 153.

The inner edge of the phelloderm seen at *e* in fig. 151, C,
passes suddenly into the inner primary region of the outer cortex
(fig. 151, A and C, *g*) which comprises two types of parenchymatous tissue, patches of isodiametric cells, *g*, *g*, alternating
with radially arranged areas consisting of tangentially elongated
elements (fig. C, *f, f*; fig. D) which extend as wedges into the
phelloderm.

The longitudinal section represented in fig. 152, B, shows an
equal bifurcation of a stem in which no secondary xylem is

present; in the lower part of the section the xylem and the outgoing leaf-traces are seen in radial section and at the upper end of each arm the leaf-traces alone, *lt*, are exposed, as in fig. 148, B. It is interesting to notice the large amount of

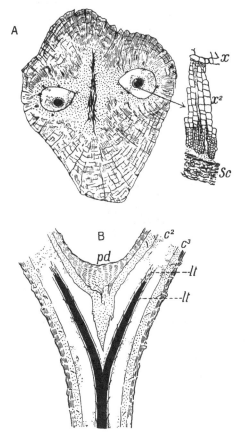

FIG. 152. *Lepidodendron vasculare.* Sections of dichotomously branched shoot.
　A.　From a section (10·5 × 9 cm.) in the Cambridge Botany School Collection.
　B.　From a section (8 cm. long) in the Cambridge Collection.

phelloderm which has been produced in the fork of the branch, at *pd*, where greater strength is required.

The section represented diagrammatically in fig. 152, A, has lost the outermost part of the cortex together with the leaf-

cushions; it consists largely of secondary cortex composed of
radially disposed phelloderm cells and tangentially placed
secretory strands (represented by the discontinuous black lines
in the drawing): the dotted region in the central part of the
axis is composed of primary cortical parenchyma, and the two
spaces surrounding the steles contain portions of the lacunar
middle cortex.   Each stele posseses a narrow crescentic zone

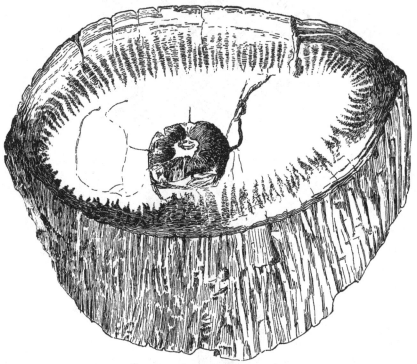

Fig. 153.   *Lepidodendron vasculare.*
(From a specimen (16 × 7·5 cm.) in the Manchester Museum.)

of secondary xylem; the amount is greater in the case of the
right-hand stele, of which a small piece is shown on a larger
scale; the striking contrast in size between the outer and more
internal secondary tracheae is no doubt the expression of some
unfavourable condition of growth.   The position of the secretory
zone beyond the secondary xylem is shown at *sc*, fig. 152, A.

An example of a large and partially decorticated stem is
afforded by the specimen (16 × 7·5 cm.) shown in fig. 153. The
irregularly ribbed surface is formed of rather thick-walled
phelloderm, in which occur tangentially arranged rows of
secretory strands. The tapered form of the secondary cortex
as it abuts internally on the primary cortex is shown very clearly
in the drawing (cf. fig. 151, C). The stele in this much older
stem consists mainly of secondary wood.

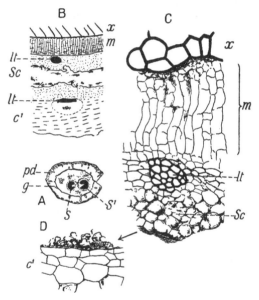

Fig. 154. *Lepidodendron vasculare.* Shoot (2·8 cm. diam.) with two steles.
(From a specimen from Halifax in the Williamson Collection,
British Museum, No. 340.)

An interesting example of a small shoot, the largest
diameter of which is 2·8 cm., is shown in fig. 154, A : the section
was cut a short distance above the bifurcation of the stele into
two approximately equal branches. The outer part of the
cortex consists of phelloderm, *pd*, with the usual rows of
secretory tracts, and primary outer cortex *g* ; the middle cortex
is represented by patches of parenchyma with a few leaf-traces.
To one of the steles, *s′* (fig. 154, A), a crescent-shaped band of

secondary xylem has been added; the other stele, *s*, possesses no fully developed secondary elements.

Fig. 154, B and C, illustrates the anatomical features immediately external to the primary xylem of the smaller stele, *s*. The comparatively broad band of radially disposed parenchyma, *m*, is connected with the outermost elements of the xylem by a few rather dark and small crushed parenchymatous cells. The band *m*, which we may speak of as the meristematic zone, clearly consists of cells in a state of division; it is in this region that the secondary xylem is produced. Beyond the leaf-trace, (fig. 154, C *lt*), occurs a portion of the secretory zone, some of the

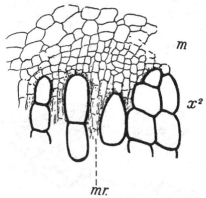

FIG. 155. *Lepidodendron vasculare*. Outer edge of secondary xylem : *m*, meristematic zone; *mr*, medullary ray. (Drawn from the section shown in fig. 168, A).

smaller cells of which show signs of disorganisation; but most of this tissue has been destroyed (fig. 154, B, *sc*). The outer edge of the secretory zone is shown in fig. 154, D abutting on the cells of the inner cortex, *c'*. The leaf-trace shown in the inner cortex in fig. 154, B illustrates the more oval or tangentially extended form of the xylem in this region, in contrast to the more circular outline which it exhibits on the inner side of the secretory zone.

The transverse section, part of which is reproduced in fig. 168, A, illustrates a characteristic feature, namely the juxtaposition of the outermost tracheae of the secondary xylem and much smaller cells of the meristematic zone. This is seen in fig. 155, which shows a small piece of fig. 168, A, on a larger scale. In plants with a normal cambium the segments cut off from the initial layer fit on to the elements of the xylem or phloem to which they are to form additions, but in *Lepidodendron* it seems to be a general rule to find each of the most external lignified elements abutting on a group of two or three much smaller cells. It is difficult to believe that the meristem shown in fig. 155, *m*, could produce secondary xylem elements equal in size to those already formed: in all probability had growth continued there would have been a marked difference between the size of the secondary tracheids, as in fig. 152, A, x^2, where there was no doubt some cause which interfered with normal cambial activity. This disparity in size between the secondary xylem elements and the adjacent parenchymatous tissue of the meristematic zone is by no means exceptional and may be described as the general rule. It is at least certain that in *Lepidodendron vasculare*, as in other species, the secondary xylem was succeeded by a broad band of parenchymatous tissue, from which new tracheae and medullary-ray elements were produced, and not by a narrow cambium such as occurs in recent plants.

v. Lepidodendron *stems as represented by casts and impressions of partially decorticated specimens.*

The differentiation of the outer cortex of a *Lepidodendron* into comparatively thin-walled and more resistant tissue has been the cause of unequal decay and the consequent formation of shrinkage cavities. In addition to the unequal resisting power of contiguous tissues, another important factor in determining the nature of casts and impressions is the existence of the cylinder of delicate cells in the outer cortex of stems and branches. As already pointed out, this meristematic cylinder or phellogen constitutes a natural line of separation, as in the case of the cambium layer between the wood and the external tissues in a fresh Sycamore twig. The result of the separation of an outer shell of bark from the rest of the stem and the

results of unequal decay in the more superficial tissues, have
necessarily led to the preservation of the same specific type
under a variety of forms.

Our knowledge of the anatomy of Lepidodendron stems
enables us to recognise in fossils of very different appearance
specimens in various conditions of preservation of one and
the same type. Such names as *Knorria, Bergeria* and *Aspidiaria* are examples of generic titles instituted before any
adequate knowledge of Lepidodendron anatomy was available.

Differences in age as well as different degrees of decortication have contributed in no small measure to the institution of
generic and specific names which more recently acquired knowledge has shown to be superfluous.

a.　*Knorria.*

The designation Knorria, after a certain G. W. Knorr of
Nürnberg, was proposed by Sternberg in 1826[1] for casts of
Palaeozoic stems of a type figured more than a century earlier
by Volkmann[2]. Goeppert, in his earlier works, published drawings of fossil stems which he referred to Sternberg's genus: one
species he at first called *Didymophyllum Schollini*. He afterwards[3] described some specimens which showed that the
features characteristic of *Knorria* may occur on partially decorticated stems with leaf-cushions of the true Lepidodendron
type. His specimens, preserved in the Breslau Museum,
demonstrate the accuracy of his drawings and conclusions.
Goeppert, and after him Balfour[4], drew attention to the different
appearances presented by branches of *Araucaria imbricata* when
preserved with the surface intact and after partial decortication,
as illustrating possible sources of error in the determination of
fossil stems.

Although it is now a well-established fact that fossils bearing the name *Knorria* are imperfect lepidodendroid stems, the
use of the term may be conveniently retained for descriptive
purposes. The specimen from the Commentry coal-field of

---

[1] Sternberg (26) A.　　　[2] Goeppert (52) A. p. 196. See also Kidston (01) p. 50.
[3] *ibid.* (52) A. p. 44. Pls. xxx. xxxi. Lief. i and ii.　　　[4] Balfour (72) A.

France, shown in fig. 156, affords some excuse for the institution of several generic names for different states of preservation
or decortication of one species. The cortical level exposed at *e*
is characterised by spirally disposed peg-like ridges with truncated apices: it is this form of cast which is usually designated
Knorria. The ridges vary in size and shape in different types

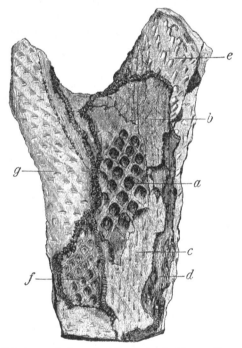

Fig. 156. A dichotomously branched Lepidodendroid stem (*Knorria mirabilis*
Ren. and Zeill.). (After Renault and Zeiller.) (¼ nat. size.) The
original specimen is in the Natural History Museum, Paris.
a—g, surface features exposed as the result of different degrees of
decortication. (See vol. I. p. 102, fig. 23).

of stem; they may be narrow as shown at *e*, fig. 156, or short
and broad with rounded distal ends. In some cases they are
forked at the apex, as in the partially decorticated specimen
of *Lepidodendron Veltheimianum* represented in fig. 185, A.

The *Knorria* state represents the impression or cast of the
outer cortical region too deep below the leaf-cushion region to

retain any indications of the cushion-form; the ridges are the casts of the spaces produced in the cortex by the decay of the sheath of delicate cells surrounding each leaf-trace and by the decay of the thin-walled cells of the parichnos. The occasional forked apex of a ridge is the expression of the fact that the cast was made at the region where the parichnos divides into two arms (cf. p. 100). In certain specimens it is possible to connect the Knorria casts with associated lepidodendroid stems which may be determined specifically; but when we have no evidence as to surface-features the fossils may be designated casts of lepidodendroid stems in the Knorria condition. Such casts are illustrated by numerous drawings in palaeobotanical literature[1].

b. Bergeria.

This is another name first used by Sternberg in his classic work, *Die Flora der Vorwelt*, for casts of lepidodendroid plants such as Steinhauer[2] had previously figured as *Phytolithus cancellatus*. Brongniart[3] recognised that the application of the generic title *Lepidodendron* should be extended to include specimens referred by Sternberg to *Bergeria*, and a few years later Goldenberg[4] realised that this name does not stand for well-defined generic characters. The correctness of these views was, however, first satisfactorily demonstrated by Carruthers[5] and by Feistmantel[6].

If a Lepidodendron stem loses its superficial layers of outer cortex and in this condition is embedded in sand or mud, the cast is distinguished from that of a perfect stem by the absence of the leaf-scars and by other features. It may, however, still show spirally disposed areas, corresponding approximately to the original leaf-cushions, which are characterised by a small depression or pit either at the apex or near the centre of each oval area: the pit marks the position of the leaf-trace and its parichnos strand. In some cases the exposed surface may be smooth without any indication of leaf-cushions, while narrow

---

[1] Good examples are given by Schmalhausen (77) Pl. III.
[2] Steinhauer (18) A. Pl. IV. fig. 5.   [3] Brongniart (49) A. p. 42.
[4] Goldenberg (55).   [5] Carruthers (73²) p. 6.
[6] Feistmantel (75) A.

spirally arranged grooves represent the obliquely ascending vascular bundles passing through the cortex to the leaves.

Fig. 185, B, shows the Bergeria state of *Lepidodendron Veltheimianum*, which differs from the Knorria condition in the fact that decortication had not extended below the level at which the form of the leaf-cushions could be recognised. It is clear that no sharp line can be drawn in all cases between the different degrees of decortication as expressed by the terms *Knorria* and *Bergeria*.

A list of synonyms of *Knorria*, *Bergeria*, and *Aspidiaria* forms of stem and a detailed treatment of their characteristic features may be found in a recent work by Potonié[1].

c. *Aspidiaria.*

In one of the earliest English books on fossil plants, the *Antediluvian Phytology* by Artis[2], a specimen from the Carboniferous sandstone of Yorkshire is figured as *Aphyllum cristatum*, and a similar fossil is described as *A. asperum*. These are impressions of Lepidodendron stems in which the characteristic leaf-cushions are replaced by smooth and slightly convex areas with a narrow central ridge. To this type of specimen Presl gave the name *Aspidiaria*[3], under the impression, shared by subsequent writers, that the supposed external features were entitled to generic recognition.

It is to Stur[4] that we owe the first satisfactory interpretation of fossils included under the name *Aspidiaria*: he showed that on the removal of the projecting convex areas from some of his specimens a typical Lepidodendron leaf-cushion was exposed (fig. 144, A, *a*). The Aspidiaria condition (fig. 144, A, *b*) represents the inner face of the detached shell of outer bark of a Lepidodendron stem, while in the Bergeria casts we have a view of the external face of a stem deprived of its superficial tissues.

In a Lepidodendron stem embedded in sediment the more delicate portions of the leaf-cushions would tend to shrink away

[1] Potonié (05) Lief. iii. 42—44.   [2] Artis (25) A. Pls. xvi. xxiii.
[3] Sternberg (38) A.   [4] Stur (75) A. Heft ii. p. 229.

from the internal and more resistant tissues of the outer cortex, thus producing spaces between each cushion; further decay would cause rupture of the leaf-traces and the superficial tissues would thus be separated from the rest of the stem. The tendency of Lepidodendron stems to split along the line of phellogen in the outer cortex is seen in fig. 148, A, *g*. The deposition of sediment on the exposed inner face of this cortical shell would result in the production of a specimen of the Aspidiaria type: the reticulum enclosing the spirally disposed convex areas is formed by the impression of the firmer tissue between the leaf-cushions.

vi.   *Lepidodendroid axes known as* Ulodendron *and* Halonia.

a.   *Ulodendron.*

This generic name was suggested by Lindley and Hutton[1] for two specimens from the English Coal-measures characterised by leaf-cushions like those of a *Lepidodendron,* but distinguished by the presence of two vertical rows of large and more or less circular cup-shaped scars. These authors, while recognising the possibility that the fossils might be identical with *Lepidodendron,* regarded them as generically distinct. The generic title *Ulodendron,* though no longer denoting generic rank, is still applied to certain shoots of lycopodiaceous plants which may belong to the genera *Lepidodendron, Bothrodendron,* and according to some authors[2], also to *Sigillaria.*

The large specimen from the Belgian coal-measures, represented in fig. 211, affords a good example of the Ulodendron form of shoot of the genus *Bothrodendron,* which is described on page 249. The specimen shown in fig. 157 shows the *Ulodendron* shoot of *Lepidodendron Veltheimianum.*

Casts of large Ulodendron scars are occasionally met with as separate fossils bearing a resemblance to an oval shell.

In Steinhauer's paper on *Fossil Reliquiae*[3] a drawing is given

---

[1] Lindley and Hutton (31) A. Pls. v. and vi.

[2] Kidston (85). In this important paper Dr Kidston gives a full account of the history of our knowledge of *Ulodendron.*

[3] Steinhauer (18) A. p. 286, Pl. vii. fig. 1.

of a Ulodendron stem under the name *Phytolithus parmatus* and a similar stem specifically identical with that shown in fig. 157 was figured by Rhode[1], one of the earliest writers on

FIG. 157. *Lepidodendron Veltheimianum.* Ulodendron condition. (From a photograph by Dr Kidston of a specimen from the Calciferous Sandstone series, Midlothian; ⅔ nat. size.) [Kidston (02) Pl. LVII.]

fossil plants, under the comprehensive designation "Schuppen-pflanze."

There has been no lack of ingenuity on the part of authors

[1] Rhode (20) Pl. III.

in offering suggestions as to the meaning of these large cup-
like depressions, and there is still difference of opinion as to
their significance.   Lindley and Hutton[1] described them as the
scars of branches or masses of inflorescence.   Sir Joseph
Hooker[2] speaks of a specimen of *Ulodendron*, shown to him by
Mr Dawes, on which a large organ, supposed to be a cone, was
inserted in one of the depressions, but he was unable to arrive
at any conclusion as to the real nature of the fossil.   While
most authors have seen in the scars pressure-areas formed by the
pressure of sessile cones against the surface of a growing branch,
others, as for example Geinitz[3], have described the depressions
as branch-scars.   Carruthers[4] regarded the scars as those of
adventitious roots and Williamson referred to them as the
scars of reproductive shoots.   The depressions vary considerably
in size.   The Belgian example shown in fig. 211 possesses scars
9 cm. in diameter.   A specimen of *Bothrodendron* in the Manchester Museum from the Lancashire Coal-Measures, to which
Williamson[5] has referred, bears two rows of scars 11—12 cm.
in diameter on a stem 112 cm. in girth and 233 cm. long.   The
scars occur in two alternate series, on opposite faces of the
axis, the distance between the successive scars in the same row
being 29 cm.   The surface-features of this large stem are not
preserved.

   Before considering the nature and origin of the scars it is
important to remember the considerable size to which they
may attain; other points of importance are the occurrence,
either in the centre of each depression or in an excentric
position, of an umbilicus or slightly projecting boss, in the
centre of which is a pit formed by the decay of an outgoing
vascular strand.   The sloping sides of the scars sometimes
bear elevations resembling leaf-cushions like those on the rest
of the stem surface.   In the specimen shown in fig. 157 the
lower margin of each cup shows indistinctly the outlines of what
appear to be leaf-cushions, while the rest of the sloping face is
characterised by radial ridges, which may be due to bracts or
leaves.

[1] Lindley and Hutton (31) A.   [2] Hooker (48), p. 427.   [3] Geinitz (55) A.
[4] Carruthers (70).   [5] Williamson (72).

It is obvious that in these cups we have the scars of some lateral organ, but the evidence afforded by specimens of which the depressions contain the remains of such organs is by no means conclusive. A Ulodendron has been figured by D'Arcy Thompson[1], in which the lower part of a lateral organ is attached by a narrow base to one of the scars, but the preservation is not sufficiently good to enable us to decide whether the organ is a cone or a vegetative shoot. Kidston[2] has described other examples showing portions of organs in connexion with the scars, but an examination of the specimens in his collection failed to convince me that his interpretation of them as strobili is correct.

The phenomenon known as cladoptosis, as shown on a stem of the Conifer *Agathis*[3] and certain Dicotyledonous trees such as *Castilloa*, suggests a possible explanation of the Ulodendron scars. This comparison was made by Shattock[4] in 1888, but he did not accept the resemblance as a real one. An objection may be urged to the cladoptosis hypothesis that in *Ulodendron* the branch, whether vegetative or reproductive, was not attached to the whole of the depressed area. On the other hand, a lateral branch originally attached by a narrow base may have continued to increase in diameter until its base became slightly sunk in the bark of the stem, thus producing a cup-like depression which, on the fall of the branch, would retain traces of the original surface-features of the stem.

Mr Watson[5] of Manchester recently published a paper on Ulodendron scars, in which he adduces fresh and, as it seems to me, satisfactory arguments in favour of the branch-scar hypothesis. Fig. 158, from one of Mr Watson's blocks, illustrates the nature of his evidence. He points out that in the obverse half of a large specimen of *Bothrodendron* in the Manchester Museum, the umbilicus consists of a cylindrical hole, 18 mm. deep and 8 mm. in diameter, surrounded by a projecting ring of mineral material which doubtless represents some portion of the original plant: on the reverse half of the

---

[1] Thompson, D'Arcy (80).   [2] Kidston (85).
[3] Seward and Ford (06) Pl. xxiii. fig. C.   [4] Shattock (88).
[5] Watson (08).

specimen the continuation of the ring is seen as a prominent
cone fitting into the cup-like depression in the obverse half: the

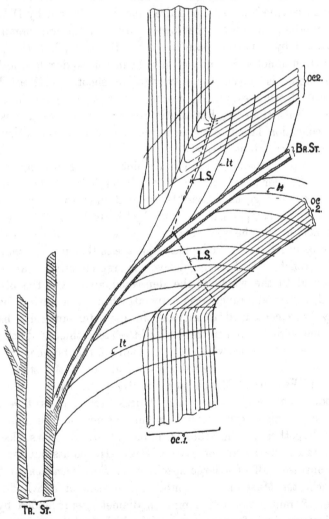

FIG. 158. Diagrammatic section through the base of a branch to illustrate the
Branch theory of the Ulodendroid scar. (After Watson.)

conical cast shows that numerous small vascular strands were
given off from this ring of tissue, and these strands have the

same arrangement and size as the dots which are found on typical Ulodendron scars. He interprets the ring surrounding the umbilicus as the remains of the primary wood and the small strands as leaf-traces supplying the branch.

In the diagrammatic section shown in fig. 158 the outer cortex of the main stem is represented by *oc* 1; this consists of secondary tissue. The corresponding tissue in the branch is seen at *oc* 2. The stele of the stem is shown at Tr. St. and that of the branch at Br. St.; *lt, lt*, mark the position of the leaf-traces. If we assume the branch to be detached along the line LS, the depression would show numerous spirally arranged dots representing the points of exit of leaf-traces and the vascular axis would be exposed in the umbilicus. This explanation appears to me to be in harmony with the surface-features of Ulodendron scars on both Bothrodendron and Lepidodendron stems. The occasional occurrence of leaf-cushions on a portion of a Ulodendron scar is a difficulty on the cladoptosis hypothesis. Assuming that true leaf-cushions occur, their presence may, as Watson suggests, be due to the folding back of a piece of the outer cortex of the branch which has been "crushed down on to the area of the scar[1]."

Since this account was written a note has been published by M. Renier[2] in which he describes a specimen of *Bothrodendron* from Liège, one face of which shows a projecting Ulodendroid scar with an excentric umbilicus. On the other face is a dichotomously branched shoot with surface-features corresponding to those on the scar; the evidence that the scar represents the base of the branch is described as indisputable.

Stur[3] held the view that the depressions on Ulodendron stems represent the places of attachment of special shoots comparable with the bulbils of *Lycopodium Selago*, or, it may be added, with the short branches occasionally produced on *Cycas* stems. If the depressions were formed by the pressure of the bases of cones, it is clear that the size of the cavity must be an index of the diameter of the cone. The larger Ulodendron scars exceed in diameter the base of any known lepidodendroid strobilus. Another obvious difficulty, which has not been over-

[1] Watson (08) p. 10.   [2] Renier (08).   [3] Stur (75) A. Heft II.

looked by Kidston who holds that the scars were produced by
sessile cones, is that in *Lepidodendron Veltheimianum* strobili were
borne at the tips of slender branches; the same difficulty is
presented by *Bothrodendron* (Fig. 213). It is unlikely that two
types of strobili were produced on the same plant, particularly
as the cone of *L. Veltheimianum* was heterosporous.

The cones of certain species of *Pinus* remain attached to
the tree· for many years and their bases become embedded in
the stem; this is particularly well shown in the drawing of a

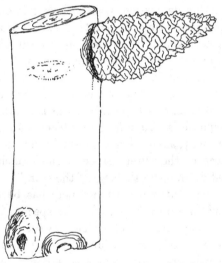

Fig. 159.   *Pinus clausa.*   ¼ nat. size.

cone of *Pinus clausa* (fig. 159), for which I am indebted to
Mr Sudworth, Dendrologist in the United States Forest Service.
Mr Sudworth has drawn my attention to *P attenuata* and *P.
muricata* in illustration of the same phenomenon[1]. The
example shown in fig. 159 cannot, however, be matched by any
known specimen of *Ulodendron*; in the case of the depressions
on the stem of a Pine the cone-base fits the circular scar,
but in the fossil stems it is practically certain that this was
not the case.

---

[1] *Garden and Forest*, vol. v., pp. 160—162, fig. 24 (April 6, 1902).

There can be little doubt that certain Palaeozoic Lycopods shed their branches by a method similar to that employed by the Kauri Pine of New Zealand and by some species of Dicotyledons. The evidence adduced in the case of *Bothrodendron punctatum* is a strong argument in favour of extending the same explanation to other Ulodendron shoots.

A   B

FIG. 160. A. *Lepidophloios scoticus* Kidst. From a specimen from the Calciferous Sandstone, Midlothian, in Dr Kidston's Collection ; rather less than ⅓ nat. size.

B. *L. scoticus* cone. From a specimen from the Calciferous Sandstone of Midlothian in Dr Kidston's Collection ; slightly reduced.

b. Halonia.

The branched axis with *Lepidophloios* leaf-cushions, represented in fig. 160, A, illustrates a special form of shoot described by Lindley and Hutton[1] under the generic name *Halonia.* The original specimens referred to this genus are

[1] Lindley and Hutton (35) A.

decorticated axes showing remains of Lepidodendroid leaf-
cushions.   The spirally disposed circular scars in the specimen
of *Halonia* (*Lepidophloios scoticus*[1]) shown in fig. 160 constitute
the characteristic feature of the genus; they may have the
form, as in fig. 160, A, of circular discs with a central umbilicus
marking the position of a vascular strand, or, as in the sandstone cast of *Halonia tortuosa* shown in fig. 161[2], they may

FIG. 161.   *Halonia tortuosa* L. and H.   From a specimen in Dr Kidston's Collection, from the Lower Coal-measures of Ayrshire (No. 1561); ⅔ nat. size.

appear as prominent tubercles.   The latter example illustrates
the condition characteristic of partially decorticated stems.

In 1883 Williamson[3] described a specimen, now in the Leeds
Museum, which convinced him that *Halonia* is merely a special
form of *Lepidodendron* concerned with the production of fertile
shoots or strobili.   Feistmantel[4] also recognised that *Halonia*

---

[1] Kidston (93) Pl. II. fig. 6.   [2] *ibid.* (02) Pl. LIII. fig. 2.
[3] Williamson (83²) A. Pl. 34.   [4] Feistmantel (75)A. p. 193, Pls. XXXIV.—XXXVII.

regularis is identical in the form of the cushions with the type known as *Lepidophloios laricinus*. It is worthy of note that under the name *Halonia*, Feistmantel[1] figured a piece of decorticated axis characterised by two rows instead of the usual spiral series of large cup-shaped scars. Recent researches have, however, tended to break down the distinction between *Ulodendron* and *Halonia* founded respectively on the biseriate and spiral arrangement of the scars or tubercles.

The interpretation of Halonial branches as cone-bearing members of Lepidodendroid plants has passed into a generally accepted statement of fact, but, so far as I know, only one specimen has been figured in which strobili are seen attached to an Halonia axis. This specimen, described by Grand'Eury[2] from the coal-field of Gard, is hardly sufficiently well-preserved to constitute a demonstration of the correctness of the generally received view, which, as is not unusual, has been repeated by one writer after another without due regard being paid to the nature of the evidence on which the statement is based. It may, indeed, be correct to describe Halonial branches as cone-bearing, but there are certain considerations which make one pause before unhesitatingly accepting this explanation. The vascular strand which passes from the central cylinder of the shoot to the tubercle or scar is composed of a solid rod of xylem distinguished from the main stele by the absence of a pith. In such petrified peduncles as have been discovered the stele is of the medullated type. The common occurrence of strobili terminating slender branches of lepidodendroid plants, though not a fatal objection to their attachment to Halonial shoots, shows that in many cases the cones were borne at the tip of leafy shoots. It may be that some of the Halonial scars are in origin like those of the Ulodendron axes of *Bothrodendron* and mark the position of deciduous vegetative branches.

The first account of the anatomy of *Halonia* we owe to Dawes[3]; this was followed by a fuller description by Binney[4] The history of our knowledge of this type of branch has been given by Carruthers[5], who expressed the opinion that *Halonia*

---

[1] Feistmantel *loc. cit.* Pl. xlvii.   [2] Grand' Eury (90) A.
[3] Dawes (48).   [4] Binney (72) ; see also Seward (99).   [5] Carruthers (73²).

is merely a fertile condition of *Lepidophloios* and possibly of other lepidodendroid plants. He was also inclined to regard the Halonial tubercles as younger stages of the larger scars characteristic of the genus *Ulodendron*. Williamson's contributions to our knowledge of *Halonia* are of primary importance; he supplied further proof of the Lepidodendroid nature of these branches and advanced our knowledge of their anatomy. In an early paper[1] he expressed the view that the differences on which *Halonia* and *Ulodendron* are separated are such as result from a difference in age and are not of generic importance. In the last memoir, of which he was sole author, published by the Royal Society[2], Williamson brought forward further evidence in support of this well-founded opinion.

That the fossils known as *Halonia* are branches of a lepidodendroid plant is at least certain, and it is probable that the lateral branches which they bore were fertile, though satisfactory proof of this is lacking. We know also that Halonia branches are characterised by the Lepidophloios form of leaf-cushion; there is, however, no sufficient reason to assume that such branches were never attached to stems with the cushions of the Lepidodendron form. The further question, namely whether Williamson was correct in his contention as to the absence of any essential distinction between *Ulodendron* and *Halonia*, does not admit of an unchallenged answer. In 1903 Weiss[3] described the anatomy of a specimen of a biseriate *Halonia* branch of *Lepidophloios*. The form of the leaf-cushions is unfortunately not very well preserved, but Weiss figures other specimens with two rows of tubercles on which the leaf-cushions are sufficiently distinct to justify a comparison with those of *Lepidophloios*. He believes with Williamson that it is the presence of tubercles in place of scars which distinguishes *Halonia* from *Ulodendron*, and that the arrangement of the tubercles or scars is a matter of little importance. He expresses the opinion justified by the evidence available that the absence or presence of tubercles is merely due to accidents of preservation or, one may add, to difference in age. Kidston[4] dissents from Weiss's description of his specimen as a biseriate

[1] Williamson (72).　　[2] *ibid.* (93).　　[3] Weiss, F. E. (03).　　[4] Kidston (05).

Halonia; he regards it as a Ulodendron branch of *Sigillaria discophora* (König). Until specimens with more clearly preserved external features are forthcoming it is impossible to settle the point in dispute, but on the facts before us there would seem to be a *prima facie* case in favour of Weiss's contention.

The designation *Halonia* may be retained as a descriptive term for Lepidodendroid shoots characterised by spirally disposed scars or tubercles and bearing leaf-cushions of the Lepidophloios type. In the case of specimens showing prominent tubercles, the superficial tissues are usually absent and, as in the fossil represented in fig. 161, the name *Halonia* does not necessarily imply the presence of leaf-cushions of a particular type.

vii. *Anatomical characters of Vegetative Lepidodendron shoots (Lepidodendron and Lepidophloios).*

The type already described under the name *Lepidodendron vasculare* differs from those dealt with in the following pages chiefly in the anatomy of the stele. The simplest and probably most primitive type of Lepidodendron stem is that in which the xylem forms a solid rod; the type of stele most frequently represented is that of *L. Harcourtii, L. fuliginosum,* and other species in which the diameter of the stele is greater and a cylinder of primary xylem encloses a comparatively large parenchymatous pith.

1. *Lepidodendron esnostense,* Renault[1].

This species was founded by Renault on petrified specimens from the Culm beds of Esnost in France. The surface of a young twig bears prominent leaf-cushions of elongated rhomboidal form similar to those of *Lepidodendron obovatum* (fig. 173) and other species. In older branches the primary cortex is replaced by a considerable thickness of radially disposed secondary cortical tissue which, as shown in tangential section, consists of a reticulum of elongated pointed elements

---

[1] Renault (96) A. p. 175, Pls. xxxiii. xxxiv.

with comparatively thick walls enclosing meshes filled with large-celled parenchyma. It is worthy of note that if such a branch were exposed to decay, the earlier destruction of the more delicate tissue in the meshes of the secondary cortex would produce a series of oval depressions, corresponding to the parenchymatous areas, separated by a projecting reticulum of the more resistant elements: a cast of this partially decayed surface would be indistinguishable from that of some types of *Sigillaria* or of a *Lyginodendron*. The inner regions of the cortex of the type-specimens have not been preserved. The xylem, which is the only part of the stele represented, has the form of a protostele or solid cylinder of scalariform tracheids with peripheral groups of narrower protoxylem elements which mark the points of exit of the leaf-traces: in a branch 1—2 cm. wide the xylem column has a diameter of 3 mm. The small leaves (fig. 143, B, C), similar to those of a *Sigillaria*, are sub-rhomboidal in section near the base and approximately circular near the apex[1]. The mesophyll consists of palisade cells having the appearance of typical chlorophyll-tissue. The heterosporous strobili attributed to this species bore microsporangia on the upper and megasporangia on the lower sporophylls; the megaspores, of which a considerable number occur in each megasporangium, are identical in size with those of another Culm form, *Lepidodendron rhodumnense*. Some of these have retained traces of prothallus tissue, and in one spore Renault figures what he regards as an archegonium: the drawing is by no means convincing.

2. *Lepidodendron rhodumnense*, Renault[2].

The species from the Culm of Combres (Loire) agrees in its solid xylem cylinder and in the differentiation of the secondary cortex, as also in the association of two kinds of spore, with *Lepidodendron esnostense*. A comparison of the leaves of the two types reveals certain differences which may be of specific rank, but, apart from minor differences, these Culm species may be classed under one anatomical type.

[1] For description of the leaf-anatomy, see pp. 98, 99.
[2] Renault (79) p. 249, Pl. x.

3. *Lepidodendron saalfeldense,* Solms-Laubach[1].

This Devonian species was founded on a specimen 3 × 2·5 cm. broad at the base, which shows the stumps of four branches recalling the dichotomously branched arms of *Stigmaria* and *Pleuromeia.* If these are in reality the remains of Stigmaria-like horizontal branches the species affords an interesting example of a Lepidodendron axis with a subterranean rhizome of the type which has been found in several Sigillarian stems. In the upper end of the axis the stele consists of a solid strand of xylem which is not sufficiently well preserved to show the position of the protoxylem groups. A transverse section taken near the base reveals a type of stele differing from that at the upper end in being composed of radially disposed tracheids and in its resemblance to the stele of *Stigmaria.*

4. *Lepidodendron fuliginosum,* Williamson. Figs. 162—172, 179, E.

1871. *Lepidodendron Harcourtii,* Binney, Palæont. Soc., p. 48, Pl. VII. fig. 6.
1872. *Halonia regularis,* Binney, Palæont. Soc., p. 89, Pl. XV.
1881. *Lepidodendron Harcourtii,* Williamson, Phil. Trans. Roy. Soc., Vol. 172, p. 288, Pls. XLIX–LII.
1887. *Lepidodendron fuliginosum,* Williamson, Proc. Roy. Soc., Vol. XLII. p. 6.
1891. *Lepidodendron Williamsoni,* Solms-Laubach, Fossil Botany, p. 226.
1893. *Lepidophloios fuliginosus,* Kidston, Trans. Roy. Soc. Edinburgh, Vol. XXXVIII. p. 548.

The name *Lepidodendron fuliginosum* was proposed by Williamson in 1887 for petrified stems previously included by him in Witham's species *L. Harcourtii,* but subsequently recognised as a distinct type characterised by "the greater uniformity in the composition of the entire cortex" and by other features some of which do not constitute distinctive characters. The species agrees with *L. Harcourtii* and with *L. Veltheimianum* in having a medullated stele; it is distinguished not

[1] Solms-Laubach (96) p. 18, Pl. x. figs. 7—11.

only by the more frequent preservation of the middle cortex, a
fact due to a difference in minute structure, but chiefly by
the peculiar structure of the secondary tissue added to the
stele; this is in part composed of radial series of parenchymatous cells and of a varying amount of tracheal tissue the
elements of which are narrower than in other species and are
characterised also by their sinuous vertical course. As is pointed
out in the sequel, the anatomical features of *L. fuliginosum,* as
at present understood, are not confined to one type of *Lepidodendron* stem. Specimens have been described with leaf-
cushions of the form characteristic of *L. aculeatum, L. obovatum*
and *Lepidophloios* combined with the anatomical features of
Williamson's species: it is possible that the two species
L. obovatum and *L. aculeatum* are not really distinct[1], but it is
certain that shoots with both the *Lepidodendron* and *Lepidophloios* cushions may have the same type of anatomical
structure.

A more detailed knowledge of the structural features of
Lepidodendron shoots may enable us to define anatomical
species with more exactness than is possible at present. There
can, however, be little doubt that well-marked anatomical
features may be associated with more than one specific form of
shoot as defined by the form of the leaf-cushions.

Solms-Laubach proposed the name *Lepidodendron Williamsoni* for the anatomical type *L. fuliginosum* of Williamson, but
the latter name has been generally adopted.

In the following account special attention is directed to the
nature and origin of the secondary stelar tissue and to the
secretory zone, as difference of opinion exists as to the interpretation of these features. Among the best examples of shoots
of *Lepidodendron fuliginosum* without secondary tissue or in
which it is feebly developed are those originally described by
Binney. The stele includes a large parenchymatous pith, the
cells of which frequently show signs of recent division, a feature
observed also in the pith of the large stem of *L. Wünschianum,*
represented in figs. 181, 182. The primary xylem cylinder has
an irregularly crenulate outer edge like that of *L. Wünschianum*

---

[1] They are regarded as identical by Fischer (04).

and *L. Harcourtii* and the protoxylem elements occupy an exarch position. Isodiametric reticulately-pitted elements are met with both on the inner and outer edge of the xylem.

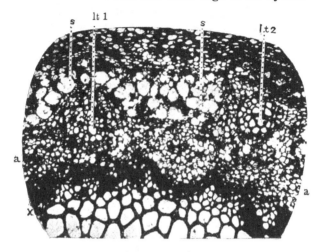

Fig. 162. *Lepidodendron fuliginosum.* Part of the stele in transverse section. (Binney Collection, Sedgwick Museum, Cambridge.)

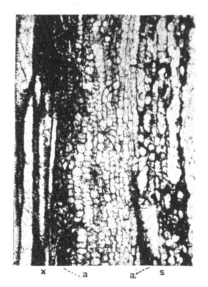

Fig. 163. *Lepidodendron fuliginosum.* Longitudinal section. (Binney Collection, Cambridge.)

Figs. 162 and 163 illustrate the structure of the outer portion
of the xylem and adjacent tissues in a section of a shoot
3·8 cm. × 2·5 cm. in diameter, which is in the act of branching,
as shown by the occurrence of two steles of equal size.
A figure of the complete section will be found in Binney's
memoir[1], and additional illustrations were published in 1899[2].

The primary xylem (figs. 162, 163, x) is succeeded by 2—3

Fig. 164.   *Lepidodendron fuliginosum.*   Leaf-trace.
(Binney Collection, Cambridge.)

rows of polygonal cells with dark contents and associated with
isodiametric tracheae: these pass into clearer parenchymatous
tissue, a, characterised by the arrangement of the cells in
vertical series, to which the term meristematic zone has been
applied. The secretory zone, s, abutting on the meristematic
zone, consists of more or less disorganised parenchymatous cells

[1] Binney (72) Pl. xiii. fig. 1.   [2] Seward (99).

and broader and more elongated spaces; it is interrupted here
and there by an outgoing leaf-trace, as at lt 1 and lt 2 in fig 162.
The secretory zone is succeeded by a homogeneous inner cortex
like that described in *L. vasculare*; part of this region is seen
at the upper edge of fig. 162. The broad middle cortex, which
is separated from the inner cortex by a sharply defined
boundary, is composed of rather small lacunar parenchymatous tissue consisting of sinuous tubular elements interspersed
among isodiametric cells of various sizes (fig. 166, p). In
the middle cortical region the leaf-traces pursue an almost

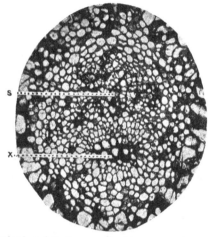

FIG. 165. *Lepidodendron fuliginosum.* Leaf-trace: x, xylem; s, secretory zone.
(Binney Collection, Cambridge.)

horizontal course; one is shown in fig. 164, in oblique longitudinal section, in a reversed position; the xylem, x, should be
on the inner side of the secretory tissue, s. The clear space
between the two parts of the vascular bundle was originally
occupied by a few layers of parenchymatous cells, as seen in the
transverse sections, figs. 165 and 166. In some specimens the
leaf-traces pass through the middle cortex in a much more
vertical course, as shown by the section represented in fig. 165.
This section illustrates the structure of a typical leaf-trace with
unusual clearness; it shows the tangentially elongated group of
xylem, the strand of tissue which occupies the position of phloem,

s (to which the term secretory zone is applied), the compact
parenchyma between the two parts of the bundle, and surrounding the whole a narrow sheath sharply contrasted by the
smaller and more uniform size of the cells from the middle
cortex, a few cells of which are seen in the photograph. The
middle cortex shows a well-defined junction with the more
compact outer cortical region, which consists of primary
parenchyma passing externally into a zone of phelloderm composed of thick-walled and more elongated cells. A noticeable
feature in many Lepidodendron shoots is the occurrence of a
circle of strands of secretory cells often surrounding fairly large

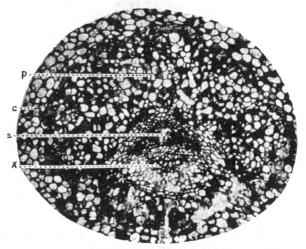

Fig. 166.   *Lepidodendron fuliginosum.*   Leaf-trace : *p*, parichnos.
(Binney Collection, Cambridge.)

ducts just internal to the edge of phelloderm : similar strands
form irregularly concentric circles, as was pointed out in the
case of *L. vasculare,* in the phelloderm itself.

Fig. 166 shows a leaf-trace in the outer cortex accompanied
by its crescent-shaped parichnos, *p*, derived from the middle
cortex and by means of which the outer cortex and the lamina
of the leaves are connected with the inner region of the shoot.
This lacunar middle cortex and parichnos doubtless constitute an
aerating tissue-system which after leaf-fall is exposed directly
to the air at the ends of the parichnos arms on the leaf-scars.

Some of the sections in the Binney Collection (Sedgwick Museum, Cambridge) show early stages in the production of secondary xylem: in the section represented in fig. 167 the secretory zone is succeeded on its inner face by a zone of radially elongated cells, *m* which are clearly in a meristematic condition. The same section shows also the more radially extended form of the xylem of a leaf-trace with its internal protoxylem, *px*, in

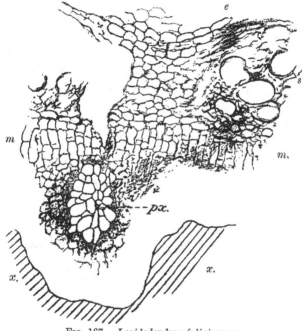

Fig. 167.　*Lepidodendron fuliginosum.*
(Binney Collection, Cambridge.)

contrast to the tangentially elongated form which is assumed during its passage through the cortex (cf. figs. 165, 166).

Some sections of *Lepidodendron fuliginosum* in the Manchester University Collection are of special interest from the point of view of the method of secondary thickening. In the section reproduced in fig. 168, B, the meristematic zone is seen to consist in part of radially elongated elements, *m*, with parallel cross-walls evidently of recent origin. The same tissue is shown also in fig. 168, C, *a*, D, *a*, and in fig. 169, A, *a*

This band of meristem, which we may speak of as the cambium, occurs in the outer region of the meristematic zone immediately internal to the secretory zone, *sc*.

The result of the activity of this cambium band is the production of secondary parenchyma and tracheal tissue. In fig. 179, E, drawn from a portion of the section represented in

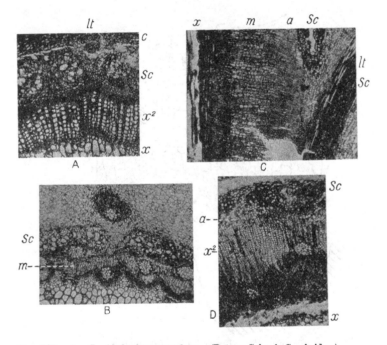

Fig. 168. A.   *Lepidodendron vasculare*. (Botany School, Cambridge.)
       B.   *Lepidodendron fuliginosum*. (From a specimen from Shore, Lancashire, in the Cambridge Botany School Collection).
       C.   *L. fuliginosum*. ("Biseriate Halonia" of Weiss No. 257, Manchester University Museum.)
       D.   *L. fuliginosum*. (Manchester Univ. Museum.)

fig. 168, B, a projecting arm of primary xylem is seen at x; this is followed by 2—3 layers of parenchymatous cells, some of which have dark contents, and beyond this is seen a group of secondary elements, *tr*, cut across somewhat obliquely, which are evidently products of the cambial cells on the inner margin

of the secretory zone, *sc*. The longitudinal section (fig. 169, D)
shows the cambial cells, *a*, next the secretory zone, *sc*, passing
internally into crushed and imperfectly preserved elongated
elements which are presumably miniature tracheae, and these
are succeeded by older and more completely lignified xylem

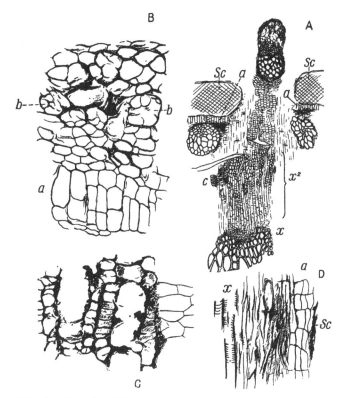

Fig. 169.   *Lepidodendron fuliginosum.*
    A, B.   (Manchester University Collection. No. Q. 645 A.)
    B, C.   (Manchester. No. 257.)
    D.   (Manchester. No. 6.)

elements, *x*. In larger shoots the amount of secondary tissue
is considerably greater; it may consist almost entirely of short-
celled parenchyma (fig. 168, C, from *x* to *sc*), or it may include
a large proportion of radially disposed and vertically elongated
tracheae (fig. 168, D, x^2, and fig, 170, A, x^2), or it may consist of

150 LYCOPODIALES [CH.

parenchyma containing scattered groups of tracheae (fig. 169, A, x^2)[1].

Fig. 169, A, is a diagrammatic sketch of the tissues—1 mm. wide—between the primary xylem, x, and the inner cortex. The primary xylem is succeeded by short parenchymatous cells followed by a zone of radially elongated elements passing occasionally into rows of narrow scalariform tracheae, some of which, owing to their sinuous longitudinal course (fig. 171, C), are seen in oblique section, as at C, fig. 169, A. At its outer edge this secondary tissue, x^2, consisting of parenchyma and tracheae, passes into the cambial band (fig. 169, B, a).

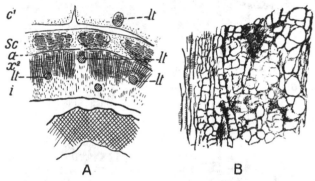

Fig. 170. *Lepidodendron fuliginosum.* (From sections in the Manchester Museum.)

The radial longitudinal section represented in fig. 168, C, is taken from the fossil described by Weiss as a biseriate *Halonia*; it agrees sufficiently closely in structure with others referred to *Lepidodendron fuliginosum* to be classed as an example of this anatomical type. A complete transverse section of the stem measures $9 \times 6\cdot3$ cm.; the breadth of the tissues between the edge of the primary xylem and the outer edge of the secretory zone is $2\cdot5$ mm. The middle cortical region, characterised by the sooty appearance, which led Williamson to choose the specific name *fuliginosum*, is traversed by the leaf-traces and is sharply differentiated from both the inner and outer cortex.

[1] As Miss Stokey (09) points out the production of parenchyma internal to the cambium of *L. fuliginosum* is a feature shared by *Isoetes*. See also Scott and Hill (00), p. 424.

The longitudinal section (fig. 168, C) shows the outer edge of
the primary xylem, x, abutting on a band of dark and small-
celled parenchyma which passes into the broad zone of secondary
tissue, m, the inner region of which consists of fairly thick-
walled elements in radial series passing externally into the
thin-walled cells of the cambial region, a, on the inner edge of
the secretory zone, sc. This section shows also the interruption

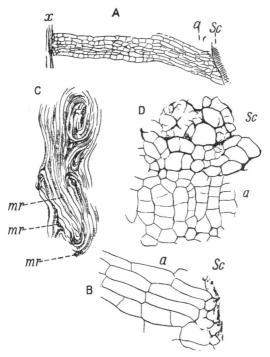

Fig. 171. *Lepidodendron fuliginosum.* (From sections in the Manchester
Museum.)

of the secretory zone by an out-going leaf-trace, lt, the lower part
of which, sc, is continued downwards into the secretory zone. The
exit of a leaf-trace produces a gap in the secretory zone of the
stem, but not in the xylem. If we applied the term phloem to the
secretory zone—a course adopted by Prof. F. E. Weiss and some
other authors, but which I do not propose to follow—we should
speak of a phloem foliar-gap as a characteristic feature of a

Lepidodendron shoot. This applies to other species of the genus as well as to *L. fuliginosum.*

Fig. 171, A, shows more clearly the broad zone of secondary parenchyma with the thinner-walled cambial region, a; the latter is represented on a larger scale in fig. 171, B. The section shown in fig. 168, D, and in fig. 170, A, affords an example of a stem in which the secondary tissue consists largely of narrow scalariform tracheae, x^2; the primary stele has a diameter of 1 cm.; the secondary xylem, x^2, forms a fairly broad zone of parenchyma and tracheal elements through which leaf-traces pass vertically, a fact of some interest in comparison with the

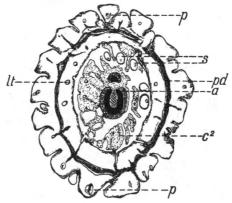

Fig. 172. *Lepidodendron fuliginosum.* From a section (4 × 3·4 cm.) in the Williamson Collection, British Museum (No. 379), figured by Williamson, *Phil. Trans. R. Soc.* 1881, Pl. 52.

horizontal course which they pursue through the medullary rays in the normal secondary wood of *L. vasculare* and *L. Wünschianum.* The secondary tracheae pass gradually into thin-walled cambial cells (*a*, fig. 168, D; 170, A) with parallel tangential walls. Fig. 171, C, shows the sinuous course of the secondary tracheae as seen in longitudinal section, and a few small groups of parenchymatous cells, *mr*, which may be of the nature of medullary rays, enclosed between the winding scalariform tracheae.

The secretory zone of *Lepidodendron fuliginosum* agrees essentially with that of other species; it usually presents the appearance shown in fig. 168, B, *sc*; fig. 169, B and C; fig. 170, B (longitudinal section); fig. 171, D, *sc*. The comparatively large clear spaces which characterise this tissue, as seen in fig. 168, B, appear to owe their origin to groups of small cells which gradually break down and give rise to spaces containing remnants of the disorganised elements, as in fig. 171, D, and fig. 169, B, *b*. The secretory tissue seen in fig. 170, B, consists of large and small parenchymatous cells without any of the broad sacs or spaces such as are shown in fig. 169, C.

Fig. 172 represents a diagrammatic sketch of a transverse section (4 × 3·4 cm. in diameter) of a young shoot from the Lower Coal-Measures of Lancashire figured by Williamson[1] in 1881 as *Lepidodendron Harcourtii*. It shows the features characteristic of *L. fuliginosum* and is of importance as affording an example of a shoot giving off a branch from the stele to supply a lateral axis of the type characteristic of *Halonia*. The exit of the branch-stele forms a gap in the main stele; a ramular gap as distinguished from a foliar gap. The outgoing vascular strand is at first crescentic, but becomes gradually converted into a solid stele. The primary xylem of the main stele (black in the figure) consists of a ring six tracheae in breadth; this is succeeded by a few layers of dark parenchymatous cells and a band of radially elongated elements, *a*, which abuts on the secretory zone. The middle lacunar cortex, c^2, with *Stigmaria* rootlets, *s*, is fairly well preserved. In the outer cortex occur several leaf-traces, *lt*, accompanied by spaces originally occupied by the parichnos strand, *p*. A band of secondary cortex, consisting chiefly of phelloderm, is seen at *pd*. The prominent leaf-cushions, some of which show the parichnos, *p*, appear to be of the Lepidophloios type.

It remains to consider the external characters of Lepidodendroid shoots possessing the anatomical features represented by the comprehensive species *Lepidodendron fuliginosum*.

Certain sections exhibiting this type of structure were described by Binney in 1872 as *Halonia regularis*[2] on evidence supplied by Mr Dawes, who stated that they were cut from a

[1] Williamson (81) A. Pl. LII. p. 288. (Will. Coll. No. 379.)   [2] Binney (72).

specimen bearing Halonia tubercles. The section represented
in fig. 172 is no doubt from an Halonia axis. In 1890 Cash
and Lomax[1] stated that they had in their possession a stem of
the *L. fuliginosum* type with the external features of *Lepidophloios*; this identification has been confirmed by Kidston[2]
and Weiss[3]. It is, however, equally clear that certain species
with the elongated leaf-cushions of *Lepidodendron* must be
included among examples of shoots with the anatomical
characters of *L. fuliginosum*.

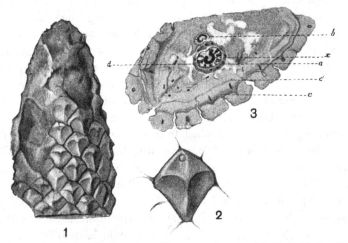

FIG. 173. *Lepidodendron obovatum.* (From a specimen lent by Dr D. H. Scott.)

Dr Scott[4] published in 1906 a short account of the structure
of a specimen from the Lower Coal-Measures of Lancashire, the
external features of which were identified by Kidston with
those of *Lepidodendron obovatum* Sternb. Dr Scott generously
allowed me to have drawings made from his specimen; these
are reproduced in fig. 173. The form of the leaf-cushion is by
no means perfect; there is a well-marked median ridge, and
the small circular scar near the upper end of some of the
cushions may represent the ligular cavity. At the base of the
leaf-cushions a cortical meristem has produced a zone of

[1] Cash and Lomax (90).   [2] Kidston (93) p. 547.
[3] Weiss, F. E. (03) p. 218.   [4] Scott, D. H. (06³).

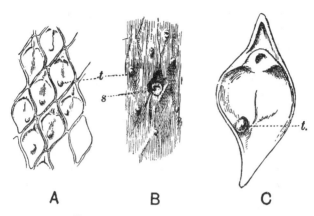

FIG. 174. *Lepidodendron aculeatum.*
(Cambridge Botany School.)

secondary cortex; at *c* a second meristem is seen in the outer
cortex: the dark dots in the cortex mark the positions of leaf-
trace bundles.   The inner cortex, *d*, is a more compact tissue

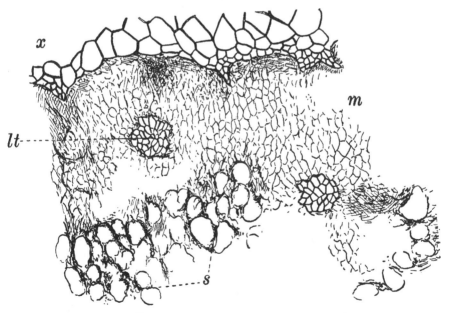

FIG. 175. *Lepidodendron aculeatum.*
(Cambridge Botany School.)

surrounding the imperfectly preserved secretory zone. From the medullated stele a lateral branch, *b*, is being given off; its crescentic form becoming changed to circular as it passes nearer to the surface.

A type of *Lepidodendron*, *L. Hickii*, founded on anatomical characters by Mr Watson[1], is believed by him to possess leaf-cushions like those of *L. obovatum*; if this is so, it is interesting, as he points out, to find two distinct anatomical types associated with one species. Watson thinks it probable that the " species" *L. obovatum* includes at least two widely different species. This merely emphasizes the importance of correlating structure and external characters as far as available data permit.

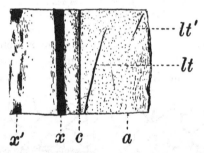

Fig. 176. *Lepidodendron aculeatum.*
(Cambridge Botany School.)

The specimen, of which part of the surface is shown in fig. 174, is in all probability *L. aculeatum* Sternb. This was described by me in detail in *The Annals of Botany* (1906) as another example of the co-existence of the *Lepidodendron fuli-ginosum* type of anatomy with a true *Lepidodendron*. The locality of the specimen is not known. The leaf-cushions are 1·5 cm. long with tapered upper and lower ends; a ligular cavity may be recognised on some parts of the fossil, also faint indications of leaf-trace scars. The tubercles (fig. 174, A—C, *t*) probably represent leaf-traces which the shrinkage of the super-ficial tissues has rendered visible in the lower part of their course. The circular scar, *s* (fig. B), on the partially decorticated surface is apparently a wound. The stele is sufficiently

[1] Watson (07) p. 18.

well preserved to justify its reference to *L. fuliginosum.* The irregularly crenulated edge of the primary xylem, x (fig. 175), is succeeded by a broad band of parenchyma (the meristematic zone), m, and beyond this are remnants of the secretory zone, s. The structure of the leaf-traces corresponds with that of other specimens of the type, but the much steeper course of these vascular strands, lt, lt' (fig. 176), is a feature in which this example differs from most of those referred to *L. fuliginosum.* Such evidence as is available would seem to point to the

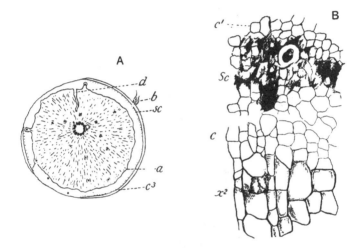

FIG. 177. *Stigmaria radiculosa* (Hick). (From sections in the Manchester University Collection.)

absence of trustworthy criteria enabling us to separate, on anatomical grounds, *Lepidophloios* and *Lepidodendron*[1].

Stigmaria radiculosa (Hick).

We have no proof of the nature of the subterranean organs of *Lepidodendron fuliginosum*, though it is not improbable that the specimens described below may be correctly assigned by Weiss to that species. Prof. Weiss[2] has made an interesting contribution to our knowledge of a type first described by Hick[3] under the name *Tylophora radiculosa*, a designation which he after-

[1] Seward (06) p. 378.   [2] Weiss, F. E. (02).   [3] Hick (93).

wards altered to *Xenophyton radiculosum*[1] and for which we may
now substitute *Stigmaria radiculosa* (Hick).   Prof. Williamson
expressed the opinion that *Xenophyton* exhibited considerable
affinity with *Stigmaria ficoides* and Weiss's further study of the
species leads him to regard Hick's plant as probably the Stigmarian organ of *Lepidodendron fuliginosum*. The diagrammatic
transverse section represented in fig. 177, A (4·5 cm. in diameter),
shows an outer cortex of parenchyma, c^3, consisting in part of

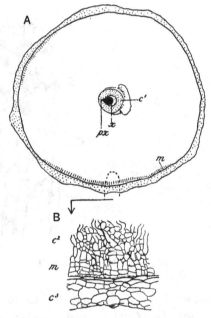

Fig. 178. *Rootlet of Stigmaria.* (From a section in the Manchester Collection.)

radial rows of secondary tissue and of a band of compact parenchyma bounded by the wavy line *a*; at *sc* is a series of secretory
strands exactly like those in a corresponding position in *Lepidodendron fuliginosum* and other species of the genus. The greater
part of the organ is occupied by a lacunar and hyphal middle
cortex identical in structure with that shown in fig. 178, B, drawn
from a rootlet. At *d*, fig. 177, A, the middle cortex has been

[1] Hick (93²).

invaded by a narrow tongue of outer cortical tissue. The stele is characterised by a large pith filled with parenchyma; in *Stigmaria ficoides*[1] the general absence of pith-tissue has led to the inference that the stele was hollow. The xylem is represented by a ring of bundles separated by broad medullary rays; each bundle contains a few small, apparently primary, elements on its inner edge but is mainly composed of radial rows of secondary tracheae x^2, fig. 177, B. On the outer face of the secondary xylem occur a few smaller and thinner walled cells, c, having the appearance of meristematic tissue; from these additional tracheae were added to the xylem. This meristematic zone occurs, as in the stems of *Lepidodendron*, immediately internal to the secretory tissue, sc; at C^1, fig. 177, B, is seen the inner cortical tissue.

In surface-view a specimen figured by Hick[2] shows a number of circular scars agreeing in shape and arrangement with the rootlet scars of *Stigmaria ficoides*. At b in fig. 177, A, the basal portion of a rootlet is shown in organic connexion with the outer cortex. The rootlet bundles are given off from the stele as in other examples of *Stigmaria*; each bundle consists of a triangular strand of xylem with an endarch protoxylem at the narrow end accompanied by a portion of the secretory tissue as in the leaf-traces. As in *Stigmaria ficoides* the rootlets are attached to the outer cortex above a cushion of small cells. It is interesting to find that rootlet-bundles, as seen in tangential section of the main axis, are associated with a parichnos strand, but this is on the xylem side of the vascular strand, whereas in the case of leaf-traces the parichnos is on the other side of the bundle.

Fig. 178, A, represents a transverse section of a rootlet (6 mm. in diameter) associated with *Stigmaria radiculosa* and probably belonging to this species. The xylem strand x is composed of a group of tracheae with a single protoxylem strand, px, at the pointed end and with small metaxylem elements at the broad end next the space originally occupied by the so-called phloem. A parenchymatous sheath, c', surrounds the bundle, and beyond this is the broad middle cortex a small portion of which is shown on a larger scale in fig. 178, B; as Weiss

[1] See p. 240.   [2] Hick (93) Pl. xvi. fig. 1.

points out, some of the outermost cells of the lacunar cortex (m) are clearly in a state of meristematic activity.

The preservation of the middle cortex and the small quantity of secondary xylem are characters which this *Stigmaria* shares with *Lepidodendron fuliginosum*, and although decisive evidence is still to seek, we may express the opinion that Weiss's surmise of a connexion between *Stigmaria radiculosa* and *Lepidodendron fuliginosum* is probably correct.

5. *Lepidodendron Harcourtii.*   Fig. 179, A—D.

In 1831 Mr Witham[1] published an anatomical description of a fragment of a *Lepidodendron* which he named *Lepidodendron Harcourtii* after Mr C. G. V Vernon Harcourt from whom the specimen was originally obtained.   The fossil was found in rocks belonging to the Calciferous series in Northumberland.   Witham reproduced the account of this species in his classic work on *Fossil Vegetables*[2], and Lindley and Hutton[3], who examined Mr Harcourt's material, published a description of it in their *Fossil Flora* in which they expressed the view that *Lepidodendron* is intermediate between Conifers and Lycopods.   Adolphe Brongniart[4] included in his memoir on *Sigillaria elegans* an account of Witham's species based on material presented to the Paris Museum by Mr Hutton and Robert Brown.   Dr Kidston[5] has shown that the actual transverse section figured by Witham is now in the York Museum; a piece of stem in the same Museum, which is not the specimen from which Witham's section was cut, supplied the transverse section figured by Brongniart.   The figures given by Lindley and Hutton do not appear to have been made from the York specimens.   In 1887 Williamson[6] published a note in which he pointed out that some of the specimens described by him as *L. Harcourtii* should be transferred to a distinct species, which he named *L. fuliginosum*. Subsequently in 1893 he gave a fuller account of Witham's species; it has, however, been shown by Dr Kidston and by

[1] Witham (31) A.   [2] Witham (33) A. Pls. XII. XIII.
[3] Lindley and Hutton (35) A. Pls. 98, 99.   [4] Brongniart (39) A.
[5] Kidston (03) p. 822.   [6] Williamson (87).

Mr Watson[1] that certain specimens identified by Williamson as
L. Harcourtii differ sufficiently from that type to be placed in
another species, for which Watson proposes the name *I. Hickii.*

A paper on *L. Harcourtii* published by Bertrand[2] in 1891
extends our knowledge of this type in regard to several anatomical details. It was recognised by Williamson that the
absence of secondary wood in shoots possessing the anatomical
characters of *L. Harcourtii* is a feature to which no great
importance should be attached. It is possible that the large
stems from the Isle of Arran described by Williamson[3] as
Lepidodendron Wünschianum, in which the secondary wood is
well developed, may be specifically identical with the smaller
specimens from Northumberland and elsewhere which are recognised as examples of Witham's type.

The diagrammatic sketch shown in fig. 179, A, was made from
a section figured by Williamson in 1893[4]; it has a diameter of
9×8.5 cm. The stele is of the medullated type like that of
L. Wünschianum, and the outer edge of the primary xylem is
characterised by sharp and prominent projecting ridges similar
to those of *L. fuliginosum* but rather more prominent. Parenchymatous cells succeed the xylem, as in other species, but in
this case there is no indication of meristematic activity; beyond
this region occur occasional patches of a partially destroyed
secretory zone. Remains of a lacunar tissue are seen in the
middle cortical region; also numerous leaf-traces, *lt,* consisting
of a tangentially elongated xylem strand accompanied by a
strand of secretory zone tissue enclosed in a sheath of delicate
parenchyma. In the inner part of the outer cortex, c^3, the leaf-
traces lie in a space originally occupied by the parichnos; in
the outer portion of the same region a band of secondary
cortex, *pd,* has been formed; immediately internal to this
occur numerous patches of secretory tissue, represented by small
dots in the drawing close to *pd*; one is shown on a larger scale
in fig. B.

The position of the phellogen is seen at *a*; external to this
are radial rows of rather large cells with dark contents.

---

[1] Kidston (03) p. 822; Watson (07).   [2] Bertrand, C. E. (91).
[3] Williamson (80) A.   [4] Williamson (93) Pl. i. fig. 3.

Fig. 179, C, *x*, shows the characteristic form of the primary xylem edge, beyond which are seen oval or circular leaf-traces with a mesarch protoxylem, *lt, px.* It is possible that this specimen may not be specifically identical with Witham's species, but it represents a very similar if not identical type; it may on the other hand be referable to *L. fuliginosum.* The importance of the specimen, apart from its precise specific position, is that

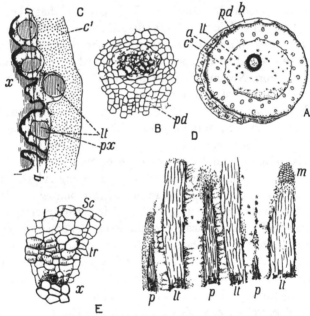

FIG. 179.   A—D.   *Lepidodendron Harcourtii,* Witham.
E.   *Lepidodendron fuliginosum,* Shore, Lancashire.
A, B.   From a specimen in the Williamson Collection, British Museum (No. 380), from Airdrie, Scotland.
C, D.   From sections in the Collection of Dr Kidston, from Shore, Lancashire.

it serves to illustrate the general appearance of the xylem surface met with in both species, *L. Harcourtii* and *L. fuliginosum.* A tangential longitudinal section, taken through the line *ab* in fig. C, is represented in fig. 179, D. The xylem of the leaf-traces *lt,* consisting chiefly of scalariform tracheae, alternates with patches of crushed and delicate parenchyma which immediately abut on the primary xylem; at p, p, the section passes through some of the projecting arms of the xylem cylinder; at m is seen a patch of meristematic zone tissue. This section together with the similar section of *Lepidodendron vasculare* described on a previous page demonstrates that the projecting ridges of the primary xylem form apparently vertical bands: they are not characterised by a lattice-work arrangement as described by Bertrand and by other authors who have accepted his conclusions. If a reticulum of intersecting ridges were present on the face of the xylem cylinder its existence would be revealed by such a section as that represented in fig. 179, D.

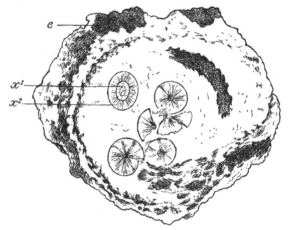

FIG. 180. *Lepidodendron Wünschianum.* From Arran. ($\frac{1}{5}$ nat. size.)
(Sedgwick Museum, Cambridge.)

6. *Lepidodendron Wünschianum* (Williamson). Figs. 180—184.

Reference was made in Volume I. to the occurrence of large stems of a Lepidodendron in volcanic beds of Calciferous sandstone age in the island of Arran[1]. These were discovered and briefly described by Mr Wünsch in 1867[2] and afterwards named by Carruthers *Lomatophloyos Wünschianus*[3]. Mr Carruthers

---

[1] Volume I. p. 89. For other references to these stems, see Seward and Hill (00) p. 918.
[2] Wünsch (67).   [3] Carruthers (69²) p. 6.

visited the locality and published an account of the peculiar
method of preservation of the plant remains[1]. It is, however, to
Williamson[2] that we owe the more complete description of these
Arran stems. Portions of large stems from the Arran beds are
preserved in the British Museum, the Sedgwick Museum, Cambridge, and in the Manchester Museum. The section of one
of these is shown in fig. 180; an outer shell of bark encloses a
mass of volcanic ash in which are embedded several woody
cylinders originally described as "internal piths[3]," and by Carruthers as young stems produced from spores which had germinated
in the hollow trunk of a large tree. The true interpretation was
supplied by Williamson who showed that a stem of the dimensions
of that represented by the outer cortex, *e*, fig. 180, must have
possessed a single stele of the size of those seen in the interior
of the hollow trunk. The additional woody cylinders, or steles,
were derived from other stems, and carried, probably by water,
into the partially decayed trunk. In addition to large Lepidodendron stems Williamson described smaller shoots as well
as an Halonial branch and made brief reference to some
cones described by Binney[4] in 1871 from the same locality.

The following account of *Lepidodendron Wünschianum* is
based on an exceptionally fine specimen discovered by Mr T.
Kerr of Edinburgh in Calciferous sandstone volcanic ashes at
Dalmeny in Linlithgowshire. The material from this locality
described by Mr Hill and myself[5] was generously placed in my
hands by Dr Kidston of Stirling. Fig. 181, A, shows a
transverse section, 33 cm. in diameter, consisting of a shell of
outer cortical tissue enclosing a core of light-coloured volcanic
ash; on the decay of the more delicate middle cortex the cylindrical stele dropped to one side of the hollow trunk. The stele,
fig. 182, has a diameter of 6·5 cm.; the centre is occupied by
concentric layers of silica, *s*, surrounded externally by the
remains of a parenchymatous pith, *p*, made up of isodiametric
and sinuous hypha-like elements like those in the middle cortex
of Lepidodendron shoots. On the inner edge of the primary

---

[1] Carruthers (69).   [2] Williamson (80) A. ; (93); (95).
[3] Wünsch *loc. cit.*   [4] Binney (71) p. 56.
[5] Seward and Hill (00).

xylem, x', occur several isodiametric tracheae with fine scalariform
and reticulate thickening bands like those in the central region

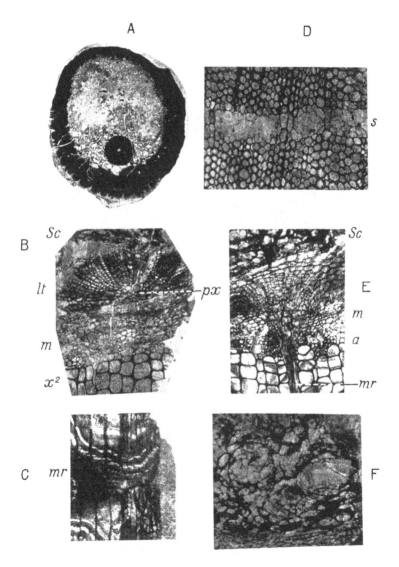

FIG. 181. *Lepidodendron Wünschianum.* Calciferous Sandstone, Dalmeny.
(A, Sedgwick Museum, Cambridge. B—F, Botany School, Cambridge.)

of the stele of *Lepidodendron vasculare*: it is probable that these
elements are vestiges of conducting tissue which in ancestral
forms formed a solid and not a medullated stele.

The primary xylem is limited externally by an unequally
fluted surface with exarch protoxylem elements; it is, however,

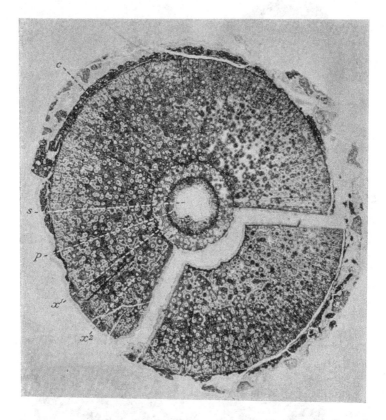

FIG. 182.   *Lepidodendron Wünschianum.*   The stele of the stem shown in
fig. 181, A.   (Cambridge Botany School.)

noteworthy that there is not always a very clearly defined
difference between the small protoxylem and the large centripetally developed tracheae.   Immediately beyond the primary
xylem occur numerous thin-walled parenchymatous cells with
spiral and reticulate pitting; beyond these is the broad zone of

secondary xylem, x^2, composed of scalariform tracheae and numerous medullary rays consisting of one, two, or several rows of radially elongated elements with spiral and reticulate pitting. In tangential sections the rays are seen to vary considerably in size, some being made up of a single row of cells while others are longer and broader; through the latter leaf-traces pass horizontally. Portions of medullary rays are seen at mr in fig. 181, C and E.

The leaf-traces given off from projecting ridges on the outer edge of the primary xylem pass upwards for a short distance and then bend outwards through a broad medullary ray; on reaching the limit of the secondary xylem they again bend sharply upwards, appearing in transverse section at lt fig. 181, B. Each leaf-trace consists at first of long tracheae accompanied by numerous thin-walled spiral and reticulate parenchymatous cells derived from the tissue in contact with the outer edge of the primary wood. Fig. 181, B, shows a leaf-trace near the edge of the secondary xylem; it consists of a group of primary tracheae, with narrower protoxylem elements, px, near the outer margin, almost completely enclosed by radially disposed series of smaller and more delicate tracheae. These secondary elements of the leaf-trace are apparently added during its passage through the medullary ray, but additions are also made to this tissue by the meristematic zone, m, fig. 181, B and E. In contact with the outermost tracheae of normal size at the edge of the secondary xylem there are some smaller lignified elements, as at a, fig. 181, E, and at T, fig 183; this juxtaposition of large and small tracheae has been referred to in the description of *L. vasculare*.

Prof. Williamson[1], in his account of the Arran specimens of this species, expressed the opinion that the trees probably perished "in consequence of the mephitic vapours which filled the atmosphere"; it may be that in the striking difference in the diameter of the conducting elements on the margin of the wood we have evidence of approaching death.

Beyond the most recently formed tracheae we have a band

of delicate parenchymatous cells (*m*, figs. B and E, 181; C, figs.
183, 184) forming the meristematic zone[1]. The longitudinal
section represented in fig. 184 shows some recently formed narrow
tracheae, T, and beyond these the meristematic zone composed of
thin-walled short cells, C, arranged in horizontal rows. It is this
small-celled tissue to which the name phloem has been applied
by some authors[2], a term which seems to me to be misleading

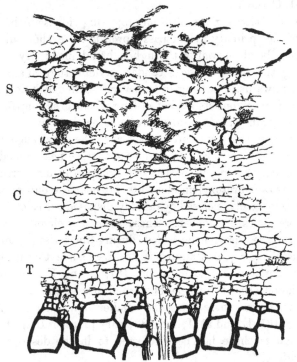

Fig. 183. *Lepidodendron Wünschianum.* From the specimen shown in fig. 181
     S, secretory zone; C, meristem; T, immature tracheae.

and inappropriate. In passing through this zone of dividing cells
the leaf-traces become surrounded by an arc of meristem from
which elements are added to the radially placed rows of secondary
tracheae. Beyond the meristematic region portions of the

---

[1] The term meristematic zone is used because some of the cells in this
region are in a state of active division, though the inner portion may consist
of permanent tissue.
[2] Scott (00) p. 131 ; (08) p. 142.

secretory zone are preserved, consisting of large sacs or spaces
and small dark cells as seen in figs. 181, B, E, *sc*, F; 183, 184.
This tissue has the same structure as in *L. vasculare* and in
L. fuliginosum: it is a striking fact that there are no indications
of any additions to the secretory zone even in stems with such
a large amount of secondary xylem as in the Dalmeny specimen
(fig. 182, x^2). If the secretory zone were of the nature of phloem
we should expect to see signs of additions made to it in the

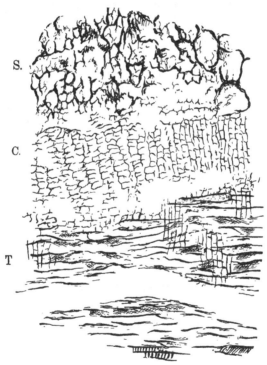

FIG. 184. *Lepidodendron Wünschianum.* Longitudinal section of the specimen
represented in transverse section in fig. 183.

course of growth. In this connexion it is worth mentioning
that in the recent fern *Botrychium* (Ophioglossaceae) secondary
xylem is formed in the stem, but apparently no additions are
made to the phloem. The structure of the secretory zone tissue
as seen in the longitudinal section fig. 184, S, is also a serious
difficulty in the way of accepting the designation phloem as

employed by Scott and Weiss. Between the secretory zone and
the outer cortical region, no tissues have been preserved. The
shell of bark consists chiefly of radial rows of elongated cells
with rather thick walls characterised by the occurrence of small
intercellular spaces and by tangentially placed bands of secretory
cells and sacs (fig. 181, D, *s*). Immediately internal to the
secondary cortex or phelloderm occur groups of secretory tissue
as shown in the section of *L. Harcourtii* (fig. 179, B).

The large tree shown in transverse section in fig. 181, A,
has lost its leaf-cushions; the bark, as seen in the lower
part of the photograph, presents a fissured appearance like
that with which we are familiar on an old Oak or Elm stem.
A radial longitudinal section through the phelloderm revealed
the existence of a crushed leaf-trace passing outwards in an
approximately horizontal course accompanied by a strand of
parenchymatous tissue[1] having the characteristic structure of a
parichnos. It is probable that the surface of this partially
decorticated stem differed in appearance from that of an old
Sigillaria (cf. fig. 198) in the much smaller and less conspicuous
parichnos strands.

In addition to the large stems of *L. Wünschianum* from
Arran and Dalmeny numerous examples of smaller axes from
the former locality are represented in the Williamson collection
(British Museum). Some of the twigs are characterised by a
solid stele (protostele) giving off numerous leaf-traces accompanied by short spirally thickened tracheids like those which
occur at the outer edge of the primary xylem in the larger stem:
these extend into the leaf where they are arranged round the
vascular bundle like the transfusion tracheids[2] in many recent
conifers. The surface of these smaller shoots bears large leaf-
cushions which are seen in longitudinal section to have the form
characteristic of *Lepidophloios*. It is worthy of note that a
section of a bifurcating axis of this species from the Calciferous
Sandstone of Craigleith (British Museum Collection[3]), although its
diameter is 19 × 14 cm., shows no signs of secondary wood. This
late appearance of secondary xylem and other anatomical features

---

[1] Seward and Hill (00) Pl. II. fig. 14.   [2] Worsdell (95); Bernard (04).
[3] No. 52, 625.

suggest the possibility of the specific identity of *L. Wünschianum* and *L. Harcourtii*[1].

In 1871 Binney[2] described a specimen of a heterosporous cone, *Lepidostrobus Wünschianus*, from Arran exhibiting the ordinary features of lepidodendroid strobili; this was probably borne by *Lepidodendron Wünschianum*.

7. *Lepidodendron macrophyllum* (Williamson). Fig. 186, C.

The diagrammatic sketch reproduced in fig. 186, C, was made from the transverse section of a small twig, slightly less than 2 cm. in its longest diameter, originally figured by Williamson[3] in 1872. Earlier in the same year Carruthers[4] published a short account of the same form based on specimens collected by Mr Butterworth from the Coal-Measures of Lancashire near Oldham, but both authors refrained from instituting a new specific name. In a later publication Williamson spoke of the type as *Lepidodendron macrophyllum*[5]. Williamson's species has nothing to do with *Lycopodites macrophyllus* of Goldenberg[6]. The most striking feature of this rare form is the large size of the leaf-cushions, which are of the *Lepidophloios* type, in proportion to the diameter of the shoot. The stele consists of a ring of xylem, all of which is primary in the sections so far described, enclosing a parenchymatous pith: a Stigmarian rootlet is shown at *s*.

8. *Lepidodendron Veltheimianum* Sternb. (General account). Figs. 157, 185, 186, A, B.

1820. "Schuppenpflanze," Rhode, Beit. zur Pflanzenkunde der Vorwelt, Pl. III. fig. 1.
1825. *Lepidodendron Veltheimianum*, Sternberg, Flora der Vorwelt, Pl. LII. fig. 5.
1836. *Pachyphloeus tetragonus*, Goeppert, Diefossilen Farnkräuter, Pl. XLIII. fig. 5.
1852. *Sagenaria Veltheimiana*, Goeppert, Foss. Flora des Übergangsgebirges, Pls. XVII—XXIV.

[1] Seward and Hill (00) p. 922.   [2] Binney (71) p. 56, Pl. XI. figs. 2a—2c.
[3] Williamson (72) p. 298, pl. XLV. fig. 35.   [4] Carruthers (72).
[5] Williamson (93) p. 30.   [6] Goldenberg (55) p. 12.

1875. *Lepidodendron Veltheimianum*, Stur, Culm Flora, p. 269, Pls. XVIII—XXII.

1886. *Lepidodendron Veltheimianum*, Kidston, Catalogue of Palaeozoic plants, British Museum, p. 160.

1901. *Lepidodendron Veltheimianum*, Potonié, Silur und 'Culm Flora, p. 116, figs. 72—76.

1904. *Lepidodendron Veltheimianum*, Zalessky, Mém. Com. Géol. Russie, Pl. IV. figs. 4, 5.

1906. *Lepidodendron Veltheimi*, Potonié, Königl. Preuss. geol. Landesanstalt, Lief. III.

The above list may serve to call attention to a few synonyms[1] of this plant, and to a selection of sources from which full information may be obtained as to the history of our knowledge of this characteristic and widely spread Lower Carboniferous type.

Lepidodendron Veltheimianum is represented by casts of stems, the largest of which hitherto described reaches a length of 5·22 metres with a maximum diameter of 63 cm.; this specimen, figured by Stur[2], consists of a tapered main axis giving off smaller lateral shoots, some of which exhibit dichotomous branching. Fig. 185, C and D, represent the external features of a well-preserved cast and impression respectively. Oblique rows of prominent cushions wind round the surface of the stem and branches: each cushion is prolonged upwards and downwards in the form of a narrow ridge with sloping sides which connects adjacent cushions by an ogee curve. At the upper limit of the broader kite-shaped portion of the cushion the ligular pit forms a conspicuous feature; immediately below this is the leaf-scar with its three small scars,— the lateral parichnos strands and the central leaf-trace. The two oval areas shown in fig. 185, D, just below the lower edge of the leaf-scars, represent the parichnos arms which impinge on the surface of the cushions on their way to the leaves, as explained on a previous page. It is possible that these areas were visible on the living stem as strands of loose parenchyma comparable with the lenticel-like pits on the stipules of *Angiopteris*[3] and the leaf-bases of Cyatheaceous ferns, or it

---

[1] See also Kidston (94), (86) A. p. 160; Potonié (05) Lief. III. 50.
[2] Stur (75) A. II. p. 330, fig. 34.   [3] Hannig (98).

may be that their prominence in the specimen before us is the
result of the decay of a thin layer of superficial cortex which hid

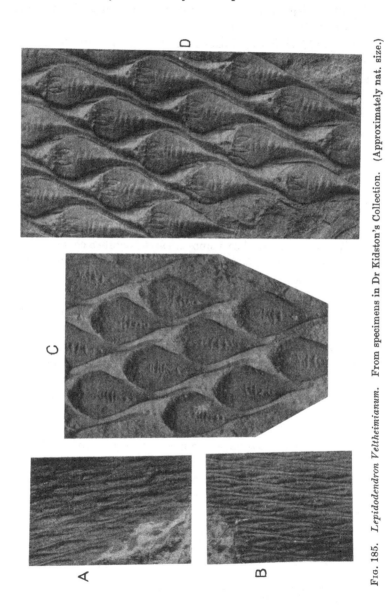

FIG. 185. *Lepidodendron Veltheimianum.* From specimens in Dr Kidston's Collection. (Approximately nat. size.)

them on the living tree. Fig. 185, B, illustrates the appearance of
a stem in a partially decorticated condition (*Bergeria* state).
A further degree of decortication is seen in fig. 185, A, which
represents the *Knorria* condition.

Fig. 157 shows a Ulodendron axis of this species; in the
lower part the specimen illustrates the partial obliteration of
the surface features as the result of the splitting of the
outer bark consequent on growth in thickness of the tree.
By an extension of the cracks, shown in an early stage in
fig. 157, the leaf-cushions would be entirely destroyed and the
surface of the bark would be characterised by longitudinal
fissures simulating the vertical grooves and ridges of a Sigillarian stem. The large stumps of trees shown in the frontispiece to Volume I. are probably, as Kidston[1] suggests, trunks
of *L. Veltheimianum* in which the leaf-cushions have been
replaced by irregular longitudinal fissures. In old stems of
Sigillaria the enlarged parichnos areas constitute a characteristic
feature (p. 205), but it does not follow that the absence of large
parichnos scars is a distinguishing feature of all *Lepidodendra*.

In this species, as in others, the form of the leaf-cushion
exhibits a considerable range of variation dependent on the
thickness of the shoot; the contiguous cushions of young
branches become stretched apart as the result of increasing
girth of the whole organ, and casts of still older branches may
exhibit very different surface-features[2]. The leaves as seen
on impressions of slender branches are comparatively short,
reaching a length of 1—2 cm. It is important to notice that
leafy twigs of this species may bear terminal cones[3] resembling
in form those of *Picea excelsa* and other recent conifers, though
differing essentially in their morphological features.

The fossil stumps of trees represented in the frontispiece to
Volume I. bear horizontally spreading and dichotomously
branched root-like organs having the characters of *Stigmaria
ficoides*[4]. Geinitz has suggested that *Stigmaria inaequalis*
Göpp. may be the underground portion of *Lepidodendron Veltheimianum*.

[1] Young and Kidston (88) A.   [2] Potonié (01²) fig. 72, p. 117.
[3] Stur (75) II. A. Pl. xxxvi. fig. 9.   [4] See Chap. xvii.

It is unfortunately seldom possible to connect petrified *Lepidodendron* cones with particular species of the genus based on purely vegetative characters, but it is practically certain that we are justified in recognising certain strobili described by Williamson[1] from the Calciferous Sandstone series of Burntisland on the Firth of Forth as those of *Lepidodendron Veltheimianum*. Williamson believed that the cone which he described belonged to the plant with shoots characterised by the anatomical features of his species *Lepidodendron brevifolium* (=*L. Veltheimianum*), a conclusion which is confirmed by Kidston[2]. The cone of *L. Veltheimianum*, which reached a diameter of at least 1 cm. and a length of 4 cm., agrees in essentials with other species of *Lepidostrobus*; the axis has a single medullated stele of the same general type as that of the vegetative shoots of *Lepidodendron fuliginosum* and *L. Harcourtii*. The sporophylls are described by Williamson as spirally disposed, and Scott notices that in some specimens they are arranged in alternate whorls; as in recent Lycopods both forms of phyllotaxis may occur in the same species. The heterosporous nature of this strobilus, to which Scott first applied the name *Lepidostrobus Veltheimianus*, is clearly demonstrated by the two longitudinal sections contributed by Mr Carruthers and figured by Williamson in 1893[3].

Each sporophyll, attached almost at right angles to the cone-axis, bears a radially elongated sporangium seated on the median line of its upper face; its margins are laterally expanded as a thin lamina; from the middle of the lower face a narrow keel extends downwards between two sporangia belonging to a lower series. From the base of a sporangium a mass of sterile tissue penetrates into the spore-producing region as in the large sporangia of *Isoetes* (cf. fig. 191, H, a, and fig. 133, H). The distal and free portion of the sporophylls is bent upwards as a protecting bract. Some of the sporangia in the upper part of the cone produced numerous microspores, while 8—16

---

[1] Williamson (72) Pl. XLIV. p. 294 : (93) (93²).

[2] Kidston (01) p. 60. See also Scott (00) p. 170, figs. 67, 68.

[3] Williamson (93), Pl. VIII. figs. 51, 52. See also figs. 67—69 given by Scott (00).

megaspores occur in the lower sporangia. The megaspores, having a mean diameter of 0·8 mm. "quite 40 times the size of the microspores[1]," are characterised by tubular capitate appendages, and by a conspicuous three-lobed projection (fig. 191, E)[2] which, as Scott suggests, may represent the outer spore-wall which has split as the result of germination. It is not improbable, as shown in fig. 191, I, that this cap was present

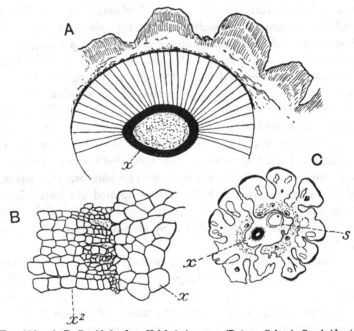

FIG. 186. A, B. *Lepidodendron Veltheimianum*. (Botany School, Cambridge.)
C. *Lepidodendron macrophyllum*. (British Museum. No. 377.)
x, Primary xylem ; x^2, secondary xylem ; s, Stigmarian rootlet.

before germination. The megaspores represented in fig. 191, I, illustrate their characteristic form as seen in a section of a megasporangium, *Sm* ; the open beak-like portion of the larger spore is probably the apical region which has split along the three-rayed lines. These lines form a characteristic feature of

[1] Scott (00) p. 173.
[2] Scott [(08) p. 187] suggests that the projection may have formed a passage for the admission of the microspores, or of the spermatozoids which they produced.

both recent and extinct spores and denote their origin in tetrads. The spore shown in fig. 191, E[1], illustrates the external features. The apical region of the prothallus of a megaspore of *Lepidodendron Veltheimianum* described by Mr Gordon[2] consists of smaller cells than those occupying the greater part of the spore-cavity, a differentiation which he compares with that of the prothallus of *Selaginella*.

There can be little doubt that the petrified shoots described by Williamson[3] from the Calciferous Sandstone beds of Burnt-island as *Lepidodendron brevifolium* are identical with specimens possessing the external features of *L. Veltheimianum*. In 1872 Dawson expressed the opinion that Williamson's species should be referred to *L. Veltheimianum*, and evidence subsequently obtained confirms this view. The stele of this species is of the medullated type, differing from that of *L. fuliginosum* and *L. Harcourtii* in the absence of prominent ridges on the external surface of the primary xylem, and from *L. vasculare* in the possession of a parenchymatous pith. In younger twigs the cortex consists of fairly homogeneous tissue, but in older branches there is a greater distinction between a delicate middle cortex and a stronger outer cortex. Fig. 186, A, represents a stem in which the vascular cylinder is composed of a primary xylem ring, x, 1·5 mm. broad, succeeded by a zone of secondary wood 1·2 cm. in breadth. The junction between the primary and secondary xylem is shown on a larger scale in fig. 186, B. The tissues abutting on the secondary xylem have not been preserved; the outer cortex, which consists chiefly of secondary elements, is divided superficially into unequal ridges corresponding to the leaf-cushions which have been more or less obliterated as the result of growth in thickness of the stem.

9. *Lepidodendron Pedroanum* (Carruthers).

In 1869 Mr Carruthers described some specimens of vegetative stems and isolated sporangia, collected by Mr Plant in Brazil, as *Flemingites Pedroanus*[4]. From a more recent account

---

[1] Bennie and Kidston (88) Pl. vi. figs. 20, *a—s*.   [2] Gordon (08).

[3] Williamson (72).   [4] Carruthers (69²).

published by Zeiller[1] it is clear that Carruthers' species is a
true *Lepidodendron*; an examination of the type-specimens in
the British Museum confirms this determination. The contiguous leaf-cushions have rounded angles similar in form
to those of *Lepidodendron Veltheimianum* and *L. dichotomum*,
but it is not unlikely that the Brazilian plant is specifically
distinct from European species. A figure of one of the specimens
on which Carruthers founded the species is given by Arber[2] in
his *Glossopteris Flora*. The Brazilian plant is chiefly interesting
as affording proof of the existence of *Lepidodendron* in the
southern hemisphere; the species has also been recognised
in South Africa from material collected by Mr Leslie at
Vereeniging[3].

As Zeiller[4] has suggested, it is not improbable that the
fossils described by Renault[5] from Brazil as *Lycopodiopsis
Derbyi* may be the petrified stems of *Lepidodendron Pedroanum*. The structure of the central cylinder of Renault's
species is of the type represented by *L. Harcourtii*; the
xylem forms a continuous ring and does not consist of separate
strands of tracheae as Renault believed.

10. *Lepidodendron australe* (M'Coy). Figs. 187, A—C.

Specimens described under this name are interesting rather
on account of their extended geographical range and geological
antiquity than on botanical grounds. The drawings reproduced
in fig. 187 illustrate the characteristic appearance of this Lower
Carboniferous and Upper Devonian type, as represented by a
specimen recently described[6] from the Lower Karroo (Dwyka)
series, which is probably of Carboniferous age, near Orange
River Station, South Africa. The surface is divided into polygonal or rhomboidal areas (figs. A and B) 8—9 mm. long and
7—8 mm. broad, arranged in regular series and representing
leaf-scars, comparable with those of *Sigillaria Brardi* and other
species, or possibly partially decorticated leaf-cushions. A short

[1] Zeiller (95). See also White (08) p. 447.
[2] Arber (05) Pl. I. fig. 2.   [3] Seward and Leslie (08).
[4] Zeiller (98).   [5] Renault (90).
[6] Seward (07³).

distance below the apex of each area there is a more or less circular prominence or depression (fig. 187, B) and on a few of the areas there are indications of a groove (fig. A, g) extending from the raised scar to the pointed base, as at g, g.

In examining the graphitic layer on the surface of the South African specimen shown in fig. 187, A, use was made of a method recently described by Professor Nathorst[1]. A few

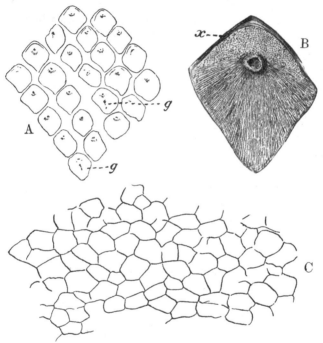

FIG. 187. *Lepidodendron australe.* Fig. A, nat. size.

drops of collodion were placed on the surface, and after a short interval the film was removed and mounted on a slide. The addition of a stain facilitated the microscopic examination and the drawing of the collodion film. The cell-outlines (fig. 187, C) on the surface of the polygonal areas may be those of the epidermis, but they were more probably formed by a sub-epidermal tissue; the scar, which interrupts the continuity of the flat surface, may mark the position of a leaf-base, or,

[1] Nathorst (07) ; Bather (07); (08).

assuming a partial decortication to have occurred prior to fossilisation, it may represent a gap in the cortical tissue caused by the decay of delicate tissue which surrounded the vascular bundle of each leaf in its course through the cortex of the stem. If the impression were that of the actual surface of a *Lepidodendron* or a *Sigillaria*, we should expect to find traces of the parichnos appearing on the leaf-scar as two small scars, one on each side of the leaf-bundle. In specimens from Vereeniging described in 1897[1] as *Sigillaria Brardi*, which bear a superficial resemblance to that shown in fig. A, the parichnos is clearly shown. On the other hand, an impression of a partially decorticated Lepidodendroid stem need not necessarily show the parichnos as a distinct feature: owing to its close association with the leaf-trace in the outer cortex, before its separation in the form of two diverging arms, it would not appear as a distinct gap apart from that representing the leaf-bundle. The absence of the parichnos may be regarded as a point in favour of the view that the impression is that of a partially decorticated stem. Similarly, the absence of any demarcation between a leaf-cushion and a true leaf-scar such as characterises the stems of Lepidodendra and many Sigillariae is also favourable to the same interpretation.

In 1872 Mr Carruthers[2] described some fossils from Queensland, some of which appear to be identical with that shown in fig. 187 under the name *Lepidodendron nothum*, Unger[3], a species founded on Upper Devonian specimens from Thuringia. The Queensland plant is probably identical with Dawson's Canadian species, *Leptophloeum rhombicum*[4]. In 1874 M'Coy[5] instituted the name *Lepidodendron australe* for some Lower Carboniferous specimens from Victoria, Australia: these are in all probability identical with the Queensland fossils referred by Carruthers to Unger's species, but as the identity of the German and Australian plants is very doubtful[6] it is better to adopt M'Coy's specific designation.

---

[1] Seward (97²) A. p. 326, Pl. xxiii.   [2] Carruthers (72²).
[3] Unger and Richter (56).
[4] Dawson (71) A. Pl. viii. See also Smith and White (05).
[5] M'Coy (74). See also Feistmantel (90) A.   [6] Kidston (86) A. p. 231.

Krasser[1] has described a similar, but probably not specifically identical, type from China; from Devonian rocks of Spitzbergen Nathorst[2] has figured, under the name *Bergeria*, an example of this form of stem, and Szajnocha[3] has described other specimens from Lower Carboniferous strata in the Argentine.

Lepidodendron australe has been recorded from several Australian localities[4] from strata below those containing the genus *Glossopteris* and other members of the Glossopteris, or, as it has recently been re-christened, the Gangamopteris[5] Flora.

viii.   *Fertile shoots of* Lepidodendron.

A.   *Lepidostrobus.*

The generic name *Lepidostrobus* was first used by Brongniart[6] for the cones of *Lepidodendron*, the type-species of the genus being *Lepidostrobus ornatus*, the designation given by the author of the genus to a Lepidostrobus previously figured by Parkinson[7] in his *Organic Remains of a Former World*. The generic name *Flemingites* proposed by Carruthers[8] in 1865, under a misapprehension as to the nature of spores which he identified as sporangia, was applied to specimens of true *Lepidostrobi*. Brongniart also instituted the generic name *Lepidophyllum* for detached leaves of *Lepidodendron*, both vegetative and fertile; the specimen figured by him in 1822 as *Filicites* (*Glossopteris*) *dubius*[9], and which was afterwards made the type-species of the genus, was recognised as being a portion of the lanceolate limb of a large single-veined sporophyll belonging to a species of *Lepidostrobus*.

In an unusually large *Lepidophyllum,* or detached sporophyll of *Lepidostrobus*, in the Manchester University Museum, the free laminar portion reaches a length of 8 cm.

It is not uncommon to find *Lepidodendron* preserved in the form of a shell of outer cortex, which has become separated along the phellogen from the rest of the stem; as the result of

---

[1] Krasser (00) Pl. ɪɪ. fig. 1.   [2] Nathorst (94) A. Pl. ɪɪ. fig. 8.
[3] Szajnocha (91) p. 203.   [4] See Etheridge (90); David and Pittman (93).
[5] White (08).   [6] Brongniart (28) A. p. 87.
[7] Parkinson (11) A. Pl. ɪx. fig. 1, p. 428.   [8] Carruthers (69²).
[9] Brongniart (22) A. Pl. ɪɪ. fig. 4.

compression the cylinder of bark may assume the appearance of a flattened stem covered with leaf-cushions. A specimen preserved in this way was described by E. Weiss as a cone of *Lomatophloios macrolepidotus* Gold., and is quoted by Solms-Laubach and other authors[1] as an example of an unusually large *Lepidostrobus*. An examination of the type-specimen in the Bergakademie of Berlin convinced me that Weiss had mistaken the partially destroyed leaf-cushions for sporophylls, and Stigmarian rootlets, which had invaded the empty space, for sporangia[2].

In external appearance some species of *Lepidostrobus* bear a superficial resemblance to the cone of a Spruce Fir (*Picea excelsa*), but the surface of a lycopodiaceous strobilus is usually covered by the overlapping and upturned laminae which terminate the more or less horizontal sporangium-bearing portion of the sporophyll.

Fig. 188 affords a good example of a long and narrow *Lepidostrobus*. This specimen from the Middle Coal-Measures of Lancashire has a length of 23 cm.; like other *Lepidostrobi* it is borne at the tip of a slender shoot. The fossil is sufficiently well preserved to show the characteristic radially elongated form of the large sporangia and the long and upturned distal portions of the sporophylls.

We may briefly describe *Lepidostrobus* as follows:—Cylindrical strobili consisting of an axis containing a single cylindrical stele which agrees generally with that of the vegetative shoots of *L. Harcourtii* and other species. The amount of parenchymatous pith varies in different forms; in some the primary xylem is almost solid. The middle cortical region, which has usually been destroyed before fossilisation, possesses the loose lacunar structure characteristic of this region in the vegetative branches. The thicker walled outer cortex is continued at the periphery into crowded, usually spirally disposed sporophylls, each of which consists of a more or less horizontal pedicel, which may be characterised by a keel-like median ridge on its lower surface, while to the central region of the upper face is attached a large radially elongated sporangium. One of the chief differences between a *Lepidodendron* cone

[1] Bower (08) p. 305.   [2] Seward (90) ; Potonié (93²).

and those of the recent genus *Lycopodium* is the greater radial elongation of the sporangia in the former. Some species of *Lepidostrobus* may have been homosporous; some are known to

Fig. 188. *Lepidostrobus*. Middle Coal-Measures, Bardsley, Lancashire. From a specimen in the Manchester Museum. (½ nat. size.)

be heterosporous. In the latter the megasporangia borne on the lower sporophylls usually contain several megaspores as in

Isoetes (cf. fig. 133, E). Beyond the distal end of the sporangium
the sporophyll becomes broader in a horizontal plane and is
bent upwards as a lanceolate limb; it may also be prolonged a
short distance downwards as a bluntly triangular expansion.

Fig. 189 is an accurate representation of a transverse
section, 6 mm. in diameter, of what is no doubt the apical
portion of a *Lepidostrobus* from the Coal-Measures of Shore,
Lancashire. The section cuts across the upturned free laminae
above the level of the apex of the cone-axis. Each lamina

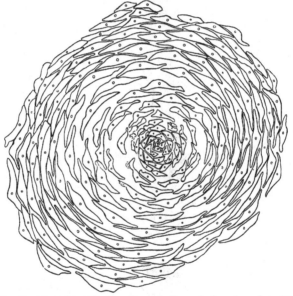

Fig. 189. *Lepidostrobus.* Section through the apical region of a cone above
the axis. (Manchester University Collection.)

contains a small vascular bundle composed of a few tracheae and
some thin-walled cells surrounded by delicate mesophyll tissue.
Immediately in front of the distal end of a sporangium a
small ligule is borne on the upper face of the sporophyll
(fig. 191, A, B, *l*) occupying the same position as in *Selaginella*
(cf. fig. 131, F). Strands of vascular tissue pass in a steeply
ascending course from the xylem to the pedicels of sporophylls, finally curving upwards and ending in the upper limb.

Each vascular bundle consists of a strand of xylem, apparently of mesarch structure, accompanied by a few layers of parenchyma on its outer face and by a group of cambiform elements, the whole being enclosed in a sheath of parenchyma continuous with the inner cortex of the cone axis. The vascular bundle is accompanied by a parichnos in the outer cortex and in the sporophyll.

There can be little doubt that the Palaeozoic *Lepidodendra*, like *Lycopodium cernuum* (fig. 123) and other recent Lycopods, usually bore their cones at the tips of slender shoots. The fertile shoot of *Lepidophloios scoticus* shown in fig. 160, B, affords one of several instances supporting this statement; similar examples are figured by Brongniart[1], Morris[2], and by more recent writers. The apparently sessile cone figured by Williamson[3] from a specimen in the Manchester Museum is certainly not *in situ*, but is accidentally associated with the stem.

The general absence of secondary wood in the steles of *Lepidostrobi* is, as Dr Kidston[4] points out, consistent with the view that the cones were shed on maturity and that fertilisation probably took place on the ground, or perhaps on the surface of the water where the slender hairs of the megaspores (fig. 191, F, I) may have served to catch the microspores.

Reference has already been made to the belief on the part of some palaeobotanists that the large scars of *Ulodendron* represent attachment-surfaces of sessile cones, and reasons have been given against the acceptance of this view.

There is considerable range in the size of *Lepidostrobi*. An incomplete specimen, 33 cm. long and 6 cm. broad, which may have been 50 cm. in length, is described by Renault and Zeiller[5] from the Commentry Coal-field. The larger cones afford a striking demonstration of the enormous spore-output of some species of *Lepidodendron*.

Among the earliest accounts of the anatomy of *Lepidostrobus* are those by Hooker[6] and Binney[7]. One of the specimens

[1] Brongniart (37) Pl. xxiv.   [2] Morris (40) Pl. xxxviii. fig. 10.

[3] Williamson (93) Pl. vi. fig. 26, A.   [4] Kidston (01) p. 62.

[5] Renault and Zeiller (88) A. Pl. lxi. fig. 4.   [6] Hooker (48²).

[7] Binney (71).

described by the former author (fig. 190) affords an interesting
example of an unusual manner of fossilisation ; a hollow stem or
Lepidodendron is filled with sedimentary material containing
several pieces of *Lepidostrobi* in an approximately vertical
position.

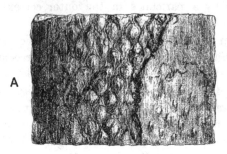

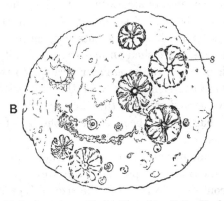

FIG. 190.   *Lepidodendron* stem with *Lepidostrobi*.   (After Hooker.)
    A.   Side-view showing leaf-cushions on the left-hand side and the
        Knorria condition on the right.
    B.   View of transverse section ; *s*, sections of *Lepidostrobi*.

The fact that *Lepidostrobi* usually occur as isolated specimens renders it impossible in most cases to refer them to
particular species of *Lepidodendron*.   Neither external features
nor anatomical characters afford satisfactory criteria by which
to correlate vegetative and fertile shoots ; in some measure this
is due to the imperfection of our knowledge as regards the
range of structure within the limits of species ; it is also due

to lack of information as to the extent to which the transition
from sterile to fertile portions of a shoot is accompanied by
anatomical differences. Prof. Williamson wrote: "I have for
many years endeavoured to discover some specific characters by
which different *Lepidostrobi* can be distinguished and identified,
but thus far my efforts have been unsuccessful[1]." In a few
cases, such as those mentioned in the description of *Lepidodendron Veltheimianum* and *L. Wünschianum*, it has been possible
to correlate cones and vegetative shoots.

The most complete account we possess of the anatomy of
Lepidodendron cones is that by Mr Maslen[2], who first demonstrated the occurrence of a ligule on the sporophylls, and thus
supplied a missing piece of evidence in support of the generally
accepted view as to the homology of the sporangium-bearing
members and foliage leaves.

i. *Lepidostrobus variabilis* (Lindley and Hutton).

1811. "Strobilus," Parkinson, Organic Remains, Vol. I. p. 428,
Pl. IX. fig. 1.

1828. *Lepidostrobus ornatus*, Brongniart, Prodrome, p. 87.

1831. *L. variabilis*, Lindley and Hutton, Foss. Flora, Pls. X. XI.

1831. *L. ornatus*, Lindley and Hutton, Foss. Flora, Pl. XXVI.

1837. *L. ornatus* var. *didymus*, *Ibid*. Pl. CLXIII.

1850. *Arancarites Cordai*, Unger, Genera et Spec. Plant. foss. p. 382.

1875. *Lepidostrobus variabilis*, Feistmantel, Palaeontographica, Vol.
LXIII. Pl. XLIV.

1886. *L. variabilis*, Kidston, Cat. Palaeozoic Plants, p. 197.

1890. *L. ornatus*, Zeiller, Flor. Valenciennes, p. 497, Pl. LXXVI.
figs. 5, 6.

—— *L. variabilis*, Zeiller, Flor. Valenciennes, p. 499, Pl. LXXVI.
figs. 3, 4.

Under this specific name are included strobili from Upper
Carboniferous rocks which, in spite of minor differences, may be
considered as one type. The cylindrical cones vary considerably
in size, some reaching a length of 50 cm. or more. The sporophylls are attached by a pedicel, 4—8 mm. long, at right angles
to the axis, while the distal portion forms an oval lanceolate
limb 10—20 mm. in length. The sporangia are 4—8 mm. long.

[1] Williamson (93) p. 26.　　　　　　　[2] Maslen (99).

The branched example figured by Lindley and Hutton[1] as a variety (*L. ornatus* var. *didymus*) illustrates a phenomenon not uncommon in both Palaeozoic and recent lycopodiaceous strobili.

ii.   *Lepidostrobus oldhamius* Williamson[2].   Fig. 191, A—D.

Williamson[3] instituted this term for strobili previously described by Binney[4], without adequate evidence, as the cones

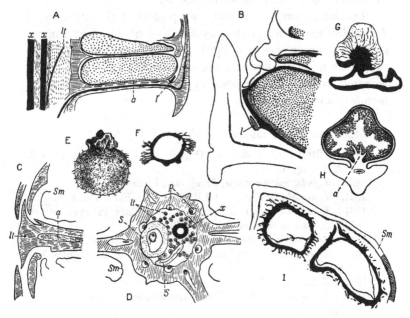

FIG. 191.   *Lepidostrobus.*

  A—D.   *L. oldhamius.*
  B, C, D.   From sections in the Binney Collection, Cambridge.
  E.   Megaspore. (After Kidston.)
  F.   Megaspore (Coal-Measures, Halifax). (After Williamson.)
  G.   Megaspore of *Lepidostrobus foliaceus*. (After Mrs Scott.)
  H.   Tangential section of sporangium. (After Bower.)
  I.   Part of sporangium wall, *Sm*, of the cone of *Lepidodendron Veltheimianum*, enclosing two megaspores. (Cambridge Botany School.)

  [1] Lindley and Hutton (37) A. Pl. 163.
  [2] For a detailed account of this type, see Maslen (99).
  [3] Williamson (93) p. 28.   [4] Binney (71).

of *Lepidodendron Harcourtii*. In shape and in the main morphological features this type resembles *L. variabilis*, which is however known only in the form of casts and impressions. A cone of *L. oldhamius*, 2—3 cm. in diameter, possesses a medullated stele consisting of a ring of primary xylem (fig. 191, D, x) with exarch protoxylem and no secondary elements. Maslen found several short tracheae at the periphery of the xylem and states that these led him to compare the cone with the vegetative shoots of *Lepidodendron vasculare*, but the common occurrence of such elements in different types of shoot renders them of little or no specific value. The inner cortex is like that of vegetative shoots of *Lepidodendron* and the middle cortex, which was no doubt of the type described in *Lepidostrobus Brownii*, is represented by a gap in the sections, beyond which is the stronger outer cortex (fig. 191, D) passing into the horizontal pedicels of the sporophylls. The section of the axis reproduced in fig. 191, D, was figured by Binney[1] as *Lepidodendron vasculare*. The leaf-traces, several of which are seen in the middle cortical region in fig. D, *lt*, consist of a strand of scalariform tracheae, with a mesarch protoxylem, succeeded by a few parenchymatous cells; beyond these there is usually a small gap which was originally occupied by a strand of thin-walled cells. It is important to note that in one sporophyll-trace figured by Maslen[2] there is a strand of thin-walled elongated elements abutting on the xylem, which he describes as phloem. This tissue is certainly more like true phloem than any which has hitherto been described in the leaf-traces of vegetative shoots. The state of preservation is not, however, sufficiently good to enable us to recognise undoubted phloem features.

In such cones as I have examined no tissue has been seen which shows the histological features characteristic of the secretory zone of vegetative shoots : the " phloem " (Maslen) occupies the position in the sporophyll bundle which in the vascular bundles of foliage leaves is occupied by a dark-celled and partially disorganised tissue in continuity with the secretory zone of the main stele. It may be that in the strobili

[1] Binney (71) Pl. VIII. figs. 2, 4.   [2] Maslen (99) Pl. XXXVI. fig. 11.

this tissue occurred in a modified form, but even assuming that
the section figured by Maslen shows true phloem, an assumption based on slender evidence, this is not sufficient justification
for the application of the term phloem to a tissue occupying
a corresponding position in vegetative shoots and distinguished
by well-marked histological features.

The sporophyll-traces, as seen in the outer cortex in fig.
191, D, are partially surrounded by a large crescentic space, p,
which was originally occupied by the parichnos. The sporangia
are attached along the middle line of the sporophyll and, as in
Lepidostrobus Brownii, a cushion of parenchyma projects into
the lower part of the sporangial cavity (fig. 191, A, a; C, a).

The diagrammatic sketch of part of a section in the Binney
Collection reproduced in fig. 191, B, shows the position of the
ligule, l. No megaspores have been discovered in any specimens of this type; the microspores, which occur both singly
and in tetrads, have a length of 0·02—0·03 mm.

The drawing shown in fig. 191, A, based on a section in the
Binney Collection, illustrates the general arrangement of the
parts of a typical *Lepidostrobus*. I have made use of this
sketch instead of that given by Maslen, as his figure conveys
the idea that the sporophylls are superposed, whereas, whether
they are verticillate or spiral, a radial longitudinal section
would not cut successive sporangia in the same plane.

iii. *Lepidostrobus Brownii* (Brongn.)

In 1843 a specimen of a portion of a petrified cone was
purchased by the British Museum, assisted by the Marquis of
Northampton and Robert Brown, for £30 from a French
dealer. This fossil, from an unknown locality, was briefly
described by Brown in 1851[1] and named by him *Triplosporites*,
but in a note added to his paper he expressed the opinion that
the generic designation *Lepidostrobus* would be more appropriate. Brongniart afterwards named the cone *Triplosporites
Brownii*[2], and Schimper[3] described it in his *Traité* as *Lepido-*

[1] Brown, R. (51).   [2] Brongniart (68).
[3] Schimper (70) A. p. 67, Pl. LXII. figs. 13—29.

strobus Brownii. The type-specimen is preserved in the British Museum and the Paris Museum possesses a piece of the same fossil.

The central axis of the cone has a stele of the type characteristic of *Lepidodendron fuliginosum* and *L. Harcourtii*, and the xylem is surrounded by a thin-walled tissue described by Bower[1] as possibly phloem; but in the absence of longitudinal sections it is impossible to say how far the tissue external to the xylem agrees with that in Lepidodendron stems. The sporophylls consist of a horizontal portion, to the upper face of which the radially elongated sporangia are attached, one to each sporophyll; beyond the distal end of the sporangium the sporophyll bends sharply upwards as a fairly stout lamina. The wall of the sporangium is composed of several layers of cells, as shown in a drawing published by Bower[2]; in the interior occur groups of microspores, and from a ridge of tissue which extends along the whole length of the sporangium irregular trabeculae of sterile tissue project into the sporangial cavity, as in *Isoetes* (fig. 191, H: cf. fig. 133, H).

Further information in regard to *Lepidostrobus Brownii* has recently been supplied by Prof. Zeiller[3], who recognises the existence of a ligule, and draws attention to some interesting histological features in the tissue of the sporophylls[4].

Spores of Palaeozoic Lycopodiales.

The calcareous nodules from the Coal seams of Yorkshire and Lancashire are rich in isolated spores, many of which are undoubtedly those of *Lepidostrobi*. Examples of spores were figured by Morris[5] in 1840, and their occurrence in coal has been described by several authors, one of the earliest accounts being by Balfour[6]. The drawings of Palaeozoic and recent spores published by Kidston and Bennie[7] demonstrate a striking similarity between the megaspores of existing and extinct Lycopods, the chief difference being the larger size of the fossils.

---

[1] Bower (93).   [2] Bower (94) Pl. xlviii. fig. 93.   [3] Zeiller (09).
[4] Zalessky has recently (08) described a large species of cone, *Lepidostrobus Bertrandi*, 5 cm. in diameter.
[5] Morris (40).   [6] Balfour (57).   [7] Kidston and Bennie (88).

The general generic name *Triletes*, originally used by Reinsch[1], is a convenient term by which to designate Pteridophytic spores which cannot be referred to definite types.

It is usual to find more than four megaspores in each megasporangium in Palaeozoic and not infrequently, as we have seen, in Mesozoic lycopodiaceous strobili, but in some Palaeozoic cones, e.g. *Bothrostrobus* (fig. 216) and *Lepidostrobus foliaceus*[2], a single tetrad only appears to have reached maturity.

The occurrence of long simple or branched and sometimes capitate hairs is a common feature of Carboniferous megaspores (fig. 191, E, F, I). It is possible that these appendages served to catch the microspores, thus facilitating fertilisation. A peculiar form of megaspore has been described by Mrs Scott[3], and assigned by her to *Lepidostrobus foliaceus*, the megasporangium of which apparently contained only four spores. As shown in fig. 191, G, a large bladder-like appendage characterised by radiating veins is attached to the thick spore-coat; it is suggested that this excrescence may be compared with the "swimming" apparatus of the recent water-fern *Azolla*. The epithet swimming which it is customary to apply to the appendages of *Azolla* megaspores would seem to be inappropriate if Campbell[4] is correct in stating that spores of *Azolla* are incapable of floating.

B.   *Spencerites.*

        Spencerites insignis (Williamson).   Fig. 192.

      1878.   *Lepidostrobus sp.*, Williamson, Phil. Trans. R. Soc., p. 340, Pl. XXII.
      1880.   *Lepidostrobus insignis*, Williamson, Phil. Trans. R. Soc., p. 502, Pl. XV. figs. 8—12.
      1889.   *Lepidodendron Spenceri*, Williamson, Phil. Trans. R. Soc., p. 199, Pl. VII. figs. 20—22; Pl. VIII. fig. 19.
      1897.   *Spencerites insignis*, Scott, Phil. Trans. p. 83, Pls. XII—XV.

Another type of lycopodiaceous strobilus, differing sufficiently from *Lepidostrobus* to deserve a special generic designation,

---

[1] Reinsch (81) A.   [2] Maslen (99) p. 373; Scott, R. (06) p. 117.
[3] Scott, R. (06).   [4] Campbell (05) p. 414.

is that originally described by Williamson[1], from the Lower
Coal-Measures of Yorkshire, as a type of *Lepidostrobus, L. insignis*, but afterwards[2] more fully investigated and assigned to
a new genus by Scott[3]. It should be pointed out that in a
later publication Williamson spoke of the lycopodiaceous axis,

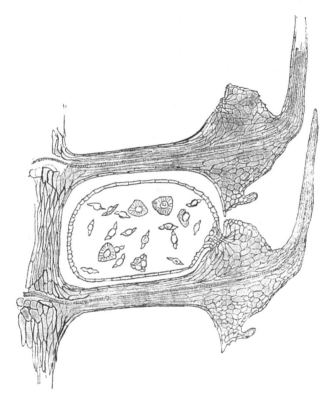

FIG. 192.   *Spencerites insignis* (Williamson).   (After Miss Berridge.)

which he suspected might belong to his *L. insignis*, as possibly
worthy of recognition as a distinct generic type.

Of the two species included by Scott in his genus *Spencerites*
only one, *S. insignis*, need be considered. Since the publica-

---

[1] Williamson (78) A. p. 340, Pl. xxii. See also the drawings in Williamson's
later papers quoted in the synonymy.

[2] Williamson (93²).   Scott, D. H. (98).

tion of Scott's paper our knowledge of this type has been
extended by Miss Berridge[1] and by Prof. Lang[2].

The axis of the strobilus has a stele characterised by a
pith of elongated elements, most of which have thin walls;
the xylem cylinder possesses about twenty protoxylem strands
forming more or less prominent exarch ridges. The cortex
exhibits a differentiation comparable with that in the shoots of
Lepidodendron. The sporophylls are arranged in alternating
verticils, each whorl consisting of ten members: the narrow
horizontal pedicel of a sporophyll, containing a single vascular
bundle, as shown in fig. 192, is expanded distally into a
prominent upper lobe bearing a cushion of small and delicate
cells, to which the sporangium is attached, and prolonged obliquely
upwards as a free leaf-like lamina. The lower blunt prolongation of the sporophylls appears to form a thick dorsal lobe, but,
as Lang has pointed out, it is highly probable that the present
form of the dorsal lobe is of secondary origin, and is " due to the
disappearance of a mucilage cavity from a large sporophyll
base[3]." As Miss Berridge remarks, the vascular bundle of the
sporophyll does not give off a branch to the ventral lobe and
sporangium. In attachment, in shape, and in the structure of
the wall the sporangia differ markedly from those of *Lepidostrobi*.
The spores, which also constitute a characteristic feature of
the genus, have a maximum diameter of 0·14 mm.; they are
described as oblate spheroids with a broad hollow wing running
round the equator (fig. 192) comparable with the air-sacs of the
pollen of *Pinus*. Scott points out that the spores of *Spencerites*
are intermediate in size between the microspores of *Lepidodendron* and the megaspores of *Lycopodium*; it is difficult therefore
to decide to which category they should be referred. *Spencerites*
is clearly distinct from *Lepidostrobus*; the absence of a ligule,
the manner of attachment of the sporangia, and the form and
size of the spores, are characteristic features.

A comparison of *Spencerites* with the strobili of *Lycopodium
cernuum* (figs. 123, 126—129) has recently been made by Lang,
who draws attention to the striking agreement as regards general
plan and even detailed structural features between the Palaeozoic

[1] Berridge (05).   [2] Lang (08).   [3] Lang (08) p. 364.

and the recent type of strobilus. It is interesting to find, as Lang points out, that in the original account of the fossil cone by Williamson, the view is expressed that the sporangiophores were confluent. An examination of the section figured by Williamson[1] led Lang to confirm this opinion. It would be out of place to enter here into a detailed comparison of *Spencerites insignis* and the cone of *Lycopodium*, but the resemblances are considered by Lang to be sufficiently close to suggest that the striking similarity may be indicative of relationship[2].

It is worthy of notice that the radial section of *Spencerites* (fig. 192) presents a fairly close resemblance to a corresponding section through a cone-scale of *Agathis* (Kauri Pine)[3]. In each case the megasporangium is attached by a narrow pedicel to the sporophyll and the latter has a similar form in the two plants, though the extent of the resemblance is somewhat lessened by Lang's more complete account of the Palaeozoic type. If the *Spencerites* sporangia possessed an integument the similarity with the *Agathis* ovule would of course be much closer: recent palaeobotanical investigations have shown that ovules and sporangia are not separated by impassable barriers.

[Since this Chapter was set up in type a paper has appeared by Dr Bruno Kubart on a new species of *Spencerites* spore, *S. membranaceus*, from the Ostrau-Karwiner Coal-basin (Austria). The spores are larger than those of *S. insignis* and in some the cells of a prothallus are preserved. Kubart figures a section of a spore containing a group of seven cells, a central cell, which he regards as an antheridial mother-cell, surrounded by six wall-cells. Kubart (90).]

---

[1] Williamson (78) A. Pl. xxii. fig. 53.

[2] Lang (08) p. 367. Since this was written a paper has been published by Mr Watson on a new type of Lycopodiaceous cone from the Lower Coal-Measures (*Mesostrobus*): in an appendix he criticises Dr Lang's views in regard to *Spencerites*. [Watson, *Annals of Botany*, Vol. xxiii. p. 379, 1909.]

[3] Seward and Ford (06) p. 395.

CHAPTER XVI.

Sigillaria.

i. *General.*

In view of the close resemblance between *Lepidodendron* and *Sigillaria*, another lycopodiaceous plant characteristic of Carboniferous and Permian floras, a comparatively brief description of the latter genus must suffice, more particularly as *Lepidodendron* has received rather an undue share of attention. *Sigillaria*, though abundantly represented among the relics of Palaeozoic floras, especially those preserved in the Coal-Measures, is rare in a petrified state, and our knowledge of its anatomy is far from complete. In external form as in internal structure the difference between the two genera are not such as enable us to draw in all cases a clearly defined line of separation.

In the *Antediluvian Phytology*, Artis[1] figured a fossil from the Carboniferous sandstones of Yorkshire which he called *Euphorbites vulgaris* on account of a superficial resemblance to the stems of existing succulent Euphorbias. Rhode[2] also compared Sigillarian stems with those of recent *Cacti*. The specimen described by Artis is characterised by regular vertical and slightly convex ribs bearing rows of leaf-scars in spiral series, like those on the cushions of *Lepidodendron*. A few years earlier Brongniart[3] had instituted the genus *Sigillaria*[4] for plants with ribbed but not jointed stems bearing "disc-like impressions" (leaf-scars) disposed in quincunx; the type-species named by the author of the genus *Sigillaria scutellata* is identical, as Kidston[5] points

---

[1] Artis (25) A. Pl. xv.   [2] Rhode (20).

[3] Brongniart (22) A. Pl. xii. fig. 4.

[4] For generic names wholly or in part synonymous with *Sigillaria*, see White (99) p. 230.

[5] Kidston (86) A. p. 186.

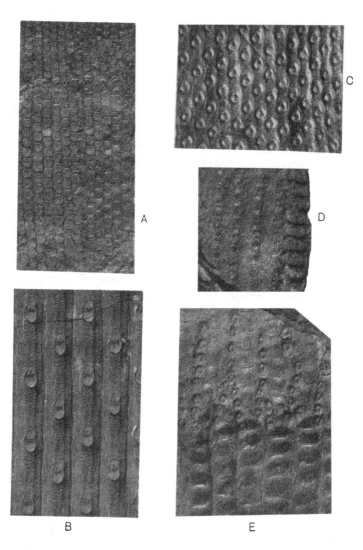

FIG. 193. A. *Sigillaria elegans* Brongn.
           B. *Sigillaria rugosa* Brongn.   Middle Coal-Measures.
           C. *Omphalophloios anglicus* Kidst.   Barnsley.
           D. *Sigillaria elegans* Brongn.
           E. *Sigillaria tessellata* Brongn.
           (A, B, C, E, about ¾ nat. size.   Dr Kidston's Collection.)

out, with *Euphorbites vulgaris* of Artis and with the plant afterwards figured by Brongniart as *S. pachyderma*[1]. Brongniart in
1822 figured another type of stem characterised by the absence
of ribs and by prominent spirally arranged cushions bearing
relatively large leaf-scars like the upper part of the specimen
shown in fig. 203; this he named *Clathraria Brardii*, a well-
known and widely distributed Carboniferous and Permian species
now spoken of as *Sigillaria Brardi* (figs. 196, A—C; 203). A
third type of stem figured by Brongniart as *Syringodendron
striatum*[2] agrees with *Sigillaria scutellata* in having ribs, but
differs in the substitution of narrow oval ridges or depressions
for leaf-scars; this is now recognised as a partially decorticated
Sigillaria, in which the vascular bundle of each leaf is represented by a narrow ridge or depression. The name *Syringodendron*, originally used by Sternberg, is conveniently applied
to certain forms of Sigillarian stems which have lost their
superficial tissues. A fourth generic name, *Favularia*, was
instituted by Sternberg[3] for Sigillarian stems with ribs covered
with contiguous leaf-scars of hexagonal form and prominent
lateral angles (fig. 193, A; fig. 200, G).

The generic or subgeneric title *Rhytidolepis*, also instituted
by Sternberg, is applied to ribbed Sigillarian stems such as
S. scutellata, *S. rugosa* (fig. 193, B), *S. mammillaris* (fig. 195), or
S. laevigata (fig. 196, D). Goldenberg[4] proposed the name
Leiodermaria for smooth Sigillarian stems with leaf-scars not
in contact with one another (fig. 196, C).

The shoot system of *Sigillaria* consisted of a stout stem
tapering upwards to a height of 100 feet[5] or more as an unbranched column, with its dome-shaped apex[6] covered with
linear grass-like leaves or, in some species, such as *Sigillaria
Brardi*[7], *S. Eugenii*[8], etc., the main trunk was occasionally divided
by apparently equal dichotomy. The younger portions of the
stem or branches were in some species clothed with leaves
separated by a narrow zigzag groove surrounding their hexa-

[1] Brongniart (37) Pl. CL. fig. 1.
[2] Brongniart (22) A. Pl. XII. fig. 3.
[3] Sternberg (23) A.   [4] Goldenberg (55).
[5] Zeiller (88) A. (*S. elegans*).   [6] Goldenberg (55).
[7] Renault (96) A. Pl. XXXV.   [8] Stur (75) II. A. Pl. XLII.

gonal bases, while in other forms each leaf was seated on a more or less prominent cushion having the form illustrated by *Sigillaria McMurtriei* (fig. 194) or by the example represented in fig. 200, H ; or as in the ribbed species shown in figs. 193, B, and

FIG. 194.   *Sigillaria McMurtriei* Kidst. From a specimen from the Upper Coal-Measures of Radstock, in the British Museum (V. 952). Nat. size.

195, the leaves in vertical series were separated from one another by longer portions of the ribs. As in *Lepidodendron* the cushions are frequently characterised by irregular transverse wrinklings and other[1] surface-ornamentation which in some instances at

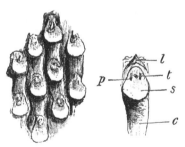

FIG. 195.   *Sigillaria mammillaris.* (Rhytidolepis form.) From a specimen in the Manchester Museum. *p*, parichnos; *l*, ligule-pit; *t*, leaf-trace; *c*, cushion ; *s*, leaf-scar.

least may have been produced as the result of *post-mortem* shrinkage of superficial tissue. From the rarity of shoots with the foliage attached, it would seem that the leaves persisted for a comparatively short time and were cut off by an absciss-layer leaving behind a well-marked leaf-scar area. The linear leaves,

[1] For an account of the various external features made use of in the classification of Sigillarias, see Koehne (04).

reaching in rare cases a length of one metre (e.g. *S. lepidoden-drifolia*) but usually much shorter, possessed a single median bundle, and the lower face was characterised by two stomatal grooves and a median keel.   It is not uncommon to find leaf-bases of *Sigillaria* detached from the stem and preserved as separate impressions.   The term *Sigillariophyllum* used by

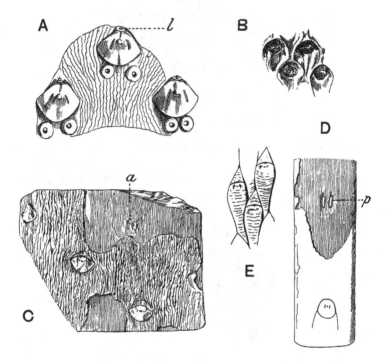

Fig. 196.   A—C.   *Sigillaria Brardi*.   (A after Germar; B, C after Zeiller.)
         D.   *Sigillaria laevigata.*
         E.   *Lepidodendron Wortheni* (D and E after Zeiller).

Grand'Eury[1] may be applied to detached leaves, though it is by no means easy to distinguish between the foliage of *Sigillaria* and *Lepidodendron*.   A comparison of a typical species of *Sigillaria*, such as *S. rugosa* (fig. 193, B) or *S. Brardi* (fig. 196, A—C) with a typical *Lepidodendron* reveals obvious

[1] Grand'Eury (90) A.

differences in the form of the leaf-cushion, but in some cases the distinction becomes purely arbitrary.

Immediately above the centre of the upper boundary of a Sigillarian leaf-scar a ligule pit may often be detected, as shown in fig. 195, *l*, and in some cases, e.g. a specimen figured by Germar[1] (fig. 196, A) as *Sigillaria spinulosa* (identical with *S. Brardi*), some circular scars with a central pit surrounded by a raised rim occur on the surface of the stem, either singly or in pairs, near the leaf-scars; these, it is suggested, may represent the position of adventitious roots or, as Germar thought, of some deciduous spinous processes. The leaf-scars are frequently hexagonal in shape, with the lateral angles either rounded (fig. 200, F) or sharply pointed (fig. 200, G, H); each scar bears three smaller scars as in *Lepidodendron*, a central circular, oval or crescentic leaf-trace scar and larger oval or slightly curved scars formed by the two parichnos arms (fig. 195, *p*). The larger size of the parichnos arms, the individual cells of which may often be detected as a fine punctation, is a distinguishing feature of the genus, but otherwise the structure is very similar to that in *Lepidodendron*. As shown in figs. 195, 200, F, G, the three scars may occur nearer the upper than the lower margin of the leaf-base area.

Lepidodendron Wortheni[2] (fig. 196, E), described from North America by Lesquereux[3], by Zeiller[4] from France, and by Kidston[5] from the Upper and Middle Coal-Measures of England, may be quoted as a *Lepidodendron* bearing a close resemblance to *Sigillaria*. The shoots bear cushions two or three times as long as broad and without the usual median division, but with numerous irregular and discontinuous transverse wrinklings. *Lepidodendron Peachii* Kidston[6] affords another example of a form agreeing both with *Sigillaria* and with *Lepidodendron*. An Upper Devonian type described by White[7] as *Archaeosigillaria primaeva* affords a striking instance of the combination on one stem of Sigillarian and Lepidodendroid leaf-cushions.

---

[1] Germar (53).
[2] Cf. *Lepidodendron Zeilleri*, Zalessky (04) Pl. IV. fig. 1.
[3] Lesquereux (79) A. Pl. LXIV.   [4] Zeiller (88) A. Pl. LXXI.
[5] Kidston (01) p. 46.   [6] Kidston (85).   [7] White, D. (07²).

The difference between the original surface of a Sigillaria stem and that of partially decorticated specimens is seen in figs. 196, C and D; in fig. C the bark of *Sigillaria Brardi* shows the characteristic wrinklings of the superficial tissue, while at a slightly lower level the leaf-scars are replaced by the parichnos casts, *a*, and fine longitudinal striations represent

FIG. 197. *Carica sp.* From the Royal Gardens, Kew. (Much reduced.) M.S.

the elongated phelloderm cells laid bare by the exfoliation of the surface-layers. Similarly, in the rib of *Sigillaria laevigata* (fig. 196, D) the parichnos arms, *p*, and the longitudinal striations are exposed at the lower level, while the surface is smooth and bears rows of widely separated leaf-scars.

The older part of a Sigillarian stem may present an appearance very different from that of the younger shoots. The leaf-cushions may be stretched apart as the result of elongation and increase in girth, while in some cases the arrangement of the leaf-scars may vary on the same axis as the result of inequalities in growth or changing climatic conditions. The contiguous arrangement of the leaf-scars and narrow cushions characteristic of the Clathrarian form of stem, as was first demonstrated by Weiss[1], and afterwards illustrated by Zeiller[2] and Kidston, may be gradually replaced (on the same specimen) by a more distant disposition of the leaf-scars separated by a smooth intervening surface of bark. The specimen of *S. Brardi* reproduced in part in fig. 203, and first figured by Kidston, affords an example of three "species" on one piece of stem, *S. Brardi* Brongn. *S. denudata* Goepp. and *S. rhomboidea* Brongn.[3]

The piece of *Carica* stem, represented in fig. 197, illustrates the danger of trusting to the disposition of leaves as a specific criterion.

Similarly, in the ribbed forms the degree of separation of the leaf-scars is by no means uniform in a single species[4]. Some authors have adopted a two-fold classification of Sigillarian stems proposed by the late Prof. Weiss[5] of Berlin, who divided the Sigillariae into (A) Sub-Sigillariae, comprising Leiodermariae and Cancellatae, and (B) Eu-Sigillariae, including *Favulariae* and *Rhytidolepis*. Grand'Eury[6] adopts the terms *Rhytidolepis* and *Leiodermaria* for ribbed and smooth stems respectively, the type to which the name *Clathraria* was applied by Brongniart being in some cases at least the young form of Leiodermarian stems. While recognising the artificial distinction implied by such terms as *Rhytidolepis*, *Leiodermaria*, and other sub-generic titles, we may conveniently speak of the two main types of *Sigillaria* stems as ribbed and smooth.

Still older stems of *Sigillaria* are not uncommon from which the leaf-scars and other superficial tissues have been exfoliated, leaving exposed a longitudinally fissured surface of secondary cortex characterised by pairs of considerably enlarged parichnos

[1] Weiss, C. E. (88).   [2] Zeiller (89).   [3] Kidston (01) p. 94.
[4] Seward (90²).   [5] Weiss, C. E. (89).   [6] Grand'Eury (90) A.

strands (fig. 198) which are sometimes partially or wholly fused into one (*Syringodendron* state of *Sigillaria*). The single or double nature of the elliptical or circular parichnos areas is doubtless due to the degree of exfoliation, which may extend sufficiently deep into the cortex to reach the level of the parichnos before the single strand has bifurcated (cf. *Lepidodendron*, p. 100). In the Museums of Manchester, Newcastle, and other places casts of large Sigillaria stems may be seen, which illustrate the differences in breadth and regularity of the vertical ribs, and in the size and shape of the parichnos areas in different regions of a partially decorticated stem. A cast of a ribbed species in the Manchester Museum, having a length of 185 cm. and a breadth of 56 cm., shows in the upper portion straight vertical grooves and broad ribs bearing pairs of parichnos scars 11 mm. long; in the lower portion the ribs tend to become obliterated and the parichnos scars, 2 cm. in length, may be partially fused and arranged in much less regular vertical series. A feature of these older ribbed Sigillarian stems is the increase in the number of the ribs from below upwards. Kidston[1] has described a specimen in the Sunderland Museum, 6 feet 6 inches long, with a circumference at the slightly bottle-shaped base of 5 feet. On the lower portion of the stem there are 29 broad ribs; about one-third the height many of these bifurcate, producing as many as 40 ribs in the upper part where the cast has a circumference of 3 feet. The increase in number of the ribs is due in part to bifurcation, but also to the intercalation of new ones. As Kidston points out, this example shows that as a stem grew in length additional leaves were developed at the apex. A similar stem, which illustrates very clearly the increase in the number of ribs from below upwards, may be seen in the Newcastle Museum.

Grand'Eury[2] has described an example of an old stem of a ribless species of *Sigillaria*, *Syringodendron bioculatum*, bearing single and double parichnos areas of nearly circular form and with a diameter of 1—2 cm. In a specimen figured by Renault and Roche[3] (*Syringodendron esnostense*) from the Culm strata

[1] Kidston (97) p. 46.   [2] Grand'Eury (90) A. Pl. XIII. fig. 8.
[3] Renault and Roche (97).

in France, the parichnos scars reach a length of 3 cm. As seen in the fragment of a ribbed *Sigillaria* represented in fig. 198, the large parichnos areas exhibit a distinct surface pitting in contrast to the fine longitudinal striation of the rib; the difference in surface-appearance is due to the nature of the tissue, which in the parichnos consists of fairly large parenchymatous elements with groups of secretory cells[1], and in the exposed cortex of elongated elements. The vertical line in the middle of fig. 198, which occurs in the middle of the rib, has probably been formed by splitting of the bark.

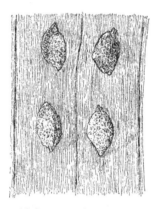

Fig. 198. *Sigillaria* with large parichnos areas. (⅓ nat. size.) M.S.

Grand'Eury's description of fossil forests of Sigillariae in the rocks of the St Étienne[2] district affords a striking picture of these arborescent Pteridophytes; he speaks of the stems of some of the trees as swollen like a bottle at the base, characterised by the Syringodendron features and terminating below in short repeatedly forked roots of the type known as *Stigmariopsis*. Other specimens of *Sigillaria* stumps show a marked decrease in girth towards the base; this tapered form is regarded by Grand'Eury as the result of the development of aerial columnar stems from underground rhizomes.

The nature of the root-like organs of *Sigillaria* is dealt with in the sequel: a brief reference may, however, be made

---

[1] Coward (07); Renault (96) A.   [2] Grand'Eury (90) A. Pl. III.

to the occurrence of stumps of vertical trunks which pass downwards into regularly forked and spreading arms. These arms
lie almost horizontally in the sand or mud like the underground

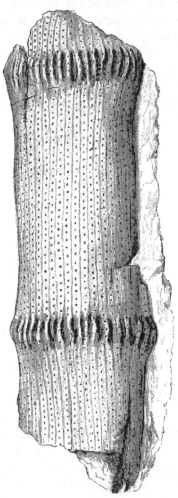

FIG. 199.   Partially decorticated stem of *Sigillaria* showing two zones of cone-
scars. From a cast in the Sedgwick Museum, Cambridge. M.S.
(⅛ nat. size.)

rhizomes of *Phragmites* and other recent plants growing in
swampy situations where water is abundant and where deeper

penetration of the soil would expose them to an insufficient supply of oxygen[1]. It is certain that *Sigillaria* had no tap-root, but was supported on spreading subterranean organs bearing spirally disposed long and slender rootlets which absorbed water from a swampy soil.

The regularity of the leaf-scar series on a Sigillarian stem may be interrupted by the occurrence of oval scars with a central scar and surrounding groove (fig. 193, E); these occur in zones at more or less regular intervals on the stem, as seen in the partially decorticated cast represented in fig. 199. Zeiller has pointed out that the rows of oval or circular scars, which mark the position of caducous stalked strobili, may occur between the leaf-scars in vertical series, each of which may include as many as 20 scars, while in other cases a single series of such cone-scars may encircle the stem[2]. The zones are usually of uneven breadth, as in *S. Brardi*, and their occurrence produces some deformation of the adjacent leaf-scars.

By the earlier writers *Sigillaria* was compared with succulent Euphorbias, Cacti, and Palms; Brongniart[3] at first included undoubted Sigillarian stems among Ferns, but after investigating an agatised stem from Autun, he referred *Sigillaria* to the Gymnosperms[4] on the ground that it had the power of producing secondary wood. It was then supposed that *Lepidodendron* possessed only primary xylem, and that the presence of a vascular meristem in *Sigillaria* necessitated its separation from the lycopodiaceous genus *Lepidodendron* and its inclusion in the higher plants. By slow degrees it was recognised, as in the parallel case of the genus *Calamites*, that the presence or absence of secondary vascular tissue is a character of small importance. Williamson, whose anatomical researches played the most important part in ridding the minds of palaeobotanists of the superstition that secondary growth in thickness is a monopoly of the Phanerogams, spoke in 1883 of the conflict as to the affinities of *Lepidodendron* and *Sigillaria* as virtually over but leaving here and there "the ground-swell of a

[1] Cf. Prof. Yapp's account (08) of Fen vegetation.
[2] Zeiller (88) A. Pl. LXXXV.   [3] Brongniart (28) A. p. 63.
[4] Brongniart (39) A. ; (49) A. p. 55.

stormy past[1]." In 1872 the same author had written: "If then
I am correct in thus bringing the Lepidodendra and Sigillariae
into such close affinity, there is an end of M. Brongniart's theory,
that the latter were gymnospermous exogens, because the crypto-
gamic character of the former is disputed by no one; we must
rather conclude as I have done that the entire series represents,
along with the Calamites, an exogenous group of Cryptogams in
which the woody zone separated a medullary from a cortical
portion[2]."

In 1879 Renault[3] expressed the opinion that Brongniart by
his investigation of the anatomy of *Sigillaria elegans* had
established in a manner "presque irréfutable" that *Sigillaria*
must be classed as a Gymnosperm showing affinity with the
Cycads.

In 1855 Goldenberg[4] described some strobili which he regarded as those of *Sigillaria* and recognised their close
resemblance to a fertile plant of *Isoetes*. He was led to the
conclusion, which had little influence on contemporary opinion,
that *Sigillaria* is related to *Isoetes* and must be classed among
Pteridophytes. To these long and narrow strobili Schimper
gave the name *Sigillariostrobus*[5]. In 1884 Zeiller[6] supplied confirmation of Goldenberg's view by the discovery of cones borne
on pedicels with Sigillarian leaf-scars, thus demonstrating the
generic identity of cones and vegetative shoots, which Goldenberg had connected on the evidence of association. Zeiller's
more recent work[7] and the still later researches of Kidston[8]
have added considerably to our knowledge of the morphology
of Sigillarian cones. Grand'Eury's remark made so recently as
1890[9] that opinion in regard to the Gymnospermous nature of
Sigillaria is losing ground every day, bears striking testimony
to the pertinacity with which old beliefs linger even in the face
of overwhelming proof of their falsity.

It is remarkable, in view of the abundance of vegetative
shoots, how rarely undoubted Sigillarian strobili have been

[1] Williamson (83).   [2] Williamson (72) p. 228.
[3] Renault (79).   [4] Goldenberg (55) p. 24.
[5] Schimper (70) A. p. 105.   [6] Zeiller (84).
[7] Zeiller (88) A.   [8] Kidston (97).   [9] Grand'Eury (90) A. Vol. II.

found; this may, however, be in part due to a confusion with Lepidostrobi which so far as we know do not differ in important respects from Sigillariostrobi[1].

There can be no doubt that *Sigillaria* usually produced its cones on slender pedicels which bore a few leaves or bracts in irregular verticils, or in short vertical series on comparatively stout stems, an arrangement reminding us of the occurrence of flowers on old stems of *Theobroma* and other recent Dicotyledons. As Renault[2] pointed out the fertile shoots are axillary in origin.

Dr Kidston[3] is of opinion that certain species of *Sigillaria* bore cones sessile on large vegetative shoots characterised by two opposite rows of cup-like depressions like those in the Ulodendron form of *Lepidodendron Veltheimianum* (fig. 157). He has described the Ulodendron condition of two species, *Sigillaria discophora* (König) and *S. Taylori* (Carr.); the cup-like depressions may have a diameter of several centimetres and are distinguished from those of *Bothrodendron* by the almost central position of the umbilicus. The specimens which he figures as *S. discophora* are identified by him with the stem figured by König as *Lepidodendron discophorum* and by Lindley and Hutton[4] as *Ulodendron minus*. We have already dealt with the nature of Ulodendron shoots, expressing the opinion that in spite of the often quoted specimen described by D'Arcy Thompson[5], in which a supposed cone occurs in one of the cups, there is no satisfactory case of any undoubted cone having been found attached to the large Ulodendron scars. It is more probable that the Ulodendron depressions represent the scars of branches, either elongated axes, or possibly in some cases deciduous tuberous shoots which served as organs of vegetative reproduction. A specimen figured by Kidston as *Sigillaria Taylori* from the Calciferous sandstone of Scotland[6] bears a row of slightly projecting "appendicular organs" attached to a Ulodendron axis; but these furnish no proof of their strobiloid nature. The main question is, are these Ulodendron shoots correctly identified by Kidston as Sigillarian?

---

[1] Kidston (97).   [2] Renault (96) A.
[3] Kidston (85).   [4] Lindley and Hutton (31) A. Pl. vi.
[5] Thompson (80).   [6] Kidston (85) Pl. vi. fig. 10.

The surface of the specimens shows crowded rhomboidal scars surrounded in some cases by a very narrow border or cushion; the general appearance is, as Kidston maintains, like that of *Sigillaria Brardi* in which the leaf-scars are contiguous (e.g. fig. 203, upper part). None of the leaf-scars exhibit the three characteristic features, the leaf-trace and parichnos scars, but only one small scar appears on each leaf-base area. In a more recent paper Kidston figures a small piece of a stem from Kilmarnock, which he identifies as *Sigillaria discophora*, showing the three characteristic scars on the leaf-base area. There is no doubt as to the Sigillarian nature of this specimen, but it is not clear if the piece figured is part of a Ulodendron shoot[1].

Prof. Zeiller[2] retains the older name *Ulodendron minus* Lind. and Hutt. in place of König's specific designation and dissents from Kidston's identification of *Ulodendron minus* and *U. majus* of Lindley and Hutton as one species; he is also inclined to refer these Ulodendron axes to *Lepidodendron*. In spite of the superficial resemblance to *Sigillaria* of the specimens described by Kidston, and which I have had an opportunity of examining, I venture to regard their reference to that genus as by no means definitely established. We must recognise the difficulty in certain cases of drawing any satisfactory distinction between *Sigillaria* and *Lepidodendron* based on external features, and while giving due weight to the conclusions of so experienced a palaeobotanist as my friend Dr Kidston, I venture to think we are not in a position to state with confidence that *Sigillaria* possessed Ulodendron shoots.

ii. *Leaves.*

The leaves of *Sigillaria* agree closely with those of *Lepidodendron*; they are either acicular (fig. 200, D) like Pine needles or broader and flatter like the leaves of *Podocarpus*. Their attachment to comparatively thick branches[3]

---

[1] Kidston (89²) p. 61 ; Pl. vi. fig. 1.
[2] Zeiller (88) A. p. 483, Pls. lxxiii. lxxiv.   [3] Zeiller (06) Pl. xlii.

shows that they persisted, in some cases at least, for several
years as in *Araucaria imbricata*. The lower surface of the
lamina was characterised by a prominent keel (fig. 142, A
and C) which dies out towards the apex; on either side of it
are well-defined stomatal grooves (figs. 142, *g, g*; 143, A; 200,
D, *g*). The upper face may be characterised by another groove
(fig. 142, B) but without stomata. The occurrence of the
stomatal grooves, the abundance of transfusion tracheae
(fig. 142, *t*) surrounding the vascular bundle, and the presence of strengthening hypodermal tissue suggest that the
leaves of *Sigillaria* were of a more or less pronounced xerophilous type and had a fairly strong and leathery lamina. The
mesophyll tissue consists either of short parenchymatous cells
or of radially elongated palisade-like elements and has the loose
or lacunar arrangement characteristic of the aerating system
in recent leaves; the slight development or absence of palisade-
tissue may indicate exposure to diffuse light of no great intensity.

In most species there is a single vein, but in others the xylem
forms a double strand (fig. 142, B). Sections of the lamina near
the apical region present a more circular form, owing to the
gradual obliteration of the upper groove and lower keel and to
the dying out of the stomatal grooves.

The transverse section of the leaf diagrammatically represented in fig. 142, A, A′, shows the two stomatal grooves, *g*,
and a prominent keel; the single vein consists of a small group
of primary tracheae, *x*, some delicate parenchyma, and a brown
patch of imperfectly preserved tissues, *a*, resembling the secretory zone tissue of a *Lepidodendron*. The whole is surrounded
by a sheath of rather wide and short thinner-walled spiral or
reticulate tracheids, which may be spoken of as transfusion
tracheae, *t*, and compared with similar elements in the leaves
of many recent Conifers. To this tissue Renault applies the
epithet "water-bearing" and it is very likely that this may have
been its function. The shaded portions of the lamina, in
fig. 142, A, represent the distribution of thicker-walled hypodermal tissue. The section of a leaf 3 mm. wide shown in
fig. 142, C, shows an almost identical structure; the transfusion
tracheae are richly developed especially on the sides and lower

surface of the vascular strand. This leaf occurs in association
with a petrified stem of *Sigillaria scutellata*[1].

Renault[2] has shown that the leaf-traces of *Sigillaria
spinulosa* (= *S. Brardi*) are accompanied in the outer cortical

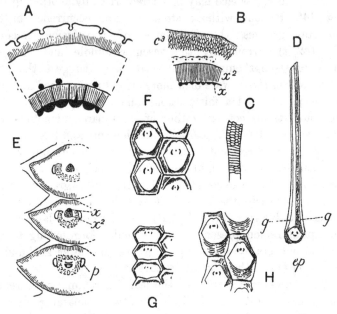

FIG. 200. A. *Sigillaria Brardi* [= *S. elegans* Brongn. (39)]. Transverse
                section of stem.
           B. *Sigillaria Brardi* [= *S. spinulosa*, Renault and Grand'Eury (75)]:
                c^3, outer cortex ; x, x^2, xylem. (After Renault and Grand'Eury.)
           C. *S. Brardi*, primary xylem element. (A and C after Brongniart.)
           D. Leaf of *Sigillaria Brardi* : g, g, stomatal grooves ; ep, piece of
                epidermis of stem. (After Renault.)
           E. *Sigillaria Brardi*. Tangential section of leaf-bases : p, parich-
                nos. (After Renault.)
           F, G, H. Sigillaria leaf-scars and cushions. (After Weiss.)

region of the stem by a fairly large amount of secondary xylem ;
in sections of the free lamina which he figures the secondary
elements are much less obvious and represented by a few
tracheae only. Similarly, in the leaf-base of *S. Brardi*

---

[1] Kidston (07[2]).   [2] Renault (96) A. Pl. xxxvii. fig. 3.

(fig. 200, E) the xylem consists of both primary and secondary elements (x, x^2), but in the lamina the latter is poorly if at all represented. In the lamina of the leaves of *S. Brardi* the primary xylem forms a narrow slightly curved band with two lateral groups of narrower, presumably protoxylem elements; this is surrounded by delicate parenchyma styled by Renault, on very slender evidence, phloem ("liber"). Some dark cells below the xylem are described as sclerous tissue, and surrounding the bundle is a sheath of transfusion tracheae (dotted area in fig. 200, E). It is possible that the elements spoken of with hesitation by Renault as secondary xylem are transfusion tracheae.

There has probably been some confusion in the minds of authors between sclerous tissue and dark secretory tissue in Sigillarian leaves; the crescentic band, a, shown in fig. 142, B, which corresponds in position with the sclerous tissue of Renault in *S. Brardi* leaves, appears to be of the nature of secretory tissue.

The diagram shown in fig. 142, B, illustrates a type of leaf very like those already described, except that there are two xylem strands, x. The difference between the double strand and the single bundle seen in figs. 142, A, C and 200 E, is comparatively small, but it is a real distinction. This type of leaf (fig. 142, B) was originally described by Renault[1] under the generic title *Sigillariopsis*. The genus was founded on a French petrified specimen consisting of part of a ribbed stem possessing a stele of the Sigillarian type and characterised by separate primary xylem strands, like those of *S. Brardi* described by Brongniart in 1839. Renault considered the presence of two xylem strands in the leaf a sufficient reason for the institution of a new genus and named the specimen *Sigillariopsis Decaisnei*. Prof. Bertrand of Lille kindly photographed for me Renault's type-specimen and sent several prints with explanatory notes. The transverse section of the leaves shows very clearly the two xylem strands; each strand consists of a triangular group of primary tracheae with the protoxylem apex pointing towards the lower surface of the lamina. Below each primary strand

---

[1] Renault (79) Pls. xii. xiii. p. 270 ; (96) A. p. 245.

of centripetal xylem is an arc composed of a few small tracheae
which Renault and Bertrand describe as secondary xylem; it
is, however, not clear from the photomicrographs that these
are of secondary origin, their position and appearance reminding
one of the primary centrifugal xylem of a cycadean foliar
bundle. Below this centrifugal xylem is another arc of imperfectly preserved elements described by Renault as a
protective sheath and by Bertrand as glandular tissue; the
latter term is probably the more correct as the tissue may well
correspond to the secretory-zone tissue of Lepidodendron stems.
Fairly large groups of transfusion tracheids occur on the flanks
of the xylem. Prof. Bertrand points out that one of his sections,
cut nearer the apex of a leaf than that figured by Renault with
a single xylem strand, contains a double strand and thus shows
the latter's description to be an incorrect interpretation of the
imperfectly preserved tissues.

The *Sigillariopsis* type of leaf was recognised by Scott[1]
in English material on which he founded the species *Sigillariopsis sulcata*. In a section which he has recently figured[2] a
lacuna below the two xylem strands is described as " representing
secretory tissue "; a band of transfusion tracheae almost encircles
the pair of bundles.

In a note published in 1907, Kidston[3] demonstrated the
association of *Sigillariopsis* leaves with an undoubted Sigillarian stem of the Rhytidolepis type and expressed his
conviction that Renault's genus is identical with *Sigillaria*.
The correctness of Kidston's conclusion has been proved by
Arber and Thomas[4] who found that the leaf-traces of *Sigillaria
scutellata* bifurcate during their course through the outer region
of the cortex and enter the leaf as two distinct strands of
primary xylem. In the section from Dr Kidston's collection
shown in fig. 142, B, the lamina, 4 mm. wide, consists mainly of
thin-walled assimilating tissue composed of radially elongated
cells abutting at the periphery on hypodermal mechanical tissue,
except at the edges of the stomatal grooves which are bounded
by the small-celled epidermis. A broad sheath of thicker-walled

[1] Scott, D. H. (04²).   [2] Scott (08) p. 230, fig. 95.
[3] Kidston (07²).   [4] Arber and Thomas (08).

elements, *s*, surrounds numerous scattered transfusion tracheae, *t*, and below the two xylem strands, *x*, which are embedded in delicate parenchyma there is a crescentic band of dark tissue, *a*, resembling the smaller strand, *a*, in fig. 142, A′, and the secretory zone tissue of a Lepidodendron stem.

iii. *Fertile shoots of* Sigillaria.

Reference has already been made to the manner of occurrence of strobili on Sigillarian stems; it remains to describe the structure of these reproductive shoots. *Sigillariostrobus*, the name given to Sigillarian strobili, may be defined in general terms as follows:

Cylindrical cones, rarely dichotomously branched[1] as in species of *Lycopodium* and *Selaginella*, which may reach a length of 30 cm. (*e.g. Sigillariostrobus nobilis* Zeill.[2]) and a diameter 2—5 cm.; peduncle long and slender, sometimes bearing acicular bracts or, after leaf-fall, characterised by leaf-cushions and leaf-scars like those on vegetative shoots (fig. 201, E). The stalked cones are borne in irregular verticils and in some species in vertical series, the fertile zones being separated by comparatively long sterile portions of the stem (fig. 199). The cones were deciduous and, in certain cases if not in all, the individual sporophylls became detached from the cone-axis on maturity. The slender axis bore spiral or verticillate imbricate sporophylls attached at right angles or more or less obliquely. The basal rhomboidal portion bore spores on its upper surface (fig. 201, F), presumably enclosed in a somewhat radially elongated sporangium (fig. B) and was prolonged distally into a narrow lanceolate free portion, in some species with a ciliate border (fig. D). The sporangia probably produced megaspores and microspores, but such spores as have been recognised appear to belong to the former category. The designation *Triletes* is applied to isolated spores of *Sigillaria* or to those of *Lepidodendron*.

Sigillariostrobus Tieghemi Zeiller[3] (figs. 201, E, F). In this

---

[1] Goldenberg (55); Kidston (97).   [2] Zeiller (88) A. Pl. xc. 1, p. 598.
[3] Zeiller (84); (88) A. Pl. LXXXIX.

species, from the Coal-field of Valenciennes, the pedicel bore
acicular leaves or bracts attached to the upper portion of leaf-
cushions arranged in vertical series (fig. E). The cones reached
a length of 16 cm. and a breadth of 2·5—5 cm.; the sporophylls
are borne in alternating verticils with 8—10 in each whorl.
Several megaspores (2 mm. in diameter) appear to have been
produced in tetrads in each sporangium.

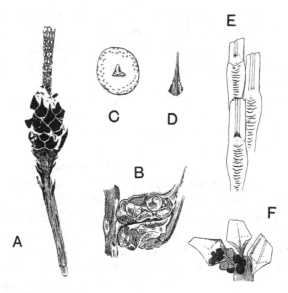

FIG. 201.   *Sigillariostrobus.*
   A, C.   *Sigillariostrobus rhombibracteatus* Kidst.   (After Kidston.)
   A.   Portion of strobilus.
   C.   Megaspore.
   B, D.   *Sigillariostrobus ciliatus* Kidst.   (After Kidston.)
   E, F.   *Sigillariostrobus Tieghemi* Zeill.   (After Zeiller.)

Sigillariostrobus rhombibracteatus Kidston[1].   Fig. 201, A, C.

   Kidston described this species from the Middle Coal-
Measures of England: it is similiar in habit and in the form of
the sporophylls to *S. Tieghemi,* but rather smaller, and the more
definitely rhomboidal sporophylls have a ciliate margin   The

---

[1] Kidston (97) Pls. I. II. p. 50.

cone was probably heterosporous, but megaspores alone have so far been discovered. The sporophylls bear a close resemblance to those of *Lycopodium cernuum* (fig. 126, C). In some of the illustrations of this type given by Kidston the naked cone-axis with its numerous sporophyll-scars is clearly shown, reminding one of the naked axes of the cones of the Silver Fir (*Abies pectinata*) or Cedar after the fall of the scales.

Our knowledge of Sigillarian cones is too incomplete to admit of a detailed comparison with the strobili of *Lepidodendron* or with those of recent Pteridophytes. There can, however, be little doubt that Goldenberg[1] was correct in his selection of *Isoetes* as the most nearly allied recent plant so far as the fertile leaves are concerned. It would seem that the sporangia were comparatively delicate structures which have left no clearly defined remains of their walls in the carbonised specimens; Kidston, indeed, speaks of the hollow bases of the sporophylls as holding the spores, but this is hardly likely to have been the case. Our knowledge of the anatomy of *Sigillariostrobus* is practically nil, but in one specimen of a *Sigillaria elegans* stem Kidston[2] describes the structure of the tissues as seen in a transverse section of a scar of a fertile shoot; from this we learn that the stele was composed exclusively of primary tracheids forming a solid strand without a pith. It is probable that the cones of *Sigillaria* were heterosporous, but in no instance have undoubted microspores been discovered; the megaspores in each megasporangium were fairly numerous as in *Isoetes* (fig. 133, E). In one species, *Sigillariostrobus major* (Germar), from Permian rocks of France and Germany, Zeiller[3] states that the whole of a single cone bore megaspores (0·8—1 mm. in diameter) only; this is, however, not opposed to the idea of heterospory, as we find instances in *Selaginella* of strobili bearing one kind of spore only (cf. p. 56).

In a few instances, it has been possible to correlate cones with certain species of *Sigillaria*, but in most cases the strobili occur as isolated fossils.

[1] Goldenberg (55)
[2] Kidston (05) Pl. III. figs. 23, 25, 26, 27.　　　　[3] Zeiller (06) p. 160.

iv. *The structure of Sigillarian stems.*

The first account of the anatomy of *Sigillaria* we owe to
Brongniart[1] who published a description of the internal structure
of an agatised stem, about 4 cm. in diameter, from Autun, which
he referred to *Sigillaria elegans.* It has, however, been shown
by Zeiller[2] and by Renault that this petrified fragment belongs
to Brongniart's species *S. Menardi*, which is probably a young
form of *S. Brardi.* Brongniart's specimen, now preserved in
the Paris Natural History Museum, is a very beautiful example
of a silicified plant : on part of the surface are preserved the
hexagonal contiguous leaf-scars, like those shown in fig. 193, A,
and on the polished transverse section is seen a relatively large
stele consisting of a ring of secondary xylem surrounding
a series of crescentic groups of primary xylem (fig. 200, A)
enclosing a wide pith occupied by concentric layers of silica. A
portion of the outer cortex is preserved, and this is separated
from the stele by a broad space filled with siliceous rock. The
main features of this type may be described in a few words.
The primary xylem differs from that of such Lepidodendron
stems as have been described in being made up of groups of
scalariform and occasionally reticulate (fig. 200, C) tracheae,
having a plano-convex or more or less crescentic form as seen in
transverse section. These primary strands, in contact with one
another laterally, have their narrowest elements on the outer
edge. The leaf-traces are given off from the middle of the
abaxial face of each xylem strand (fig. 202, C, *lt*); these pass
obliquely outwards through medullary rays and then, as in
Lepidodendron, turn sharply upwards before bending outwards
again on their way to the leaves. Each leaf-trace consists of a
group of primary tracheae to which a few secondary tracheae are
added during the passage through the secondary wood. The
secondary xylem forms a continuous cylinder of tracheae with
scalariform bands on both radial and tangential walls; the
medullary rays are numerous and consist of long and narrow

[1] Brongniart (39); Renault (96) A.
[2] Zeiller (88) A. p. 586; Kidston (05) p. 534.

series, usually one cell broad, of parenchymatous cells with occasional short rays one or more cells in depth.

The slightly greater breadth of the rays between each primary xylem strand tends to divide the secondary wood into bundles corresponding in breadth to the primary groups. The outer cortex closely resembles that of *Lepidodendron*; it consists internally of radial series of secondary, elongated and rather stout, elements abutting on the parenchymatous tissue of the leaf-cushions.

The next contribution to our knowledge of the anatomy of *Sigillaria* was made by Renault and Grand'Eury[1] who described the structure of *Sigillaria spinulosa* Germar[2], a species now recognised as the Leiodermarian condition of *S. Brardi*, and probably, therefore, not specially distinct from the specimen described by Brongniart in 1839 as *S. elegans*. In Brongniart's fossil the leaf-cushions are in contact (Clathrarian form of *S. Brardi*: fig. 203, upper part) whereas in the specimen now under consideration the leaf-scars are further apart (Leiodermarian form of *S. Brardi*, fig. 203, lower part, and fig. 196, C). It may be, as Scott suggests, that these two specimens are not specifically identical but closely allied, an opinion based on certain anatomical differences[3]; we may, however, include both under the comprehensive name *S. Brardi*.

The primary xylem (fig. 200, B, x), is in some regions separated into distinct strands, in others it forms a continuous band equal in length to several of the separate groups. This type of stele, in which the primary xylem consists in part of separate strands and in part of a continuous cylinder, forms a transition between that represented in fig. 200, A, and the steles of *Sigillaria elegans* (fig. 202, A) and most species of *Lepidodendron*. The tendency of the primary xylem strands to become united laterally, forming broader bands, was first described by Solms-Laubach[4] in a French specimen of *Sigillaria spinulosa* in the Williamson collection. The leaf-traces arise from the middle of the concave outer face of the primary xylem groups. The inner cortex is composed of small parenchymatous cells as in *Lepido-*

[1] Renault and Grand'Eury (75); Renault (96) A.   [2] Germar (44) A.
[3] Scott (08) p. 219.   [4] Solms-Laubach (91) A. p. 253.

dendron, and it is noteworthy that traces of partially disorganised tissue, described as large canals, in the region external to the secondary wood, bear a resemblance[1] to the secretory tissue of *Lepidodendron.*

Other interesting features are presented by the structure of the outer cortex and the parichnos. The outer cortex in the leaf-scar region is composed of parenchyma, but for the most part it consists of radially elongated groups of thin-walled

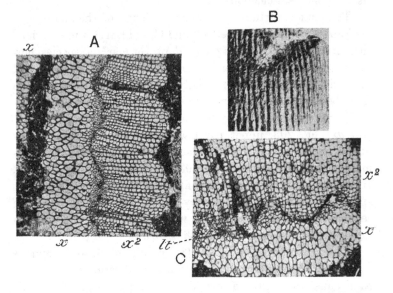

FIG. 202. A. *Sigillaria elegans* Brongn. (Section in Dr Kidston's Collection cut from the specimen shown in fig. 193, D.)
B, C. *Sigillaria elongata* Brongn. *lt,* leaf-trace. (From specimens in the collection of Prof. Bertrand.)

parenchyma enclosed in a framework of thicker-walled and elongated elements (fig. 200, B, c^3). This type of cortex, to which Brongniart applied the name *Dictyoxylon,* would produce a cast in the case of a partially decorticated stem characterised by a surface formed of irregularly oval and raised areas bounded by narrow grooves; the greater prominence of the former being due to the more rapid decay of the softer tissue, which would

[1] Renault and Grand'Eury (75) Pl. I. fig. 5.

produce depressions on the exposed face of the dead stem. Casts of this type are not uncommon in Carboniferous rocks, and while some may belong to the Pteridosperm *Lyginodendron*, others may be those of Sigillarian stems.

The large parichnos-strands, produced as in *Lepidodendron*, by the forking of a single strand arising in the middle cortical region, consist in part of tissue containing secretory canals, a structure like that recently described by Miss Coward[1] in the large parichnos strands of *Syringodendron* stems.

An example of a decorticated specimen is described by Renault[2] as *Sigillaria xylina*. This stem is presumably referred to *Sigillaria* because the primary xylem consists of separate strands. It is characterised by the unusually large development of secondary wood and by the relatively small size of the pith. The xylem cylinder has a diameter of 4—5 cm. and the pith is only 4—5 mm. in breadth.

Another example of a petrified Sigillaria stem has been described by Kidston[3] as *S. elegans* Brongn.[4] (fig. 193, D), a species characterised by vertical rows of sub-hexagonal and contiguous leaf-scars and by the presence of verticils of cone-scars. Fig. 193, D, represents Kidston's specimen in surface-view; one row of leaf-scars is shown, but most of the superficial tissues have been destroyed. The crushed stele, 13 mm. in its longest diameter, has a continuous cylinder of primary xylem, (fig. 202, A, x) characterised by a regularly crenulate outer margin with the smallest elements at the edge; the prominent ridges separating the sinuses are rounded. The leaf-traces arise from the bottom of each sinus; the leaf-bundles are mesarch, and consist exclusively of primary elements. The secondary xylem, x^2, like that of the primary xylem, has a crenulate outer edge. The most interesting feature of the outer cortex is afforded by a tangential section which, in addition to the leaf-scars, cuts through a cone-scar showing a solid primary stele surrounded by the cortex of the cone-peduncle.

Another type of *Sigillaria*, probably *S. elongata* Brongn. (fig. 202, B, C), which is very similar to *S. scutellata* has been

[1] Coward (07).   [2] Renault (96) A. p. 237, Pl. xxxviii. figs. 1—4.
[3] Kidston (05).   [4] Brongniart (28) A. Pls. cxlvi. clv. clviii.

briefly described by Prof. Bertrand[1], to whom my thanks are
due for the two photographs reproduced in fig. 202, B. C.   His
specimen, from the Pas de Calais Coal-field, shows a ribbed
Rhytidolepis form of surface (fig. 202, B).   The stele (fig. 202, C)
agrees closely with that of *S. elegans* as described by Kidston,
but the ridges on the fluted surface of the primary xylem are
more pointed.   "In the immediate neighbourhood of the origin
of a leaf-trace, the spiral elements form a median band in the
middle of a sinus" and from this the leaf-traces are given off.
No secondary xylem was found in the leaf-traces at any part of
their course.

Bertrand compares the stele of *S. elongata* with that of the
type of *Lepidodendron* represented by the Burntisland species
named by Williamson *L. brevifolium* (fig. 186) and now usually
referred to *L. Veltheimianum*; the chief distinguishing features
are the greater prominence in the French species of the surface-
ridges or teeth of the primary xylem, a feature which occurs in
L. Wünschianum, and the detachment of the leaf-traces from
the bottom of each sinus (fig. 202, C, *lt*) instead of from the
sides of the sinus.   It is, however, not clear how far this latter
distinction is a real one; in *Lepidodendron Wünschianum* the
leaf-traces appear to arise, as in *Sigillaria*, from the middle of
each sinus.

Other types of ribbed Sigillaria stems have been briefly
described by Scott[2], Kidston[3], and more recently, by Arber
and Thomas[4].

The specimen described by Scott agrees in the main with
S. elegans of Kidston and with *S. elongata* of Bertrand.

Kidston's sections of *S. scutellata* show a continuous primary
xylem cylinder with a slightly and irregularly crenulate outer
margin.   It would seem that one important diagnostic character
in Sigillarian stems is afforded by the degree and form of the
crenulations on the outer surface of the primary xylem.   *S. scutellata* has been described also by Arber and Thomas; these
authors were the first to demonstrate the presence of a ligule
and ligular pit on the leaf-base in a petrified stem, and they

[1] Bertrand (99).   [2] Scott (08) p. 227, fig. 93.
[3] Kidston (07[2]).   [4] Arber and Thomas (07).

also contribute the important fact that the leaf-traces in passing through the phelloderm bifurcate and enter the leaf as two distinct vascular strands. This double bundle has been referred to in the description of Sigillaria leaves. (Page 214.)

Although our knowledge of the anatomy of *Sigillaria* has been considerably extended since Williamson[1] drew attention to our comparative ignorance of the subject, there are several points on which information is either lacking or very meagre. As regards the stele, it is in all types so far investigated, of the medullated type and constructed on the same plan as that of *Lepidodendron Wünschianum, L. Veltheimianum,* and other species. Secondary xylem was developed at an early stage of growth, and its relation to the primary xylem, from which as Kidston points out in his description of *S. elegans,* it may be separated by a few parenchymatous elements, is like that in *Lepidodendron.* The tendency of the outer face of the secondary xylem to present a crenulate appearance in transverse sections may, as Scott thinks[2], be a feature of some diagnostic importance, but this is not a constant character in the genus. In origin and in their mesarch structure, the leaf-traces closely resemble those of *Lepidodendron.* The earlier account of the structure of the leaf-traces of *Sigillaria,* which were described as possessing both centrifugal and centripetal wood, led Mettenius[3] to draw attention to an important anatomical resemblance between this genus and modern Cycads. This comparison was, however, based on a misconception; the Cycadean leaf-trace, consisting solely of primary wood, is not strictly comparable with those of some species of *Sigillaria,* in which one part of the xylem is primary and another secondary. The occasional presence of secondary xylem in Sigillarian leaf-traces is matched in some *Lepidodendra*[4], and cannot be accepted as a distinguishing feature.

The origin of the leaf-traces from the middle of the sinuses on the edge of the primary xylem is regarded as a difference; in *Lepidodendron* the leaf-traces are said to arise in some species from the sides of the crenulations; but, as already pointed out,

---

[1] Williamson (72).   [2] Scott (08) p. 227.
[3] Mettenius (60).   [4] e.g. *L. Wünschianum* (fig. 181, B, *lt*).

this is a distinction of doubtful value. The division of the primary xylem into separate strands in some stems of *Sigillaria* of the Clathrarian and Leiodermarian forms is a characteristic peculiarity; but *S. spinulosa* forms a connecting link between this type and the continuous arrangement of the xylem in *S. elongata* and *S. elegans.* Kidston[1] has shown that the discontinuous primary xylem occurs in Lower Permian species, a fact consistent with the view that the greater abundance of the centripetally developed wood, characteristic of the older species, represents a more primitive feature. This is not merely a conclusion drawn from a consideration of geological age, but it is in harmony with the view expressed by Scott[2] that as plants achieved greater success in producing secondary centrifugal wood, the retention of any considerable quantity of primary xylem became superfluous. As yet we know very little of the structure of the perixylic tissues of *Sigillaria,* but there is no sufficient reason for supposing that these differ in essentials from those in *Lepidodendron.* The middle and outer cortical tissues are practically identical in the two genera. The parichnos is of the same type, except that in *Sigillaria* it reached greater dimensions in the outer part of its course.

v.   *Sigillaria Brardi*[3] Brongniart.

Figs. 196, A—C; 200; 203.

1822.   *Clathraria Brardi,* Brongniart, Classif. Vég. foss., Pl. xii. fig. 5.
1828.   *Sigillaria Brardi,* Brongniart, Hist. Vég. foss. p. 430, Pl. clviii. fig. 4.
        S. Menardi, ibid. Pl. clviii.
1836.   *Lepidodendron Ottonis,* Goeppert, Fossil Farnkr. Pl. xlii.
1839.   *S. elegans,* Brongniart, Arch. Mus. Nat. Hist. Paris, Vol. i. p. 406, Pl. xxv.
1849.   *S. spinulosa,* Germar, Verstein. Wettin und Löbejün, p. 59, Pl. xxv.
1893.   *S. mutans,* Weiss, Abhand. Preuss. Geol. Anst. [N.F.] Heft 2, Pl. viii.

---

[1] Kidston (05) p. 547.   [2] Scott, D. H. (02).
[3] For fuller synonymy, see Kidston (86) A. p. 179; and Zeiller (06) p. 160; Koehne (04) p. 62.

The aerial shoots of this species are occasionally branched dichotomously[1], the apical portions bearing short crowded leaves[2]; the surface of the bark is either completely covered with contiguous leaf-scars without definite leaf-cushions or with projecting cushions forming a narrow sloping surface surrounding each

FIG. 203.   *Sigillaria Brardi* Brongn. (¾ nat. size).   From a photograph of a specimen in Dr Kidston's collection, from the Upper Transition Series of Staffordshire. Published by Kidston (02) Pl. LIX. fig. 1.

leaf-scar. Other parts of the plant may possess cushions similar in their kite-shaped form to those of *Lepidodendron*, but without a median vertical groove, or the leaf-scars may be spirally disposed at varying distances apart on a comparatively smooth and longitudinally wrinkled bark. The species exhibits striking instances of a transition between the Favularian, Clathrarian, and Leiodermarian forms of stems. The leaf-scars, which are hexagonal in

[1] Renault (96) A. Pl. XXXV.; Zeiller (06) Pl. XLII.   [2] Grand'Eury (90) A.

outline,—the lateral angles pointed and transversely elongated, the upper and lower angles rounded,—bear three scars, the central leaf-trace and two straight or curved lateral parichnos scars; a ligular pit occurs immediately above the centre of the upper edge of the leaf-scar and occasionally circular elevations with a central pit occur singly or in pairs below a leaf-scar (fig. 196, A). The linear leaves, which may persist on shoots having a fairly large diameter[1], have a single median vein and two stomatal grooves on the lower surface[2] (fig. 200, D).

Partially decorticated and younger shoots are characterised by the occurrence of pairs of elliptical parichnos areas and a smaller median leaf-trace scar. The surface of older stems, which may show signs of longitudinal splitting (*Syringodendron* state), bears pairs of parichnos scars reaching a length of 2—2·5 cm. and a breadth of 10—13 mm. The regularity of the leaf-scar arrangement is interrupted at intervals by the occurrence of more or less regular verticils of scars marking the position of deciduous shoots. Grand'Eury[3] has figured cones which he believes to be those of this species, and Zeiller refers the large strobili, *Sigillariostrobus major*, to *Sigillaria Brardi*[4].

The subterranean axes were characterised by spirally disposed rootlet-scars like those of *Stigmaria ficoides* (figs. 204, 205) and by a cortical surface with the features of *Stigmaria rimosa* Gold.[5]

The anatomy of the stele and leaves has already been described (p. 219). The stele of the Stigmarian portion of the plant consists of a band of centripetal primary xylem and a cylinder of centrifugally formed secondary wood with medullary rays containing vascular bundles passing out to the rootlets[6].

Sigillaria Brardi occurs not uncommonly in Permian rocks; it is recorded from France[7], Germany[8], Pennsylvania[9], and elsewhere. It is found in the Upper, Middle, and Lower Coal-Measures of England[10] and in Permo-Carboniferous strata in Africa[11] and Brazil[12].

[1] Zeiller (06) Pl. xlii.   [2] Renault (96) A. Pls. xxxvii. xli.
[3] Grand'Eury (90) A. Pl. xi.   [4] Zeiller (06) p. 176.
[5] Goldenberg (55) Pl. xii.   [6] Renault (96) A. Pl. xxxix.
[7] Zeiller (92) A.; (06).   [8] Potonié (96) A.   [9] Fontaine and White (80).
[10] Kidston (94) p. 252.   [11] Seward (97²) A.
[12] White (08) p. 450, Pl. v. fig. 12.

CHAPTER XVII.

UNDERGROUND RHIZOMES AND ROOTS OF PALAEOZOIC LYCOPODIACEOUS PLANTS.

Stigmaria.

Stigmaria ficoides is the name given to cylindrical casts met with in Palaeozoic rocks, from the Devonian[1] to the Permian[2], characterised by a smooth or irregularly wrinkled surface bearing

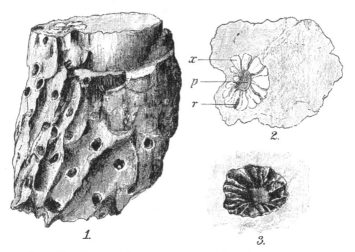

FIG. 204. *Stigmaria ficoides* Brongn. M.S. (See Vol. i. p. 73.)

spirally disposed circular scars bounded by a raised rim and containing a small central pit. It is not uncommon to find evidence of a partial collapse of the substance of the plant as

[1] Potonié (01[2]).   [2] Goeppert (64) A.

seen in fig. 204; this is doubtless the expression of a shrinkage of the middle cortical region, which was composed of a delicate and lacunar system of cells. There can be no reasonable doubt that Stigmaria grew in water or in swampy ground. Specimens are occasionally met with in which the cast terminates in a bluntly rounded apex; such are, perhaps, young branches which have not grown far from the base of the aerial stem from which they arose (cf. fig. 207, B, C). Other examples occur, such as Goeppert[1] figured and Gresley[2] has more recently described,

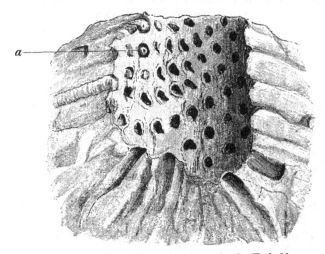

Fig. 205. *Stigmaria ficoides.* From a specimen in the York Museum, from Bishop Auckland. *a*, base of rootlet showing vascular bundle scar. M.S.

which are twisted and distorted as though obstacles had been encountered in the ground in which they grew.

The circular scars mark the bases of long single and occasionally forked appendages (rootlets) which spread on all sides into the surrounding medium (figs. 205, 208). The occurrence of rootlets radiating through the shale or sandstone affords proof that the Stigmarias are often preserved in their position of growth. This was recognised by Steinhauer[3] and Logan[4], and has been more recently emphasised by Potonié[5] as

[1] Goeppert, *loc. cit.*   [2] Gresley (89) Pl. ii.
[3] Steinhauer (18) A.   [4] Logan (42).   [5] Potonié (93³).

an argument in favour of the view that the beds containing such specimens are old surface-soils.

Stigmaria usually shows regular dichotomous branching, the arms spreading horizontally or slightly downwards and always arising from four main branches in the form of a cross (fig. 207). The most remarkable specimens found in England are described by Williamson[1] in his monograph of *Stigmaria*. One of two large casts found near Bradford in Yorkshire, and now in the Manchester Museum, shows four large primary arms radiating from the base of an erect stump 4 feet in diameter. Each arm divides a short distance from its base into two, and the smaller branches extend almost horizontally for several feet[2].

An illustration published by Martin in 1809[3] shows a characteristic feature of Stigmarian casts, namely the presence of a smaller axis, usually occupying an eccentric position inside the larger. This represents the cast of the fairly broad parenchymatous pith which, on decay, left a space subsequently filled by sand or mud: at a later stage the surrounding wood and cortex were removed and the cavity so formed was similarly filled. A thin layer of coal formed by the carbonisation of some of the tissues frequently surrounds the medullary cast, and Steinhauer, whose account of the genus is much fuller and more scientific than those of other earlier and many later writers, recognised the true nature of this internal cast. Artis[4] regarded it as the remains of a young plant, which he described as "perforating its parent," at length bursting it and assuming its place, a gratuitously drastic interpretation.

In 1838[5] Lindley and Hutton figured a partially petrified specimen of *Stigmaria* obtained by Prestwich from Carboniferous rock of Shropshire. This example showed a fairly broad cylinder of secondary wood penetrated by medullary rays. The medullated stele consisted of a pith surrounded by a small amount of primary xylem and by a cylinder of

[1] Williamson (87) A.
[2] A similar example, now in the Bergakademie of Berlin, has been described by Potonié (90) A.; see also a note on the German specimen by Seward (91).
[3] Martin (09) A. Pl. xii.   [4] Artis (25) A.
[5] Lindley and Hutton (38) A. Pl. clxvi.

secondary scalariform tracheae. The preservation of the tissues
abutting on the edge of the wood is usually very imperfect, and
the middle cortex of lacunar parenchyma has practically in every
case eluded the action of mineralising agents; the outer cortex,
on the other hand, consists of more resistant elements and is
frequently well preserved. As in *Lepidodendron* and *Sigillaria*
stems, meristematic activity produced a broad band of secondary
cortex; and beyond this were attached to cushion-like pads the
numerous appendages, each supplied with a single vascular
bundle which arose from the primary xylem and passed outwards through a medullary ray. There is abundant evidence
that the appendages were hollow, a fact in striking accord with
the aquatic and semi-aquatic habitat (cf. *Isoetes* root, fig. 133, G).

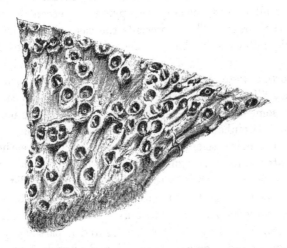

Fɪɢ. 206.   *Cyperus papyrus.*   Piece of rhizome showing rootlet-scars.   Nat. size.
M.S.

The piece of dried rhizome of *Cyperus papyrus* shown in fig.
206 is an almost exact counterpart of *Stigmaria ficoides*; the
wrinkled and shrivelled surface and the circular root-scars
containing the remains of a vascular bundle are striking
features in common and, it may be added, the two plants, though
very different in structure and in systematic position, illustrate
anatomical adaptations to a similar environment.

Stigmaria ficoides Brongniart[1].   Figs. 204, 205, 207, 208.

1809.   *Phytolithus verrucosus*, Martin, Petrifact. Derb. Pls. XI–XIV.
1818.   *Phytolithus verrucosus*, Steinhauer, Trans. Phil. Soc. America,
          [N.S.] Vol. I. p. 268, Pl. IV.
1820.   *Variolaria ficoides*, Sternberg, Flora der Vorwelt, p. 22,
          Pl. XII.
1822.   *Stigmaria ficoides*, Brongniart, Mem. Mus. d'hist. nat. Paris,
          Pl. XII. fig. 7, p. 228.
1825.   *Ficoidites verrucosus*, Artis, Antediluvian Phytology, Pl. X.
1840.   *Stigmaria anabathra*, Corda, Flor. der Vorwelt, Pl. XIV.

The first figure of *Stigmaria* is said to be by Petver in 1704;
Volkmann published illustrations of this common fossil in 1720
and Parkinson in 1804[2].   Binney, whose researches may be
said to have inaugurated a new era in the investigation of fossil
plants, wrote in 1844: "Probably no fossil plant has excited
more discussion among botanists than the *Stigmaria*.   It is the
most common of the whole number of plants found in the Coal-
Measures, but there has hitherto been the greatest uncertainty
as to its real nature[3]."   This uncertainty still exists, at least in
the minds of some who know enough of the available data to
realise that our knowledge is imperfect.

To pass to the questions of the affinity and nature of
Stigmaria: Brongniart[4] at first compared his genus with recent
Aroideae, but he afterwards[5] spoke of it as probably the root of
Sigillaria.   Other writers regarded *Stigmaria* as a dicotyledonous
plant comparable with Cacti and succulent Euphorbias.   For
many years opinion was divided as to whether *Stigmaria*
represents an independent and complete plant or the underground system of *Sigillaria*.

Artis[6], Lindley and Hutton[7], as well as Goldenberg[8], believed
it to be a prostrate plant unconnected with any erect aerial stem.
Goldenberg figured one of the slender rootlets terminating in
an oval body described as a reproductive organ.   This seed-like
impression is either some extraneous body or an abnormal

---

[1] For a fuller synonymy, see Kidston (03) p. 757.
[2] Goldenberg (55) p. 6.   [3] Binney (44) p. 165.
[4] Brongniart (22) A. p. 228.   [5] *ibid.* (49) A. p. 456.   [6] Artis (25) A. Pl. x.
[7] Lindley and Hutton (31) A. Pl. xxxi.   [8] Goldenberg (55).

development at the end of a rootlet. In 1842 Logan drew attention to the almost complete monopolisation by *Stigmaria* of the underclays, the rock which as a general rule occurs below a seam of coal. He wrote: "The grand distinguishing feature of the underclays is the peculiar character of the vegetable organic remains; they are always of one kind (*Stigmaria ficoides*) and are so diffused throughout every part of the bed, that by their uniform effect alone the clay is readily recognised by the eye of the miner[1]." This fact, which has played a very conspicuous part in the perennial discussions on the origin of coal, led to the almost general recognition of the underclays as surface-soils of the Coal period forests.

The next step was the discovery of *Stigmaria* in the Coal-Measures of Lancashire and in the Carboniferous rocks of Cape Breton, Nova Scotia, forming the basal branches of erect stems identified by Binney[2], Bowman[3] and Richard Brown[4] as undoubted Sigillariae. In one case Brown found what he considered to be convincing evidence of the continuity between *Stigmaria* and *Lepidodendron*.

In 1842 Hawkshaw[5] described certain fossil trees, the largest of which had a circumference at the base of 15 ft., discovered, in the course of excavations for a railway in Lancashire, in soft shale at right angles to the bedding. The surface features were not sufficiently clear to enable him to decide with certainty between *Sigillaria* and *Lepidodendron*, but while inclining to the former, it is interesting to note that the occurrence of numerous Lepido-strobi near the root led him to recognise the possibility of a connexion between the Stigmarian roots and Lepidodendron stems. In 1846 Binney gave an account of similar trees found at Dukinfield near Manchester: he spoke of one stem as unquestionably a *Sigillaria* with vertical ribs, furrows, and scars, about 15 inches high and 4 ft. 10 inches in circumference. He expressed his conviction that "*Sigillaria* was a plant of an aquatic nature[6]." Similar descriptions of rooted stems in the Coal-Measures of Nova Scotia were published by Brown in

---

[1] Logan (42) p. 492.   [2] Binney (44) ; (46).   [3] Bowman (41).
[4] Brown (45); (46); (47); (49).   See also Dawson (66).
[5] Hawkshaw (42).   [6] Binney (46) p. 393.

1845, 1846 and 1849; in the last paper he figured a specimen, which has become famous, showing a Syringodendron stem terminating in branching Stigmarian (or possibly Stigmariopsis) roots bearing on the lower surface a series of what he called conical tap roots[1]. A similar specimen discovered in Central France nearly fifty years later demonstrated the accuracy of Brown's description.

Despite these discoveries the root-like nature of *Stigmaria* was not universally accepted. It was, however, generally agreed that *Stigmaria* formed the roots of *Sigillaria*; it was, moreover, held by some that *Lepidodendron* stems also possessed this type of root, an opinion based on Brown's record and on the occurrence of *Stigmaria* in beds containing *Lepidodendron* but no *Sigillaria* stems, as in the volcanic beds of Arran and else-where, and on observations of Geinitz and others[2]. There is now general agreement that *Lepidodendron* and *Sigillaria* had the same type of "root," though the connexion of *Stigmaria* with the former was not so readily admitted, and indeed the evidence in support of it is still very meagre. Goeppert and other authors were unable to believe that the numerous species of *Sigillaria* possessed roots of so uniform a type, but Goeppert, by his recognition of several varieties of *Stigmaria*, supplied a partial answer to this objection.

Messrs Mellor and Leslie[3] have described and figured some large casts of roots exposed in Permo-Carboniferous rocks in the bed of the Vaal river at Vereeniging (Transvaal) which exhibit certain features suggesting comparison with *Stigmaria*. Some of these reach a length of 40—50 feet and, when complete, were probably not less than 100 feet long: in some of them the centre of the cast from which forked arms spread almost horizontally shows a depression in the form of a cross indicating a regular dichotomous branching like that of *Stig-maria*. The authors incline to the belief that the roots belong to *Noeggerathiopsis* and not to a lycopodiaceous plant, though Lepidodendroid stems are abundant in the sandstone a few feet

---

[1] Brown (49). This figure is reproduced by Williamson (87) A. p. 16.

[2] Williamson (87) A. p. 3. Solms-Laubach (91) A. p. 284.

[3] Mellor and Leslie (06).

higher in the series. Despite the absence of any Stigmarian
scars on the surface of the fossil it is probable that these fine
specimens are the rhizomes of some lycopodiaceous plant,
possibly *Bothrodendron*, which is not uncommon in the Vereeniging beds.

Admitting that *Stigmaria* is part of *Sigillaria*, the next
question is, is *Stigmaria* a root in the ordinary sense, the underground system formed on germination of the spore and of equal
age with the shoot, or did it bear a different relation to the Sigillarian stems ? To this question different answers would still be
given. Goeppert[1] discussed evidence in favour of the view that
aerial Sigillarian shoots were produced as vegetative buds
on pre-existing Stigmarian axes, like young moss plants on
a protonema. At a later date Renault[2] developed a similar
view as regards *Sigillaria*; but we may pass on to consider the
more recent and complete observations of Grand'Eury[3] and
Solms-Laubach[4].

The recognition of two distinct types of Stigmariae in the
Coal-Measures of Central France led Grand'Eury[5] to institute a new genus, *Stigmariopsis*. This type which is characterised by a difference in habit as well as by other distinguishing
features, is represented by such specimens as those figured by
Goldenberg as *Stigmaria abbreviata*, bearing lenticular scars
spirally disposed on a cortical surface characterised by irregular
longitudinal wrinklings. *Stigmariopsis* has frequently been found
in direct continuity with Sigillarian stems of the Leiodermarian-
Clathrarian type, spreading obliquely downwards in the form of
rapidly narrowing arms clothed with slender and usually simple
appendages; and from the under surface of these arms short
conical outgrowths are given off. It is probable, as Solms-
Laubach believes, that *Stigmariopsis* was represented also by
long horizontally creeping rhizomes[6] of uniform breadth from
which ribless Sigillarian aerial shoots arose as bud-like outgrowths. Grand'Eury, the author of the genus, confined the
term to the shorter and more rapidly tapered organs spreading

---

[1] Goeppert (64) A. p. 197, Pls. 34—36.   [2] Renault (81).
[3] Grand'Eury (90) A.   [4] Solms-Laubach (94).
[5] Grand'Eury (77) A. p. 171.   [6] For figures see Grand'Eury (87) A.; (90) A.

from the base of erect stems; the horizontal rhizomes of all
Sigillarian stems he refers to *Stigmaria*. The pith-casts of
Sigillariopsis may be recognised by their long vertical ridges
and grooves, a feature readily understood by reference to the
stem structure. The *Stigmariopsis* rhizomes though rare in
England have been recognised by Dr Kidston[1] in the Middle
Coal-Measures of Yorkshire; he has figured a pith-cast very
like that illustrated in Solms-Laubach's Memoir as *Stigmariopsis
anglica*.

The surface-features of a *Stigmariopsis* pith-cast are clearly
shown on a specimen from St Étienne in the Williamson
collection[2].

The most complete account of Grand'Eury's views in regard
to the anchoring and absorbing organs of *Sigillaria* is given in
his monograph on the Coal-field of Gard[3], St Étienne, and these
are clearly stated also by Solms-Laubach[4] who confirms the
conclusions of the French author as to the manner of development of the aerial shoots. Grand'Eury believes that both
Stigmaria and *Stigmariopsis* are rhizomes and not true roots.
The surface-features of *Stigmaria* have already been described.
This type Grand'Eury speaks of as characterised by the uniform
diameter and considerable horizontal elongation of the bifurcated axes; he thinks they grew both as floating rhizomes and
on the ground: they may frequently be traced for a considerable distance without showing any signs of connexion with
aerial shoots, but occasionally they have been seen in organic
union with Sigillarian stems. He believes that these rhizomes
were produced as the result of germination under water of the
spores of *Sigillaria* or *Lepidodendron* and developed as long and
branched aquatic rhizomes capable of independent existence.
Under certain conditions, as he thinks in shallower water, the
rhizomes produced bulb-like outgrowths which grew into erect
stems having the surface-features of *Sigillaria*. This method
of origin is practically the same as that described by Goeppert
in 1865. The vascular medullated cylinder of these erect

[1] Kidston (02) Pl. LI. fig. 4.   [2] British Museum, No. 870 F.
[3] Grand'Eury (90) A.   [4] Solms-Laubach (94).

branches was in direct continuity with that of the Stigmarian
rhizomes.

The next stage is that in which the undifferentiated bulb
becomes swollen at the base and developes four primary roots
(fig. 207 B, C) which grow obliquely downwards and produce
numerous rootlets. Meanwhile the parent rhizome gradually
decays, finally setting free the aerial stems which are now
provided with spreading and forked roots (fig. 208) such as we

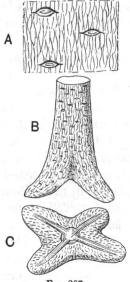

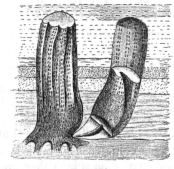

Fig. 208.

Fig. 207.

Fig. 207. An early stage in the development of *Sigillaria*.
           A. Surface-features enlarged. (After Grand'Eury.)

Fig. 208. Later stage in the development of *Sigillaria*; *Syringodendron* with
           Stigmariopsis. (After Grand'Eury.)

are familiar with in English specimens as *Stigmaria ficoides*,
but which in the French specimens show the features of *Stigmariopsis*. At this later stage conical outgrowths are formed
from the under surface of the Stigmariopsis arranged in a
more or less regular series surrounding the centre of the forked
and spreading roots (fig. 209). These conical and positively
geotropic organs were long ago described by Richard Brown as

tap-roots.   Grand'Eury's conclusions are briefly as follows:
Sigillaria, and we may add *Lepidodendron*, had no true roots
and in this respect are comparable with *Psilotum* (fig. 118):
the organs which are described by Grand'Eury as roots are
correctly so named in a physiological sense, but morphologically
they do not strictly conform, either in origin or in the arrangement of their appendages, to true roots.   The question as to
whether they are entitled to the designation root is one which
it is needless and indeed futile to discuss in detail; it would be
conceding too much to a formal academic standpoint to refrain
from applying to them the term root, as that best describes
their share in the life of the Sigillarian stems.   The horizontal
Stigmarian axes are rhizomes in the ordinary sense of the term
and from these were developed Sigillarian shoots, characterised

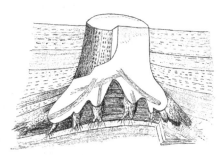

FIG. 209.   *Stigmariopsis* and "tap-roots."   (After Grand'Eury.)

in the lower portions by large parichnos strands.   From
the base of the young bulbous shoots roots were formed: these
roots being, in the French specimens, of the *Stigmariopsis* type.
    These conclusions require some modification when applied
to British representatives of the arborescent Lycopodiales.
The long spreading and dichotomously branched root-like organs
attached to the base of Sigillarian and Lepidodendron stems
are true examples of *Stigmaria ficoides* or other species.   *Stigmariopsis* occurs but rarely.   This marked difference between
French and English specimens may be explained if we adopt
the opinion of Solms-Laubach, who believes that the true
Stigmaria represents both the parent rhizome and the later-
formed roots of the Rhytidolepis Sigillarian species and of

Lepidodendron, the *Stigmariopsis* form having the corresponding relation to the Leiodermarian-Clathrarian species.

The opinion expressed by Williamson[1] in 1892 that Grand'-Eury's hypothesis "appears to be identical with the vague and speculative guesses that were prevalent among us in the early years of the present [nineteenth] century" illustrates the strength of conviction based on English specimens as to the root-nature of *Stigmaria*.

There is undoubtedly considerable confusion, which can be cleared up only by further research, as to the precise relation between *Stigmaria* and *Stigmariopsis* on the one hand and the different types of *Sigillariae* on the other. The main contention, and this is the most important point, of Renault, Grand'Eury and Solms-Laubach as to the manner of formation of the aerial shoots from rhizomes and the subsequent production of forked roots and their ultimate separation from the parent rhizome is, as I believe, correct. Williamson held that *Stigmaria* must be regarded as a true root; he found no evidence to support the view that the large rooted stem discovered by Hawshaw, Binney, and others had been originally produced from aquatic rhizomes. It must, however, be remembered that Grand'Eury's opinion is based on evidence afforded by the exceptionally well displayed Sigillarian forests of St Étienne, on a scale such as English strata have not as yet afforded. Moreover, the absence of any parent-rhizome in association with the rooted stumps described by Williamson and by others is not a serious argument against their rhizome origin.

The specimen represented in fig. 209, which was examined *in situ* by Solms-Laubach and Grand'Eury, shows a Sigillarian stem in the Syringodendron condition bearing rows of paired parichnos scars; from the base forked and rapidly tapering arms radiate through the surrounding rock and, as shown by other specimens, these bear numerous appendages like those of the English Stigmarias. The surface-features of the arms are those of *Stigmariopsis* and the centre of each, as seen on the broken face, is occupied by a pith-cast characterised by parallel longitudinal ridges resembling those on the medullary casts of

---

[1] Williamson (92).

Calamites. It is noteworthy that the petrified rhizome originally described by Renault as *Stigmaria flexuosa*, and afterwards identified by him as the subterranean system of *Sigillaria Brardi,* possesses a vascular cylinder composed of primary xylem strands of crescentic transverse section lining the pith; a cast of the pith, after the removal by decay of its delicate parenchymatous tissue, would exhibit the surface-features of *Stigmariopsis*. *Stigmaria flexuosa* no doubt represents a true *Stigmariopsis* rhizome. On the other hand, as Williamson has shown, the inner surface of the wood of *Stigmaria ficoides* consists of a reticulum of xylem with meshes of medullary-ray tissue; a cast of such a surface presents a very different appearance from that of *Stigmariopsis*.

Returning to fig. 209: from the lower surface of the *Stigmariopsis* arms numerous conical outgrowths, reaching a length of several centimetres, project vertically downwards; these also possess Stigmariopsis pith-casts and are identical with the "tap-roots" of Richard Brown. The stump seen in fig. 209 shows the characteristic hollow base of the erect stem: this is the region which, it is believed, represents the position of the Stigmarian rhizome from which the aerial shoot was developed. Although no remains of the parent rhizome were found, traces of the rootlets which probably belonged to it were found in the neighbourhood. The absence of the actual rhizome is, however, not surprising as it would not persist after its aerial Sigillarian branches had attained independence by the development of their own dichotomously branched absorbing and holdfast organs.

The Stigmarian axes of Palaeozoic Lycopods are compared by Miss Thomas[1] with the prop-roots of certain recent flowering plants which grow in tropical tidal swamps; their roots grow downwards from the stem at an angle of 50°—60° before spreading out horizontally. This author also makes some interesting suggestions in regard to the evidence afforded by anatomical structure as to the habitat of *Sigillaria* and *Lepidodendron*.

---

[1] Thomas, E. N. (05) p. 187.

Anatomy.

The more important anatomical features of *Stigmaria* must be dealt with briefly. Williamson's monograph, published in 1887[1], is considerably in advance of the work of that of any of the numerous writers who had previously dealt with the subject. The diagrammatic transverse section reproduced in fig. 210, H, illustrates the general arrangement of the tissues. The medullated stele was described by Williamson as consisting entirely of centrifugally developed secondary xylem and distinguished, therefore, from the stele of a *Lepidodendron* or *Sigillaria* by the absence of a centripetally produced primary xylem zone. The secondary xylem tracheae are characterised by scalariform pits on both radial and tangential walls and, as shown in a figure given by Solms-Laubach[2], the spaces between the transverse bars are bridged across by fine threads, as in the tracheae of *Lepidodendron*.

One of the largest specimens of a petrified *Stigmaria* which I have seen is one lent to me by Mr Lomax from the Coal-Measures of Halifax in which the flattened transverse section measures 18 cm. × 3·5 cm., the cylinder of wood being 1·1 cm. × 7 mm. in diameter.

In French examples of *Stigmaria* or *Stigmariopsis* it has been demonstrated by Renault[3] that primary xylem strands occur very like those in the stem of some species of *Sigillariae* (see p. 219). If a well-preserved section of an English *Stigmaria* is examined it will be seen that the edge of the secondary wood consists of a few narrower elements which do not exhibit the radial seriation characteristic of secondary elements.

A type of *Stigmaria* characterised by centripetal primary wood has been described by Weiss[4] and referred by him to *Bothrodendron mundum*; the main results of his observations are stated in the account of *Bothrodendron* on a subsequent page. This discovery is of considerable interest not only as rendering our knowledge of *Bothrodendron* remarkably complete but as confirmatory of Renault's account of French Stigmarian axes in

[1] Williamson (87) A.
[2] Solms-Laubach (92).
[3] Renault (96) A. Pl. xl. fig. 5.
[4] Weiss, F. E. (08).

which centripetal primary wood is well developed between the
secondary xylem and the centre of the stele. The Stigmarian
axis of Bothrodendron was originally figured by Williamson
as *Lepidodendron mundum*[1]. The chief difference between

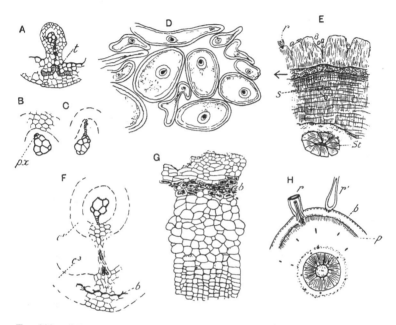

FIG. 210. *Stigmaria.*

    A.   Transverse section of vascular bundle of rootlet and part of outer
        cortex. *t*, tracheae. (After F. E. Weiss.)

    B, C. Vascular bundle of rootlet; in C a series of small tracheae are
        shown extending from the protoxylem.

    D.   Rootlets from the outer cortex of E.

    E.   Part of a large *Stigmaria*: *St*, stele ; *s*, intruded rootlet.

    F.   Vascular bundle and tracheae passing obliquely towards the
        outer cortex, *c*³.

    G.   Outer cortex of *Stigmaria.*

    H.   Diagrammatic section of *Stigmaria*: *p*, phelloderm ; *r*, rootlets.

Weiss's specimen and those described by Renault[2] as the
Stigmarian axes of *Sigillaria Brardi*, is that in the English plant
the centripetal wood forms a cylinder of uniform breadth

---

       [1] Williamson (89) A.   [2] Renault (96) A.

instead of a band with a crenulated inner margin as figured by Renault.

An interesting agreement between the French and English specimens is the occurrence in the cortex of groups of reticulate elements: in Weiss's section these are short and wide and occur in the middle cortex; in Renault's plant they are more fusiform and occur in the secondary cortical tissue. These elements appear to have been arranged as an interlacing network in the middle cortex and were in close connexion with the rootlet-bundles, comparable, as Weiss points out, with the transfusion tracheids accompanying Lepidodendron leaf-traces.

It is probable that these short and wide tracheal elements served for water-storage and thus afford another indication of the xerophilous character of the Carboniferous Lycopods, a feature possibly connected with a salt-marsh habitat.

The presence of conspicuous medullary rays gives the secondary xylem of *Stigmaria* the appearance of being divided into several more or less distinct groups (fig. 210, E, *St*). In tangential longitudinal section the xylem assumes the form of a broad reticulum with lenticular meshes filled with medullary-ray tissue through which strands of xylem are cut across in a transverse direction as they pass outwards from the inner edge of the wood to supply the rootlets. In addition to these broader or primary medullary rays, there were numerous secondary rays composed of narrow plates of parenchymatous cells one or several elements in depth. As Williamson pointed out, the medullary-ray tissue consists in part of radially elongated tracheal elements with spiral or scalariform thickening bands like those described in the same position in Lepidodendron stems.

Our knowledge of the minute structure of the tissues abutting on the secondary xylem is far from complete.

The xylem is succeeded by a zone of delicate cells which was the seat of meristematic activity. It is noteworthy that in a section figured by Williamson[1] there is the same disparity in size between the outermost elements of the xylem and the adjacent cells of the meristematic zone as in Lepidodendron stems. Beyond this region an imperfectly preserved lacunar

---

[1] Williamson (87) A. Pl. IV. fig. 20.

tissue occurs like that which I have called the secretory zone in
Lepidodendron stems; but information as to the structure of
this part of *Stigmaria* is much more incomplete than in the
case of the aerial shoots. The middle cortex was of the same
lacunar type as in the stems, and the fact that it is never well
preserved in large Stigmarian axes suggests that it may have
been even more richly supplied than in the aerial stems with an
aerating system of spaces. The outer cortex, consisting in young
examples of large-celled parenchyma, became at an early stage
of growth the seat of cambial activity which resulted in the
production of radially placed series of secondary elements (fig.
210, H, *p*). The outer and older elements of this secondary
cortex are more tangentially stretched than the inner cells, a
necessary result of the position of the phellogen on the internal
edge of the tissue and of the increasing girth of the axis.

In comparatively young Stigmarian axes the outer cortex
already possesses a band of secondary radially disposed cells
characterised by the greater tangential extension of the more
external elements; usually this tissue terminates abruptly on
the inner edge and the line of separation no doubt marks the
position of the phellogen. Occasionally some delicate secondary
elements are preserved internal to the phellogen, and these in
young specimens form a narrow cylinder composed in part of
radially elongated cells showing signs of recent tangential
divisions. In its earlier stage of activity the phellogen seems to
form a greater amount of secondary tissue on the outside, but
this is clearly not of the nature of cork, the tissue which occupies
a corresponding position in recent plants. The primary cortex
shows no signs of shrinkage or collapse as would be the case
were it cut off from the vascular system by a zone of imper-
meable cork.

Fig. 210, G, represents a piece of the external tissue of a
specimen in which the slightly flattened xylem cylinder
measures 1·4 × 1 cm.; the inner cortex has disappeared and
fragments only of the middle cortex are preserved. The outer
cortex, with an average breadth of 2 mm., consists superficially of
primary parenchyma with a somewhat uneven surface and with
a rootlet attached here and there; a short distance below the

surface is a band of conspicuous cells, *b*, characterised by dark contents suggesting very imperfectly preserved fungal hyphae, but the nature of the substance filling the cells cannot be made out with certainty. It is, however, interesting to find that this dark band constitutes an obvious feature (fig. H, *b*); its position is comparable with that of the dark-walled cells in the outer cortex of rootlets. A short distance internal to this dark band tangentially elongated cells form the outermost elements of the secondary cortex; these become gradually narrower towards the interior and pass into radial series of smaller cells of uniform size, as seen on the inner edge of fig. 210, G. At the inner boundary of this tissue, just below the region shown at the bottom of the drawing, was situated the phellogen. Such traces of tissue as occur on the inner side of the line where splitting has usually occurred, consist of thinner elements with recently formed tangential walls and probably represent an early stage in the development of phelloderm.

A much older section is shown in part in fig. 210, E. The secondary xylem cylinder, *St*, is shown in the lower part of the section; beyond this is a band of secondary tissue which reaches in some places a breadth of 6 cm. The greater part of this tissue consists of phelloderm of very uniform structure made up of radial series of cells: this is interrupted in most parts of the section by a gap crowded with intruded rootlets (a portion of this is enlarged in fig. 210, D). Beyond this gap the secondary tissue consists of radial series of cells characterised by the considerable tangential elongation of many of the elements, precisely like the tissue figured by Williamson. In all probability the gap represents a line of weakness due to the phellogen, and if this is the case it is clear that in an old *Stigmaria* the phelloderm exceeded in amount the tissue formed external to the phellogen. The secondary tissue on the inner side of the phellogen is characterised by numerous irregular concentric lines superficially resembling rings of growth in the wood of a Conifer: these are, however, not the result of any periodic change in external conditions, but are apparently due to crushing of the tissue and are possibly, to some extent, the result of the presence of secretory strands like those in the phelloderm of *Lepidodendron*. The surface of this older rhizome retains patches of primary tissue, and an occasional rootlet, as at *r*, fig. 210, E, is seen in connexion with the cortex ; the cortex has been vertically fissured as the result of secondary growth and presents an appearance like that shown in *Lepidodendron Wünschianum* and *L. Veltheimianum* (figs. 181, A, and 186, A).

The form in which a Stigmarian rootlet is usually preserved is shown in fig. 210, D ; the single vascular bundle strand with its endarch protoxylem (fig. 210, B, *px*) is enclosed by a ring of inner cortical parenchyma (fig. 210, F c^1) ; the cells in immediate contact with the xylem having usually disappeared. Beyond the middle cortical space a second cylinder of parenchyma represents the outer cortex (F, c^3) in which a layer of dark-walled cells (*b*, fig. 210, F) may be compared with the hypodermal band in the main Stigmarian axis (G, *b*). These Stigmarian rootlets, usually less than 1 cm. in diameter, are the commonest objects in sections of the calcareous nodules from English coal-seams. A good example of their abundance is shown in fig. 210, D and E ; here they have invaded the space formed by the splitting of the secondary cortical tissues along the line of the phellogen and a few are seen here and there in the deeper layers of the phelloderm (*s*, fig. 210, E). Not infrequently the close contact of these ubiquitous rootlets with the tissues of the plant which they have invaded leads to confusion between invader and invaded. Partially decayed tissues lying, probably, under water were penetrated by Stigmarian rootlets in exactly the same way as the roots of recent plants bore through vegetable substances which happen to be in their path. The rootlet bundles are in the first instance composed of the primary tracheae which line the inner edge of the secondary xylem ; these receive additions from the meristematic zone, and thus, when seen in the cortex outside the stelar region, are found to consist in part of primary and in part of a fan-shaped group of secondary tracheae. On the other hand, the monarch bundle as it appears in a free rootlet is usually composed entirely of primary elements (fig. 210, A—C, F). It has been shown by Weiss[1] that in the Stigmarian rhizome of what is probably

[1] Weiss, F. E. (02).

Lepidodendron fuliginosum, the rootlet bundle is accompanied
by a parichnos strand, but this has not been detected in the
ordinary *Stigmaria ficoides*. When free from the parent axis a
rootlet usually consists of an outer cylinder of cortex enclosing
a broad space in which remnants of lacunar tissue are sometimes seen. The relation of the external features of a well-
preserved Stigmarian rootlet-scar to the internal structure of a
petrified rootlet is very clearly seen on comparing such sections
as those represented in fig. 210, D, with the form of the scar on
a Stigmarian cast. A specimen figured by Hooker[1] in 1848
affords a good illustration of the structure of a rootlet-base as
seen in an unusually complete cast; this correlation of anatomical and surface features is clearly described also by
Williamson[2] and by Solms-Laubach[3]. It is probable that
even during life the rootlets were hollow for a part at least
of their length as are the roots of *Isoetes* (fig. 133, G).

An interesting discovery was made a few years ago which
confirmed a statement by Renault which Williamson was
unable to accept, namely that the xylem bundle of a rootlet
occasionally gives off a delicate tracheal strand at right angles
to the long axis of a rootlet. In some rootlets Weiss[4] found
obliquely running delicate strands of xylem, surrounded by a
layer of parenchymatous tissue, in the space between the
vascular bundle and the outer cortical cylinder. It is clear
that a few spiral tracheids are occasionally given off from the
protoxylem of a rootlet bundle : these follow an oblique course
to the outer cortex, where in some cases they have been traced
into connexion with short and spirally marked cells resembling
transfusion tracheae (fig. 210, A). This arrangement may serve
as a means of facilitating the passage of water absorbed by the
superficial cells into the xylem strand. It should be noticed
that, like roots of recent water-plants, the rootlets of *Stigmaria*
had no root-hairs. Fig. 210, F, shows a transverse section of
part of a rootlet in which the outer cortical cylinder, c^3, is
connected, as in the roots of *Isoetes*, with the sheath surround-

---

[1] Hooker (48²) Pls. I. II. The sections of *Stigmaria* figured by Hooker are
in the British Museum (V. 8754).
[2] Williamson (87) A. Pl. XII.   [3] Solms-Laubach (91) A.   [4] Weiss, F. E. (02).

ing the vascular bundle. A few obliquely cut tracheae are seen in this section traversing the connecting band of paren- chyma *t*, fig. 210, A.

A point of biological interest in connexion with Stigmaria rootlets is the occasional presence of hypertrophied cells, the large size of which is due to the attacks of a fungus named by Weiss[1] *Urophlyctites stigmariae*.

In addition to *Stigmaria ficoides*, which is by far the commonest form, a few other species have been founded on external characters. One of these is represented by *Stigmaria stellata*, Goepp.[2], characterised by the presence of radially dis- posed ridges and small tubercles surrounding each rootlet-scar. Kidston refers to Goeppert's species as a Lower Carboniferous type. We have no evidence as to the meaning of the stellate ridges and tubercles, nor have we any reason to suppose that this form differed essentially in structure from *Stigmaria ficoides*.

[1] Weiss, F. E. (04).
[2] Goeppert (41) Pl. x. Lief. i. ii.; Williamson (87) A. Pl. xiii. fig. 78; Eich- wald (60) Pl. xv. ; Kidston (94) p. 254.

CHAPTER XVIII.

Bothrodendreae.

Bothrodendron. Figs. 211—216.

ALTHOUGH in many respects the genus *Bothrodendron* agrees very closely in habit and in its anatomical features with *Lepidodendron,* there are reasons for referring it to a distinct family of Palaeozoic Lycopods. As the following description shows, the external features do not differ in any essential points from those of certain types of the genus *Sigillaria,* particularly such a species as *S. rimosa,* Gold.[1], which has recently been refigured and described by Nathorst[2] from Goldenberg's type-specimen in the Stockholm Museum. The small size of the leaf-scars is, however, a characteristic feature of *Bothrodendron* (fig. 212, F); but a more important point is the fact that in a recently described[3] English example of a cone of *Bothrodendron* (fig. 216), the sporangia are very like those of recent Lycopods, and differ from the radially elongated sporangia of *Lepidostrobus.* On the other hand, a French cone described by Zeiller[4] as *Lepidostrobus Olryi,* which is probably a strobilus of *Bothrodendron,* has the radially elongated type of sporangium (fig. 212, E). The comparative abundance of *Bothrodendron* in Lower Carboniferous and Devonian rocks points to the greater antiquity of this member of the Lycopodiales as compared with *Lepidodendron.*

---

[1] Goldenberg (55) Pl. vi. figs. 1—4.   [2] Nathorst (94) A. Pl. xvi. fig. 9.
[3] Watson (08).   [4] Zeiller (88) A. Pl. lxxvii. fig. 1.

The name *Bothrodendron* was instituted by Lindley and Hutton[1] for impressions of stems from the English Coal-Measures, characterised by two opposite rows of large depressions like those shown in fig. 211 and, in one of the specimens, by "a considerable number of minute dots, arranged in a quincuncial manner." The minute dots were recognised as leaf-scars and the cup-like cavities were described as probably connected with the occurrence of large cones. On very slender evidence this Palaeozoic plant, which was named *Bothrodendron punctatum*, was considered by these authors as probably a member of the Coniferales. The large stem from the Coal-Measures in the neighbourhood of Mons, Belgium, shown in fig. 211, affords a good illustration of *Bothrodendron* in a partially decorticated condition, exhibiting a row of depressions similar to those on the Ulodendron form of *Lepidodendron Veltheimianum* (fig. 157), but distinguished by the eccentric position of the scar at the bottom of each cup-shaped cavity : in the Belgian specimen, which is partially decorticated and shows the leaf-traces as small dots, the depressions have a diameter of 9 cm. It is believed by some authors that these Ulodendron shoots of *Bothrodendron* and *Lepidodendron* owe their characteristic appearance to the pressure of large cones, but, as I have already stated, there are reasons for preferring the view that these crater-like hollows are the scars of deciduous branches. Our knowledge of the strobili borne by Bothrodendron stems is still meagre, but we have no reason to assume the existence of any cones large enough to produce by the pressure of their bases such depressions as those shown in fig. 211. In one species at least the strobili were borne terminally on slender shoots (fig. 213). The Ulodendron condition has so far been recognised in one species only, *B. punctatum*.

In his catalogue of Palaeozoic plants, Kidston[2] included *Bothrodendron punctatum* as a synonym of *Sigillaria discophora* König, a mistake which he afterwards rectified[3] : the generic name *Bothrodendron* was generally ignored by authors in the belief that the specimens described by Lindley and Hutton

---

[1] Cf. Lindley and Hutton (35) A. Pls. 80, 81.
[2] Kidston (86) A. p. 175.   [3] *ibid.* (86⁴) p. 65.

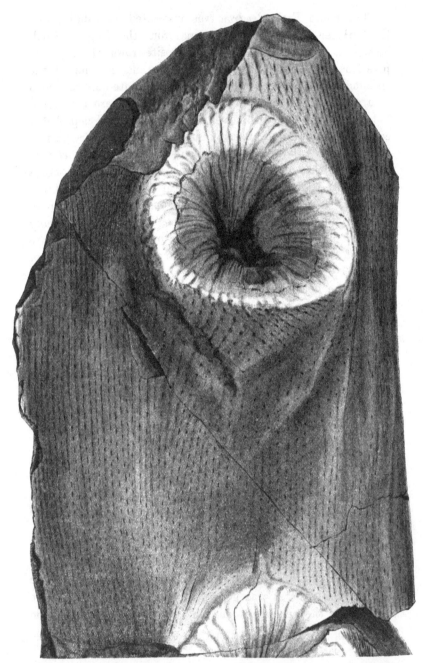

Fig. 211. *Bothrodendron punctatum.* Part of a specimen from near Mons (Hainaut), in the Brussels Museum. (Reduced.)

were not generically distinct from the fossils originally figured by Rhode as *Ulodendron*. It was Prof. Zeiller who first demonstrated that the English authors were justified in their choice of a new designation for stems with large depressions in association with minute leaf-scars. In 1859 Haughton[1] proposed a new family name Cyclostigmaceae for some Upper Devonian plants from County Kilkenny, Ireland: he described three species of his new genus *Cyclostigma*, *Cyclostigma kiltorkense* C. *minutum*, and C. *Griffithsi*; these are now generally recognised as a single species of *Bothrodendron*, though, as Nathorst suggests, the Irish plant should perhaps be separated as a sub-genus *Bothrodendron* (*Cyclostigma*) by reason of certain minor differences which distinguish it from other species of the genus.

Another generic name, *Rhytidodendron*, was instituted by Boulay in 1876 for stems characterised by a finely wrinkled bark and small spirally disposed leaf-scars. A short description of this type, which occurs in the Middle and Lower Coal-Measures, may serve to illustrate the external features of the commonest British example of the genus.

a. *Bothrodendron minutifolium* (Boulay.)   Figs. 212, A, C, D; 213.

1875. *Lycopodium carbonaceum* (*Lycopodites carbonaceus*), Feistmantel, Palaeontographica XXXIII., Pl. XXX. figs. 1, 2 ; p. 183.

1876. *Rhytidodendron minutifolium*, Boulay, Terr. Houill. Nord France, p. 39, Pl. III. fig. 1.

1886. *Bothrodendron minutifolium*, Zeiller, Bull. Soc. Géol. France [iii] XIV. p. 176, Pl. IX. figs. 1, 2.

1888. *Lepidostrobus Olryi*, Zeiller, Flor. Valenciennes, p. 502, Pl. LXXVII. fig. 1.

1889. *Bothrodendron minutifolium*, Kidston, Trans. R. Soc. Edinburgh, Vol. XXXV. Pt ii.

1893. *Sigillaria (Bothrodendron) minutifolia*, Weiss and Sterzel, K. Preuss. Geol. Landesanstalt, Heft 2, p. 49, Pl. I. figs. 3 and 4 ; Pl. II. figs. 8 and 9.

1904. *Bothrodendron minutifolium*, Zalessky, Mém. Com. Géol. Russie, Pl. VI. fig. 6.

In habit a plant of *Bothrodendron* recalls *Lepidodendron* and recent species of *Lycopodium*; the slender dichotomously

[1] Haughton (59).

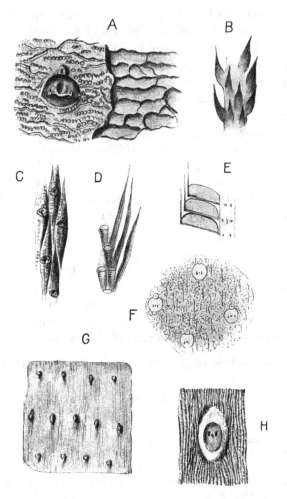

FIG. 212. *Bothrodendron.*

    A. *Bothrodendron minutifolium,* var. *rotundata* Weiss. After Weiss and Sterzel.

    B. *B. punctatum.* After Zeiller.

    C. *B. minutifolium.* After Weiss and Sterzel.

    D. *B. minutifolium.* After Zeiller.

    E. *Lepidostrobus Olryi.* After Zeiller.

    F. *Bothrodendron punctatum.* After Zeiller.

    G, H. *B. kiltorkense.* G, after Nathorst; H, after Weiss and Sterzel.

branched twigs bearing numerous leaves (fig. 212, D), have been mistaken for shoots of *Lycopodium*, and fragments of branches might well be identified as impressions of Mosses. The leaf-scars on the smaller shoots occur on elongated cushions (fig. 212, C, D) with a transversely wrinkled surface; on the older branches the leaf-scars are separated by fairly large areas of bark characterised by sinuous transverse grooves and narrow ridges bearing numerous small pits, as shown on an enlarged scale in fig. 212, A. The original surface-features are shown on the left of the drawing, and a slightly deeper level in the cortex is represented on the right-hand side. The absence of leaf-cushions on the older shoots is probably the result of secondary thickening, which also alters the size and shape of the leaf-scars. Each scar has three pits on its surface, as in *Lepidodendron*; a central leaf-trace scar and lateral parichnos scars. The circular pit above the leaf-scars, which occurs in most species, marks the position of the ligule. The relation of the short leaves, 5 mm. long, to the leaf-cushions is shown in fig. 212, D. The absence of leaves, except in impressions of slender twigs, may be interpreted as an indication that they were shed at an early stage and did not persist many years. The leaf-cushions of the smaller shoots of *Bothrodendron minutifolium* closely resemble those figured by Weiss on a Devonian plant, *Lepidodendron Losseni*[1].

One of the few examples so far discovered of a Bothrodendron cone is shown in fig. 213 ; this specimen, at least 10 cm. long, was found by Mr Hemingway in the Middle Coal-Measures of Yorkshire and described by Dr Kidston. Numerous sporophylls are attached at right angles to the axis, the surface of which is protected by their upturned distal portions; the arrangement of the parts appears to be the same as in *Lepidostrobus*. A specimen figured by Zeiller as *Lepidostrobus Olryi*, which Kidston is probably correct in identifying with *Bothrodendron minutifolium*, shows that each sporophyll carries a horizontally elongated sporangium (fig. 212, E).

---

[1] Weiss, C. E. (84) Pl. vi. figs. 6, 7.

b.   Bothrodendron punctatum Lindley and Hutton[1].   Figs. 211,
212 B, F.

This species, which is less abundant than *B. minutifolium,*

Fig. 213.   *Bothrodendron minutifolium* Cone.   From a specimen in Dr Kidston's
Collection.   (Slightly reduced.   Kidston (02) Pl. LIX.)

in British Coal-Measures, has been described by several authors
as *Ulodendron* on account of the occurrence of large depressions,
like those shown in fig. 211, on certain branches of the plant.

[1] Lindley and Hutton (35) A. Pls. 80, 81.   For synonymy, see Kidston (93)
p. 344.

At the suggestion of Dr Kidston, Prof. Zeiller[1] figured an English specimen of this species, presented to the Paris Museum by Mr Hutton, in which the leaf-scars are preserved on the bark of a stem with Ulodendron scars. The surface of the bark is characterised by numerous small pits and discontinuous vertical lines in contrast to the transverse lines of *B. minutifolium* (cf. fig. 212, A and F). The leaf-scars on the smaller shoots may have a diameter of only 0·3—0·5 mm., while on the larger branches they reach a breadth of 1 mm. The ligule-pit may be in contact with the upper edge (fig. 212, F) of the leaf-scar or separated from it by a short distance.

c.   *Bothrodendron kiltorkense* (Haughton).   Fig. 212, G, H.

> 1859.   *Cyclostigma kiltorkense*, Haughton, Journ. R. Soc. Dublin, Vol. II. p. 418, Pls. XIV.–XVII.
> *C. minutum*, Haughton, Journ. R. Soc. Dublin, Vol. II. p. 418, Pls. XIV.–XVII.
> *C. Griffithsi*, Haughton, Journ. R. Soc. Dublin, Vol. II. p. 418, Pls. XIV.–XVII.
> 1870.   *Lepidodendron Veltheimianum*, Heer (*ex parte*), K. Svensk. Vet. Akad. Handl. Vol. IX. Pl. IX. figs. 2–4.
> *Cyclostigma kiltorkense*, ibid. Pl. XI. figs. 1–5.
> *Calamites radiatus* (*ex parte*), ibid. Pl. III. fig. 2 a : Pl. IX. fig. 2 b.
> *Stigmaria ficoides minuta*, ibid.   Pl. IX. fig. 2 c.
> *Knorria imbricata*, ibid.   Pl. X. fig. 4.
> 1889.   *Bothrodendron kiltorkense*, Kidston, Ann. Mag. Nat. Hist. [VI.], Vol. IV. p. 66.
> 1894.   *Bothrodendron kiltorkense*, Nathorst, K. Svensk, Vet. Akad. Handl. Vol. XXVI. No. 4, p. 65, Pls. XIV. XV.
> 1902.   *Bothrodendron (Cyclostigma) kiltorkense*, ibid.   Vol. XXXVI. No. 3, p. 31, Pls. X.—XIV.

The specimens from the Upper Devonian rocks of Co. Kilkenny on which Haughton founded this and two other species may be regarded as representing one specific type. He described the circular leaf-scars as arranged in alternating whorls. In habit the Irish species agrees with *Bothrodendron minutifolium*, but the leaf-scars are more elliptical (fig. 212, H)

---

[1] Zeiller (86) Pl. IX. figs 1—3.

and the ligule-pit is usually absent. The leaf-scar shown in
fig. H is 1·2 mm. broad and 1·4 mm. in height. The large
collection obtained during the visit of a Swedish expedition to
Bear Island in 1898 under the leadership of Dr Nathorst has
materially increased our knowledge of this ancient type. The
form of the leaf-scars varies according to the age of the branch
and their disposition is far from constant even on the same
specimen; in some cases the scars are in fairly regular
whorls (fig. 212, G; an Irish specimen) while in others they are
in regular spirals. This irregularity of arrangement, which is
well illustrated by Nathorst's figures of Bear Island and Irish
specimens, finds its counterpart, though in a less marked form,
in recent species of *Lycopodium, e.g. L. Selago.* Partially
decorticated stems may present a superficial resemblance to
Calamites, the fissured bark simulating the ribs of a Calamitean
cast. Such stems, as Nathorst has pointed out, were mistaken
by Heer for *Calamites radiatus.* The smaller branches are
characterised by a smooth surface, and older shoots resemble
Bothrodendron minutifolium in the presence of fine vertical
lines. The preservation of only one pit on the leaf-scars of
many examples led authors to conclude that the species is
peculiar in this respect, but Nathorst has shown that in more
perfectly preserved specimens each leaf-scar bears three small
dots. A specimen from Ireland in the British Museum[1] illustrates the dichotomous branching and the longitudinal wrinkling
of the bark; the leaf-scars are 2 mm. broad and 2·5 mm. deep.

Nathorst[2] has described some examples in which the leaf-
scars occur on the lower instead of on the upper end of the
leaf-cushions; these and other specimens with obscure surface-
features he suggests may be underground axes, comparable
in habit with *Stigmaria* though not identical as regards
details. It is pointed out that the absence or scarcity of
Stigmaria in the Bear Island beds renders it unlikely that
Bothrodendron bore typical Stigmaria branches. F. E. Weiss[3]
has recently described root-bearing organs possessing primary
xylem identical with that of *Bothrodendron mundum*; while
closely resembling *Stigmaria ficoides* in certain anatomical

[1] No. 52524.   [2] Nathorst (02) Pl. x. figs. 4, 5.   [3] Weiss, F. E. (08).

characters, they clearly represent a distinct type. This discovery of a Stigmaria-like axis almost certainly belonging to *Bothrodendron* is consistent with Nathorst's views on some of the Bothrodendron impressions from Bear Island.

Information as to the cones of this species is restricted to a description by Schimper[1] of a specimen in the Dublin Museum as *Lepidostrobus Bailyanus*; this has sporophylls with a subtriangular base bearing several megaspores and terminating distally in a slender lamina 12 cm. in length.

An example of a *Bothrodendron* with more prominent leaf-cushions than those already mentioned is afforded by a species from Bear Island described by Heer[2] as *Lepidodendron Wükianum* and afterwards referred by Nathorst[3] to *Bothrodendron*. The same type is recorded also by Schmalhausen[4] from Lower Carboniferous or Devonian strata of Siberia. Certain Scotch specimens from the Calciferous Sandstone, which Kidston[5] referred to Heer's species, are regarded by Nathorst and, in part at least, by Weiss[6] and Sterzel as representing a distinct species which these authors designate *Bothrodendron Kidstoni*[7].

Without attempting the hopeless task of discriminating between the various Carboniferous and Devonian specimens described under the names *Cyclostigma* or *Bothrodendron*, reference may be made to the following records as illustrating the wide distribution of the genus. Schmalhausen[8] records *Cyclostigma kiltorkense* from Siberian rocks assigned to the Ursa stage (Devonian or Lower Carboniferous). The fossil described by Dawson[9] from the Devonian of Gaspé as *Cyclostigma densifolium* probably represents a badly preserved example of *Bothrodendron*: Weiss's species *Cyclostigma hercynium*[10] from Lower Devonian rocks of the Hartz district may be identical with *Bothrodendron kiltorkense*. The supposed identity of the latter species with *Dechenia Roemeriana* Goepp., as de-

---

[1] Schimper (70) A. p. 71.   [2] Heer (71) Pl. vi. fig. 11; Pl. ix. fig. 1.
[3] Nathorst (94) A. p. 67, Pl. xv. figs. 14, 15.
[4] Schmalhausen (77) p. 281, Pl. i. fig. 5.
[5] Kidston (89[2]) Pl. iv. figs. 2—4, p. 65.
[6] Weiss and Sterzel (93) p. 56.   [7] Kidston (03) p. 823.
[8] Schmalhausen (77) p. 290, Pl. i. figs. 7—12.
[9] Dawson (71) A. Pl. viii.   [10] Weiss, C. E. (84) Pl. vii.

scribed by Potonié[1], appears to require confirmation[2], but if this author is correct the connexion demonstrates the continuity of Bothrodendron shoots and Stigmaria-like subterranean organs. The specimens described from South Africa, from strata which may be correlated with the Upper or possibly with the Lower Carboniferous series of Europe, as *Bothrodendron Leslei*[3] in all probability represents a species closely allied to the Irish and Bear Island type. *Bothrodendron Leslei* named after Mr Leslie whose discoveries in the Carboniferous Sandstone of Vereenig-

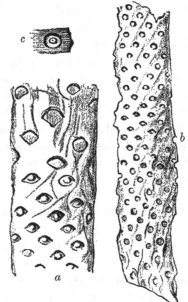

Fig. 214. *Bothrodendron Leslei* Seward.
    b. Natural size.
    a, c. Slightly enlarged.

ing (Transvaal) have added considerably to our knowledge of the South African Palaeozoic types, is represented by imperfectly preserved casts characterised by more or less circular scars displaying the same irregularity of arrangement as in *Bothrodendron kiltorkense.* The leaf-scars appear to

[1] Potonié (01²) figs. 25—27.   [2] Nathorst (02) p. 35.
[3] Seward (03) Pl. xi. figs. 1—6, p. 87 ; Arber (05) p. 166.

have only one small pit, but this may not be an original feature. The identification of this plant as *Bothrodendron* receives support from the discovery of rather more satisfactory specimens at Witteberg sent to me for examination by Dr Schwarz[1]. These fossils bear a striking resemblance to *B. kiltorkense.* *Cyclostigma australe*[2] Feist. described from the Lower Carboniferous rocks of New South Wales, though too imperfectly preserved to refer with confidence to *B. kiltorkense*, is no doubt a closely allied type.

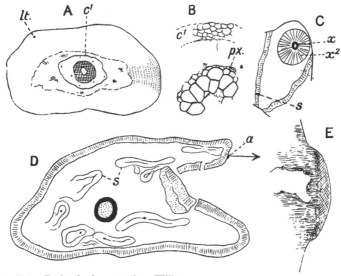

Fig. 215. *Bothrodendron mundum* (Will.).
A, B.   From a specimen (No. 26) in the Cambridge Botany School.
C.   British Museum, Williamson Collection.   (No. 416 b.)
D, E.   From a section in Dr Kidston's Collection.

Reference was made in Volume I. (p. 133) to the so-called paper coal of Carboniferous age from Central Russia, which consists of masses of thin strips of cuticle of Bothrodendron stems. The figures published by Zeiller[3] show that the plant possessed an epidermis consisting of polygonal cells interrupted by spirally disposed gaps marking the position of leaves; the gaps measure 0·5—1·5 mm. in breadth and agree, therefore, with the size of

[1] Seward (09).   [2] Feistmantel (90) A.   [3] Zeiller (80²) A.

the leaf-scars of the smaller forms of *Bothrodendron*. The specimens from the Russian mines were first figured by Trautschold and Auerbach[1] as *Lepidodendron tenerrimum* and afterwards referred by Zeiller to *Bothrodendron punctatum*[2]. Nathorst[3], however, states that an examination of the Russian material leads him to retain the name originally proposed; he records the same type from Upper Devonian rocks of Spitzbergen. The chief interest of these Russian specimens is their manner of preservation, which Renault has described as the result of bacterial action; he claims to have recognised the actual bacteria associated with the cuticular membranes[4].

Anatomy of vegetative shoots of Bothrodendron.

In 1889 Williamson[5] described several specimens of petrified shoots from the Coal-Measures of Halifax which he named *Lepidodendron mundum*: these are now known to be branches of a *Bothrodendron*. The discovery was made by Mr Lomax[6] who found specimens showing the external characters of *Bothrodendron* and the anatomical characters of *Lepidodendron mundum*. In some of the smaller twigs, the stele consists of a solid core of xylem with external protoxylem; but in the majority of specimens the centre of the xylem is replaced by parenchymatous tissue, either as a small axial strand or, as in the specimen shown in fig. 215, D, a wide pith, the elements of which are arranged in regular vertical series. A diagrammatic section of a small axis is represented in fig. 215, A : this branch, 2 mm. in diameter, is composed of a broad outer cortex consisting exclusively of primary tissue the outer cells of which are smaller and have thicker walls than the more internal elements. The leaf-traces, *lt*, are accompanied by a strand of delicate tissue, the parichnos. The stele is almost solid; the tissues in contact with the xylem have not been preserved but the inner cortex is represented by a few layers of small paren-

---

[1] Trautschold and Auerbach (60) Pl. III.
[2] Zeiller (82) A. ; (86).   [3] Nathorst (94) A. Pls. x. xi.
[4] Volume I. p. 134.   [5] Williamson (89) A. p. 197.
[6] I am indebted to Mr Lomax for photographs of his specimens. For former references to Mr Lomax's discovery, see Kidston (05); Weiss, F. E. (08); Scott, D. H. (08) p. 200.

chymatous cells, c^1. The larger section shown in fig. 215, D, was cut from a specimen from Dulesgate of which the smooth surface exhibits the characteristic leaf-scars of *Bothrodendron*. The section measures 3 cm. in its longest diameter and the stele has a breadth of 3 mm. The outer cortex has a smooth surface and is composed of rather thick-walled cells succeeded by a zone of secondary elements. The middle cortex has disappeared and the space is partially occupied by Stigmarian rootlets, *s*, and crushed patches of cortical tissue. The position of a leaf-scar is seen at *a*; this is more clearly shown in the enlarged drawing fig. E.

In his account of *Lepidodendron mundum*, Williamson[1] described a section in which the primary wood is surrounded by a considerable thickness of secondary xylem; a diagram of this is shown in fig. 215, C. An examination of the section led me to compare the structure of the outer cortical cells, characterised by radial rows of tangentially elongated elements, with the outer cortex of *Stigmaria*. It has recently been shown by Weiss[2] that this and other similar sections present several points of agreement with *Stigmaria*, particularly with *Stigmaria Brardi* as described by Renault. At *s* in fig. 215, C, a vascular strand is seen passing through the outer cortex; this is almost certainly the bundle of a rootlet: in the sections described by Weiss rootlets are shown in a similar position. The chief anatomical features of the Stigmaria-like organs of *Bothrodendron* are :—the considerable development of secondary xylem, the structure of the outer cortex, which is practically identical with that of *Stigmaria ficoides*, and the association of groups of short transfusion tracheids with the bundles of the rootlets. It is very probable that the absence of secondary xylem in the vegetative shoots of *Bothrodendron* is merely an accident and not a real distinction between the aerial and subterranean branches of the plant; a supposition rendered probable by the occurrence of secondary xylem in the axis of the cone described by Watson. As Weiss points out, there are certain differences between the true *Stigmaria* and the corresponding organ of *Bothrodendron*; the secondary xylem in *Bothrodendron* is not

[1] Williamson (89) A.   [2] Weiss, F. E. (08).

broken up by broad medullary rays as in the common *Stigmaria,* and in *Bothrodendron* the occurrence of a ring of primary xylem is another peculiarity.

In the vegetative shoots of *Bothrodendron mundum* the stele differs from those of *Lepidodendron* in the narrower primary xylem ring and in the large size of the metaxylem tracheae; from *Lepidodendron Harcourtii* and *L. fuliginosum* the xylem is distinguished by its smoother outer face which consists of numerous narrow xylem elements.

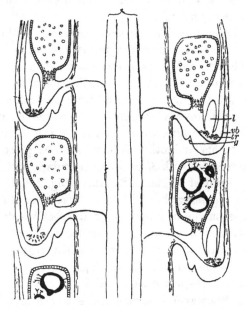

Fig. 216. *Bothrostrobus. l,* ligule. (After Watson.)

Cones of Bothrodendron (*Bothrostrobus*[1]).

The long and narrow cones referred to *Bothrodendron minutifolium* from English and French Coal-Measures are known only as impressions and it is not possible to say whether they were heterosporous or homosporous; the drawing given by Zeiller (Fig. 212, E) shows that the sporangia were of the same

[1] Nathorst (94) A. p. 42.

form as those in *Lepidostrobus*, but we have no more exact information as to their morphology. A recently published description of a petrified strobilus by Mr Watson affords a welcome addition to our knowledge. There is little doubt that this cone was borne by a species of *Bothrodendron*; the evidence for this conclusion is supplied by the agreement of the anatomical characters of the stele with that of the vegetative shoots originally described by Williamson as *Lepidodendron mundum* and by the constant association of the cones and vegetative shoots. In 1880 Williamson described a crushed cone containing both megaspores and microspores which he spoke of as " a diminutive organism, reminding us more of the dwarfed fruits of many living Selaginellas than of the large *Lepidostrobi*[1]." Watson's specimens enable us to give a more complete account of this type. The axis of the strobilus bears short sporophylls bent upwards into a distal limb with a conspicuous ligule in a deep pit beyond the shortly stalked sporangium. The length of the strobilus is estimated at 10 mm.; the stele is of the same type as that of *Bothrodendron mundum*, but it differs from the specimens of the vegetative shoots so far found in having some secondary xylem. As shown in the sketch reproduced in fig. 216 each sporophyll is characterised by two tangentially placed grooves, *g*, on the lower face, and by numerous transfusion tracheids, *tr*, above the vascular bundle, *vb*, immediately below the ligule, *l*. Megasporangia and microsporangia occur on the same cone, the megasporangia being on the lower sporophylls and containing a single tetrad of megaspores. Fig. 219, E, shows a radial longitudinal section of a microsporophyll bearing a sporangium on the adaxial side of the ligule, *l*, below which is the single vascular bundle and a group of short tracheids at *t*. The sporangia closely resemble those of species of *Selaginella* and *Lycopodium* and, as pointed out by Watson[2], they also recall the sporangia of the Palaeozoic genus *Spencerites*. *Bothrostrobus* is distinguished from *Spencerites* by the presence of a ligule, by the structure of the axis, and by the different form of the sporophylls. The occurrence of four spores only in the megasporangia is another character in which the extinct type

---

[1] Williamson (80) A. p. 500, Pl. xv. 8.　　　[2] Watson (08[2]) p. 12.

resembles recent Lycopods. It is impossible to decide whether
Watson's cone represents a more or a less primitive type than
Lepidostrobus: if we accept Professor Bower's views in regard
to the evolution of vegetative organs by the sterilisation of
sporogenous tissue, we should probably place *Lepidostrobus* lower
in the series than *Bothrostrobus*; but the greater resemblance
between the fertile and vegetative shoots of *Bothrodendron*, as
compared with the more pronounced difference in the case of
Lepidodendron, may be regarded as an argument in favour of
recognising *Bothrodendron* as the more primitive type.

Another possible example of a *Bothrodendron* cone has been
described by Nathorst from Spitzbergen as *Lepidostrobus
Zeilleri*[1]; this appears to consist of an axis bearing spirally
disposed sporangia without any indication of sporophylls. This
strobilus may belong to *Bothrodendron tenerrimum*.

Pinakodendron.

The name *Pinakodendron*[2] was instituted by the late Prof.
Weiss for a type of stem closely resembling *Bothrodendron* but
differing in the presence of a fine reticulation on the outer bark
and in the form of the leaf-scars. Weiss's genus has been recognised by Kidston in Dumfriesshire but our knowledge of the
plant is as yet based solely on a few small specimens.

Omphalophloios (a genus of uncertain systematic position).
Figs. 193, C, 217.

This generic name was instituted by White[3] for certain
specimens of large stems originally described by Lesquereux
from the Coal-Measures of North America as *Lepidodendron
mammillatum* and *L. cyclostigma*. The photograph reproduced
in fig. 193, C, for which I am indebted to Dr Kidston[4], represents
a specimen described by him from the Upper Coal-Measures
of Somerset as *Omphalophloios anglicus*, and identified with
Lepidodendron anglicum of Sternberg.

The surface of the impression shown in fig. 193, C, is

---

[1] Nathorst (94) A. p. 42, Pl. XII. figs. 8—10.   [2] Kidston (03) p. 797.
[3] White (98); (99) p. 218, Pls. LXV.—LXVIII.   [4] Kidston (02) pp. 358, 359.

characterised by clearly defined rhomboidal areas or cushions (fig.
217, E) like those of *Lepidodendron*, except in the absence of a
median keel, and similar to those on some forms of *Sigillaria
Brardi*. A short distance above the centre of each cushion is
an oval or subcordate region bounded by a rim-like margin and
containing a small oval scar, presumably that of a vascular strand.
A triangular elevation which also shows a small pit (Fig. 217,
E, *a*) occurs below the oval area. The appearance of the
surface-features varies considerably on different parts of a single
specimen. Fig. 217, D, represents one of the numerous figures
published by White in his detailed account of the American
material. Each cushion bears a widely open V-shaped ridge,
which is described as a leaf-scar; above this is an oval area (2·5

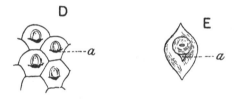

Fig. 217. *Omphalophloios.*
D. After White.   E.   After Kidston.

mm. × 1·75 mm.), the surface of which is bounded by a narrow
rim. Within the rim is a smaller concave oval region with a
small pit near its upper end.

We cannot, in the absence of petrified material, arrive at
any satisfactory conclusion as to the meaning of these surface-
features. White considers that *Omphalophloios* is probably a
rhizome of one of the arborescent Lycopods, but whether or
not this is its true nature must be left for future discoveries.
The fact that the rootlet bundles of some Stigmarian axes are
accompanied by a parichnos strand, as Weiss has shown, may
prepare us for the discovery of surface-features on *Stigmariae*
not unlike those of *Omphalophloios*. (Fig. 193, C.)

A possible comparison may be suggested also with *Sigillaria
Brardi* as figured by Germar (fig. 196, A) in which circular
scars, which may be the scars of rootlets, occur below the leaf-

base areas. It is not impossible that in the surface-features of *Omphalophloios* we have both leaf and rootlet scars represented.

General considerations.

The solid xylem core characteristic of the stele of some species of Palaeozoic Lycopodiales (e.g. *Lepidodendron esnostense* and *L. rhodumnense*) may probably, as Tansley and Chick[1] point out, be regarded as the lineal descendant of a primitive axial strand of water-conducting elements. In the course of evolution the centre of the tracheal column became partially converted into parenchymatous tissue, as in *Lepidodendron vasculare.* The arrangement of the short cells in regular vertical series is reminiscent of an early stage in the development of tracheae : instead of forming tubular conducting elements the central part of the stelar meristem acquired the short-celled form ; some of the cells became lignified as isodiametric storage tracheae while others persisted as thin-walled parenchyma.

The production of secondary xylem and an increase in the girth of the whole stem led to reduction in the amount of centripetally developed conducting channels. Some of these assumed a new rôle and a shape in harmony with their functions. A later stage is represented by a further encroachment of the central parenchyma on the cylinder of centripetal xylem, as seen in *Lepidodendron Harcourtii* and other species. The next stage is afforded by ribless species of *Sigillaria* in which the primary xylem is broken up into separate conducting strands. As Kidston[2] reminds us, it is in the geologically more recent species of *Sigillaria*, such as *S. Brardi*, which persist into the Permian era, that this more extreme case of reduction occurs. The older genus *Lepidodendron* seems to have retained to the last the complete cylinder of primary xylem. In the stele of *Stigmaria*, the rhizome of *Sigillaria* and of *Lepidodendron*, reduction of the centripetal xylem has passed beyond the stage represented by the broken cylinder of the ribless Sigillarias. With the exception of the examples described by Renault[3] and by Weiss[4], *Stigmaria* is characterised by little or no centripetal

[1] Tansley and Chick (01) p. 36.   [2] Kidston (05) p. 547.
[3] Renault (96) A.   [4] Weiss, F. E. (08).

primary xylem. It is, however, noteworthy that Renault's *Stigmaria*, in which centripetal xylem forms a prominent feature, is attributed to *Sigillaria Brardi*, a species in which the vascular cylinder of the aerial stem illustrates a later and not an earlier phase in the replacement of centripetal by centrifugal wood.

It would seem, as Lady Isabel Browne[1] says, that most Stigmarian axes had reached a more advanced stage in specialisation than is shown in the stelar structure of the aerial shoots. The relatively greater and probably the more precocious development of secondary xylem in *Stigmaria* than in *Lepidodendron* or *Sigillaria* may have some significance in relation to the smaller amount of "old wood[2]" (in a phylogenetic sense) in their steles.

As is pointed out in a later chapter, recent researches into the anatomy of extinct members of the Osmundaceae by Kidston and Gwynne-Vaughan have brought to light a striking parallelism in evolutionary sequence between the Lepidodendreae and the ancestors of *Osmunda* and *Todea*, the two surviving genera of one of the most ancient families of ferns.

There can be little doubt as to a very close relationship between *Sigillaria, Lepidodendron*, and *Bothrodendron*. *Sigillaria* seems to have outlived *Lepidodendron* and *Bothrodendron*. The two latter genera are recorded from Upper Devonian rocks in several localities, *Bothrodendron* being particularly abundant in the pre-Carboniferous floras of Bear Island and other parts of the world. A remarkable stem described by Dr White[3] as *Archaeosigillaria primaeva* from Upper Devonian shales of New York is spoken of by him as "one of the most highly developed representatives of a fairly distinct archaic group foreshadowing the later genera *Bothrodendron, Sigillaria, Lepidodendron* and *Lepidophloios.*" The type-specimen, when first discovered, consisted of an apparently unbranched stem reaching a length of 5 metres. From the swollen basal part

---

[1] Browne (09) p. 25.

[2] Scott (02) uses the terms old and new wood in discussing the evolutionary sequence in plant steles.

[3] White (07).

Stigmaria-like rootlets spread into the surrounding shale. At a higher level the fissured bark shows indistinctly defined leaf-cushions which pass gradually upwards into cushions and scars arranged in closer order on regular vertical ribs. The surface-features in this region are practically those of a ribbed *Sigillaria*. Traced farther upwards the vertical ribs die out and cushions of the Lepidodendroid form cover the surface of the bark. The leaf-scars, with a supraposed ligular pit and two vertically elongated parichnos-scars, are said to bear a closer resemblance to those of *Sigillaria* and *Bothrodendron* than to the leaf-areas of *Lepidodendron*. Nothing is known as to the anatomy of this stem, nor have fertile shoots been discovered. In the absence of more trustworthy evidence than is available conclusions of a phylogenetic nature must be accepted at their true value. It is however legitimate to describe *Archaeosigillaria primaeva* as one of the oldest examples of a lycopodiaceous plant which shows well-preserved external features, and these are of exceptional interest as indicating a combination of generic characters. This Devonian type lends support to the view that *Lepidodendron* and *Sigillaria* are offshoots, differing from one another in comparatively unimportant points, from a common ancestral type.

The generally accepted statement that arborescent Palaeozoic Lycopodiales bore their sporangia on specially modified leaves (sporophylls) grouped in cones which were usually produced at the tip of slender branches, has recently shared the fate of most rules. Prof. Bower in his *Origin of a Land Flora* mentions a Belgian specimen of *Pinakodendron musivum* Weiss from the Westphalian series (Middle Coal-Measures), to be described by Dr Kidston, which bore its sporangia " associated with the leaves of certain portions of the stem, without any cone-formation. The fertile and sterile portions are distinguished only by the presence or absence of sporangia[1]."

Lepidodendron and *Sigillaria* can hardly be claimed as the direct ancestors of any existing type of Lycopodiales, but while exhibiting points of contact with *Lycopodium*, *Selaginella*, and *Psilotum* they are perhaps more closely allied to *Isoetes*.

[1] Bower (08) p. 305.

Lady Isabel Browne[1], who has recently published an excellent summary of the evidence on the relation of the Lepidodendreae to *Isoetes*, concludes her examination of the arguments by expressing the opinion that there is a strong probability of the correctness of the view that *Isoetes* may be derived "from the Lepidodendraceae in the widest sense of the word." This decision seems to me to accord best with the facts.

The further question as to the relation of these Palaeozoic genera to plants higher in the scale must be reserved for fuller consideration in another volume. An attempt will also be made to consider how far anatomical structure may be used as a guide to the conditions under which *Lepidodendron* and *Sigillaria* as well as other members of the Permo-Carboniferous floras passed their lives. The secondary xylem of *Lepidodendron* and *Sigillaria* affords a striking example of water-conducting tissue of homogeneous structure comparable with the wood of Conifers rather than with that of Angiosperms. It was presumably formed, for the most part, under uniform climatic conditions: the absence of rings of growth points to uninterrupted supply to evergreen shoots exposed to no alternation of activity and arrested growth. Attention has already been called to the absence of any tissue corresponding to secondary phloem. Even in young shoots of *Lepidodendron,* no tissue has been found external to the meristematic zone agreeing in the form of its elements with the channels through which the elaborated food is conveyed from the leaves of recent plants to the regions of cell-building. That the ' secretory zone ' may have served this purpose, at least in young stems, is not improbable. On the other hand, it is difficult to understand why older Lepidodendron stems show no indication of additions to the secretory zone. If this tissue served for the transport of proteids we should expect to find provision made for its constant renewal *pari passu* with the secondary growth of the xylem. The conclusion seems to me inevitable that the supply of building-material was otherwise provided for than in recent vascular plants. The physiological division of labour may have been less complete in the tissue-systems of the Palaeozoic Lycopods than in the more highly

[1] Browne (09) p. 37.

specialised organs of such an extinct genus as *Lyginodendron* or
than in recent plants. Our knowledge of the anatomical structure
of many extinct types has already reached a stage when we
should take greater heed of the *modus operandi* of the complex
machinery revealed by a study of petrified stems. From the
known we proceed to interpret the unknown; but there is a
danger of neglecting the possibilities of evolution during the
countless ages which separate the forests of the Coal period from
those of the present era. We may easily allow preconceived
ideas to warp our judgment in attempting to distribute the
manifold activities which made up the life of a *Lepidodendron*
among the structural units of the plant-body.

CHAPTER XIX.

Seed-bearing plants closely allied to members of the Lycopodiales.

i. *Lepidocarpon.*

IN 1877 Williamson[1] published an account of some fossil seeds which he referred to Brongniart's genus *Cardiocarpon*[2], a generic title for certain Gymnospermous seeds. Some of these he identified, on the authority of the author of the species, with *Cardiocarpon anomalum* Carruthers[3]. Several years later Wild and Lomax described a new type of strobilus from the Lower Coal-Measures of Lancashire[4]. The result of this discovery and of the subsequent examination by Scott of additional material, was to establish the fact that the seeds described by Williamson and generally accepted as Gymnospermous, are in reality sporangia belonging to a Lycopodiaceous cone. The seeds to which Carruthers gave the name *Cardiocarpon anomalum* are, however, distinct from those described under the same name by Williamson and are those of a true Gymnosperm. For this seed-bearing strobilus Scott[5] instituted the generic name *Lepidocarpon*, which he thus defined: "Strobili, with the characters of *Lepidostrobus*, but each megasporangium was inclosed, when mature, in an integument, growing up from the superior face of the sporophyll-pedicel. Integument, together with the lamina of the sporo-

---

[1] Williamson (77) and (80) A. [2] Brongniart (28) A. p. 87.
[3] Carruthers (72³). [4] Wild and Lomax (00).
[5] Scott (01).

phyll, completely enveloping the megasporangium, or nucellus, leaving only an elongated, slit-like micropyle above. A single functional megaspore or embryo-sac developed in each mega-sporangium, occupying almost the whole of its cavity. Megaspore ultimately filled by the prothallus or endosperm. Sporophyll, together with the integumented megasporangium and its contents, detached entire from the axis of the strobilus, the whole forming a closed, seed-like, reproductive body. Seed-like organ horizontally elongated, in the direction of the sporophyll-pedicel, to which the micropylar crevice is parallel."

Lepidocarpon Lomaxi, Scott.   Fig. 218.

An immature cone of *L. Lomaxi* is practically identical with a *Lepidostrobus*; its sporangia are naked and only acquire their integuments at a later stage. A mature strobilus has a diameter of at least 3 cm. and is about 4 cm. in length. As in typical *Lepidostrobi*, the axis bears spirally disposed sporophylls, and each sporophyll has a long narrow pedicel approximately at right angles to the cone axis with its distal end expanded into a broad and thick lamina (fig. 218, B).

At the distal end the pedicel has a thin marginal wing (fig. 218, C, right-hand half) continuous with the upturned protective lamina. To the upper face of each sporophyll is attached along the whole length as far as the ligule, a single large sporangium; on each side of the base of the sporangium the sporophyll forms a supporting cushion. The relation of the sporangium to the ligule, *l*, is shown in fig. 218, B, and in the tangential section, C, which illustrates the triangular form of the sporangium near its distal end.

In mature cones, the sporangia assumed the form of seeds, the change being due to the growth of an investing integument from the upper face of the sporophylls on each side of the sporangia. Fig. 218, A, illustrates the form of a sporangium as shown in tangential sections; the vascular bundle is seen below the base of the sporangium and the gaps right and left of it probably mark the position of parichnos strands. On each side of the sporangium, *b*, a fairly thick wall of tissue has grown

up from the sporophyll, forming an integument which overtops
the apical ridge of the sporangium, leaving a narrow micropyle in
the form of a long crevice (*m*, fig. 218, B). At the proximal end
of the sporangium the integument forms an enclosing wall; at
the distal end it abuts on and is continuous with the upturned
end of the sporophyll. It is clearly established by Scott that the

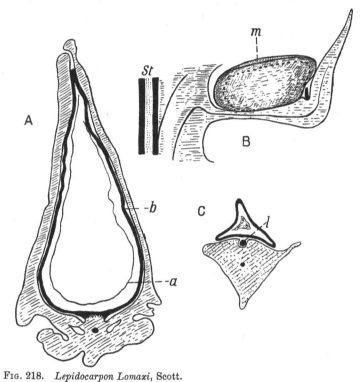

Fig. 218. *Lepidocarpon Lomaxi*, Scott.
           A and C. After Scott.
           B.   Diagram of a single sporophyll : *m*, micropyle ; *St*, stele.

tissue which invests the sporangia is not the upturned margins
of the sporophyll, but a new formation fully entitled to the
designation integument. It is noteworthy that the integument is
not developed until a late stage in the ontogeny of the strobilus ;
it is not formed until after the production of the prothallus[1].

---

[1] Letter from D. H. Scott (March 30, 1908).

The diagrammatic sketch, fig. 218, B, shows the relation of the integument to the sporophyll and sporangium, the outline of the latter being indicated by a broken line. The columnar wall of the sporangium (fig. 218, A, *b*) forms a closed beak within the micropylar crevice, and in the interior of the sporangial cavity the slightly shrivelled membrane, *a*, represents the single megaspore; traces of the aborted sister-cells of the megaspore are occasionally met with. Scott describes a specimen in which the megaspore is filled with tissue agreeing in appearance with the prothallus in a megaspore of *Isoetes* or *Selaginella*; no undoubted archegonia or female organs have been discovered, nor has any spore been found containing an embryo.

The axis of *L. Lomaxi* has a medullated stele constructed on the same plan as that of some species of *Lepidodendron* and *Lepidostrobus*; the vascular bundles supplying the sporophylls pass obliquely upwards and outwards from the stele, *St*, fig. 218, B, and bend slightly downward just before entering the pedicel of a sporophyll.

Dr Scott has also described a strobilus containing microsporangia partially enclosed by a rudimentary integument. It is, however, of considerable interest to find a partial development in the case of a male flower of an integumentary outgrowth, which it would seem could only be of real functional importance in the female shoot.

It is important to notice that specimens of a second species of *Lepidocarpon*, *L. Wildianum*, are recorded from Lower Carboniferous beds of Scotland, a fact which points to a considerable antiquity for this seed-bearing Lycopodiaceous type[1].

The most important question to consider in regard to *Lepidocarpon* is—are we justified in applying to the integumented sporangia the term seed? The megaspore was not set free as it is in recent Pteridophytes, such as *Azolla* and other genera with which *Lepidocarpon* may be compared; it was on the other hand retained in the sporangium, as may sometimes happen even in recent species of *Selaginella* (cf. fig. 131, D). Moreover, the megaspore is characterised by a thin

---

[1] Scott (01) 314.

enclosing membrane in contrast to the thick coat of a spore which is destined to be shed. The peculiar slit-like form of the micropyle is a distinguishing feature, but this may be readily explained as a convenient form in the case of a radially elongated sporangium. The absence of an embryo, though a distinguishing feature of *Lepidocarpon*, cannot be held to be a serious obstacle to the use of the term seed; in recent Cycads the embryo, as Scott points out, may not begin to develope until the seed has been shed. It is possible that the seeds of *Lepidocarpon* were not pollinated on the parent plant.

The lesson which this extinct type teaches, is that certain Lycopodiaceous plants of the Palaeozoic era had reached an important stage in the evolution of a seed. The morphological essentials of true seeds had been acquired; but we do not know the biological conditions under which pollination and fertilisation were effected. Another point of considerable interest is the value of this discovery as an argument in favour of the view that some Gymnosperms are derived from Lycopod ancestors. Leaving the general question until later, it may at any rate be stated that in *Lepidocarpon* we have a demonstration of the fact that the Lycopodiales were not always distinguished from Gymnosperms by the absence of seeds. There are certain features in *Lepidocarpon* shared by the seeds of Araucarieae[1] which may well mean something more than mere parallel development in two distinct phyla of the plant-kingdom[2].

ii. *Miadesmia*.

In 1894 Prof. Bertrand[3] published an account of certain fragments of petrified leaves and twigs of a small herbaceous Lycopodiaceous plant, under the name *Miadesmia membranacea*, which he discovered in English material in association with *Lepidodendron Harcourtii*. Subsequently Scott recognised the megasporophylls of the same plant, and microsporophylls have also been discovered. The most complete account of

[1] Seward and Ford (06).   [2] For a contrary opinion, see Scott (09) p. 656.
[3] Bertrand, E. (94).

Miadesmia so far published we owe to Dr Benson[1], whose description is based on specimens from several sources.

Miadesmia membranacea, Bertrand.   Fig. 219, A—D.

The slender stem, characterised by unequal dichotomy, has a single protostele composed of scalariform tracheids with 3—6 peripheral protoxylem groups.   A zone of delicate tissue sur-

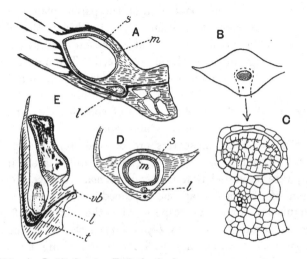

Fig. 219.   A—D *Miadesmia*; E *Bothrodendron*.
A.   Radial section of megasporophyll: *s*, sporangium; *m*, megaspore; *l*, ligule. (From a drawing kindly lent by Mrs D. H. Scott.)
B, C.   Leaf with ligule. (From a section in Dr Kidston's Collection.)
D.   Transverse section of sporophyll. (After Scott.)
E.   Radial section of microsporophyll of *Bothrodendron*. (From a section in the Manchester Museum; Hick Collection R. 406.)

rounds the xylem; this is described as phloem, but it is not clear whether the designation is based on histological characters or primarily on its position.   The cortex consists of an inner lacunar tissue and an outer region limited by a small-celled superficial layer sharply contrasted with the underlying layers

[1] Benson (08).

of larger cells. The stem of *Miadesmia* is not uncommon in sections of the Lancashire calcareous nodules, and may be recognised by the delicate crushed tissue of which it mainly consists and by large hypodermal parenchyma. The spirally disposed leaves bear a conspicuous and relatively large ligule, 3 mm. long, in a deep pit (fig. 219, B and C) roofed over by a few layers of tissue corresponding to the velum in *Isoetes* (cf. fig. 133, E, v). The fairly thick central region of the lamina is expanded laterally into thin wings, which in the living state probably bore delicate hairs. These delicate leaves, apparently without stomata, were attached to the stem at an acute angle, and Miss Benson suggests that their form ano arrangement may have enabled them to hold water by surface-tension. As seen in fig. 219, B, C, which represents part of a transverse section near the leaf-base, the ligule is a very characteristic feature, and the size of the single vein is in keeping with the almost filmy nature of the lamina.

In addition to the sections in British collections, I have been enabled by the kindness of Prof. Bertrand to see photo-micrographs of the sections on which he founded the genus. One of these sections, transverse to the stem and leaves, illustrates in a striking manner the relatively large size of the leaves and ligules in proportion to the delicate axis of the shoot.

The megasporangiate cone has an axis which agrees in its structure with that of the vegetative stem and bears several megasporophylls approximately at right-angles. As in the foliage leaves, the ligule is prominent and large, and lies in a groove which contains also the megasporangium; both ligule, *l*, and sporangium, *s*, as seen in the transverse section represented in fig. 219, D, are covered by an integument or velum which arises in the proximal part of the leaf and leaves a circular micropylar opening at the beak-like apex of the sporangium. The circular micropyle is surrounded by numerous hairs borne on the integument and which presumably played the part of a feathery stigma. A single megaspore with a thin membrane, *m*, abuts on the fairly strong sporangial wall, *s*; in some cases the sporangium and megaspore walls may be indistinguishable,

a feature suggesting comparison with seed-structure. Some
megaspores have been found filled with a prothallus. The
longitudinal section shown in fig. 219, A, illustrates the characteristic horizontal position of the megasporophyll, as also the
relation of the ligule, *l*, to the sporophyll with its single vascular
bundle, and to the hairy integument, which overarches both
sporangium and ligule; the line *m* shows the position of the
megaspore-membrane, detached from the sporangial wall on the
upper side but in contact with it below. The microsporophyll
shown in 219, E, was originally referred to *Miadesmia* but has
since been recognised by Watson[1] as that of a *Bothrostrobus*.

Miadesmia affords an example of a Palaeozoic plant comparable with *Isoetes* and *Selaginella*; it agrees also with
Lepidocarpon in possessing true seeds, and with Watson's
Bothrodendron cone in the shape of the sporangia, which are
more like those of *Selaginella* than the radially elongated
sporangia of *Lepidostrobus*. *Miadesmia* agrees with *Selaginella*,
e.g. *S. spinosa*, in its stelar structure, in the form of the
sporangia, and in the presence of a ligule. It is distinguished
by having only one instead of four megaspores in a sporangium,
in the possession of an integument which formed a close investment to the spore and served as a stigma (comparable with
the stigma-like integument of the male flower of *Welwitschia*),
and in the shedding of the megasporophylls, which have been
aptly compared with winged seeds.

On the ground of their general anatomical features *Lepidocarpon* and *Miadesmia* are clearly entitled to be included among
extinct representatives of the Pteridophyta. These plants had,
however, crossed what it has been customary to regard as the
boundary between Pteridophytes and Phanerogams: they
possessed megasporangia with the attributes of seeds. It has
been suggested by Lester Ward[2] that Pteridophytic seed-bearing
plants shall be recognised as a distinct phylum for which he
proposes the name Pteridospermaphyta, a designation implying
exclusion from the Spermatophyta as usually understood. For
seed-bearing Lycopodiaceous genera he suggests the name
Lepidospermae. As knowledge of the Palaeozoic seed-plants

[1] Watson (08²) p. 12.   [2] Ward (04).

increases revision of existing classifications and group names will become necessary, but as yet we are hardly in a position to draw up a satisfactory scheme of grouping; we know little of *Lepidocarpon* as a whole and it would be premature to commit ourselves, even provisionally, to a classification which is based on such meagre evidence as we possess. Moreover the value to be attached to the seed-habit as a basis of classification can hardly be estimated until fuller information is obtained.

CHAPTER XX.

FILICALES.

THIS division of the Pteridophyta includes both the true ferns (Filicineae) and the less familiar water-ferns or Hydropterideae. The almost complete absence of satisfactory evidence in regard to the geological history of the latter renders this group of secondary importance from a palaeobotanical standpoint, but, on the other hand, we possess a wealth of material bearing on the past history and relative antiquity of the true ferns.

The study of extinct types has so far rendered no substantial help towards bridging the wide gap between the Filicales and the lower plants. As Mr Tansley[1] says in his admirable lectures on *The Evolution of the Filicinean Vascular System,* "The biggest gap in the plant kingdom at the present time is undoubtedly that which separates the Pteridophytes from the plants definitely below them in organisation, and directly we try to step behind the ferns we tumble into this abyss." Resemblances long ago recognised between certain ferns and the cycads, a section of the Gymnosperms, were regarded by a few botanists as indications of blood-relationship, and the results of recent researches into the morphological characters of extinct Palaeozoic types are generally held to confirm these surmises. Prof. Chodat[2] of Geneva has recently challenged the validity of the arguments on which the affinity of cycads and ferns has been accepted by the great majority of botanists. Whether or not his criticisms stand the

---

[1] Tansley (08) p. 3. Cf. Braun (75) p. 267.　　　　[2] Chodat (08).

test of unbiassed examination, they must at least lead us to substitute a critical consideration of the facts for a mere repetition of conclusions which appeal to our imagination. Despite Prof. Chodat's warning, we may still quote with confidence a phrase used in another connexion—ferns "are links in a chain and branches on the tree of life, with their roots in a past inconceivably remote[1]."

Transitional forms which are regarded as pointing to a common origin for ferns and cycads are known in abundance; other types have also been discovered which lead some authors to go so far as to derive the whole of the seed-bearing plants from an ancestry the descendants of which are represented by existing ferns. While hesitating to allow the ferns or fern-like plants the peculiar position of universal ancestors, we must admit that there is no group of plants with a history of greater importance from an evolutionary standpoint than that with which we are now concerned.

There are, however, some difficulties to face in attempting to decipher the history of the Filicineae as recorded in the earth's crust. Few fossil plants are so familiar as the well-preserved carbonaceous impressions of compound leaves on the shales of our Coal-Measures, which were referred by older authors to recent genera and species of ferns and accepted by later writers as undoubted examples of Palaeozoic ferns. The common belief in the dominance of ferns in Palaeozoic floras is reflected in the novelist's description of the Carboniferous period, "when the forms of plants were few and often of the fern kind[2]" We now know that very many of these Carboniferous leaves belonged to plants differing widely in morphological characters from the modern genera to which they exhibit so deceptive a resemblance. These pseudo-ferns, recently christened Pteridosperms or seed-bearing fern-like plants, are dealt with in a later chapter. The discovery of this extinct group has added enormously to our knowledge of plant-evolution and at the same time has rendered much more difficult the task of unravelling the past history of the true ferns. As soon as it was demonstrated that many familiar Palaeozoic "ferns" are not

[1] Hudson (92) p. 29.　　[2] Hardy, *Return of the Native*, II. p. 153.

ferns, some authors went far towards concluding that however close might be the agreement between fossil and recent leaves suspicion of close relationship must be set aside. Like the earlier writers who described fossils as *lusus naturae* fashioned by devilish agency to deceive too credulous man, the discovery of seed-bearing plants with the foliage of ferns threatened to disturb the mental balance of palaeobotanists. The fact is, we cannot in some cases determine from leaf-form alone whether or not a fossil is a true fern; we may, as Professor Bower[1] suggests, regard all fern-like fossils as ferns until they are proved to be Pteridosperms, or in a spirit of scientific scepticism, we may at once admit that many Palaeozoic fern-like leaves must await further evidence before their true position can be determined. It is impossible, as Zeiller[2] says, in the present state of our knowledge to range fern-like Palaeozoic plants in two groups, one referred to Filicineae and the other to the Pteridosperms.

The following classification of the Filicales is based on that adopted by Prof. Engler in the latest edition of his *Syllabus*[3] and on the results of Bower's[4] excellent work on the spore-bearing members of recent ferns.

The members of the Filicales are characterised by the same well-marked physiological division of labour in their vegetative parts as are the Lycopods; the plant is the asexual generation (sporophyte), while the sexual generation (gametophyte) is small and inconspicuous, either an independent green prothallus or a tissue more or less completely enclosed in the spore. The large size of the leaves, which in the young state are usually coiled like a crozier (fig. 220, A), is a striking characteristic of the ferns; they are megaphyllous in contrast to the microphylly of the Lycopods.

---

[1] Bower (08).   [2] Zeiller (06) p. 8.
[3] Engler (09).   [4] Bower (00).

I. Leptosporangiate Filicales.

In these homosporous and heterosporous plants the sporangia are developed from single epidermal cells.

(a) *Eufilicineae.* The sporangia bear spores of one kind only ; the wall of a sporangium consists of one layer of cells. In

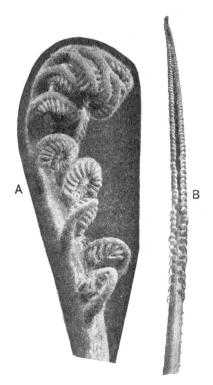

FIG. 220. Young fronds of (A) *Angiopteris evecta* and (B) *Cycas revoluta.*
(Reduced.)

the great majority of cases the sporangia are characterised by the possession of a conspicuous row of thick-walled brown cells, the annulus[1], which serves as a mechanism for dehiscence and spore-dispersal. The fertile leaves, identical in form with

[1] For an account of the mechanism of spore-dispersal, see Goebel (05) p. 587 ; Atkinson (94); Leclerc du Sablon (85); and Bower (00).

the sterile, or more or less sharply contrasted, usually bear the
sporangia on the under surface of the lamina in definite groups
or sori, and not on the upper surface or grouped in strobili as
in the Lycopodiales. The stem is dorsiventral or radial in
structure, creeping or erect, frequently clothed with chaffy
scales (ramenta) and less often with multicellular hairs. The
sexual generation is represented by a small green prothallus
which lives for a short period only and dies after nursing the
fern-plant through its earliest stages.

(b) *Hydropterideae.* Heterosporous water-ferns differing
considerably in habit from the true ferns. Each megasporangium contains a single megaspore and several microspores are
produced in each microsporangium. The gametophyte is represented by tissue more or less enclosed in the spore. [Genera
Salvinia, Azolla, Marsilia, Regnellidium, Pilularia. See
Chapter XXVI.]

(a) *Eufilicineae.* The classification of the true ferns in
common use is based almost exclusively on the structure of the
sporangium, the form and position of the sori, and on the
presence or absence of an indusium (the tissue which in some
ferns partially or completely covers each sorus). In recent
years there has been considerable activity in the investigation
of fern anatomy with a view to elucidating the natural relationship between recent families or genera. The results of these
researches are on the whole consistent with the scheme and
grouping adopted in the *Synopsis Filicum* of Hooker and Baker
and in general harmony with the main conclusions arrived at
by Bower from an intensive study of the development of fern
sporangia. The following classification is based on that of
Bower who takes as a basis (i) the relative time of appearance
of the sporangia in a single sorus, (ii) the structure of the
sporangia and their orientation relative to the whole sorus,
(iii) the productiveness of sporangia (spore-output).

Osmundaceae ⎫
Schizaeaceae ⎬ *Simplices* (Bower). The sporangia are relatively
Gleicheniaceae ⎪ large and all the sporangia in a sorus have a
Matonineae ⎭ simultaneous origin: the annulus is oblique.

Loxsomaceae
Hymenophyllaceae
Cyatheaceae
Dennstaedtiinae

Gradatae (Bower). Sporangia arise in basipetal succession on a more or less elongated receptacle (portion of the leaf lamina which projects as a cushion or column on which the sporangia are borne); annulus oblique; indusium, if present, in the form of a cup or flap of tissue arising from the base of the sorus.

Polypodiaceae
Parkeriaceae

Mixtae (Bower). This division includes the Polypodiaceae, by far the largest family of ferns. The sporangia are characterised by their relatively small size, the presence of a slender stalk, the absence of regular orientation or sequence in development, and by the presence of a vertical annulus.

Dipteridinae

The Dipteridinae include species with the characters of the *Mixtae*, and one species in which the sporangia develope simultaneously (*Simplices*).

Osmundaceae[1]. (*Osmunda, Todea.*)

Sporangia large and rather stouter than those of other Leptosporangiate ferns, borne in small groups (filmy species of *Todea*) in linear and frequently confluent sori (*Todea barbara*; fig. 221, D) or clustered round the axis of modified fertile pinnae with much reduced lamina (*Osmunda*). The annulus is represented by a group of thicker-walled cells a short distance below the apex (fig. 221, C). This family stands apart among the ferns; in some respects, e.g. in the more robust sporangia occasionally forming synangia, and in the presence of stipular wings, it forms a transitional series between the Leptosporangiate and Eusporangiate ferns. The only European species of *Osmunda*, *O regalis*, is almost cosmopolitan in range; other species occur in North and South America, in the Far East, the Malay Peninsula, and in other regions, more especially in the temperate zones. *Todea* is represented by (i) the South African and Australian species, *T. barbara*, a fern with a stem, which may reach a height of several feet, thickly covered with adventitious roots and bearing large and somewhat leathery

[1] For a fuller account of recent ferns, see Engler and Prantl (02), Christ (97), Hooker and Baker (68), and Bower (00) (08).

fronds; (ii) filmy species in New Zealand, New South Wales, New Caledonia, and elsewhere. A plant of the small tree-fern *Todea Wilkesiana* (Fiji, Samoa, and other islands) in the filmy-fern house at Kew, to which my attention was drawn by my friend Mr A. W. Hill, has a slender stem with the characteristic leaf-scars exposed; it presents a striking similarity to some of the fossil species of Osmundaceae described in a later chapter.

Schizaeaceae. (*Schizaea, Aneimia, Lygodium, Mohria.*)

Sporangia borne singly and not in groups (sori), readily

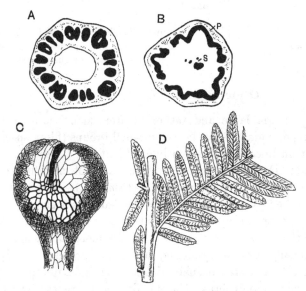

FIG. 221. A. *Osmunda cinnamomea* (after Faull).
B. *Todea barbara*, p, phloem; s, sclerenchyma.
C. *Osmunda regalis* (after Luerssen).
D. *Todea barbara* (½ nat. size).

recognised by the complete transverse apical annulus usually one layer of cells deep, but occasionally two layers in depth on the side opposite the line of dehiscence[1] (fig. 224, B). *Schizaea* (fig. 222) with the exception of one species in North America

---

[1] Prantl (81) Pl. VII. fig. 104, C; Zeiller (97) p. 215, figs. 7—10.

(*S. pusilla*) is characteristic of Northern India, the Malay region, Australia, New Caledonia, S. Africa, and elsewhere south of the Equator. *Aneimia* (figs. 223, 224, A, B), characterised by the fertile segments with reduced lamina, is chiefly American : the monotypic genus *Mohria*, resembling in habit the Poly-

Fig. 222. *Schizaea elegans*. (Slightly reduced.) A few of the segments terminate in narrow fertile lobes.

podiaceous genus *Cheilanthes*, occurs in S. Africa and Madagascar, while species of *Lygodium* are widely spread tropical ferns, with one species in temperate North America. This family has disappeared from Europe.

Gleicheniaceae [*Gleichenia, Platyzoma* (= *G. microphylla*)].

Sporangia form circular naked sori composed of a variable number of sporangia, usually not more than ten and frequently

FIG. 223. *Aneimia rotundifolia.* (From the Royal Gardens, Kew. ⅓ nat. size.)

fewer, characterised by an obliquely horizontal and almost complete annulus (fig. 224, I). In some species of *Gleichenia* (sect. *Eugleichenia*) the ultimate segments are very small and

semi-circular in form (fig. 226, C), in others (sect. *Mertensia*[1])
the segments are linear (fig. 226, D), and in many species the
fronds are distinguished by the regular dichotomous branching

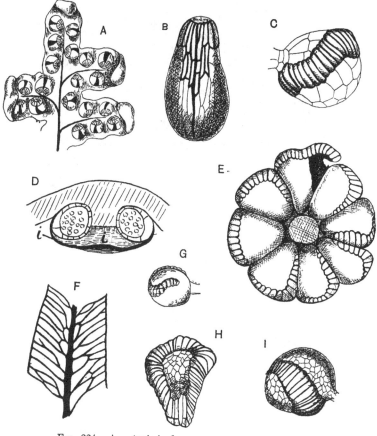

FIG. 224.　A.　*Aneimia flexuosa.*
　　　　　B.　*A. phyllitidis.*
　　　　　C.　*Hymenophyllum dilatatum.*
　　　　　D, E, F, G.　*Matonia pectinata*; *i*, indusium.
　　　　　H.　*Thyrsopteris elegans.*
　　　　　I.　*Gleichenia circinata.*
　　　　(A, B, after Prantl; C, G, H, I, after Bower.)

[1] Underwood (07), p. 243, has adopted Bernhardi's genus *Dicranopteris* in
place of *Mertensia* on the ground that the latter was used as early as 1793 for a
Boraginaceous plant.

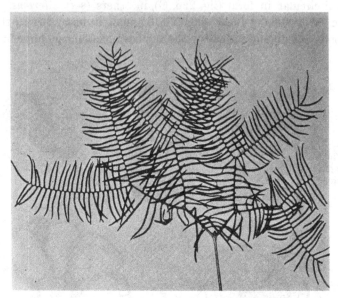

Fig. 225. *Gleichenia dicarpa.* (⅓ nat. size.)

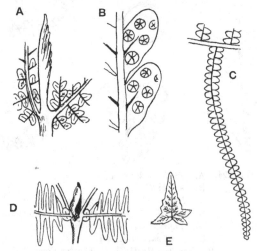

Fig. 226. A, B. *Gleichenites Rostafinskii*, Raciborski.
C. *Gleichenia dicarpa.* (Nat. size.)
D, E. *Gleichenia dichotoma.* (Reduced.)
(A, B, after Raciborski; C, after Hooker; D, E, after Goebel.)

(fig. 225), frequently showing an arrested rachis bud in the forks[1] protected by modified pinnules (fig. 226, D, E). In *Platyzoma* the leaves are simple, reaching a length of 20—30 cm., and bear small revolute oval segments.

Gleichenia is represented by several species in the tropics and extends to south temperate and Antarctic latitudes. The species *G. dichotoma* (= *G. linearis*) is one of the more successful tropical ferns, while *G. moniliformis* (by some authors recognised as a distinct genus, *Stromatopteris*) is peculiar to New Caledonia. The monotypic genus *Platyzoma* is a xerophilous Australian fern. The Gleicheniaceae are unrepresented in existing north temperate floras.

Matonineae. (*Matonia.*)

The genus *Matonia*, placed in the Cyatheaceae by Sir William Hooker and compared by other authors also with the Gleicheniaceae, is now included in a special family. The sori are circular and consist of 5—11 large sporangia (fig. 224, E, G) sessile on a central columnar receptacle which spreads out into an umbrella-like indusium (D, *i*) with its incurved margin tucked in below the ring of sporangia. The indusium is detached when the sporangia are ripe. The annulus is oblique and incomplete and often slightly sinuous; it agrees in the main with that of *Gleichenia*. The species *Matonia pectinata* is characterised by dichotomously branched fronds (figs. 227, 228) with long and slender petioles; the pinnae bear linear pinnules with forked lateral veins and occasional lateral anastomoses (fig. 224, F). The only other living representative is *M. sarmentosa*, discovered by Mr Charles Hose at Niah, Sarawak[2]: this species has long pendulous leaves apparently very different from those of *M. pectinata*, but the branching of the frond may be regarded as a modification of a primitive form of dichotomy[3]. A small bud occurs in the angle between the forked linear segments and the rachis, as in some species of *Gleichenia*[4]. *Matonia* is confined to

---

[1] Goebel (05) p. 318. [2] Baker (88).
[3] Diels, in Engler and Prantl (02) pp. 343, 344. [4] Compton (09).

the Malay region : *M. pectinata* grows in Western Borneo and in various localities in the Malay peninsula, while *M. sarmentosa*

FIG. 227.   *Matonia pectinata.*   (⅓ nat. size.)   M.S.

has been found in one locality only; the latter species has recently been transferred to a new genus *Phanerosorus*, but in view of the practical identity in anatomical structure and the close agreement as regards the sori of the two species there would seem to be no justification for this change of name[1].

Loxsomaceae.

The New Zealand genus *Loxsoma* has marginal sori with a cup-like indusium surrounding an elongated receptacle bearing

Fig. 228. *Matonia pectinata*. From a photograph by Mr Tansley of a group of plants in a wood on Gunong Tundok, Mount Ophir.

pear-shaped sporangia provided with a complete oblique annulus. The genus is chiefly interesting because of its isolated position; it agrees with *Trichomanes* (Hymenophyllaceae) in the structure of the sorus and with species of *Dicksonia* and *Davallia* in habit; it shows some resemblance also to Gleicheniaceae and Schizaeaceae[2]. A new type of fern described by Christ[3] from Costa Rica as *Loxsomopsis costaricensis* affords a striking instance of discontinuous distribution and emphasises the antiquity and generalised features of the family.

[1] Copeland (08) p. 344.   [2] Bower (00) p. 47; Gwynne-Vaughan (01).
[3] Christ (04).

Hymenophyllaceae. (*Hymenophyllum, Trichomanes.*)

The sporangia, which are attached to a columnar receptacle or prolongation of a vein beyond the margin of the lamina, are characterised by an obliquely transverse annulus (fig. 224, C). A cup-like indusium surrounds the lower portion of the receptacle which is two-lipped in *Trichomanes* and entire in *Hymenophyllum* (fig. 270, C, D). These two filmy ferns have a wide distribution both in tropical and extra-tropical regions; they

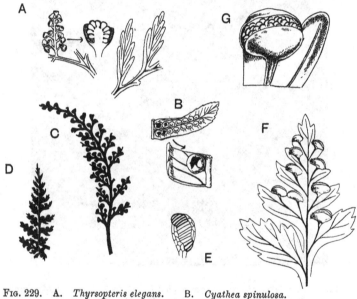

Fig. 229. A. *Thyrsopteris elegans.*   B. *Cyathea spinulosa.*
           C. *Davallia concinna.*   D. *Dicksonia coniifolia.*
           E. *Alsophila excelsa.*   F, G. *Dicksonia culcita.*
 (A, after Diels and Kunze; B, D, F, G, after Hooker; E, after Bower.)

are represented in the British Isles by *Hymenophyllum tunbridgense*, *H. Wilsoni*, and *Trichomanes radicans*.

Cyatheaceae. (*Cyathea, Hemitelia, Alsophila, Dicksonia, Thyrsopteris.*)

The sporangia occur in indusiate or naked sori and have an obliquely vertical and incomplete annulus (fig. 229, E). In the great majority of cases the fronds are large and highly compound, but *Cyathea sinuata* Hook, a rare Ceylon species, bears simple narrow linear leaves. This family includes, with few exceptions, all the tree ferns[1]. The sori of *Dicksonia* are enclosed in a two-valved indusium (fig. 229, F, G); in the species represented in fig. 230 the fertile segments, which terminate in cup-like indusia, are characterised by the absence of a lamina and closely resemble those of *Thyrsopteris* (fig. 229, A).

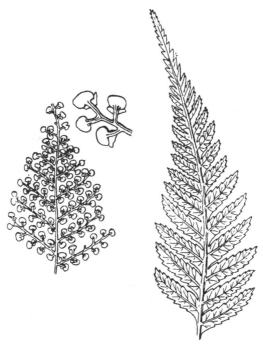

Fig. 230. *Dicksonia Bertercana* Hook. Fertile and sterile pinnae. (Nat. size. British Museum Herbarium.)

In *Cyathea* the indusium has the form of a cup which is at first closed and afterwards opens at the apex (fig. 229, B); in *Hemitelia* the indusium is much reduced and in *Alsophila* the sori are naked. *Thyrsopteris* is characterised by the reduced fertile pinnules bearing stalked sori in deep cups (fig. 229, A). The appearance of this fern "is very remarkable, for the cup-shaped

[1] Scott, J. (74); Hannig (98).

sori hang down from the fronds in masses, looking just like masses of millet seed[1]." The sporangia are described by Bower[2] as large and of rather peculiar form. As seen in fig. 224, H, the annulus is continuous; it forms a twisted loop of cells which vary in shape and in the thickness of the walls. The Cyatheaceae are for the most part tropical ferns with a wide geographical range, usually in moist regions; they are, however, able to flourish under widely different temperature conditions. In Tasmania, as Diels[3] points out, tree ferns may occasionally be seen laden with snow, and on the west coast of New Zealand they overhang the edge of a glacier[4]. The monotypic genus *Thyrsopteris* is confined to Juan Fernandez. The Cyatheaceae no longer exist in Europe.

Dennstaedtiinae. (*Microlepia, Dennstaedtia.*)

This sub-tribe, instituted by Prantl, has been revived by Bower on the ground that the sori present features intermediate between those of Cyatheaceae and the Polypodiaceous genus *Davallia*. The sporangia have a slightly oblique annulus.

Polypodiaceae.

This section of the Leptosporangiate ferns, including several sub-tribes, comprises the great majority of recent genera. The sporangia form naked or indusiate sori and have a vertical incomplete annulus. In *Plagiogyria*[5] the oblique annulus and soral features suggest comparison with the Cyatheaceae. A more intimate acquaintance with Polypodiaceous ferns will undoubtedly demonstrate the existence of other generalised types[6].

From the point of view of the identification of fossil ferns it is important to bear in mind the very close resemblance presented by some Polypodiaceous species, e.g. species of *Davallia* (fig. 229, C), to Cyatheaceous ferns (cf. fig. 229, D).

[1] *Challenger Reports* (85) p. 827. (Narrative, Pl. II.)
[2] Bower (00) p. 68.   [3] Diels (02) p. 117.
[4] Seward (92) p. 45.   [5] Bower (00) p. 80.
[6] Prof. Bower informs me that he is now at work on *Plagiogyria* and other Polypodiaceae.

Parkeriaceae. (*Ceratopteris.*)

The almost spherical and scattered sporangia are character-ised by the peculiar form of the vertical annulus, which is composed of numerous cells differing in their greater breadth

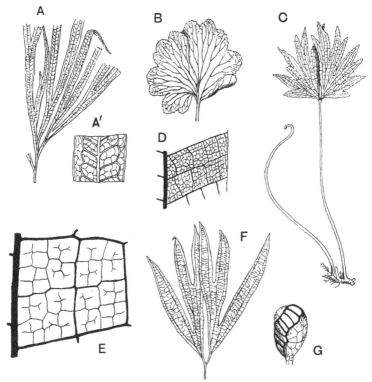

FIG. 231. A, A'. *Dipteris quinquefurcata* (type-specimen in the Kew Herbarium).
B, C, E, G. *D. conjugata.* (C, ⅛ nat. size.)
D. *Polypodium quercifolium.*
F. *Dipteris Wallichii.*
(D, after Luerssen.)

and smaller depth from those of a typical annulus. Exannulate sporangia have been described, while others occur showing different stages between a rudimentary and a complete ring. The single species of *Ceratopteris, C. thalictroides,* is an annual aquatic fern widely spread in tropical countries[1].

[1] Kny (75); Ford (02); Goebel (91).

Dipteridinae. (*Dipteris.*)

The genus *Dipteris*, formerly included in the Polypodiaceae, has been assigned to a separate family partly on account of the slight obliquity of the vertical annulus (fig. 231, G) and on other grounds[1]. The four species *Dipteris conjugata, D. Wallichii, D. Lobbiana* (= *D. bifurcata*), and *D. quinquefurcata* (fig. 231) are characterised by a creeping rhizome bearing fronds reaching a length of 50 cm.; in *D. conjugata* and *D. Wallichii* the lamina is divided by a median sinus into two symmetrical halves, while in other species the leaf is dissected into narrow linear segments. The main dichotomously branched ribs are connected by lateral branches and these by tertiary veins, the delicate branches of which end freely within the square or polygonal areolae (fig. 231, A', E). The naked sori are composed of numerous sporangia and filamentous hairs : while in some species the soral development conforms to that characteristic of the Mixtae, it has been shown that in one species, *D. Lobbiana* (= *D. bifurcata*[2]), the sporangia develope simultaneously as in the Simplices. *Dipteris* occurs in company with *Matonia* on Mt Ophir and elsewhere in the Malay peninsula; it extends to the Philippines, Samoa, New Caledonia, China, New Guinea, and the sub-tropical regions of Northern India.

The impossibility of drawing a hard and fast line between the divisions adopted in any system of classification is well illustrated by the ferns. In the main, the three-fold grouping suggested by Bower is probably consistent with the order of evolution of the true ferns. The Polypodiaceae, which are now the dominant group, are in all probability of comparatively recent origin, while the Gradatae and Simplices represent smaller sub-divisions with representatives in remote geological epochs. The genera *Loxsoma, Matonia* and *Dipteris* afford examples of ferns exhibiting points of contact with more than one of Bower's sub-divisions : they are generalised types which, like many relics of the past, are now characterised by a restricted geographical range.

[1] Seward and Dale (01).   [2] Armour (07).

It is noteworthy that while certain vegetative features may in some cases be cited as family-characters, such features are not usually of much value from a taxonomic point of view.

FIG. 232. *Davallia aculeata*. (⅔ nat. size.)

While the typical tree ferns are practically all members of the Cyatheaceae, a few members of other families, e.g. *Todea barbara* (Osmundaceae) and the monotypic Indian genus *Brainea* (Poly-

podiaceae), form erect stems several feet in height; but these
differ in appearance from the Palm-like type of the Cyatheaceous
tree ferns. On the other hand, the thin, almost transparent,
leaf of *Hymenophyllum tunbridgense* and other filmy ferns is
a character shared by several species of *Todea*, *Asplenium re-
sectum*, and *Danaea trichomanoides* (Marattiaceae); the filmy
habit is essentially a biological adaptation.

The form of frond represented by certain species of
Gleichenia, characterised by a regular dichotomy of the axis and
by the occurrence of arrested buds, is on the whole a trustworthy
character, though *Davallia aculeata* (bearing spines on its rachis)
(fig. 232) and *Matonia sarmentosa* have fronds with a similar
mode of branching and also bear arrested radius-buds. A limited
acquaintance with ferns as a whole often leads us to regard a
certain form of leaf as characteristic of a particular species, but
more extended enquiry usually exposes the fallacy of relying
upon so capricious a feature. The form of leaf illustrated by
Trichomanes reniforme is met with also in *Gymnogramme reni-
formis* and is fairly closely matched by the leaf of *Scolopendrium
nigripes*. The fronds of *Matonia pectinata* (figs. 227, 228) bear a
close resemblance to those of *Gleichenia Cunninghami*, *Adiantum
pedatum*, and *Cheiropteris palmatopedata*[1].

The habit, leaf-form, and distribution of Ferns.

The full accounts of the structure and life-history of the
common Male Fern, given by Scott in his *Structural Botany* and
by Bower in the *Origin of a Land Flora*, render superfluous
more than a brief reference to certain general considerations in
so far as they may facilitate a study of fossil types.

In size Ferns have a wide range: at the one extreme we
have the filmy fern *Trichomanes Goebelianum*[2], growing on tree
stems in Venezuela, with leaves 2·5 to 3 mm. in diameter, and
at the other the tree ferns with tall columnar stems reaching a
height of 40 to 50 feet and terminating in a crown of fronds
with a spread of several feet. A common form of stem is
represented by the subterranean or creeping rhizome covered

---

[1] Diels (02) fig. 98, p. 188.   [2] Giesenhagen (92) p. 179, fig. 3.

with ramental scales or hairs: the remains of old leaves may
persist as ragged stumps, or, as in *Oleandra, Polypodium vulgare*
and several other species, the leaf may be cut off by the formation of an absciss-layer[1] leaving a clean-cut peg projecting
from the stem. As a rule the branches bear no relation to the
leaves and are often given off from the lower part of a petiole,
but in a few cases, e.g. in the *Hymenophyllaceae*, it is noteworthy
that true axillary branching is the rule[2]. In the typical tree-
fern the surface resembles that of a Cycadean trunk covered
with persistent leaf-bases and a thick mass of roots. Among
epiphytic ferns highly modified stems are occasionally met with,
as in the Malayan species *Polypodium (Lecanopteris) carnosum*
and *P. sinuosum*[3].

The leaves of ferns are among the most protean of all plant
organs; as Darwin wrote, "the variability of ferns passes all
bounds[4]." The highly compound tri- or quadripinnate leaves
of such species as *Pteris aquilina, Davallia* and other genera
stand for the central type of fern frond; others exhibit a well-
marked dichotomy, e.g. *Lygodium, Gleichenia, Matonia*, etc., a
habit in all probability associated with the older rather than
with the more modern products of fern evolution. Before
attempting to determine specifically fossil fern fronds, it is
important to familiarise ourselves with the range of variability
among existing species and more especially in leaves of the
same plant. A striking example of heteromorphy is illustrated
in fig. 233. Reinecke[5] has figured a plant of *Asplenium multi-
lineatum* in which the segments of the compound fronds assume
various forms. In *Teratophyllum aculeatum var. inermis* Mett.,
a tropical climbing fern believed by Karsten[6] to be identical
with *Acrostichum (Lomariopsis) sorbifolium*,—an identification
which Goebel[7] questions,—the fronds which stand free of the
stem supporting the climber differ considerably from the
translucent and much more delicate filmy leaves pressed against
the supporting tree. From this fern alone Fée is said to have
created 17 distinct species. In this, as in many other cases,

---

[1] Bäsecke (08).   [2] Boodle (00).
[3] Yapp (02).   [4] Darwin (03) II. p. 381.   [5] Reinecke (97).
[6] Karsten (95); Christ (96); Bommer (03).   [7] Goebel (05) p. 347.

differences in leaf-form are the expression of a physiological
division of labour connected with an epiphytic existence. Some
tropical species of *Polypodium* (sect. *Drynaria*), e.g. *P. quercifolium* (fig. 234 and fig. 231, D), produce two distinct types of
leaf, the large green fronds, concerned with the assimilation of
carbon and spore-production, being in sharp contrast to the small

Fig. 233. *Polypodium Billardieri* Br. (¼ nat. size.) Middle Island, New
Zealand. From specimens in the Cambridge Herbarium.

slightly lobed brown leaves which act as stiff brackets (fig.
234, M) for collecting humus from which the roots absorb raw
material. Similarly in *Platycerium* the orbicular mantle-leaves
differ widely from the long pendulous or erect fronds fashioned
like the spreading antlers of an elk. In *Hemitelia capensis*, a
South African Cyatheaceous species, the basal pinnae assume

the form of finely divided leaves identified by earlier collectors as those of a parasitic *Trichomanes* (fig. 235). In a letter written by W. H. Harvey in 1837 accompanying the specimen shown in fig. 235, he says, " Apropos of *Hemitelia*, be it known abroad that supposed parasitical *Trichomanes*...is not a parasite, but a part of the frond of *Hemitelia*." The delicate reduced pinnae remain on the stem and form a cluster at the base of the fronds [1].

In many species the sporophylls are distinguished from the

FIG. 234. *Polypodium quercifolium.* (Much reduced : M, Mantle-leaves.)

sterile fronds by segments with little or no chlorophyllous tissue, as in *Onoclea struthiopteris*[2] in which, each year, the plant produces a funnel-shaped group of sterile leaves followed later in the season by a cluster of sporophylls; or, as in many other genera, the fertile leaves are distinguished also by longer petioles and thus serve as more efficient agents of spore-dissemination. In *Ceratopteris* the narrow segments of the taller fertile leaves

[1] A striking example of these so-called Aphlebiae of *Hemitelia* may be seen at the Royal Gardens, Kew.
[2] Luerssen, in Rabenhorst (89) A. p. 483, fig. 164.

are in striking contrast to the broader pinnules of the submerged foliage leaves. Leaf-form is in many cases obviously the expression of environment; the xerophilous fern *Jamesonia*[1] from the treeless paramos of the Andes[2] is characterised by its minute leaflets with strong revolute margins and a thick felt of hairs on the lower surface; in others, xerophilous features take

Fig. 235. *Hemitelia capensis* R. Brown. Nat. size. *a*, Pinna of normal frond. [From a specimen in the British Museum. M.S.]

the form of a covering of overlapping scales (*Ceterach*), or a development of water-tissue as in the fleshy leaves of the Himalayan fern *Drymoglossum carnosum*. In the Bracken fern Boodle[3] has shown how the fronds may be classed as shade and

[1] Goebel (91) Pl. xiii.　　　[2] Spruce (08) ii. p. 232.　　　[3] Boodle (04).

sun leaves; the former are spreading and softer, while the latter
are relatively smaller and of harder texture (fig. 236, *a* and *b*).
Even in one leaf six feet high, growing through a dense bush
of gorse and bramble, the lower part was found to have the

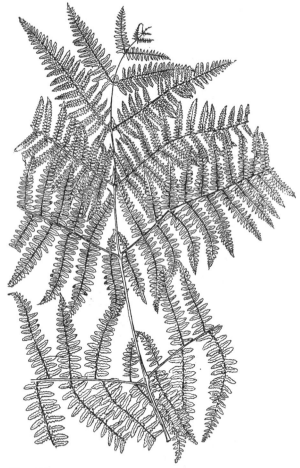

FIG. 236 *a*.   *Pteris aquilina.*
Part of leaf from greenhouse.   (¼ nat. size.)   After Boodle.

features of a shade leaf, while the uppermost exposed pinnae
were xerophilous.

The resemblance between some of the filmy Hymenophyll-

aceae and thalloid Liverworts[1] is worthy of mention as one of
the many possible pitfalls to be avoided by the palaeobotanical
student.　The long linear fronds of such genera as *Vittaria* and
Monogramme might well be identified in a fossil state as the
leaves of a grass-like Monocotyledon, or compared with the

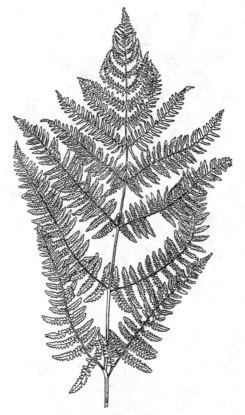

Fig. 236 *b*.　*Pteris aquilina.*
Leaf from the same plant grown out of doors.　(¼ nat. size.)
After Boodle.

foliage of *Isoetes* or *Pilularia.*　The resemblance of some fern
leaves with reticulate venation to those of Dicotyledons has led
astray experienced palaeobotanists; it is not only the anastomosing venation in the leaves of several ferns that simulates

[1] Goebel (05) ; Baker (67).

dicotyledonous foliage, but the compound leaves of many dicotyledons, e.g. *Paullinia thalictrifolia* (Sapindaceae) and species of Umbelliferae, may easily be mistaken for fronds of ferns. The dichotomously lobed lamina of some Schizaeas, e.g. *S. dichotoma* and *S. elegans* (fig. 222), bears a close resemblance to the leaves of *Baiera* or *Ginkgo*[1]. The original description by Kunze[2] of the South African Cycad *Stangeria paradoxa* as a Polypodiaceous fern illustrates the difficulty, or indeed impossibility, of distinguishing between a sterile simply pinnate fern frond and the foliage of some Cycads. The deeply divided segments of *Cycas Micholitzii*[3] simulate the dichotomously branched pinnae of *Lygodium dichotomum*, and the leaves of *Aneimia rotundifolia* (fig. 223) and other species are almost identical in form with the Jurassic species *Otozamites Beani*, a member of the Cycadophyta.

There are certain facts in regard to the geographical distribution of ferns to which attention should be directed. Mr Baker in his paper on fern distribution writes: " With the precision of an hygrometer, an increase in the fern-vegetation marks the wooded humid regions[4]." If in a collection of fossil plants we find a preponderance of ferns we are tempted to assume the existence of such conditions as are favourable to the luxuriant development of ferns at the present day. On the other hand, we must bear in mind the wonderful plasticity of many recent species and the fact that xerophilous ferns are by no means unknown in present-day floras.

Ferns are admirably adapted to rapid dispersal over comparatively wide areas. Bower[5] estimates that in one season a Male Fern may produce about 5,000,000 spores: with this enormous spore-output are coupled a thoroughly efficient mechanism for scattering the germs and an unusual facility for wind-dispersal. When Treub[6] visited the devastated and sterilised wreck of the Island of Krakatau in 1886, three years after the volcanic outburst, he found that twelve ferns had already established themselves; the spores had probably been

[1] Seward and Gowan (00).   [2] Hooker (59).
[3] Thiselton-Dyer (05).   [4] Baker (68) p. 305.
[5] Bower (08) p. 18.   [6] Treub (88) A. ; Ernst (08).

carried by the wind at least 25 to 30 miles. It is not surprising, therefore, to find that many ferns have an almost world-wide distribution; and, it may be added, in view of their efficient means of dispersal, wide range by no means implies great antiquity. Prof. Campbell[1] has recently called attention to the significance of the wide distribution of Hepaticae in its bearing on their antiquity; the spores are incapable of retaining vitality for more than a short period, and it is argued that a world-wide distribution can have been acquired only after an enormous lapse of time. If we apply this reasoning to the Osmundaceae among ferns, it may be legitimate to assume that their short-lived green spores render them much less efficient colonisers than the great majority of ferns; if this is granted, the wide distribution of Osmundaceous ferns in the Mesozoic era carries their history back to a still more remote past, a conclusion which receives support from the records of the rocks.

The Bracken fern which we regard as characteristically British is a cosmopolitan type; it was found by Treub among the pioneers of the New Flora of Krakatau; in British Central Africa, it greets one at every turn "like a messenger from the homeland[2]"; it grows on the Swiss Alps, on the mountains of Abyssinia, in Tasmania, and on the slopes of the Himalayas. The two genera *Matonia* (fig. 228) and *Dipteris*, which grow side by side on Mount Ophir in the Malay Peninsula, are examples of restricted geographical range and carry us back to the Jurassic period when closely allied types flourished abundantly in northern latitudes. Similarly *Thyrsopteris elegans*, confined to Juan Fernandez, exhibits a remarkable likeness to Jurassic species from England and the Arctic regions.

The proportion of ferns to flowering plants in recent floras is a question of some interest from a palaeobotanical point of view; but we must bear in mind the fact that the evolution of angiosperms, effected at a late stage in the history of the earth, seriously disturbed the balance of power among competitors for earth and air. The abundance of ferns in a particular region is, however, an unsafe guide to geographical or climatic conditions. Many ferns are essentially social plants; the wide stretches of

1 Campbell (07).   2 Davy (07) p. 263.

moorland carpeted with *Pteris aquilina* afford an example of the monopolisation of the soil by a single species. In Sikkim Sir Joseph Hooker speaks of extensive groves of tree ferns, and in the wet regions of the Amazon, Bates[1] describes the whole forest glade as forming a " vast fernery." In a valley in Tahiti *Alsophila tahitiensis* is said to form " a sort of forest almost to the exclusion of other ferns[2]." In the abundance of *Glossopteris* (figs. 334, etc.) fronds spread over wide areas of Permo-Carboniferous rocks in S. Africa, Australia, and India, we have a striking instance of a similar social habit in an extinct fern or at least fern-like plant.

Acrostichum aureum, with pinnate fronds several feet long, is an example of a recent fern covering immense tracts, but this species[3] is more especially interesting as a member of the Filicineae characteristic of brackish marshes and the banks of tropical rivers in company with Mangrove plants and the " Stemless Palm " *Nipa*. This species exhibits the anatomical characters of a water-plant and affords an interesting parallel with some Palaeozoic ferns (species of *Psaronius*) which probably grew under similar conditions.

The Anatomy of Ferns.

The text-book accounts of fern-anatomy convey a very inadequate idea of the architectural characters displayed by the vascular systems of recent genera. When we are concerned with the study of extinct plants it is essential to be familiar not only with the commoner recent types, but particularly with exceptional or aberrant types. The vascular system of many ferns consists of strands of xylem composed of scalariform tracheae associated with a larger or smaller amount of parenchyma, surrounded either wholly or in part (that is concentric or bicollateral) by phloem: beyond this is a pericycle, one layer or frequently several layers in breadth, limited externally by an endodermis, which can usually be readily recognised. The vascular strands are embedded in the ground-tissue of the stem

---

[1] Bates (63) A. p. 30.   [2] *Challenger Reports* (85) p. 785.
[3] Tansley and Fritsch (05) p. 43 ; Thomas, E. N. (05).

consisting of thin-walled parenchyma and, in most ferns, a
considerable quantity of hard and lignified mechanical tissue.
The narrow protoxylem elements are usually characterised by a
spiral form of thickening, but in slow-growing stems the first-
formed elements are frequently of the scalariform type.

A study of the anatomy of recent ferns both in the adult
state and in successive stages of development from the embryo
has on the whole revealed "a striking parallelism[1]" between
vascular and sporangial characters in leptosporangiate ferns.
For a masterly treatment of our knowledge of fern anatomy

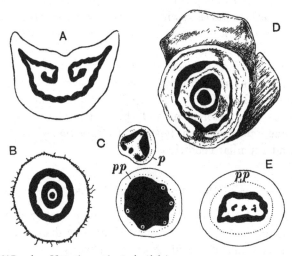

Fig. 237. A. *Matonia pectinata* (petiole).
          B. *M. pectinata* (stem).
          C. *Gleichenia dicarpa* (stem) : *p*, petiole ; *pp*, protophloem ;
              position of protoxylem indicated by black dots.
          D. *Matonidium*.
          E. *Trichomanes reniforme* : *pp*, protophloem.
              (C, E, after Boodle ; D, after Bommer.)

from a phylogenetic point of view reference should be made to
Mr Tansley's recently published lectures : within the limits of
this volume all that is possible is a brief outline of the main
types of vascular structure illustrated by recent genera.

To Prof. Jeffrey[2] we owe the term protostele which he applied

---

[1] Tansley (08) p. 27.   [2] Jeffrey (98).

to a type of stele consisting of a central core of xylem surrounded by phloem, pericycle, and endodermis. While admitting that steles of this type may sometimes be the result of the modification of less simple forms, we may confidently regard the protostele as representing the most primitive form of vascular system. The genus *Lygodium* affords an example of a protostelic fern; a solid column of xylem tracheae and parenchyma is completely encircled by a cylinder of phloem succeeded by a multi-layered pericycle and an endodermis of a single layer of cells. In this genus the stele is characterised by marginal groups of protoxylem ; it is exarch. An almost identical type is

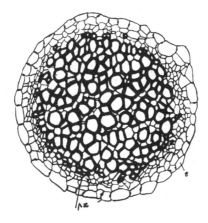

Fɪɢ. 238.　Stele of *Trichomanes scandens* : *px*, protoxylem ; *s*, endodermis. From Tansley, after Boodle.

represented by species of *Gleichenia*, but here the stele is mesarch, the protoxylem being slightly internal (fig. 237, C). *Trichomanes scandens* (fig. 238) has an exarch protostele like that of *Lygodium*; but, as Boodle[1] has suggested, the protostelic form in this case is probably the result of modification of a collateral form of stele such as occurs in *Trichomanes reniforme* (fig. 237, E). A second type of stele has been described in species of *Lindsaya*[2] in which the xylem includes a small group of phloem near the dorsal surface. This *Lindsaya* type is often passed through in the development of " seedling " ferns and may

[1] Boodle (00).　　　　　　[2] Tansley and Lulham (02).

be regarded as a stage in a series leading to another well-marked type, the solenostele. The solenostele[1], a hollow cyclinder of xylem lined within and without by phloem, pericycle, and endodermis, occurs in several genera belonging to different families, e.g. *Dipteris*, species of *Pteris*, species of *Lindsaya*, *Polypodium*, *Jamesonia*, *Loxsoma*, *Gleichenia* and other genera. In a smaller number of ferns the stele consists of what may be called a medullated protostele similar to the common form of stele in *Lepidodendron*: this type is found in species of *Schizaea* and in *Platyzoma* (fig. 239). It is important to notice that in the

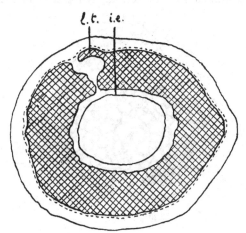

Fig. 239. *Platyzoma microphylla. l.t.,* leaf-trace ; *i.e.,* internal endodermis. (After Tansley; modified from Boodle.)

solenostele and as a rule in the medullated protostele when a leaf-trace passes out from the rhizome stele the vascular cylinder is interrupted by the formation of a foliar gap (*Platyzoma*[2], fig. 239, is an exception). This fact has been emphasized by Jeffrey[3] who draws a distinction between the Lycopodiaceous type of stele, which is not broken by the exit of leaf-traces, and the fern stele in which foliar gaps are produced: the former he speaks of as the cladosiphonic type (*Lycopsida*) and the latter as the phyllosiphonic (*Pteropsida*).

The transition to a hollow cylinder of xylem from a protostele

---

[1] Gwynne-Vaughan (01); (03).   [2] Boodle (01) p. 735.   [3] Jeffrey (00); (03).

may be described as the result of the replacement of some of the
axial conducting tracheae by parenchyma or other non-vascular
tissue consequent on an increase in diameter of the whole stele
and the concentration of the true conducting elements towards
the periphery[1].

The occurrence of the internal cylinder of phloem, pericycle,
and endodermis in a solenostele is rendered intelligible by a
study of fern seedlings and by a comparative examination of
transitional types connecting protosteles and solenosteles through
medullated protosteles and steles of the *Lindsaya* type.   A

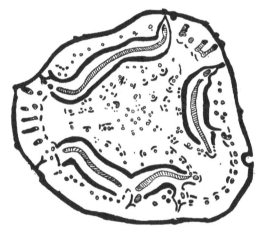

FIG. 240.   *Cyathea Imrayana.*   (From Tansley after de Bary.)   (Sclerenchyma
represented by black bands.)

further stage in stelar evolution is illustrated by what is termed
the dictyostele, the arrangement of vascular tissue characteristic
of *Nephrodium Filix-mas, Cyathea* (fig. 240), *Polypodium vulgare*
and many other common ferns.

If a solenostele is interrupted by leaf-gaps at intervals
sufficiently close to cause overlapping, a transverse section at
any part of the stele will show apparently separate curved
bands of concentrically arranged xylem and phloem, which on
dissection are seen to represent parts of a continuous lattice-

[1] For an account of the probable methods by which this has been effected
and of the factors concerned, see Tansley (08).

work or a cylinder with the wall pierced by large meshes. The manner of evolution of the dictyostele has been ably dealt with by Gwynne-Vaughan[1] and other authors. In a few ferns, e.g. *Matonia pectinata*[2], a transverse section of the stem (fig. 237, B) reveals the presence of two or in some cases three concentric solenosteles with a solid protostele in the centre: this *polycyclic* type may be regarded as the expression of the fact that in response to the need for an adequate water-supply to the large fronds, ferns have increased the conducting channels by a method other than by the mere increase of the diameter of a single stele. Fig. 237, A, shows the vascular tissue of a petiole of *Matonia* in transverse section.

The two genera of Osmundaceae, *Todea* and *Osmunda*, are peculiar among recent ferns in having a vascular cylinder composed of separate strands of xylem varying considerably in shape and size, from U-shaped strands with the concavity facing the centre of the stem and with the protoxylem in the hollow of the U, to oval or more or less circular strands with a mesarch protoxylem or without any protoxylem elements (fig. 221, A, B). These different forms are the expression of the change in contour or in structure which the parts of the lattice-work undergo at different levels in the stem[3]. Beyond this ring of xylem bundles is a continuous sheath of phloem of characteristic structure. A transverse section of a stem of *Osmunda regalis* may show 15 or more xylem strands; in *O. Claytoniana* there may be as many as 40. In *Todea barbara* (fig. 221, B) the leaf-gaps are shorter, and in consequence of the less amount of overlapping the xylem cylinder becomes an almost continuous tube. The recent researches of Kidston and Gwynne-Vaughan[4] have resulted in the discovery of fossil Osmundaceous stems with a complete xylem ring, the stele being of the medullated protostele type; in another extinct member of the family the stele consists of a solid xylem core. The Osmundaceous type of stele is complicated in *O. cinnamomea* (fig. 221, A) by the occurrence of local internal phloem and by

---

[1] Gwynne-Vaughan (03).   [2] Seward (99[2]); Wigglesworth (02).
[3] Seward and Ford (03); Jeffrey (03); Faull (01).
[4] Kidston and Gwynne-Vaughan (07); (08); (09).

an internal endodermis, a feature which leads Jeffrey to what
I believe to be an incorrect conclusion that the vascular arrangement found in *Osmunda regalis* has been evolved by reduction
from a stele in which the xylem was enclosed within and without
by phloem. New facts recently brought to light enable us to
derive the ordinary Osmundaceous type from the protostele
and solenostele. It is worthy of remark that the Osmundaceae
occupy a somewhat isolated position among recent ferns; their
anatomy represents a special type, their sporangia differ in
several respects from those of other leptosporangiate ferns and in
some features *Osmunda* and *Todea* agree with the Eusporangiate
ferns. The possession of such distinguishing characters as these
suggests antiquity; and the 'facts of palaeobotany, as also the
present geographical range of the family, confirm the correctness
of this deduction.

Before leaving the stelar structure of leptosporangiate fern
stems, a word must be added in regard to a type of structure
met with in the Hymenophyllaceae. In this family *Trichomanes
reniforme* (fig. 237, E) may be regarded, as Boodle suggests, as
the central type: the stele consists of a ring of metaxylem
tracheae, the dorsal portion having the form of a flat arch
and the ventral half that of a straight band. This flattened ring
of xylem encloses parenchymatous tissue containing scattered
tracheae some of which are protoxylem elements. In *Trichomanes radicans* the rhizome is stouter than in *T. reniforme* and
the stele consists of a greater number of tracheae. The stele
is cylindrical like that shown in fig. 238, but the centre is occupied by two groups of protoxylem and associated parenchyma.
In *Hymenophyllum tunbridgense* the stele is of the subcol-
lateral type; the ventral plate of the xylem ring has disappeared
leaving a single strand of xylem with endarch protoxylem and
completely surrounded by phloem. *Trichomanes muscoides*
possesses a still simpler stele consisting of a slender xylem
strand with phloem on one side only. Reference has already
been made to the occurrence in this family of the protostelic
type. The Hymenophyllaceae afford a striking illustration
of the modification in different directions of stelar structure
connected with differences in habit, and of the correlation of

demand and supply as shown in the varying amount of conducting tissue in the steles of different species.

The leaf-trace in a great number of ferns is characterised by
its C-shaped form[1] as seen in transverse section: this in some
genera, e.g. *Matonia* (fig. 237, A), is complicated by the spiral
infolding of the free edges of the C; in other ferns (e.g.
some Cyatheaceae) (fig. 278, C) the sides of the C are incurved,
while in some species the xylem is broken up into a large
number of separate strands.

An elaborate treatment of the leaf-traces of ferns was
published a few years ago by MM. Bertrand and Cornaille[2]
in which the authors show how the various systems of vascular
tissue in the fronds of ferns may be derived from a common type.
As Prof. Chodat[3] justly remarks this important work has not
received the attention it deserves, the neglect being attributed
to the strange notation which is adopted[4].

The roots of ferns are characterised by a uniformity of plan
in marked contrast to the wide range of structure met with in
the stem and to a less extent in the leaves. The xylem may
consist of a plate of scalariform tracheae with a protoxylem
group at each end, or the stele may include six or more alternating strands of xylem and phloem.

II. Marattiales (Eusporangiate isosporous Filicales).

The Marattiaceae, the single family of ferns included in the
Marattiales, comprise the genera *Angiopteris*, *Archangiopteris*,
Marattia, *Danaea*, and *Kaulfussia*, which are for the most part
tropical in distribution. These genera are characterised by
eusporangiate sori or synangia, the presence of stipules at the
base of the petioles, and by the complex arrangement of the
vascular tissue. In view of the fact that many fossil ferns
show a close resemblance to the recent Marattiaceae, the surviving genera are briefly described. The prothallus is green
and relatively large.

[1] Gwynne-Vaughan (08).
[2] Bertrand and Cornaille (02).   [3] Chodat (08) p. 15.
[4] See also Pelourde (09) for an account of the anatomy of fern petioles.

Angiopteris. This genus occurs in Polynesia, tropical Asia, and Madagascar ; it is characterised by a short and thick fleshy stem bearing large bipinnate leaves which occasionally show a forking of the rachis[1], a feature reminiscent of some Palaeozoic fern-like fronds. One of the large plants of *Angiopteris evecta* in the Royal Gardens, Kew, bears leaves 12 feet in length with a stalk 6 inches in diameter at the base. The sessile or shortly stalked and rather leathery linear or broadly lanceolate pinnules have a prominent midrib and dichotomously branched lateral veins. The surface of an old stem is covered with the thick stumps of petioles enclosed by pairs of fleshy stipules (fig. 241, A) and bears numerous fleshy roots, which hang free in the air or penetrate the soil. The young fronds (fig. 220, A) exhibit very clearly the characteristic circinate vernation. The proximal part of each primary pinna is characterised by a pulvinus-like swelling. The sporangia, in short linear

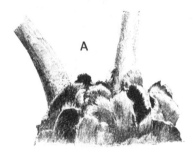 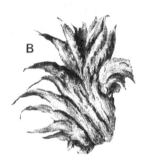

FIG. 241. A. *Angiopteris evecta.* (Considerably reduced.)
            B. *Marattia fraxinea.* Stipule. M.S.

elliptical sori near the edge of the pinnules, consist of free sporangia (fig. 242, A—D) provided with a peculiar type of "annulus"[2], in the form of a narrow band of thicker-walled cells, which extends as a broad strip on either side of the apex. An examination of sections through the sporangia of *Angiopteris* in different planes[3] illustrates the difficulty of determining the precise nature of the annulus in a petrified sporangium which is seen only in one or two planes. Many of the sporangia from the English Coal-Measures, compared by authors with those of Leptosporangiate ferns, are in all probability referable to the Marattiaceous type.

The vascular system[4] of the stem constitutes a highly complex dictyostelic or polycylic type which may consist of as many as nine concentric series of strands of xylem surrounded by phloem, with large sieve-tubes

[1] Observed in plants in the Botanic Gardens of Brussels and Leipzig. A.C.S.
[2] For an account of the spore-producing members of the Marattiaceae, see Bower (97).
[3] Zeiller (90) p. 19.   [4] Shove (00); Tansley (08).

and a pericycle which abuts on the parenchymatous ground-tissue without
any definite endodermal layer. A peculiarity in the vascular strands is
that the first-formed elements of the phloem lie close to the edge of the
xylem, the metaphloem being therefore centrifugal in its development.
The ground-tissue is devoid of mechanical tissue and is penetrated by
roots, a few of which arise from the outer vascular strands while others
force their way to the surface from the more internal dictyosteles. Leaf-
traces, consisting of several strands, are given off from the outermost

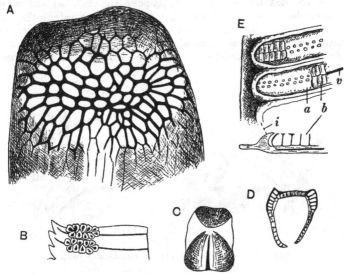

FIG. 242. A—D. *Angiopteris evecta.*
     A. Apex of sporangium showing " annulus,"
     B. Sori.
     C. Sporangium.
     D. Section of sporangium, showing the two lateral bands of thick-
        walled cells.
     E. *Danaea*: *a*, roof of synangium, with pores; *b*, sporangial
        cavities; *v*, vascular bundle; *i*, indusium.
       · (D, after Zeiller.)

cylinder and a segment of the second dictyostele moves out to fill the
gap formed in the outermost network, while the gap in the second cylinder
receives compensating strands from the third. A few layers below the
surface of the petiole there is a ring of thick-walled elements (*s*, fig. 243),
and in both petiole and stem numerous mucilage ducts and tannin-sacs
occur in the ground-tissue. It has been shown by Farmer and Hill[1] that
in some of the vascular strands in an *Angiopteris* stem a few secondary

[1] Farmer and Hill (02) Pl. xviii. figs. 26, 28.

tracheae are added to the primary xylem by the activity of the adjacent parenchyma. The vascular bundles in the petiole form more or less regular concentric series ; they have no endodermis and are characterised also by the large size of the sieve-tubes (*st*, fig. 243).

The roots of Marattiaceous ferns (fig. 244) are characterised by the

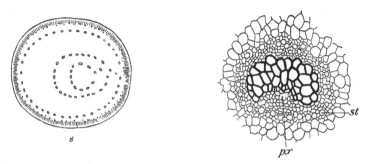

FIG. 243. *Angiopteris evecta.* Section of petiole (considerably reduced) and of a single vascular bundle (magnified) : *px*, protoxylem ; *st*, sieve-tubes.

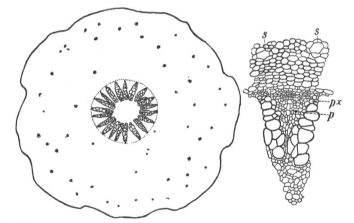

FIG. 244. *Angiopteris evecta.* Transverse section of root, with part of the stele magnified: *s*, sieve-tubes ; *p*, phloem ; *px*, protoxylem.

larger number of xylem and phloem groups ; the stele is polyarch and not diarch, tetrach or hexarch as in most Leptosporangiate ferns.

Archangiopteris. This monotypic genus, discovered by Mr Henry in South Eastern Yunnan, was described by Christ and Giesenhagen in 1899[1]. The comparatively slender rhizome has a fairly simple vascular system[2]. The simply-pinnate leaves bear pinnules like those of *Danaea*,

---

[1] Christ and Geisenhagen (99).   [2] Gwynne-Vaughan (05).

but the sori agree with those of *Angiopteris* except in their greater length
and in the larger number of sporangia.

Marattia. This genus, which extends "all round the world within the
tropics[1]," includes some species which closely resemble *Angiopteris*, while
others are characterised by more finely divided leaves with smaller
ultimate segments. The fleshy stipules occasionally have an irregularly
pinnatifid form (fig. 241, B). The sporangia are represented by oval
synangia[2] (fig. 245, A; the black patches at the ends of the lateral veins)

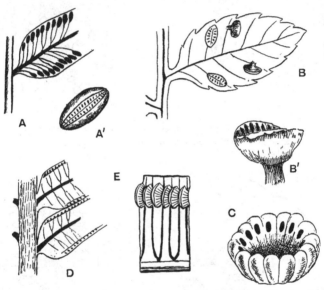

Fig. 245. A. *Marattia fraxinea.* A'. A single synangium showing the two
valves and pores of the sporangial compartments.
B, B'. *M. Kaulfussii.*
C. *Kaulfussia* (synangium showing pores of sporangial compart-
ments).
D, E. *Marattiopsis Münsteri.*
(C, after Hooker; D, E, after Schimper.)

composed of two valves, which on ripening come apart and expose
two rows of pores formed by the apical dehiscence of the sporangial compartments (fig. 245, A', B). In *Marattia Kaulfussii* the sori are attached
to the lamina by a short stalk (fig. 245, B, B') and the leaf bears a close

[1] Hooker and Baker (68) p. 440.
[2] The term synangium is applied to sporangia more or less completely
united with one another and producing spores in groups separated by walls of
sterile cells. A synangium may be regarded as a spore-forming organ produced
by partial sterilization of sporogenous tissue or as a group of coalescent sporangia.

resemblance to those of the Umbelliferous genera *Anthriscus* and *Chaerophyllum*. The vascular system is constructed on the same plan as that of *Angiopteris* but is of simpler form.

Danaea. Danaea, represented by about 14 species confined to tropical America, is characterised by simple or simply pinnate leaves with linear segments bearing elongated sori extending from the midrib almost to the margin of the lamina. Each sorus consists of numerous sporangia in two parallel rows united into an oblong mass partially overarched by an indusium (fig. 242, E, *i*) which grows up from the leaf between the sori. In the portion of a fertile segment shown in fig. 242, E, the apical pores are seen at *a*; and at *b*, where the roof of the synangium has been removed, the spore-bearing compartments are exposed. The vascular system[1] agrees in general plan with that characteristic of the family.

Kaulfussia. The form of the leaf (Vol. I. p. 97, fig. 22) closely resembles that of the Horse Chestnut; the stem is a creeping dorsiventral rhizome with a vascular system in the form of a "much perforated solenostele[2]." The synangia are circular, with a median depression; each sporangial compartment opens by an apical pore on the sloping sides of the synangial cup (fig. 245, C)[3].

Copeland has recently described a Marattiaceous leaf which he makes the type of a new genus, *Macroglossum alidae*. The sori are nearer the margin than in *Angiopteris* and are said to consist of a greater number of sporangia. The photograph[4] of a single pinna which accompanies the brief description hardly affords satisfactory evidence in support of the creation of a new genus. The structure of a petiole which I have had an opportunity of examining, through the kindness of Mr Hewitt of Sarawak, shows no distinctive features.

III. Ophioglossales. (Isosporous and Eusporangiate.)

The three genera, *Ophioglossum*, *Botrychium*, and *Helminthostachys*, are characterised by the division of the leaves into a sterile and a fertile lobe. The fertile lobe in *Ophioglossum* bears two rows of spherical sporangia sunk in its tissue; in *Botrychium* and *Helminthostachys* the spores are contained in large sporangia with a stout wall[5]. The prothallus is subterranean and without chlorophyll. In the British species of *Ophioglossum*, *O. vulgatum* (the adder's tongue fern), an almost cosmopolitan species, the sterile part of the frond is of oval form and has reticulate venation. In *O. pendulum* and *O.*

---

[1] Brebner (02); Rudolph (05).   [2] Tansley (08) p. 90; Kühn (90).

[3] Pelourde (08) has recently dealt with the anatomy of recent and fossil Marattiaceous ferns.

[4] Copeland (08) Pl. i. (09) Pl. v.   [5] Bower (96).

palmatum the lamina is deeply lobed. In the genus *Botry-chium*, represented in Britain by *B. Lunaria,* both sterile and fertile branches of the frond are pinnately divided, while in *Helminthostachys* the sporangia are borne on sporangiophores

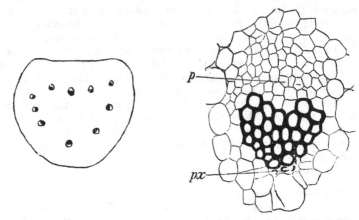

Fig. 246. *Ophioglossum vulgatum.* Transverse section of petiole and single bundle: *p*, phloem; *px*, endarch protoxylem.

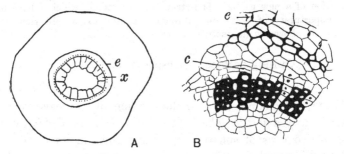

Fig. 247. *Botrychium virginianum:* *e*, endodermis; *c*, cambium; *x*, xylem. A, diagrammatic section of stem; B, portion of the stele and endodermis enlarged.

(A, after Campbell; B, after Jeffrey.)

given off from the margin of the fertile branch of a frond similar in habit to a leaf of *Helleborus.*

The stem of *Ophioglossum* is characterised by a dictyostele of collateral bundles with endarch protoxylem: the vascular system of the leaf-stalk is also composed of several separate strands (fig. 246). In *Botrychium* the stele is a cylinder of xylem surrounded externally by phloem. This genus affords the only instance among ferns of a plant in which the addition of secondary tracheae occurs on a scale large enough to produce a well-defined cylinder of secondary xylem traversed by radial rows of medullary-ray cells[1] (fig. 247). The unsatisfactory nature of the evidence in regard to the past history of the Ophioglossales renders superfluous a fuller treatment of the recent species.

[1] Jeffrey (98). For an account of the anatomy of *Helminthostachys*, see Farmer and Freeman (99).

CHAPTER XXI.

FOSSIL FERNS.

Osmundaceae.

FROM the Culm of Silesia, Stur[1] described impressions of sterile fronds which he named *Todea Lipoldi* on the ground of the similarity of the finely divided pinnules to those of *Todea superba* and other filmy species of the genus. The type-specimen of Stur (in the Geological Survey Museum, Vienna) affords no information as to sporangial characters and cannot be accepted as an authentic record of a Lower Carboniferous representative of the family. Another more satisfactory but hardly convincing piece of evidence bearing on the presence of Osmundaceae in pre-Permian floras has been adduced by Renault[2], who described petrified sporangia from the Culm beds of Esnost in France as *Todeopsis primaeva* (fig. 256, F). These pyriform sporangia are characterised by the presence of a plate of large cells comparable with the subapical group of "annulus" cells in the sporangia of the recent species (fig. 221). Zeiller[3] has published a figure of some sporangia described by Renault from Autun resembling the Osmundaceous type in having a plate of thick-walled cells instead of a true annulus, but the plate is larger than the group of cells in the recent sporangia, and both sporangia and spores are smaller in the fossil. The sporangia from Carboniferous rocks described by Weiss as *Sturiella*[4] bear some resemblance to those of recent

---

[1] Stur (75) A. p. 77, Pl. XI. fig. 8.   [2] Renault (96) A. p. 21.
[3] Zeiller (90) p. 16.   [4] Zeiller (90) p. 48.

Osmundaceae, but there is no adequate reason for referring them to this family.

The generic name *Pteridotheca* is employed by Scott as a convenient designation for unassigned petrified sporangia of Palaeozoic age with an annulus and other characters indicating fern-affinity. In the species *P. Butterworthi*[1] the sporangia are characterised by a group of large cells suggesting comparison with the annulus, or what represents the annulus, in Osmundaceae and Marattiaceae. Scott has also described a sporangium from the Coal-Measures containing germinating spores[2]; the structure is similar to that of recent Osmundaceous sporangia, and it is interesting to note that germinating spores have been observed in the recent species *Todea hymenophylloides*[3].

Additional evidence of the same kind is afforded by fertile specimens of a quadripinnate fern with deeply dissected oval-lanceolate pinnules described by Zeiller from the Coal-Measures of Heraclea in Asia Minor as *Kidstonia heracleensis*[4] (fig. 256, E). Carbonised sporangia were found at the base of narrow lobes of the ultimate segments and, as seen in fig. 256, E, the sporangial wall is distinguished by a plate of larger cells occupying a position like that of the " annulus " of recent Osmundaceae. Zeiller regards the sporangia as intermediate between those of Osmundaceae and Schizaeaceae. From the same locality Zeiller describes another frond bearing somewhat similar sporangia as *Sphenopteris (Discopteris) Rallii* (fig. 256, D)[5]: the term *Discopteris* was instituted by Stur for fertile fronds referred by him to the Marattiaceae[6].

It is by no means safe to assume that these and such Upper Carboniferous sporangia as Bower[7] compared with those of *Todea* were borne on plants possessing the anatomical characters of Osmundaceae rather than those of the extinct Palaeozoic family Botryopterideae. This brings us to the important fact, first pointed out by Renault, that the Botryopterideae are essentially generalised ferns exhibiting many points of contact with the Osmundaceae[8]. It is clear that whether or not we

[1] Scott, D. H. (08) p. 292.   [2] Scott (04) p. 18.
[3] Boodle (00) p. 484.   [4] Zeiller (99) Pl. II. figs. 5, 6.
[5] *Ibid.* Pl. II. fig. 10.   [6] See p. 402.
[7] Bower (91) Pl. VII.   [8] Scott, D. H. (09).

are justified in tracing the Osmundaceae as far back as the
Lower Carboniferous period, some of the characteristics of the
family were already foreshadowed in rocks of this age.

Through a fortunate accident of preservation, unequivocal
evidence of the existence of Osmundaceae in the Palaeozoic era
is supplied by the Russian Upper Permian genera *Zalesskya*
and *Thamnopteris*.

Zalesskya.

This generic title has been instituted by Kidston and
Gwynne-Vaughan[1] for two Russian stems of Upper Permian
age, one of which was named by Eichwald[2] *Chelepteris gracilis*,
but the probability that the type of the genus *Chelepteris* is
generically distinct from Eichwald's species necessitated a new
designation for the Permian fern.

In habit the stem of *Zalesskya* resembles that of an
Osmunda or a *Todea*, but it differs in the possession of a stele
composed of a continuous cylinder or solid column of xylem
surrounded by phloem, and by the differentiation of the xylem
into two concentric zones. The leaves are represented by
petiole-bases only; the sporangia are unknown. The stem and
leaf-base anatomy fully justifies the inclusion of *Zalesskya* in
the Osmundaceae.

Zalesskya gracilis (Eichwald). Fig. 248.

The type-specimen is a partially decorticated stem, from
Upper Permian beds in Russia, provided with a single stele,
13 mm. in diameter, surrounded by a broad thin-walled inner
cortex containing numerous leaf-traces and occasional roots: this
was doubtless succeeded by a sclerotic outer cortex. In its
main features *Zalesskya gracilis* agrees closely with *Z. diploxylon*
represented in fig. 249. The stele consists of a continuous
cylinder of xylem exhibiting a fairly distinct differentiation
into two zones, (i) a broader outer zone of narrower scalariform
tracheae ($x\,ii$, fig. 248) in which 20 to 25 protoxylem strands (px)

---

[1] Kidston and Gwynne-Vaughan (08).   [2] Eichwald (60).

occur just within the edge, (ii) an inner zone of broader and shorter tracheae (fig. 248, $x\,i$). The protoxylem elements (px, fig. 248) are characterised by a single series of scalariform pits, while the metaxylem elements have multiseriate pits like

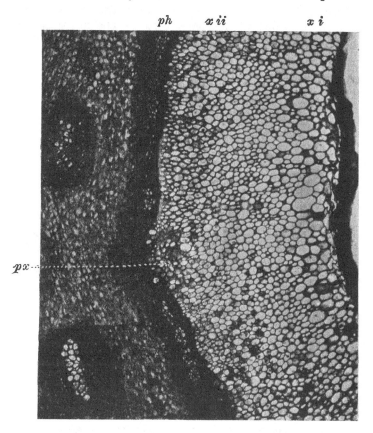

Fig. 248.   *Zalesskya gracilis* (Eich.).   Transverse section of part of the stele: ph, phloem; $x\,i$, $x\,ii$, xylem; px, protoxylem. (After Kidston and Gwynne-Vaughan.   × 20.)

those on the water-conducting elements of recent Osmundaceae. The tracheae show an interesting histological character in the absence of the middle substance of their walls, a feature recognised by Gwynne-Vaughan[1] in many recent ferns. External to

[1] Gwynne-Vaughan (08).

the xylem and separated from it by a parenchymatous sheath
is a ring of phloem, *ph*, composed of large sieve-tubes and
parenchyma separated from the inner cortex by a pericycle
4 to 5 layers in breadth. The occurrence of a few sclerotic
cells beyond the broad inner cortex points to the former
existence of a thick-walled outer cortex. The leaf-traces are

ph

Fig. 249.   *Zalesskya diploxylon.*   Kidston and Gwynne-Vaughan.   Transverse
       section of stem. *ph*, phloem. (After Kidston and Gwynne-Vaughan.
       × 2½.)

given off as mesarch strands from the edge of the xylem; they
begin as prominences opposite the protoxylem and become
gradually detached as xylem bundles, at first oblong in transverse section, then assuming a slightly crescentic and reniform
shape, while the mesarch protoxylem strand takes up an
endarch position. As a trace passes further out the curvature

increases and the protoxylem strands undergo repeated bifurcation; it assumes in fact the form and general type of structure met with in the leaf-traces of *Todea* and *Osmunda*. Numerous diarch roots, given off from the stele at points just below the out-going leaf-traces, pass outwards in a sinuous horizontal course through the cortex of the stem.

In *Zalesskya gracilis* the xylem cylinder was probably wider in the living plant than in the petrified stem. In *Zalesskya diploxylon*[1], in all probability from the same Russian locality, there can be little doubt that the xylem was originally solid to the centre (fig. 249). In this species also the phloem forms a continuous band (*ph*, fig. 249) consisting of four to six layers of sieve-tubes.

Thamnopteris.

Thamnopteris Schlechtendalii (Eich.).   Figs. 250, 312, A, Frontispiece.

In 1849 Brongniart[2] proposed the name *Thamnopteris* for a species of fern from the Upper Permian of Russia originally described by Eichwald as *Anomopteris Schlechtendalii*. A new name was employed by Brongniart on the ground that the fossil was not generically identical with the species previously named by him *Anomopteris Mougeotii*[3]. Eichwald's specimen has been thoroughly investigated by Kidston and Gwynne-Vaughan[4]. The stem (Frontispiece) agrees in habit with those of *Zalesskya* and recent Osmundaceae; on the exposed leaf-bases the action of the weather has etched out the horse-shoe form of the vascular strands and laid bare numerous branched roots boring their way through the petiole stumps. The centre of the stem is occupied by a protostele 13 mm. in diameter consisting of solid xylem separated by a parenchymatous sheath from a cylinder of phloem. The xylem is composed mainly of an axial column of short and broad reticulately pitted tracheae (fig. 250 *b* and Frontispiece), distinguished from the sharply contrasted peripheral zone of normal scalariform elements, *a*, by their thinner

---

[1] Kidston and Gwynne-Vaughan (08) p. 226.
[2] Brongniart (49) A. p. 35.   [3] Brongniart (28) A. Pl. LXXX.
[4] Kidston and Gwynne-Vaughan (09).

walls and more irregular shape. The protoxylem, *px*, is represented by groups of narrower elements rather deeply immersed
in the peripheral part of the metaxylem. A many-layered
pericycle, *per*, and traces of an endodermis, *en*, succeed the
phloem, *ph*, which is characterised by several rows of large
contiguous sieve-tubes; beyond the endodermis is a broad thin-
walled inner cortex. The leaf-traces arise as in *Zalesskya*,
but the protoxylem in *Thamnopteris* is at first central; as the

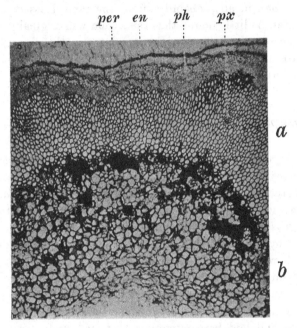

F<small>IG</small>. 250. *Thamnopteris Schlechtendalii* (Eich.). Part of stele: *a*, outer xylem;
b, inner xylem. (After Kidston and Gwynne-Vaughan.   × 13.)

trace passes outwards a group of parenchyma appears immediately internal to the protoxylem elements and gradually
assumes the form of a bay of thin-walled tissue on the inner
concave face of the curved xylem. The next stage is the
repeated division of the protoxylem strand until, in the sclerotic
outer cortex, the traces acquire the Osmundaceous structure
(fig. 312, A, p. 453). The petiole bases have stipular wings as
in *Todea* and *Osmunda*.

The striking feature exhibited by these Permian plants is the structure of the protostele, which in *Thamnopteris* and probably in *Zalesskya diploxylon* consists of solid xylem surrounded by phloem : this may be regarded as the primitive form of the Osmundaceous stele. In *Osmunda regalis* and in other recent species of the genus the xylem cylinder has the form of a lattice-work; in other words, the departure of each leaftrace makes a gap in the xylem and the overlapping of the foliar-gaps results in the separation of the xylem into a number of distinct bundles. In *Zalesskya gracilis* the continuity of

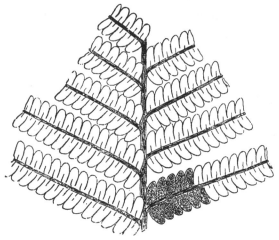

Fig. 251. *Lonchopteris virginiensis.* (After Fontaine. ½ nat. size.)

the xylem is not broken by overlapping gaps; in this it agrees with *Lepidodendron.* In *Thamnopteris* the centre of the stele was occupied by a peculiar form of xylem obviously ill-adapted for conduction, but probably serving for water-storage and comparable with the short and broad tracheae in *Megaloxylon*[1]. There is clearly a well-marked difference in stelar anatomy between these two Permian genera and *Todea* and *Osmunda* : this difference appears less when viewed in the light of the facts revealed by a study of the Jurassic species *Osmundites Dunlopi.*

[1] Seward (99).

As possible examples of Triassic Osmundaceae reference
may be made to some species included in Stur's genus
Speirocarpus[1]. *S. virginiensis* was originally described by
Fontaine[2] from the Upper Triassic rocks of Virginia as
Lonchopteris virginiensis (fig. 251) and has recently been figured
by Leuthardt[3] from the Keuper of Basel. The sporangia, which
are scattered over the lower surface of the pinnules, are
described as globose-elliptical and as having a rudimentary
apical annulus; no figures have been published. In habit
the frond agrees with *Todites Williamsoni*, but the lateral
veins form an anastomosing system like that in the Palaeozoic
genus *Lonchopteris* (fig. 290, B). There would seem to be an *a
priori* probability of this species being a representative of the
Osmundaceae and not, as Stur believed, of the Marattiaceae.
Seeing that *Lonchopteris* is a designation of a purely provisional
kind, it would be convenient to institute a new generic name
for Triassic species having the Lonchopteris venation, which
there are good reasons for regarding as Osmundaceous ferns.

Similarly *Speirocarpus tenuifolius* (Emmons) (=*Acrostichites
tenuifolius* Font.), which resembles *Todites Williamsoni* (see
p. 339) not only in habit and in the distribution of the sporangia
but also in the venation, is probably an Osmundaceous species.

Osmundites.

Osmundites Dunlopi, Kidston and Gwynne-Vaughan[4], fig. 252.

This species was found in Jurassic rocks in the Otago district
of New Zealand in association with *Cladophlebis denticulata*[5]
(fig. 257). The type-specimen forms part of a stem 17 mm. in
diameter surrounded by a broad mass of crowded leaf-bases.
The stele consists of an almost continuous xylem ring (fig. 252)
enclosing a wide pith: the phloem and inner cortex are not
preserved but the peripheral region of the stem is occupied by
a sclerotic outer cortex. The mass of encasing leaf-bases
resolves itself on closer inspection into zones of foliage-leaf

---

[1] Krasser (09) p. 10.   [2] Fontaine (83) Pls. xxviii. xxix.
[3] Leuthardt (04) Pl. xviii.   [4] Kidston and Gwynne-Vaughan (07).
[5] See p. 343.

petioles and the petioles of scale-leaves with an aborted lamina.
A similar association of two forms of leaf is seen in the existing
American species *Osmunda Claytoniana* and *O.cinnamomea*. The
cortex and armour of leaf-bases are penetrated by numerous
diarch roots. The xylem cylinder, six to seven tracheae broad,
is characterised by the narrower diameter of its innermost elements and—an important point—by the fact that the detachment
of a leaf-trace does not break the continuity of the xylem
cylinder (fig. 252). Each leaf-trace is at first elliptical in
section; it then becomes curved inwards and gradually assumes
the horse-shoe form as in *Zalesskya* and in the recent species.
The single endarch protoxylem becomes subdivided until in
the petiole it is represented by 20 or more strands.

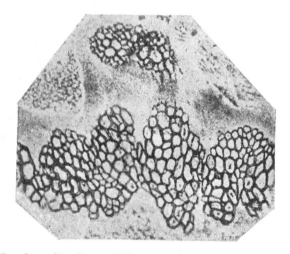

Fig. 252. *Osmundites Dunlopi* Kidst. and G.-V. Portion of xylem showing the
departure of a leaf-trace. (After Kidston and Gwynne-Vaughan; × 36.)

In the continuity of the xylem cylinder this species of
Osmundites shows a closer approach to *Todea barbara* or
T. superba (fig. 221, B) than to species of *Osmunda*; it differs
from *Zalesskya* in having reached a further stage in the reduction of a solid protostele to one composed of a xylem
cylinder enclosing a pith. This difference is of the same kind
as that which distinguishes the stele of *Lepidodendron rhodumnense* from *L. Harcourtii*. In *Lepidodendron* short tracheae
occasionally occur on the inner edge of the xylem cylinder, and

in recent species of *Todea* the same kind of reduced tracheae are met with on the inner edge of the xylem[1]. In both cases the short tracheae are probably vestiges of an axial strand of conducting elements which in the course of evolution have been converted into parenchymatous cells. In *Lepidodendron vasculare* the mixed parenchyma and short tracheae in the centre

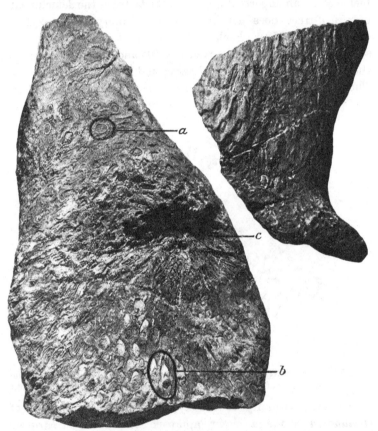

FIG. 253. *Osmundites Kolbei* Sew. (⅓ nat. size.)

of the stele represent an intermediate stage in xylem reduction, and the arrangement in vertical rows of the medullary parenchyma in *Lepidodendron* is precisely similar to that described by Kidston and Gwynne-Vaughan in *Thamnopteris*. In both cases

[1] Seward and Ford (03).

the rows of superposed short cells have probably been produced by the transverse septation of cells which began by elongating as if to form conducting tubes and ended by assuming the form of vertical series of parenchymatous elements.

In another Jurassic species, *Osmundites Gibbiana*[1], the xylem is of the *Osmunda* type and consists of about 20 strands instead of a continuous or almost continuous cylinder.

Osmundites Kolbei Seward, figs. 253—255.

This species was founded on a specimen obtained by Mr Kolbe from the Uitenhage series of Cape Colony[2]. The fossil

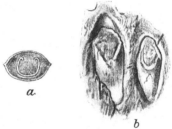

FIG. 254. *Osmundites Kolbei.* (Leaf-scars.)

flora and fauna of this series point to its correlation with the Wealden or Neocomian strata of Europe[3]. The type-specimen consists of several pieces of a stem (fig. 253) which reached a length of about 90 cm. On the weathered surface the remains of petiole-bases are clearly seen and on the reverse side of the smaller piece shown in the figure numerous sinuous roots are present in association with the leaf-stalks. The depression *c* in the larger specimen may mark the position of a branch: at *a* fig. 253 (enlarged in fig. 254, *a*) the vascular strand of a petiole is exposed as a broad U-shaped band and at *b* (fig. 254, *b*) the form of the petiole-bases is clearly shown[4]. With the stem were found imperfectly preserved impressions of fronds referred to *Cladophlebis denti-*

---

[1] Kidston and Gwynne-Vaughan (07).

[2] Seward (07³) p. 482, Pls. xx. xxɪ.

[3] Seward (03) ; Kitchin (08).

[4] Cf. *Todea Wilkesiana* (p. 286).

FIG. 255. *Osmundites Kolbei* Sew. Transverse section, from a photograph supplied by Dr Kidston and Mr Gwynne-Vaughan. (2½ nat. size.)

culata, a common type of leaf which was found also in association with the slightly older New Zealand stem, *Osmundites Dunlopi.*

An examination of the internal structure of the South African stem by Dr Kidston and Mr Gwynne-Vaughan has revealed many interesting features, which will be fully described in Part IV. of their Monograph on fossil Osmundaceous stems. I am greatly indebted to these authors for allowing me to publish the following note contributed by Dr Kidston:—

" The section of *Osmundites Kolbei* Seward, shown in fig. 255, presents the usual appearance of an Osmundaceous stock. The parts contained in this section are the stele, inner and outer cortex and a portion of the surrounding mantle of concrescent leaf-bases. The whole specimen has suffered much from pressure, but if restored to its original form the xylem ring must have been about 19 mm. in diameter. The number of xylem strands is about fifty-six and several of them are more or less joined as in the modern genus *Todea.* The tracheae are of the typical Osmundaceous type, that is to say, the pits are actual perforations and several series of them occur on each wall of the larger tracheae.

The most interesting structural characteristic of *Osmundites Kolbei* is not well seen in the figure owing to the compression of the xylem ring. This consists in the occurrence of tracheae in the pith. In fact, we have here a mixed pith, composed of parenchyma and true tracheae, a condition which connects the *Osmundaceae* with a parenchymatous medulla with those possessing a solid xylem stele like *Zalesskya* and *Thamnopteris* and so completes the series of transitions extending from the older and solid-steled forms to the modern medullated members of the *Osmundaceae.*"

Osmundites skidegatensis, Penhallow.

This lower Cretaceous Canadian species, first described by Penhallow[1] and more recently by Kidston and Gwynne-Vaughan[2], is remarkable for the large size of the stem, the stele alone having a diameter of 2·4 cm. Penhallow figures a fragment of a leaf

---

[1] Penhallow (02).   [2] Kidston and Gwynne-Vaughan (07).

bearing a superficial resemblance to that of *Osmunda Claytoniana*,
which may be the foliage borne by *Osmundites skidegatensis*.
The xylem cylinder is broken by the exit of leaf-traces into 50
or more strands varying in size and shape, and it is noteworthy
that the phloem is also interrupted as each leaf-trace is given
off. In recent species the xylem cylinder is almost always
interrupted, but the phloem retains its continuity. In the
Canadian fossil an internal band of phloem occurs between the
xylem and the pith, and this joins the external phloem at each
leaf-gap. This internal phloem finds an interesting parallel in
certain recent species[1], but in these the internal and external
phloem do not meet at the foliar gaps as they do in the extinct
type. In *Osmunda cinnamomea* the internal phloem occurs
only at the regions of branching of the stem stele; in the fossil
it is always present.

It is clear that *Osmundites skidegatensis* represents the most
complex type of stem so far recognised in the Osmundaceae; it
illustrates a stage in elaboration of the primitive protostele in
advance of that reached by any existing species.

The primitive Osmundaceous stele was composed of solid
xylem surrounded by phloem (*Thamnopteris* and *Zalesskya*);
at a later stage the xylem cylinder lost its inner zone of
wide and short tracheae and assumed the form seen in
Osmundites Kolbei, in which the centre of the stele consists of
parenchyma with some tracheae. Another type is represented
by *O. Dowkeri* in which the pith is composed wholly of parenchyma and the xylem ring is continuous. From this type,
by expansion of the xylem ring and by the formation of overlapping leaf-gaps, the form represented by *Osmunda regalis* was
reached. *Osmunda cinnamomea*, with internal phloem in the
regions of stelar branching, probably represents a further stage,
as Kidston and Gwynne-Vaughan believe, in increasing complexity due to the introduction of phloem from without through
gaps produced by the branching of the stele. In *Osmundites
skidegatensis* the leaf-gaps became wider and the external phloem
projected deeper into the stele until a continuous internal

[1] See p. 314. Also Jeffrey (03); Faull (01); Seward and Ford (03).

phloem zone was produced. This most elaborate type proved less successful than the simpler forms which still survive.

Osmundites Sturii.

Impressions of fertile pinnae with narrow linear segments bearing exannulate sporangia described by Raciborski from Lower Jurassic rocks in Poland as *Osmunda Sturii*[1] may with some hesitation be included in the list of Mesozoic Osmundaceae.

Osmundites Dowkeri.

Under this name Carruthers[2] described a petrified stem from Lower Eocene beds at Herne Bay, which in the structure of the stele agrees closely with the Jurassic species *O. Gibbiana* and conforms to the normal Osmundaceous type. It is possible, as Gardner and Ettingshausen[3] suggested, that the foliage of this species may be represented by some sterile *Osmunda*-like fragments recorded from the Middle Bagshot beds of Bovey Tracey and Bournemouth as *Osmunda lignitum*.

Todites.

This generic name[4] has been applied to fossil ferns exhibiting in the structure of the sporangia and in the general habit of the fertile fronds a close resemblance to the recent species *Todea barbara* (fig. 221, D, p. 286).

Todites Williamsoni (Brongniart) figs. 256, B, C, G.

1828. *Pecopteris Williamsonis*, Brongniart, Prodrome, p. 57 ; Hist. vég. foss., p. 324, Pl. cx. figs. 1 and 2.
—   *P. whitbiensis*, Brongniart, Hist. vég. foss. p. 321, Pl. cix. figs. 2—4.
—   *P. tenuis, ibid.* p. 322, Pl. cx. figs. 3, 4.
1829. *Pecopteris recentior*, Phillips, Geol. Yorks. p. 148, Pl. viii. fig. 15.
—   *P. curtata, ibid.* Pl. viii. fig. 12.
1833. *Neuropteris recentior*, Lindley and Hutton, Foss. Flora, Vol. i. Pl. lxviii.
—   *Pecopteris dentata, ibid.* Vol. iii., Pl. clxix.
1836. *Acrostichites Williamsonis*, Goeppert, foss. Farn. p. 285.

[1] Raciborski (94) A. p. 19, Pls. vi. xi.
[2] Carruthers (70) A. ; Kidston and Gwynne-Vaughan (07) p. 768; see also Seward, Vol. i. p. 212.
[3] Gardner and Ettingshausen (82) pp. 22, 48, Pl. iv. figs. 1—3.
[4] Seward (00) p. 86.

1841. *Neuropteris Goeppertiana,* Muenster, in Goeppert, Gattungen foss.
       Pflanz. Lief. 5 and 6, p. 104, Pls. VIII.—X.
1856. *Pecopteris Huttoniana,* Zigno, Flor. foss. Oolit. Vol. I. p. 133.
1867. *Acrostichites Goeppertianus,* Schenk, Foss. Flor. Grenzsch. p. 44,
       Pl. V. fig. 5, Pl. VII. fig. 2.

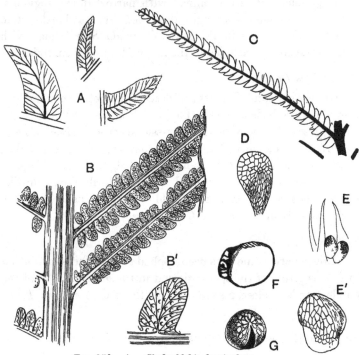

FIG. 256. A.   *Cladophlebis denticulata.*
           B, B'. *Todites Williamsoni* (fertile).
           C.   *T. Williamsoni* (sterile pinna).
           D.   *Discopteris Rallii.*
           E, E'. *Kidstonia heracleensis.*
           F.   *Todeopsis primaeva.*
           G.   *Todites Williamsoni* (sporangium).

[B, C, from specimens (13491; 39234) in the British Museum (B, very
slightly reduced; C, ½ nat. size); D, E, after Zeiller; F, after Renault; G,
after Raciborski.]

1883. *A. linnaeaefolius,* Fontaine, Older Mesoz. Flora Virginia, p. 25,
       Pls. VI.—IX.
  —   *A. rhombifolius, ibid.* Pls. VIII. XI.—XIV.
1885. *Todea Williamsonis,* Schenk, Palaeont. Vol. XXXI. p. 168, Pl. III.
       fig. 3.
1889. *Cladophlebis virginiensis,* Fontaine, Potomac Flora, p. 70, Pl. III.
       figs. 3—8; Pl. IV. figs. 1, 4.

It is hopeless to attempt to arrive at satisfactory conclusions in regard to the applicability of the name *Todites Williamsoni* to the numerous fronds from Jurassic and Rhaetic rocks, agreeing more or less closely with Brongniart's type-specimen. Specimens from the Rhaetic may not be specifically identical with those from the Jurassic; the main point is that, whether actually identical or not, both sets of fossils clearly represent the same general type of Osmundaceous fern[1] and may for present purposes be included under the same designation. The above synonymy, though by no means complete[2], serves to illustrate the confusion which has existed in regard to this widely spread type of Mesozoic fern.

Todites Williamsoni may be briefly described as follows:—

Frond bipinnate; long linear pinnae (20—30 cm.) of uniform breadth arise at an acute angle, or in the lower part of a frond, almost at right angles, from a stout rachis. Closely set pinnules attached by a broad base; slightly falcate, the side towards the rachis strongly convex and the outer margin straight or concave and bulged outwards towards the base of each segment, margin usually entire, or it may be slightly lobed. Fertile pinnules similar to the sterile; sporangia of the Osmundaceous type and often scattered over the whole lower suface of the lamina (fig. 256, B, B', G). Venation of the *Cladophlebis* type (cf. fig. 256, A).

It is not always easy to distinguish *Todites Williamsoni* from *Cladophlebis denticulata*, another common Jurassic fern, but in the latter the pinnules are usually longer and relatively narrower and the rachis is more slender (cf. fig. 256, B and 257). Schenk[3] and Raciborski[4] have shown that the sporangia of *Todites* conform in the absence of a true annulus to those of *Todea* (fig. 256, G) and *Osmunda*. Nathorst[5] has recently figured a group of spores of *Todites Williamsoni* in illustration of the use of the treatment of carbonised impressions with nitric acid and potassium chlorate. This species, though widely distributed in Jurassic rocks, is hardly distinguishable from the German Rhaetic fronds figured by Schenk from Bayreuth as *Acrostichites Goeppertianus*[6],

---

[1] Seward and Ford (03) p. 251.
[2] For a more complete list, see Seward (00) p. 87.
[3] Schenk (85) Pl. iii. fig. 3.   [4] Raciborski (94) A. Pl. vi.
[5] Nathorst (08) Pl. i. fig. 7.   [6] Schenk (67) A.

or from other fossils referred to an unnecessarily large number of species by Fontaine[1] from Upper Triassic rocks of Virginia[2].

It would seem from the paucity of later records of Osmundaceae that the family reached its zenith in the Jurassic era. When we pass to the later Tertiary and more recent deposits

Fig. 257. *Cladophlebis denticulata.* (From a specimen in the British Museum from the Inferior Oolite rocks of Yorkshire. Slightly reduced.)

evidence is afforded in regard to the geographical range of *Osmunda regalis.* It has been shown to occur in the Pliocene forest-bed of Norfolk[3] as well as in Palaeolithic and Neolithic deposits[4].

A fertile frond from the Molteno (Rhaetic) beds of South

---

[1] Fontaine (83).

[2] The geographical distribution of *Todites* and other genera will be dealt with in Volume III.

[3] Carruthers (70) A. p. 350.　　　　[4] Reid (99).

Africa referred to *Cladophlebis* (*Todites*) *Roesserti* (Presl)[1] represents in all probability an Osmundaceous fern closely allied to *Todites Williamsoni*. The same species is described by Zeiller[2] from Rhaetic rocks of Tonkin and very similar types are figured by Leuthardt[3] from Upper Triassic rocks of Basel as *Pecopteris Rutimeyeri* Heer, and by Fontaine[4] from rocks of the same age in Virginia.

Cladophlebis.

The generic name *Cladophlebis* was instituted by Brongniart for Mesozoic fern fronds characterised by ultimate segments of linear or more or less falcate form attached to the pinnae by the whole of the base, as in the Palaeozoic genus *Pecopteris*, possessing a midrib strongly marked at the base and dividing towards the distal end of the lamina into finer branches and giving off secondary forked and arched veins at an acute angle. The term is generally restricted to Mesozoic fern fronds which, on account of the absence or imperfection of fertile pinnae, cannot be safely assigned to a particular family. In the case of the species described below, the evidence in regard to systematic position, though not conclusive, is sufficiently strong to justify its inclusion in the Osmundaceae.

Cladophlebis denticulata Brongniart.   Figs. 256, A; 257, 258.

1828. *Pecopteris denticulata*[5], Brongniart, Prodrome, p. 57 ; Hist. vég.
        foss. p. 301, Pl. XCVIII. figs. 1, 2.
——   *P. Phillipsii*, Brongniart, Hist. p. 304, Pl. CIX. fig. 1.

This species is often confused[6] with *Todites Williamsoni*. The name *Pecopteris whitbiensis* has been used by different writers for Jurassic fronds which are undoubtedly specifically distinct : specimens so named by Brongniart should be referred to *Todites Williamsoni*, while *P. whitbiensis* of Lindley and

[1] Seward (08) Pl. VIII. p. 98.   [2] Zeiller (03) Pls. II. IV.
[3] Leuthardt (04) Pl. XV.   [4] Fontaine (83) Pls. XI.—XIV.
[5] For synonymy and figures, see Seward (00) p. 134; (04) p. 134.
[6] E.g. by Yokoyama (06) who identifies specimens of *Cladophlebis denticulata* from Jurassic rocks of China as *Todites Williamsoni*.

Hutton[1] is Brongniart's *Cladophlebis denticulata*. It is impossible to determine with accuracy the numerous examples
described as *Pecopteris whitbiensis, Asplenium whitbiense,
Cladophlebis Albertsii* (a Wealden species[2]), *Asplenium*, or
Cladophlebis, nebbense[3], etc., from Jurassic and Rhaetic strata.
The *Cladophlebis denticulata* form of frond is one of the
commonest in recent ferns; it is represented by such species as
*Onoclea Struthopteris, Pteris arguta, Sadleria sp., Gleichenia
dubia, Alsophila lunulata, Cyathea dealbata*, and species of
Polypodium. It is, therefore, not surprising to find records of
this Mesozoic species from many localities and horizons. All
that we can do is to point out what appear to be the most
probable cases of identity among the numerous examples of
fronds of this type from Mesozoic rocks, particularly Rhaetic
and Jurassic, in different parts of the world. The name *Cladophlebis denticulata* may be employed in a comprehensive sense
for fronds showing the following characters:—

Leaf large, bipinnate, with long spreading pinnae borne on a comparatively slender rachis. Pinnules, in nearly all cases, sterile, reaching a
length of 3—4 cm., acutely pointed, finely denticulate or entire, attached
by the whole of the base (fig. 257). In the apical region the pinnules
become shorter and broader. Venation of the *Cladophlebis* type (fig. 256, A).
Fertile pinnules rather straighter than the sterile, characterised by
linear sori parallel to the lateral veins (fig. 258).

In endeavouring to distinguish specifically between fronds
showing a general agreement in habit with *C. denticulata*,
special attention should be paid to venation characters, the
shape of the pinnules, the relation of the two edges of the
lamina to one another, and to the amount of curvature of the
whole pinnule. Unless the material is abundant, it is often
impossible to distinguish between characters of specific value
and others which are the expression of differences in age or of
position on a large frond, to say nothing of the well-known
variability which is amply illustrated by recent ferns. It is
remarkable that very few specimens are known which throw
any light on the nature of the fertile pinnae. Fig. 258 repre-

---

[1] Lindley and Hutton (34) A. Pl. cxxxiv.   [2] Seward (94[2]) A. p. 91.
[3] Nathorst (78).

sents an impression from the Inferior Oolite rocks of the
Yorkshire coast in which the exposed upper surface of the
pinnules shows a series of parallel ridges following the course
of the lateral veins and no doubt formed by oblong sori on the
lower surface. There can be little doubt that the specimen
figured by Lindley and Hutton and by others as *Pecopteris
undans*[1] is, as Nathorst suggests, a portion of a fertile frond of
C. denticulata. A fertile specimen of a frond resembling in
habit *C. denticulata,* which Fontaine has described from the

FIG. 258. Fertile pinnae of *Cladophlebis denticulata.* (From a Yorkshire
specimen in the Sedgwick Museum, Cambridge.)

Jurassic rocks of Oregon as *Danaeopsis Storrsii*[2], exhibits, as
that author points out, a superficial resemblance to the specimen
named by Lindley and Hutton *Pecopteris undans.* There is,
however, no adequate reason for referring the American
fragment to the Marattiaceae. In the absence of sporangia we
cannot speak confidently as to the systematic position of this
common type; but there are fairly good grounds for the assertion
that some at least of the fronds described under this name
are those of Osmundaceae. The English specimen shown in

[1] Lindley and Hutton (34) A. Pl. cxx.
[2] Fontaine, in Ward (05) Pl. xv. figs. 6—9.

fig. 258 is very similar to some Indian fossils figured by Feistmantel as *Asplenites macrocarpus*[1], which are probably identical
with *Pecopteris australis* Morris[2], a fern that is indistinguishable
from *Cladophlebis denticulata*. Renault[3] figured a fertile specimen of the Australian fossil as *Todea australis*, which agrees
very closely with that shown in fig. 258, and the sporangia
figured by the French author are of the Osmundaceous type.
Another example of a fertile specimen is afforded by a Rhaetic
fern from Franconia, *Asplenites ottonis*, which is probably
identical with *Alethopteris Roesserti* Presl [= *Cladophlebis
(Todites) Roesserti*], a plant closely resembling *Cladophlebis
denticulata*. Another argument in favour of including *C. denticulata* in the Osmundaceae is supplied by the association of
pinnae of this type with the petrified stem of *Osmundites
Dunlopi* recorded by Kidston and Gwynne-Vaughan.

Schizaeaceae.

Evidence bearing on the existence of this family in Carboniferous floras is by no means decisive. The generic name
Aneimites proposed by Dawson[4] for some Devonian Canadian
plants resembling species of the recent genus *Aneimia*, and
adopted by White[5] for a species from the Potsville beds of
Virginia, is misleading. The Canadian plants give no indication of the nature of the reproductive organs, and the fronds
described by White are, as he shows, those of a Pteridosperm
and bore seeds.

An examination of the suspiciously diagrammatic drawings
published by Corda[6] of the small fertile pinnules of a Carboniferous fern from Bohemia, which he named *Senftenbergia
elegans*, leads us to conclude that the sporangia are almost
certainly those of a Schizaeaceous species. The small linear
pinnules bear two rows of sessile sporangia, singly as in recent
Schizaeaceae and not in sori, characterised by 4—5 rows of
regular annular cells (fig. 270, A) surrounding the apex. It
has already been pointed out that the apical annulus of recent

---

[1] Feistmantel (77) Pls. xxxvi. xxxvii.
[2] Morris (45) Pl. vii.   [3] Renault (83) p. 81, Pl. xi.
[4] Dawson (61).   [5] White (04).   [6] Corda (45) A. Pl. lvii.

Schizaeaceae, though normally one row deep, may consist in part at least of two rows. Zeiller[1] examined specimens of Corda's species and decided in favour of a Schizaeaceous affinity; he describes the sporangia as 0·85—0·95 mm. in length, with 3 to 5 and occasionally only two rows of cells in the apical annulus. Zeiller's figures (fig. 270, A) confirm the impression that Corda's drawings are more beautiful than accurate. Stur[2], on the other hand, who first pointed out that the type-specimens of *Senftenbergia* came from the Radnitz beds of Bohemia and not from the Coal-Measures, convinced himself that the sporangia have no true annulus (fig. 270, E). He describes them as characterised by a comparatively strong wall and by the presence of a band of narrow vertical cells marking the line of dehiscence, features which lead him to assign the plant to the Marattiales, a group which seems to have exercised a dominating influence over his judgment. In a later publication Zeiller[3] replies to Stur's criticism but adheres to his original opinion. Solms-Laubach[4], while expressing himself in favour of Marattiaceous affinity, recognises that Zeiller's arguments cannot be set aside.

The question must remain open until further evidence is forthcoming; but it would seem that this Carboniferous type, not as yet recognised in Britain, possessed sporangia having a distinct resemblance to those of the Schizaeaceae, though this similarity does not amount to proof of the existence of the family in the Palaeozoic era.

Palaeozoic floras may be described as rich in generalised types types foreshadowing lines of evolution, which in the course of ages led to a sorting and a redistribution of characters. It may be that *Senftenbergia* is one of these generalised types.

It is not until we ascend the geological series as far as the older Jurassic rocks that we meet with a type which can with confidence be classed with the Schizaeaceae, as least so far as sporangial characters are concerned. The species *Klukia exilis*

---

[1] Zeiller (83) p. 188, Pl. x. figs. 1—5.   [2] Stur (85) A. p. 64.
[3] Zeiller (88) A. p. 50.   [4] Solms-Laubach (91) A. p. 147.

is selected as the best known and most widely-spread representative of Jurassic Schizaeaceae.

Klukia exilis (Phillips)[1].   Fig. 259.

The generic name *Klukia* was proposed by Raciborski[2] for a species originally described by Phillips[3] from the Inferior Oolite of the Yorkshire coast as *Pecopteris exilis*. Bunbury's[4] discovery (supplemented by additional evidence obtained by Raciborski) of well-preserved sporangia justified the substitution of a distinctive designation for the provisional term *Pecopteris*.

FIG. 259.   *Klukia exilis* (Phillips).   (Figs. 1—3, ×40; fig. 4, ×3|; fig. 5, nat. size.)

The species may be defined as follows:—

Frond tripinnate, of the *Cladophlebis* type; pinnae linear, lanceolate, attached to the rachis at a wide angle. Ultimate segments short and linear, entire or, in the lower part of a frond, crenulate, 5 mm. long or occasionally longer. Sporangia 0·5 mm. in length, borne singly on the lower surface of the lamina in a row on each side of the midrib.

A re-examination[5] of the specimen described by Bunbury confirmed his account of the structure of the sporangia. The pinna shown in fig. 259 is characterised by unusually small fertile pinnules some of which bear 10 sporangia in two rows; the annulus includes about 14 cells. Fertile specimens of this and similar forms are figured by Raciborski[6] from Jurassic rocks of Poland, and good examples of the English species may be seen

---

[1] For synonymy, see Seward (00) p. 130.   [2] Raciborski (91).
[3] Phillips (29) A. p. 148.   [4] Bunbury (51) A.   [5] Seward (94²) A.
[6] Raciborski (94) A.

in the Leckenby collection, Cambridge, in the British Museum, the museums of Manchester, Scarborough, and other places.

It is possible that specimens referred to *K. exilis* by Yokoyama[1] from Wealden strata in Japan may afford evidence of the persistence of the species beyond the Jurassic era, but in

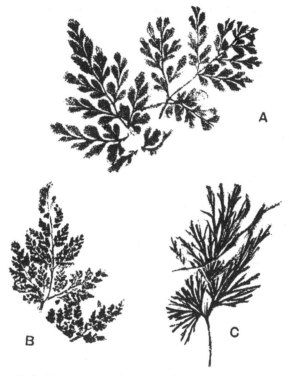

FIG. 260.   *Ruffordia Goepperti.*   (A, C, sterile; B, fertile; slightly reduced. Specimens from the Wealden of Sussex; British Museum; V. 2333, V. 2160, V. 2166.)

view of the close resemblance of the sterile fronds described from Wealden strata as *Cladophlebis Brownii*[2] and *C. Dunkeri*[2] to those of *Klukia exilis*, identity can be established only by an examination of fertile specimens. A Jurassic fern recently described by Yabe[3] from Korea as *Cladophlebis koraiensis* may be identical with *K. exilis* and there is little doubt as to the existence of the species in Jurassic Caucasian strata[4].

[1] Yokoyama (89).   [2] Seward (94²) A.   [3] Yabe (05) Pl. III.
[4] Seward (07⁴) Pls. I. III.

Ruffordia Goepperti (Dunk.). Fig. 260.

This Wealden fern[1] has been doubtfully assigned to the
Schizaeaceae on the ground of the resemblance of the sterile
fronds to those of some species of *Aneimia*, and because of the
difference between the sterile and fertile pinnae (Fig. 260).
Ruffordia cannot be regarded as a well authenticated member
of the Schizaeaceae.

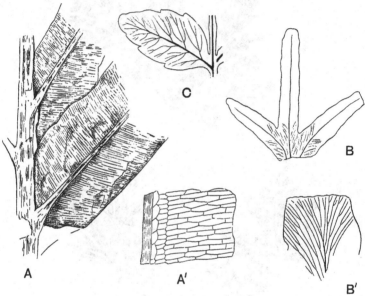

Fig. 261.   A, A'.   *Chrysodium lanzaeanum.*
B, B'.   *Lygodium Kaulfussi.*
C.   *Marattia Hookeri.*
(After Gardner and Ettingshausen ; A, B, ¾ nat. size.)

Lygodium Kaulfussi, Heer.   Fig. 261, B, B'.

Fragments of forked pinnules, agreeing very closely in venation and general appearance with recent species of *Lygodium*,
have been identified by Gardner and Ettingshausen[2] from
English Eocene beds and by Knowlton from the Miocene beds of
the Yellowstone Park[3] as *Lygodium Kaulfussi* Heer (fig. 261, B).
Despite the absence of sporangia it is probable that these

[1] Seward (94²) A. p. 75.
[2] Gardner and Ettingshausen (82) p. 47, Pls. vii. x. ; Heer (55) A. Pl. iii. p. 41.
[3] Knowlton (99) Pl. lxxx.

fragments are correctly referred to the Schizaeaceae. The sterile and fertile specimens figured by Heer[1] from Tertiary beds of Switzerland agree very closely with recent examples of *Lygodium*. Similar though perhaps less convincing evidence of the existence of this family in Europe is furnished by Saporta[2], who described two Eocene species from France.

Gleicheniaceae.

The application by Goeppert[3] and other earlier writers of the generic name *Gleichenites* to examples of Palaeozoic ferns was not justified by any satisfactory evidence. One of Goeppert's species, *Gleichenites neuropteroides*, is identical with *Neuropteris heterophylla*[4], a plant now included in the Pteridosperms.

The resemblance of sporangia and sori, whether preserved as carbonised impressions or as petrified material, from Carboniferous rocks, to those of recent species of Gleicheniaceae is in many cases at least the result of misinterpretation of deceptive appearances. Williamson[5] drew attention to the Gleichenia-like structure of some sections of sporangia from the English Coal-Measures, but he did not realise the ease with which sections of Marattiaceous sporangia in different planes may be mistaken for those of annulate (leptosporangiate) sporangia. In the regular dichotomous habit of Carboniferous fronds described as species of *Diplothmema* (Stur) and *Mariopteris* (Zeiller)[6] we have a close correspondence with the leaves of *Gleichenia*, but the common occurrence of dichotomous branching among ferns is sufficient reason for regarding this feature as an untrustworthy criterion of relationship. It is, however, interesting to find that in addition to the existence of some Upper Carboniferous ferns with sori like those of recent Gleichenias, the type of stelar anatomy illustrated by *Gleichenia dicarpa* (fig. 237, C, p. 310) and other species is characteristic of the primary structure of the stem of the Pteridosperm *Heterangium*. We find in Carboniferous types undoubted indications of anatomical and other features which in succeeding ages became the marks of Gleicheniaceae.

[1] Heer (55) A. Pl. xiii.   [2] Saporta (72) A. Pl. i. figs. 13, 14.
[3] Goeppert (36²) A. Pls. iv. v.   [4] Zeiller (88) A. p. 261.
[5] Williamson (77) Pl. vii.   [6] See Ch. xxvii.

Some Carboniferous fronds with short and small pinnules of the *Pecopteris* type, bearing sori composed of a small number of sporangia, have been assigned by Grand'Eury and other authors to the Gleicheniaceae; the same form of sorus is met with also on fronds with Sphenopteroid segments. The former is illustrated by. *Oligocarpia Gutbieri*[1] and the latter by *O. Brongniarti* described by Stur and by Zeiller[2]. Zeiller has described the circular sori of *Oligocarpia* (fig. 270, B) as consisting of three to ten pyriform sporangia borne at the ends of lateral veins and possessing a complete transverse annulus, but Stur[3] believes that the annulus-like appearance is due to the manner of preservation of exannulate sporangia. In this opinion Stur is supported by Solms-Laubach[4] and by Schenk[5]. Despite an agreement between *Oligocarpia* and *Gleichenia*, as regards the form of the sori and the number of sporangia, it is not certain that the existence of a typical Gleicheniaceous annulus has been proved to occur in any Palaeozoic sporangia[6].

From Upper Triassic beds of Virginia, Fontaine has figured several fronds for which he instituted the genus *Mertensides*[7] The habit, as he points out, is not dichotomous, but the sori are circular and are said to be composed in some species of four to six sporangia. No satisfactory evidence is brought forward in support of the use of a designation implying a close relationship with recent Gleichenias (sect. *Mertensia*). One of the species described by Fontaine was originally named by Bunbury *Pecopteris bullatus*[8], the imperfect type-specimen of which is now in the Museum of the Cambridge Botany School. In the form of the frond, the thick rachis, and in the pinnules this Triassic species resembles *Todites Williamsoni*, but the resemblance does not extend to the sori. Two of Fontaine's species are recorded by Stur from Austria[9], but he places them in the genus *Oligocarpia* and includes them in the Marattiaceae.

[1] Goeppert (41) Pl. IV. figs. 1, 2.　　[2] Zeiller (88) A. Pl. XI. figs. 3—5.
[3] Stur (85) A. p. 128.　　[4] Solms-Laubach (91) A. p. 146.
[5] Schenk (88) A. p. 30.
[6] Dr Scott tells me that an examination of Dr Zeiller's specimens led him to agree with the latter's description of the annulus of *Oligocarpia*. (A. C. S.)
[7] Fontaine (83) Pls. XV.—XIX.
[8] Bunbury (47) Pl. II. fig. 1; Seward (94²) A. p. 189.　　[9] Krasser (09) p. 16.

Leuthardt[1] figures what appears to be a Gleicheniaceous fern from the Upper Triassic beds of Basel as *Gleichenites gracilis* (Heer) showing sori composed of five sporangia (fig. 265, C) with a horizontal annulus. A Rhaetic species *Gleichenites microphyllus* Schenk[2] from Franconia agrees in the form of its small rounded pinnules with *Gleichenia*, but no sporangia have so far been found.

An impression of a frond from Jurassic rocks of northern Italy figured by Zigno as *Gleichenites elegans*[3] closely resembles in habit recent species of *Gleichenia*; though no sporangia have been found, the habit of the frond gives probability to Zigno's determination.

A Jurassic species from Poland, *Gleichenites Rostafinskii*, referred by Raciborski[4] to *Gleichenia*, exhibits a close agreement in habit and in the form of the soral impressions to some recent species of *Gleichenia*.

As we pass upwards to Wealden and more recent rocks it becomes clear that the Gleicheniaceae were prominent members of late Mesozoic floras in north Europe and reached as far north as Disco Island. In English Wealden beds portions of sterile fronds have been found which were assigned to a new genus *Leckenbya*[5], but it is probable that these specimens would be more correctly referred to *Gleichenites*. Similarly fragments of Gleichenia-like pinnae with very small rounded pinnules occur in the Wealden rocks of Bernissart, Belgium[6], in north Germany[7], and elsewhere. Conclusive evidence has been obtained by Prof. Bommer of the existence of *Gleichenites* in Wealden beds near Brussels, where many plant remains have been found in a wonderful state of preservation. The specimens, which I had an opportunity of seeing some years ago, might easily be mistaken for rather old and brown pieces of recent plants. Some of the Belgian fragments, of which Prof. Bommer has kindly sent me drawings and photographs, are characterised by an arrange-

[1] Leuthardt (04) p. 40, Pl. xviii. fig. 3.
[2] Schenk (67) A. p. 86, Pl. xxii. figs. 7, 8.
[3] Zigno (56) A. Pl. x.
[4] Raciborski (94) A. p. 43, Pl. xiii. figs. 15—20.
[5] Seward (95) A. p. 225.   [6] Seward (00) Pl. iv.   [7] Schenk (71).

ment of vascular tissue identical with that in the petioles and
rhizomes of some protostelic Gleichenias. The stele of one of
the Belgian rhizomes appears to be identical with that of
Gleichenia dicarpa (fig. 237, C. p. 310).

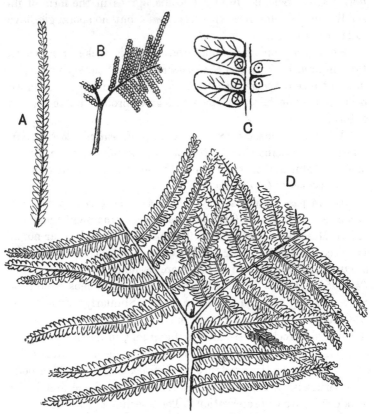

FIG. 262. A. *Gleichenites longipennis* Heer.
B. *G. delicatula* Heer.
C. *G. Nordenskioldi* Heer.
D. *G. Zippei.* (Corda.)
(After Heer ; A, B, D, very slightly reduced.)

Gleichenites Zippei (Corda).   Fig. 262, D.

This species, originally described by Corda as *Pecopteris
Zippei*[1] and afterwards figured by Heer[2] as *Gleichenia Zippei*

[1] Corda, in Reuss (46) p. 95, Pl. XLIX.   [2] Heer (75), p. 44, Pls. IV.—VII.

(fig. 262, D) from Urgonian rocks of Greenland, affords a striking example of a Mesozoic member of the *Gleicheniaceae*. It is characterised by the dichotomous branching of the frond and by the occurrence of arrested buds in the forks. The long and slender pinnae, reaching a length of 9 cm. and a breadth of 6—8 mm., bear small crowded pinnules occasionally with circular sori which are described by Heer as consisting of a small number of sporangia (cf. fig. 262, C). Several other Lower Cretaceous species are recorded by Heer from Greenland, some of which are probably unnecessarily separated from *Gleichenites Zippei*. Examples of these are represented in fig. 262, A, B, C.

A Gleicheniaceous species described by Debey and Ettingshausen from Lower Cretaceous rocks of Aix-la-Chapelle as *Didymosorus comptonifolius*[1] is very similar in habit to some of Heer's Greenland species: this should probably be referred to the genus *Gleichenites*.

Gleichenites hantonensis, Wank. Fig. 263.

From the Eocene beds of Bournemouth, Gardner and Ettingshausen[2] have described under the name *Gleichenia hantonensis* what is in all probability a true *Gleichenia* (fig. 263). This species, originally recorded by Wanklyn[3], is characterised by a slender forked rachis showing what may be traces of arrested buds between the arms of the branches, by circular sori of six or eight sporangia and by the presence of peculiar tendril-like appendages on the pinnae. If the description of the tendrils is correct, this British species affords one of the few instances of ferns adapted for climbing and may be compared with the recent species *Davallia aculeata* (fig. 232, p. 299).

Matonineae.

The genera *Laccopteris* and *Matonidium* may be described as examples of Mesozoic ferns exhibiting a very close agreement with *Matonia*.

---

[1] Debey and Ettingshausen (59) Pl. I.
[2] Gardner and Ettingshausen (82), pp. 43, 59, Pls. VI. X.
[3] Wanklyn (69).

Laccopteris. This genus, founded by Presl[1], may be described as follows:—

Frond pedate, in habit resembling *Matonia pectinata*, with pinnate or pinnatifid pinnae; ultimate segments linear, provided with a well-marked midrib giving off numerous dichotomously branched secondary veins which are in places connected by lateral anastomoses. Sori circular, forming a single row on each side of the midrib (fig. 278, B); sporangia 5—15 in each sorus, with an oblique annulus and tetrahedral spores. The presence of an indusium is not certainly established.

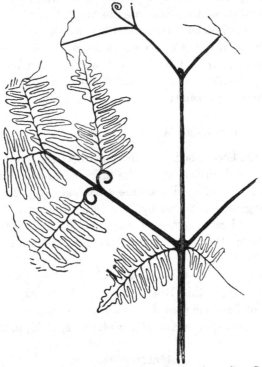

FIG. 263. *Gleichenites hantonensis* Wank. (Restoration, after Gardner and Ettingshausen.)

Schenk[2], who described several specimens of *Laccopteris* from Rhaetic rocks of Germany, compared the genus with *Gleichenia* but he also recognised the close resemblance to

[1] Presl, in Sternberg (38) A. p. 115.　　[2] Schenk (67) A.

Matonia pectinata. Zeiller[1] first established the practical identity of the sori and sporangia of *Laccopteris* and *Matonia.* The Rhaetic species, such as *L. Muensteri, L. elegans,* and *L. Goepperti,* agree very closely with *L. polypodioides* and need not be described in detail.

Fɪɢ. 264. *Laccopteris elegans* (Presl). (From a specimen in the British Museum; from the Lower Keuper of Bayreuth, Germany. Nat. size; part of pinnule × 3.)

The Rhaetic species *Laccopteris elegans,* represented in fig. 264, illustrates the characteristic habit of the genus and shows a feature usually overlooked[2], namely the occurrence of anastomoses between the lateral veins. The form of the sorus

[1] Zeiller (85).   [2] Seward (99²) p. 194.

of another Rhaetic species is shown in fig. 265, E. Schenk
figures an interesting series of fronds of *L. Goepperti* in different
stages of growth[1]; one of the younger leaves is seen in fig. 265, D.
An examination of Rhaetic specimens of *Laccopteris* in the
Bergakademie of Berlin convinced me of the correctness of the
published descriptions of the sori.

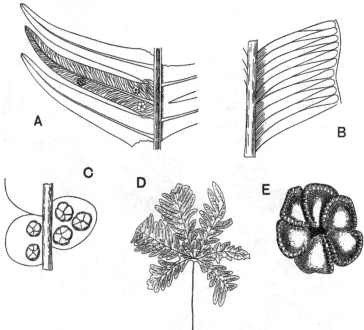

FIG. 265.　A.　*Matonidium Wiesneri.*　(Slightly enlarged.)
　　　　　　B.　*Marattiopsis marantacea.*　(Slightly enlarged.)
　　　　　　C.　*Gleichenites gracilis.*　(Slightly enlarged.)
　　　　　　D.　*Laccopteris Goepperti.*　(Slightly reduced.)
　　　　　　E.　*L. Muensteri.*　(Enlarged.)
　　　(A, after Krasser ; B, C, after Leuthardt ; D, E, after Schenk.)

Laccopteris polypodioides (Brongniart).　Figs. 266—268;
278, A.

1828.　*Phlebopteris polypodioides*[2], Brongniart, Hist. vég. foss. p. 372,
　　　　Pl. LXXXIII. fig. 1.
　—　　*P. propinqua, ibid.* Pls. CXXXII. fig. 1, CXXXIII. fig. 2.

　　　　　　[1] Schenk (67) A. Pls. XXIII. XXIV.
　　　　　　[2] For a more complete list, see Seward (00) p. 78.

1829. *Pecopteris caespitosa*, Phillips, Geol. Yorks. p. 148, Pl. VIII. fig. 10.
— *P. crenifolia, ibid.* Pl. VIII. fig. 10.
— *P. ligata, ibid.* Pl. VIII. fig. 14.

In habit this species closely resembles *Matonia* and *Matonidium*, the long petiole divides distally into several spreading

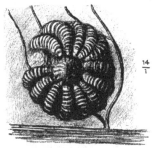

FIG. 266. *Laccopteris polypodioides* (Brongn.). (× 14.) (Brit. Mus.)

pinnatifid pinnae with linear ultimate segments (fig. 278, A). Circular sori (indusiate?) occur in a single row on each side of the midrib containing 12—14 large sporangia (fig. 266) characterised by an obliquely vertical annulus. The midrib of the

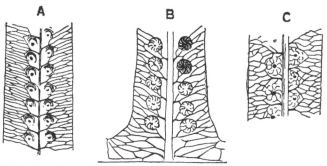

FIG. 267. Pinnules of *Laccopteris*. (Enlarged.)
A, B. From the Inferior Oolite of Yorkshire.
C. From the Inferior Oolite of Stamford. (British Museum.)

pinnules gives off secondary veins at a wide angle and these form a series of elongated meshes parallel to the median rib, as in the recent genus *Woodwardia*; forked and anastomosing branches are given off from these to the edge of the lamina (fig. 267).

The specimen shown in fig. 268 is probably a young frond of this species.

A very similar, possibly a specifically identical plant, was described by Leckenby from English Jurassic rocks as *Phlebopteris Woodwardi*[1], the distinguishing features of which are the greater number of lateral veins and the smaller sori (fig. 267, A).

The name *Microdictyon* was proposed by Saporta[2] for pinnules

Fig. 268.   *? Laccopteris polypodioides.*   Nat. size.   From a specimen in the Whitby Museum (Brit. Mus.).

differing slightly from those of *Laccopteris* in venation characters: he included *Laccopteris Woodwardi* in this genus, but such differences as are recognisable in the venation hardly justify the use of a distinct generic title.   Similarly, specimens described by Debey and Ettingshausen[3] from Lower Cretaceous rocks of Aix-la-Chapelle as species of *Carolopteris* may also be included in *Laccopteris*.

[1] Leckenby (64) A. p. 81, Pl. viii. fig. 6.   (Type-specimen in the Sedgwick Museum, Cambridge.)

[2] Saporta (73) A. p. 306.   [3] Debey and Ettingshausen (59) Pl. iii.

Laccopteris Dunkeri (Schenk)[1]

This species is represented in several Wealden localities by fragments of fertile pinnae similar to those of *L. polypodioides*. It is almost impossible to distinguish small specimens of the Wealden fern from Heer's genus *Nathorstia* (Marattiaceae) unless the sori are well preserved. This species occurs in Wealden beds in England, Germany, Belgium, and elsewhere and has been discovered by Dr Marcus Gunn in Upper Jurassic plant-beds of Sutherlandshire (N.E. Scotland).

Laccopteris is widely spread in Rhaetic, Jurassic and Lower Cretaceous floras. It affords evidence of the former abundance in northern latitudes of a family now represented by the two species of *Matonia* confined to a restricted area in the southern hemisphere.

Matonidium.

Schenk[2] instituted this convenient term for fossil fern fronds agreeing in habit and in their sori with *Matonia pectinata* (figs. 227, 228, p. 292). Zeiller[3] has drawn attention to the fact that the Mesozoic species differ from the surviving types in the greater number of sporangia in each sorus, and, it may be added, in *Matonidium* the fertile pinnules are more richly supplied with sori than are those of *Matonia*. Unfortunately our knowledge of the structure of the sporangia of *Matonidium* is less complete than in the case of *Laccopteris*, but such evidence as is available justifies the conclusion that *Matonia* is a direct descendant of ferns which formed a prominent feature in European Jurassic and Wealden floras. It is interesting to find that in a Cretaceous species, described by Krasser (fig. 265, A) since the publication of Zeiller's paper, the sori appear to be identical in distribution and in appearance with those of the recent species.

I am indebted to Prof. Bommer for permission to reproduce the unpublished drawing represented in fig. 237 D (p. 310) of a section of the rhizome of *Matonidium* from the Belgian Wealden

---

[1] See Seward (94²) A. and (00) for an account of this fern.
[2] Schenk (71) p. 219. [3] Zeiller (85).

beds of Hainaut ("Flore Bernissartienne"). The section shows an
arrangement of vascular tissue identical with that in the recent
species: there may be two solenosteles and in addition a solid
axial strand. The form of the leaf-trace in the fossil appears to
be identical with that in *Matonia pectinata* (fig. 237, A, p. 310).

Matonidium Goepperti (Ettingshausen)[1].   Fig. 269.

Under this name are included specimens from Inferior

Fig. 269. *Matonidium Goepperti* (Ettings.). (A, B, ½ nat. size; C, approximately nat. size.)

Oolite and Wealden strata in Britain and elsewhere. It is,
however, not impossible that if more information were available,
we should find adequate reasons for recognising two specific
types. Fontaine[2], adhering rigidly to the rules of priority,
speaks of this species as *Matonidium Althausii* (Dunker), but
Ettingshausen's specific term is better known.

[1] Ettingshausen (52) p. 16, Pl. v. For synonymy, see Seward (94²) A; (00).
[2] Fontaine, in Ward (05) p. 230.

Fronds pedate and apparently identical in habit with those of *Matonia pectinata*; ultimate segments linear, slightly falcate and bluntly pointed. Sori circular or oval, numerous, containing 15 to 20 sporangia with an oblique annulus, in two rows on the lower surface of the pinnules; indusium as in *Matonia*.

The English examples have so far afforded no information in regard to sporangial structure, but Schenk[1] has recognised a distinct annulus in German material. In his description of fossil plants from Lower Cretaceous rocks in California, Fontaine[2] doubtfully identifies two very small fragments as *Matonidium Althausii*; the evidence is, however, wholly inadequate.

Matonidium Wiesneri, Krasser[3]. Fig. 265, A.

This Cenomanian (Cretaceous) species from Moravia appears to be identical in habit with the older type. The pinnules are larger and bear fewer sori. Krasser's figures of the sterile pinnules show no lateral anastomosing between the secondary veins, but the small vascular network below each sorus (fig. 265, A) is identical with that in *Matonia pectinata*. The indusiate sori contain about six sporangia with an oblique annulus.

The very wide geographical distribution of the Matonineae during the Mesozoic era affords a striking contrast to the limited range of the Malayan survivals.

Hymenophyllaceae.

The frequent use of the generic name *Hymenophyllites* as a designation of Palaeozoic ferns, more particularly in the older literature, is another instance of the undue importance which palaeobotanists have always been prone to attach to external resemblances of vegetative organs. The fragment of lamina described by Stur for the Culm Measures of Austria as *Hymenophyllum waldenburgense*[4] has no claim to consideration as evidence of Palaeozoic Hymenophyllaceae. On the other hand,

[1] Schenk (71) p. 19. [2] Fontaine, *loc. cit.* Pl. LXV. figs. 22, 23.
[3] Krasser (96) p. 119, Pls. XI. XII. XIV.
[4] Stur (75) A. p. 284, Pl. XXXIII. fig. 15.

there are a few records of fertile fronds which, though not to be accepted without reserve, are worthy of more careful examination. Some petrified sporangia described by Renault[1] from the Culm of Esnost are referred to *Hymenophyllites* on account of the position of the annulus, which appears to encircle about two-thirds of the circumference; it is, however, not certain that the annulus is horizontal as in the recent genus.

The Culm species *Rhodea patentissima* described by Ettingshausen[2] as *Hymenophyllites patentissima* and subsequently

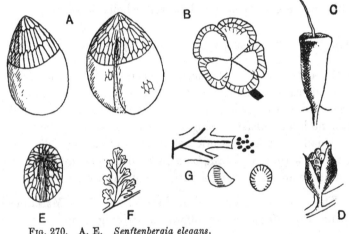

Fig. 270. A, E. *Senftenbergia elegans.*
B. *Oligocarpia Brongniartii.*
C. *Trichomanes* sp.
D. *Hymenophyllum tunbridgense.*
F, G. *Sphenopteris (Hymenophyllites) quadridactylites.*
(A, B, F, G, after Zeiller; D, after Hooker; E, after Stur.)

referred by Stur[3] to *Rhodea*, is regarded by these authors as closely allied to *Hymenophyllum* simply on the ground of the finely divided and delicate sterile fronds; another species, *Rhodea moravica* (Ett.), which Ettingshausen referred to *Trichomanes*, is compared with recent species of that genus. In neither case do we know anything of sporangial characters.

A fertile sphenopteroid frond figured by Schimper as *Hymenophyllum Weissi*[4] from the Coal-Measures of Saarbrücken

[1] Renault (96) A. p. 19. [2] Ettingshausen (66) Pl. VII. fig. 4.
[3] Stur (75) A. p. 36, Pl. IX. figs. 1—9. [4] Schimper (74) A. Pl. XXVIII. fig. 4—7.

bears some resemblance to recent Hymenophyllaceae, but the figures are by no means convincing: an examination of the type-specimens in the Strassburg Museum led Solms-Laubach[1] to express dissent from Schimper's determination. A more satisfactory example is that afforded by the fertile pieces of a frond described by Zeiller[2] from French Coal-Measures as *Hymenophyllites quadridactylites* (Gutbier). Some of the ultimate segments with a truncated tip are preserved in close association with a group of oval sporangia with a complete transverse annulus (fig. 270, F, G). The position of the sporangia is such as to suggest their separation from a terminal columnar receptacle like that in *Trichomanes* and *Hymenophyllum*. In his account of this species from the Coal-Measures of the Forest of Wyre, Kidston[3] states that Zeiller informed him that he had noticed traces of what appeared to be a columnar receptacle in the French specimens.

The records of Hymenophyllaceae from the Mesozoic and Tertiary formations are not such as need detain us. The facts bearing on the geological history of this family are singularly meagre. There is no evidence which can be adduced in favour of regarding the Hymenophyllaceae as ferns of great antiquity, which played a prominent part in the floras of the past.

It is interesting to find that the genus *Ankyropteris*[4], one of the Botryopterideae (a group of Palaeozoic Ferns for which I propose the name Coenopterideae), has a morphological character in common with *Trichomanes*, namely the production of axillary buds: there are also features in the stelar anatomy shared by the Botryopterideae and Hymenophyllaceae[5]. These resemblances, though by no means amounting to proof of near relationship, point to a remote ancestry for certain features retained by existing members of the Hymenophyllaceae.

Cyatheaceae.

The specimens from the Culm rocks of Moravia on which

[1] Solms-Laubach (91) A. p. 153.   [2] Zeiller (83) p. 155 ; (88) A. Pl. VIII. figs. 1–3.
[3] Kidston (84²) p. 593.   [4] See p. 450.   [5] Scott (08) p. 343.

Stur founded the species *Thyrsopteris schistorum*[1] are too imperfectly preserved to warrant the use of this generic name. Goeppert[2] in 1836 instituted the genera *Cyatheites*, *Hemitelites*, and *Balantites* for species of Carboniferous ferns believed to be closely allied to recent Cyatheaceae, but a fuller knowledge of these types has clearly demonstrated that in all cases the reference to this family had no justification.

The Upper Carboniferous species *Dicksonites Pluckeneti*, of which Sterzel[3] described fertile specimens in 1886 as possessing circular sori, has since been shown by Grand'Eury[4] to be a Pteridosperm bearing small seeds. In *Sphenopteris* (*Discopteris*) *cristata* (Brongn.) Zeiller[5] has described sori very like those of *Cyathea* and *Alsophila*, but differing in the exannulate sporangia: this species, like so many of the Palaeozoic ferns, is probably more akin to the Marattiaceae than to the Cyatheaceae.

We have as yet no satisfactory evidence of the existence of the Cyatheaceae in Palaeozoic floras. It is not until we reach the Jurassic period that trustworthy data are obtained. Raciborski[6] has identified as Cyatheaceous fertile Jurassic fronds from Poland, but his figures are inconclusive. In *Alsophila polonica* it is not clear whether the annulus is vertical or oblique, and in another supposed member of the family, *Gonatosorus Nathorsti*, in which the indusium is described as bivalvate, there is no proof of affinity to Cyatheaceae.

In attempting to decipher the past history of the Cyatheaceae it is important to remember the close resemblance between the fertile segments of some species of *Davallia* (Polypodiaceae) and those of *Dicksonia* (fig. 229, C, D, p. 294). Unless the sporangia are well enough preserved to show the position of the annulus, it is frequently impossible to feel much confidence in the value of the grosser features, such as the reduced lamina of the fertile segments and the form of the sori. It is, however, probable that the widely-spread Jurassic species *Coniopteris hymenophylloides* is correctly referred to the

---

[1] Stur (75) A. p. 19, Pl. x. figs. 1, 2.
[2] Goeppert (36²) A. pp. 319, 320, 329.
[3] Sterzel (86).   [4] Grand'Eury (05).
[5] Zeiller (06) Pls. ii. iii.   [6] Raciborski (94) A. Pl. ix.

Cyatheaceae, but even in the case of this species the evidence of external form needs confirmation by an examination of individual sporangia.

Coniopteris.

This genus was instituted by Brongniart[1] for fossil fronds characterised by pinnules more or less intermediate between the *Pecopteris* and *Sphenopteris* type and agreeing in the form of the sori with the leaves of recent species of *Dicksonia*. It should be noted that Stur included in this genus a species, *Coniopteris lunzensis*[2] from the Upper Trias of Lunz, which he regarded as a Marattiaceous fern.

Coniopteris hymenophylloides, Brongn. Figs. 271, 272, 275, B.

1828. *Sphenopteris hymenophylloides*, Brongniart, Hist. vég. foss. p. 189, Pl. LVI. fig. 4.

1829. *S. stipata*, Phillips, Geol. York. p. 147, Pl. x. fig. 8.

1835. *Tympanophora simplex*, Lindley and Hutton, Foss. Flor. Pl. CLXX. A.

— *T. racemosa*, ibid. Pl. CLXX. B.

— *Sphenopteris arguta*, ibid. Pl. CLXVIII.

1836. *Hymenophyllites Phillipsi*, Goeppert, Foss. Farn. p. 256.

1849. *Coniopteris hymenophylloides*, Brongniart, Tableau, p. 105.

— *Coniopteris Murrayana*, ibid.

1851. *Sphenopteris nephrocarpa*, Bunbury, Quart. Journ. Geol. Soc. Vol. VII. p. 129, Pl. XII. fig. 1.

1876. *Thyrsopteris Murrayana*, Heer, Flor. Foss. Arct. Vol. IV. (2) p. 30, Pls. I. II. VIII.

The above list represents a small selection of the names applied to Jurassic ferns from different localities which there are good grounds for regarding as referable to a single type[3].

Frond tripinnate; pinnae linear acuminate, attached to the rachis at a wide angle; the pinnules vary considerably in size and shape; in some the lamina is divided into a few broad and rounded lobes (fig. 275, B) while in others the leaflets are dissected into narrow linear segments. The sori are borne at the ends of veins; the fertile pinnules have a much reduced lamina and, in extreme cases, bear a close resemblance to those of

[1] Brongniart (49) A. p. 26. [2] Krasser (09).
[3] For fuller synonymy see Seward (00) p. 97.

Thyrsopteris elegans (fig. 229, A, p. 294).   The sori are partially enclosed in
a cup-like indusium and the sporangia appear to have an oblique annulus.
Venation and habit of frond of the *Sphenopteris* type.

The pinna shown in fig. 271 is the type-specimen of
Sphenopteris arguta Lind. and Hutt. from the Yorkshire
Inferior Oolite and is indistinguishable from the English
examples on which Brongniart founded his species *S. hymenophylloides*.   Fig. 272 shows a specimen from the York Museum
illustrating the difference between the sterile and fertile pinnae.
The resemblance of some fertile pinnae of *Coniopteris hymeno-*

Fɪɢ. 271.   *Coniopteris hymenophylloides* (Brongn.).   Nat. size.   From a specimen
in the Manchester Museum.

phylloides to those of *Thyrsopteris elegans* has led to a frequent
use, without any solid justification, of the generic name of the
Juan Fernandez fern for Jurassic and Wealden plants.   It is not
impossible that some of the fossils described by Heer from
Jurassic rocks of Siberia[1] as species of *Thyrsopteris* are Cyatheaceous ferns, but it is impossible to say with certainty that
they are generically identical with the recent species.   In his
monograph of the Potomac flora of Virginia[2] and Maryland,
Fontaine has described as species of *Thyrsopteris* several speci-

[1] Heer (76).   [2] Fontaine (89).

mens of fronds which afford no evidence as to the nature of the
sori or sporangia. Some of the fronds referred by this author
to *Thyrsopteris rarinervis*[1], which I examined in the Washington
Museum, are in all probability examples of *Onychiopsis*, a genus
included in the Polypodiaceae. The fragments described by
Lester Ward[2] as species of *Thyrsopteris* from the Lower
Cretaceous of the Black Hills of North America afford no

Fig. 272. *Coniopteris hymenophylloides*. Specimen from the Inferior Oolite,
        Scarborough; in the York Museum. [M.S.]

satisfactory evidence of relationship to the recent type. Similarly
Velenovsky has described a Lower Cretaceous *Onychiopsis* from
Bohemia[3] as a species of *Thyrsopteris,* although the fertile
segments bear little or no resemblance to those of the Cyathe-
aceous genus. Some fertile portions of fronds described by
Heer[4] as *Asplenium Johnstrupi* and afterwards as *Dicksonia*

---

[1] Fontaine (89) p. 123, Pls. xxvi. xliii. etc.   [2] Ward (99) Pl. clxi.
[3] Velenovský (88).   [4] Heer (75) A. Pl. i. figs. 6, 7.

Johnstrupi[1] from the Cretaceous beds (Kome series) of Greenland
are very similar to *Coniopteris hymenophylloides*.

Coniopteris quinqueloba (Phillips).   Fig. 273.

This species, originally described by Phillips[2] as *Sphenopteris
quinqueloba*, is very similar in habit to *C. hymenophylloides*,
differing chiefly in the smaller size of the leaf and in the
narrower ultimate segments.   The specimen shown in fig. 273, B,
illustrates the form of the sorus and sporangia.

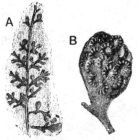

Fig. 273.   *Coniopteris quinqueloba* (Phillips).   A, ×2; B, considerably enlarged.   From drawings supplied by Dr Nathorst.

Coniopteris arguta (Lind. and Hutt.[3]).   Figs. 274, 275, A.

The sterile pinnae of this species bear pinnules of a type
met with in various species of ferns from different horizons;
the smaller ones are entire and slightly falcate, while on the
lower part of a frond the ultimate segments are longer and
have a crenulate margin.   The fertile pinnae bear pinnules
reduced to a midrib with a narrow border, and terminating
in a cup-like indusium (fig. 275, A).   In habit the sterile leaf
(fig. 274) of this species is similar to the Jurassic Schizaeaceous
fern *Klukia exilis*.

Protopteris.

Presl[4] instituted this genus for a Lower Cretaceous tree-

[1] Heer (82) A. Pl. ii. fig. 2.   [2] Phillips (75) A. p. 215.
[3] Seward (00).   [4] Sternberg (38) A. p. 169.

fern from Bohemia originally figured as *Lepidodendron punc-tatum*[1] and assigned to a Palaeozoic horizon; it was afterwards

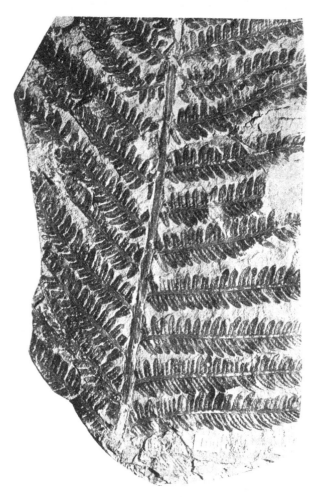

Fɪɢ. 274. *Coniopteris arguta.* (Nat. size. From a specimen in the Sedgwick Museum, Cambridge.)

named by Corda[2] *Protopteris Sternbergii* and referred by Brongniart[3] to *Sigillaria.* The genus *Protopteris* stands for

---

[1] Sternberg (20) A. Pl. ɪᴠ.
[2] Corda (45) A. Pl. ɪɪ. fig. 5.   [3] Brongniart (28) A. Pl. xʟɪ.

fossil fern-stems with the habit and, in the main, the structural features of recent tree-ferns. Persistent leaf-bases and sinuous adventitious roots cover the surface of the stems: the vascular system is of the dictyostelic type characteristic of *Cyathea* (fig. 240, p. 313) and *Alsophila*. It is by the pattern formed by the vascular tissue on the exposed surface of the leaf-bases that *Protopteris* is most readily recognised: the leaf-trace has a horse-shoe form with the ends curled inwards and the sides more or less indented (fig. 277). The generic name *Caulopteris* is used by some authors in preference to Presl's genus; but *Protopteris* is more conveniently restricted to Mesozoic Cyathe-aceous stems and *Caulopteris* to Palaeozoic stems, with the internal structure of *Psaronius* (see Chap. XXIII.). Stenzel

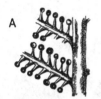

FIG. 275. A. *Coniopteris arguta.* (Fertile pinnae; nat. size.)
  B. *C. hymenophylloides.*

A, from the Inferior Oolite of Yorkshire (British Museum); B, from Jurassic rocks in Turkestan.

applies *Caulopteris* to Mesozoic stems in which the leaf-trace consists of several separate strands and not of a continuous band.

Lower Cretaceous casts of tree-fern stems in the Prague Museum have been described under the names *Alsophilina* and *Oncopteris*; the figures of the latter (fig. 276) given by Feist-mantel[1] and by Velenovský[2] show the petiole-bases arranged in vertical rows and characterised by leaf-traces consisting of two separate strands in the form of two Vs lying on their sides.

Tree-fern stems described under various generic names are not infrequently found in European Lower Cretaceous rocks: their comparative abundance affords an example of striking changes in geographical distribution since the latter part of the Mesozoic epoch. The Cyatheaceae no longer exist in Europe and the

[1] Feistmantel (72). [2] Velenovský (88).

arborescent species of the genus have retreated to more southern regions.

FIG. 276. *Oncopteris Nettvalli*. (After Velenovský; ¾ nat. size.)

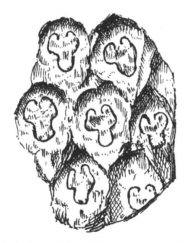

FIG. 277. *Protopteris punctata*. (After Heer; very slightly reduced.)

Protopteris punctata (Sternb.). Fig. 277.

The earliest information in regard to the anatomy of this widely spread Lower Cretaceous fern we owe to Corda, who showed that the species agrees in essentials with existing

tree-ferns. The English example described by Carruthers[1]
from Upper Greensand beds in Dorsetshire (now in the British
Museum) shows only the external features. The sandstone cast

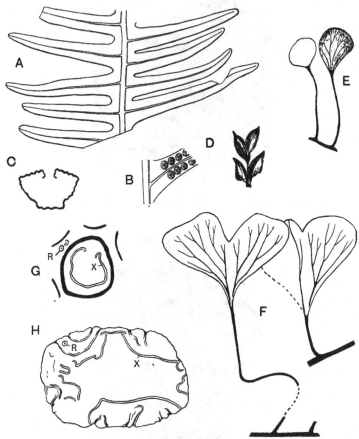

FIG. 278. A. *Laccopteris polypodioides*, Brongn. [From a specimen (39275) in
              the British Museum ; slightly reduced.]
          B. *L. Muensteri.*
          C. *Dicksonia* (petiole stele).
          D. *Onychiopsis Mantelli* (fertile segments).
          E. *Hausmannia Sewardi* Richt.
          F. *H. Kohlmanni* Richt.
       G, H. *Protopteris Witteana*, Schenk. (x, xylem ; R, roots.)
              (B, after Schenk ; E, F, after Richter.)

[1] Carruthers (65) Pl. XIII.

(14 cm. in diameter), of which a portion is seen in fig. 277, was described by Heer from Disco Island (Greenland) as a Carboniferous species[1], but afterwards correctly assigned to the Cenomanian series[2]. This species is recorded also from the Lower Cretaceous of Bohemia by Frič and Bayer[3]. Among examples of petrified stems exhibiting a general agreement with *Protopteris punctata* are those described by Stenzel[4] from Turonian rocks in Germany. In one of these, *Rhizodendron oppoliense* Göpp., attention is drawn to branches given off from the stem stele which have a solenostelic structure in contrast to the dictyostele of the stem; also to the minute structure of the tracheae which appear to have their ends perforated, a feature shown by Gwynne-Vaughan[5] to be characteristic of the xylem elements of many ferns.

Protopteris Witteana Schenk[6] (fig. 278, G, H), a Wealden species recorded from Germany and England, represents a closely allied or possibly an identical type. The section of the stem (fig. H) shows the narrow vascular bands, x, of a dictyostele similar to that of recent Cyatheaceous tree-ferns and a form of meristele (fig. G, x) resembling that of *P. punctata*. Adventitious roots are seen in section at R (figs. G and H).

Polypodiaceae.

Sections of petrified sporangia from the English Coal-Measures (*Pteridotheca* sp.) occasionally exhibit a striking resemblance to those of recent Polypodiaceae[7], but in the absence of material in which it is possible to recognise the true orientation of the sporangia, the exact position of the annulus is almost impossible to determine. We have as yet no satisfactory evidence of the existence of true Polypodiaceae in the Palaeozoic era. It is noteworthy that apart from the absence of ferns which can reasonably be included in this family, the anatomical features of the Botryopterideae (Coenopterideae) and of the Cycadofilices or Pteridosperms do not foreshadow those

[1] Heer (75).   [2] Heer (82) Pl. xlvii.
[3] Frič and Bayer (01) p. 76.
[4] Stenzel (86). See also Stenzel (97).   [5] Gwynne-Vaughan (08).
[6] Schenk (71) Pl. xxx. ; Seward (94²) A. Pl. xi.   [7] Scott (08) p. 293.

of Polypodiaceous ferns. On the other hand, as we have already noticed, anatomical characters of such families as the Gleicheni-aceae, Hymenophyllaceae, and Schizaeaceae are met with in certain generalised Palaeozoic types. These facts are perhaps of some importance as supplying collateral evidence in favour of the relatively more recent origin of the dominant family of ferns in modern floras.

The use of the generic name *Adiantites* for fern-like fronds of Lower Carboniferous age characterised by cuneate

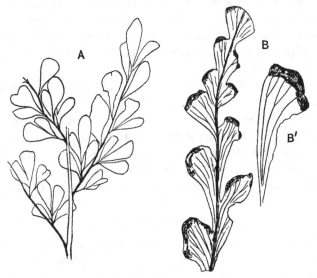

Fig. 279. A. *Adiantides antiquus* (Ett.). (½ nat. size.)
          B. *A. Lindsayoides* (Sew.). (B′ nat. size.)
                  (A, after Kidston.)

pinnules like those of species of *Adiantum*, suggests an affinity which is in all probability non-existent. It has been pointed out that this generic name was applied in the first instance to the leaves of the Jurassic plant *Ginkgo digitata*[1] and should, therefore, be discarded. Schimper[2] used the designation *Adiantides*, and Ettingshausen[3], more rashly than wisely, pre-ferred *Adiantum*. The specimens described by Kidston[4] as

[1] Goeppert (36) A. p. 217.   [2] Schimper (69) A. p. 424.
[3] Ettingshausen (66).   [4] Kidston (89³) Pl i.

XXI]   POLYPODIACEAE   377

Adiantides antiquus (Ett.) (fig. 279, A) from the Carboniferous limestone of Flintshire are portions of tripinnate fronds bearing cuneate segments with numerous forked veins radiating from the contracted base of the lamina. It is not improbable, in view of Dr White's[1] discovery of seeds on a very similar plant from the Pottsville beds of North America, that this characteristic Lower Carboniferous genus is a Pteridosperm.

From Jurassic rocks in various parts of the world numerous fossils have been described under the generic names *Aspidium*, *Asplenium*, *Davallia*, *Polypodium*, and *Pteris*. In the great majority of cases such records leave much to be desired from the point of view of students who appreciate the dangers of relying on external similarity between vegetative organs, and on resemblances founded on obscure impressions of sori. The generic term *Woodwardites*[2], which suggests affinity with the recent genus *Woodwardia*, has been used for Rhaetic plants belonging to the Dipteridinae.

A plant described as *Adiantides Lindsayoides* from Jurassic rocks of Victoria[3], characterised by marginal sori which appear to be protected by the folded-over edge of the leaflets, and by the resemblance of the pinnules to those of recent species of *Lindsaya*, may be a true Polypodiaceous fern ; but in this case, as in many similar instances, nothing is known of the structure of the sporangia. Some sterile pinnae described by Yabe from Jurassic rocks of Korea as *Adiantites Sewardi*[4] may perhaps be identical with the Australian species.

In such a species as *Polypodium oregonense* Font., from Jurassic rocks of Oregon, the generic name is chosen because the " fructification seems near enough to that of *Polypodium* to justify the placing of the plant in that genus[5]." But the fact that no sporangia have been found is a fatal objection to this identification.

Onychiopsis.

This generic name was instituted by Yokoyama[6] for a

---

[1] White (04).   [2] Schenk (67) A. Pl. xiii. ; Zeiller (03) p. 91, Pl. xvii.
[3] Seward (04²) p. 162, Pl. viii. fig. 5.   [4] Yabe (05) p. 39, Pl. i. figs. 1-8.
[5] Fontaine, in Ward (05) p. 64.   [6] Yokoyama (89), p. 26.

Japanese Wealden species, previously described by Geyler[1] as *Thyrsopteris elongata*, on the ground that, in addition to a similarity in habit of the sterile fronds, the fertile pinnae present a close agreement to those of the recent genus *Onychium*.

Onychiopsis Mantelli[2] (Brongn.). Figs. 278, D; 280, A and B.

The Japanese species *Onychiopsis elongata* may perhaps be identical with this common Wealden fern which, as Fontaine points out, should be called *O. psilotoides* if the rule of priority is to be observed irrespective of long usage.

> 1824. *Hymenopteris psilotoides*, Stokes and Webb, Trans. Geol. Soc. [ii.], Vol. I. p. 423, Pl. XLVI. fig. 7.
> 1828. *Sphenopteris Mantelli*, Brongniart, Hist. vég. foss. p. 170, Pl. XLV. figs. 3—7.
> 1890. *Onychiopsis Mantelli*, Nathorst, Denksch. Wien Akad. Vol. LVII. p. 5.

Onychiopsis Mantelli may be defined as follows:—

> Frond bipinnate, ovate lanceolate, rachis winged; pinnae approximate, given off at an acute angle; pinnules narrow, acuminate, with a single vein; the larger segments serrate and gradually passing into pinnae with narrow ultimate segments. Fertile segments sessile or shortly stalked, linear ovate, sometimes terminating in a short awn-like prolongation.

The fertile segments (fig. 278, D) bear so close a resemblance to those of species of *Onychium* that it would seem justifiable to regard the plant as a member of the Polypodiaceae. This fern is one of the most characteristic members of the Wealden floras: it occurs in abundance in the English Wealden, in Portugal, Germany, Belgium, Japan, Bohemia, South Africa, and elsewhere. A piece of rhizome figured from the English Wealden[3] is very similar to the creeping rhizomes of recent species of Polypodiaceae. The English Wealden specimens

---

[1] Geyler (77) Pl. XXXI. fig. 4.
[2] For synonymy, see Fontaine, in Ward (05) p. 155; Richter (06) p. 6; Seward (94) A. p. 41; (03) p. 5.
[3] Seward (94) A. p. 52.

shown in fig. 280, A and B, illustrate the difference in form presented by leaves of this species; the smaller pinnae reproduced in fig. A are more characteristic of the species than are those of the slightly enlarged example represented in fig. 280, B.

A

B

FIG. 280. *Onychiopsis Mantelli.* (From Wealden specimens in the British Museum; No. 13495 and No. V. 2615. A, natural size; B, very slightly enlarged.)

Among British Tertiary species referred to Polypodiaceae, it is interesting to find what may well be an authentic record of a fern closely allied to the recent tropical species *Acrostichum (Chrysodium) aureum*. This Eocene species from Bournemouth is described as *Chrysodium lanzaeanum*[1]. The frond is simply pinnate and apparently coriaceous in texture, with lanceolate or oblong lanceolate pinnules (fig. 261, A, A′, p. 350), differing from those of *Acrostichum aureum* in being sessile. A prominent

[1] Gardner and Ettingshausen (82) Pls. I. II.

midrib gives off numerous anastomosing veins. No fertile pinnules have been found.

Specimens described by Forbes from the Eocene beds of the Island of Mull as *Onoclea hebraidica*[1] bear a strong likeness to the North American and Japanese recent species *Onoclea sensibilis*. Fertile specimens referred to the latter species are recorded by Knowlton[2] from Tertiary beds of Montana.

A species described by Saporta[3] from the Eocene of Sézanne as *Adiantum apalophyllum* is recorded by Gardner and Ettingshausen from Bournemouth; an identification which is based on somewhat meagre evidence.

The following remarks by Gardner and Ettingshausen are worthy of repetition as calling attention to circumstances often overlooked in analyses of fossil floras. They speak of ferns as relatively rare in British Eocene rocks and add,—" the floras consist principally of deciduous dicotyledonous leaves, which... fell into the water and were tranquilly silted over. Ferns, on the other hand, would require some violence to remove them from the place of their growth, and their preservation would consequently be exceptional, and they would be mutilated and fragmentary. This may account for their rarity. Few as the British ferns are in the number of species, they nevertheless form the largest and most important series of Eocene ferns, even of Tertiary ferns, yet described from one group of beds[4]."

Dipteridinae.

Dictyophyllum.

This genus was founded by Lindley and Hutton for a pinnatifid leaf from the Jurassic rocks of Yorkshire which they regarded as probably dicotyledonous and named *D. rugosum*[5]. Several ferns of this genus have since been found with well-preserved sori which demonstrate a close similarity to the recent fern *Dipteris*. *Dictyophyllum* may be defined as follows:—

[1] Forbes (51) ; Gardner and Ettingshausen (82).
[2] Knowlton (02) Pl. xxvi.
[3] Saporta (68) A. ; Gardner and Ettingshausen (82) Pl. x. fig. 1.
[4] Gardner and Ettingshausen (82) p. 21.
[5] Lindley and Hutton (34) A. Pl. civ.

Fronds large and palmate, characterised by the equal dichotomy of the main rachis into two arms which curve outwards and then bend inwards (fig. 281); from the surface of each arm are given off numerous spreading pinnae with a lamina more or less deeply dissected into lobes varying in breadth and in the form of the apex.   Each lobe has a median vein, from which branches are given off approximately at right angles and then subdivide into a reticulum, in the meshes of which the veinlets end blindly (fig. 282, A and E). Sori composed of annulate sporangia are crowded on the lower surface of the lamina. In habit and in sporangial characters the genus closely resembles *Dipteris*, and in the branching of the frond suggests comparison with *Matonia*.   The rhizome (*Rhizomopteris*) is creeping and dichotomously branched, bearing leaf-scars with a horseshoe form of vascular strand.

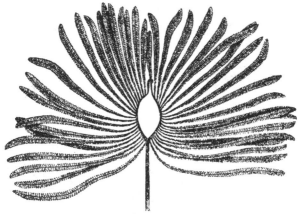

Fig. 281.   *Dictyophyllum exile*.   (After Nathorst ; much reduced.)

Dictyophyllum is represented by several types to which various specific names have been assigned, the distinguishing features being the form of the pinna lobes, the degree of concrescence between the basal portions of the pinnae, and similar features which in some cases can only be safely used as criteria when large specimens are available for comparison.

Dictyophyllum exile (Brauns).   Figs. 281, 282, D, E.

1862.   *Camptopteris exilis*, Brauns, Palaeontograph. ix. p. 54.
1867.   *Dictyophyllum acutilobum*, Schenk, Foss. Flor. Grenz. p. 77, Pls. xix. xx.
1878.   *D. exile*, Nathorst, Flora vid Bjuf, i. p. 39, Pl. v. fig. 7.
—   *D. acutilobum, ibid.* Pl. xi. fig. 1.

The restoration, after Nathorst[1], shown in fig. 281 illustrates
the habit of this striking fern, examples of which or of closely
allied species are recorded from Rhaetic rocks of Germany, Scania,
Persia, Bornholm, Tonkin, China, and elsewhere[2]. The petiole,
reaching a length of 60 cm., forks at the apex into two equal
arms leaving between them an oval space and occasionally
crossing one another. The axes of these branches are twisted
so that the pinnae, which may be as many as 24 on each arm,

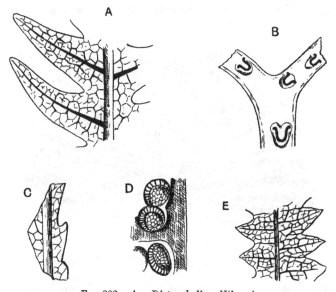

Fig. 282. A. *Dictyophyllum Nilssoni.*
B. *Rhizomopteris Schenki.*
C. *Camptopteris spiralis.*
D, E. *Dictyophyllum exile.*
(After Nathorst; A, B, C, E, ⅔ nat. size.)

and arise from the inner side, by torsion of the axes assume an
external position. An interesting analogy as regards the
twisted rachis of *Dictyophyllum exile* and *Camptopteris* is
afforded by the leaves of the Cycads, *Macrozamia Fawcettiae* and
M. corallipes, which are also characterised by the torsion of
the rachis. The habit, justly compared by Nathorst with

[1] Nathorst (06³).   [2] Seward and Dale (01) p. 505.

that of *Matonia pectinata*, affords another illustration of the
common occurrence in older ferns of a dichotomous system
of branching. The pinnae, characterised by circinate vernation,
reach a length of 60 cm. and are divided into linear lobes
inclined obliquely or at right angles to the pinna axis. The
whole of the under surface of the lamina may be covered with
sporangia, 4—7 sporangia in each sorus; the annulus is
incomplete and approximately vertical (fig. 282, D). The
rhizome is probably represented by the dichotomously branched
axis described by Nathorst from Scania as *Rhizomopteris
major*; the leaf-scars show a horse-shoe leaf-trace.

Dictyophyllum Nathorsti Zeiller[1].

This type, represented by a splendid series of specimens
from the Rhaetic beds of Tonkin, agrees very closely with
D. exile. It differs, however, in the basal parts of the pinnae
which are concrescent for a length of 5 to 8 cm. instead of free as
in *D. exile*; and, to a slight degree, in the form of the ultimate
segments. In habit and in soral characters the two species are
practically identical. Each sorus contains 5 to 8 sporangia,
which are rather larger than those of *Dipteris*.

Dictyophyllum rugosum, Lind. and Hutt. Fig. 283.

1828. *Phlebopteris Phillipsii*, Brongniart, Hist. vég. foss. p. 377,
Pl. cxxxii. fig. 3; Pl. cxxxiii. fig. 1.
1829. *Phyllites nervulosis*, Phillips, Geol. Yorks. p. 148, Pl. viii.
fig. 9.
1834. *Dictyophyllum rugosum*, Lindley and Hutton, Foss. Flor. ii.
Pl. civ.
1836. *Polypodites heracleifolius*, Goeppert, Foss. Farn. p. 344.
1849. *Camptopteris Phillipsii*, Brongniart, Tableau, p. 105.
1880. *Clathropteris whitbyensis*, Nathorst, Berättelse, p. 83.

This species, which is characteristic of Jurassic rocks, is
less completely known than the two types described above,
but in the form and venation of the pinnae there is little
difference between the Rhaetic and Jurassic plants. The leaves
of the Jurassic species appear to have been smaller and more

[1] Zeiller (03) p. 109, Pls. xxiii.—xxviii.

like those of *Dipteris conjugata* (fig. 231); there are no indications of the existence of the two curved arms at the summit
of the petiole which form so striking a feature in *D. exile* and

FIG. 283. *Dictyophyllum rugosum* (Lind. and Hutt.). (Brit. Mus. Nat. size.)

D. Nathorsti.   No sporangia have been found on English specimens, but it is safe to assume their agreement with those of other species.   A more complete list of records of *D. rugosum* is given in the first volume of the British Museum Catalogue of Jurassic plants[1].

Nathorst[2] has recently drawn attention to certain differences between *Dictyophyllum* and *Dipteris*.   The pinnate division of the pinnae is not represented in the fronds of the recent species, but this method of lobing, which is a marked characteristic of *Dictyophyllum*, is less prominent in *Clathropteris*; and in *Camptopteris lunzensis* Stur[3], an Austrian Upper Triassic species, the pinnae are entire.   In *Dictyophyllum* the sori cover the whole lower surface of the leaf; in *Dipteris* they are more widely separated and the sporangia have a diameter of 0·02 mm., but in *Dictyophyllum* the diameter is 0·4—0·6 mm.   Moreover in *Dictyophyllum* the sori contain 5 to 8 sporangia, whereas in *Dipteris* they are much more numerous.   Despite these differences it is clear, as Nathorst says, that *Dictyophyllum*, *Clathropteris*, and *Camptopteris* are existing types very closely allied to *Dipteris*.   It is a matter of secondary importance whether we include all in the Dipteridinae or follow Nathorst's suggestion and refer the fossil genera to the separate family Camptopteridinae.

Thaumatopteris.

This genus, founded by Goeppert[4] for a Rhaetic plant from Bayreuth, is by some authors[5] regarded as identical with *Dictyophyllum*, but it has recently been resuscitated by Nathorst[6] for specimens which he names *T. Schenki*, formerly included by Schenk in his species *T. Brauniana*[7].   It bears a close resemblance, in the long linear pinnules with an entire or crenulate margin, to *Dictyophyllum Fuchsi* described by Zeiller[8] from Tonkin, and it would seem hardly necessary to adopt a distinctive generic designation.   The sporangia have

[1] Seward (00) p. 122.
[3] Krasser (09) p. 111.
[5] Seward and Dale (01) p. 503.
[7] Schenk (67) A. Pl. xviii.

[2] Nathorst (06³).
[4] Goeppert (41).
[6] Nathorst (07²).
[8] Zeiller (03).

a vertical or slightly oblique annulus and the rhizome is similar to that of *Dictyophyllum exile*. The habit of the genus is shown in fig. 284, which represents one of the German Rhaetic species.

Fig. 284. *Thaumatopteris Münsteri*. (From a specimen in the Bergakademie, Berlin ; ⅓ nat. size.)

Clathropteris.

Clathropteris meniscoides, Brongn. Fig. 285.

Clathropteris, founded by Brongniart[1] for Rhaetic specimens from Scania, agrees very closely with some species of *Dictyophyllum*, but in view of the more rectangular form of the venation-meshes it is convenient to retain both names. The type-species was originally named *Filicites meniscoides*[2] and afterwards transferred to *Clathropteris*. An examination of

[1] Brongniart (25).　　[2] Brongniart (25).

Brongniart's specimens has convinced Nathorst of the specific
identity of *C. meniscoides* and *C. platyphylla*. The Tonkin
leaves described by Zeiller[1] under the latter name should,
therefore, be included in *C. meniscoides*, which may be thus
defined:

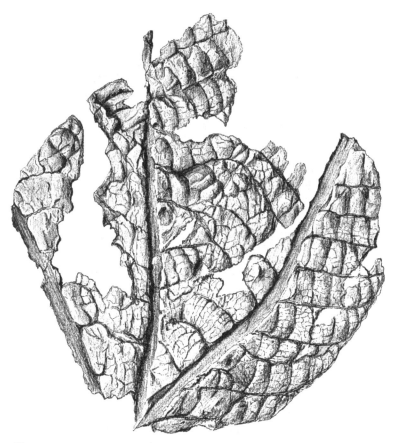

FIG. 285. *Clathropteris meniscoides.* From Rhaetic rocks near Erlangen. [M.S.]

The petiolate frond is characterised by an equal dichotomy of the
rachis, as in *Dictyophyllum*; each branch bore 5—15 pinnae, disposed
en éventail, reaching a length of 20—30 cm. and fused basally as in
D. Nathorsti Zeill. Pinnae linear lanceolate, slightly contracted at the

[1] Zeiller (03).

lower end and gradually tapered distally. The lamina, 3—14 cm. broad, is characterised by obtusely pointed marginal lobes. From the midrib of each pinna lateral veins are given off at a wide angle, and adjacent veins are connected by a series of branches which divide the lamina into a regular reticulum of rectangular and polygonal meshes (fig. 285). The sori are abundant and contain 5—12 sporangia like those of *Dictyophyllum*.

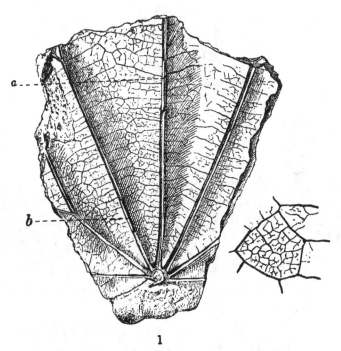

1

FIG. 286. *Clathropteris egyptiaca.* (Nat. size.) *a, b,* pieces of main ribs in grooves.

What is probably the rhizome of this species has been described by Nathorst (*Rhizomopteris cruciata*); it is similar to that of *Dictyophyllum*, but the leaf-scars are more widely separated. This species occurs in Upper Triassic, Rhaetic or Lower Jurassic rocks of Scania, France, Germany, Switzerland, Bornholm, North America, China, Tonkin, and Persia and is represented by fragments in the Rhaetic beds of Bristol[1].

[1] Seward (04) pp. 18, 164.

Clathropteris egyptiaca Sew.[1]   Fig. 286.

The specimen on which this species was founded was discovered in the Nubian Sandstone east of Edfu; the age of the beds is uncertain, but the presence of *Clathropteris* suggests a Lower Jurassic or Rhaetic horizon[2]. Seven strong ribs radiate through the lamina from the summit of the petiole; at *a* and *b* small pieces of the projecting ribs are shown in the grooves.

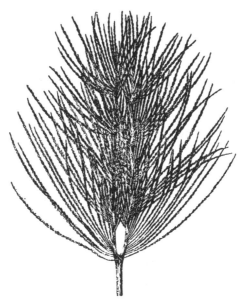

FIG. 287.   *Camptopteris spiralis*.   (After Nathorst.   Much reduced.)

From the main veins slender branches are given off at right angles and, as seen in the enlarged drawing, these again subdivide into a delicate reticulum with free-ending veinlets.

Camptopteris.

Camptopteris spiralis, Nath.   Figs. 282, C; 287.

Nathorst proposed this generic name for Rhaetic fronds[3]

---

[1] Seward (07).

[2] The evidence of the shells is stated by Mr R. B. Newton (09) to be in favour of the Cretaceous age of the Nubian Sandstone.

[3] Nathorst (78) p. 33.

resembling those of *Clathropteris* and *Dictyophyllum*, but differ-ing in the form of the pinnae and in habit. The habit of the type-species, *C. spiralis*, is shown in fig. 287. An examination of the specimens in the Stockholm Museum convinced me of the correctness of Nathorst's restoration[1]. Each of the forked arms of the rachis bore as many as 150—160 long and narrow pinnae characterised by an anastomising venation (fig. 282, C) and by a spiral disposition due to the torsion of the axes. The sporangia agree in essentials with those of *Dictyophyllum*.

Hausmannia.

A critical and exhaustive account of this genus has been given by Prof. Von Richter[2] based on an examination of specimens found in the Lower Cretaceous rocks of Quedlinburg in Germany. The name was proposed by Dunker[3] for leaves from the Wealden of Germany characterised by a deeply dissected dichotomously branched lamina. Andrae subsequently instituted the genus *Protorhipis*[4] for suborbicular leaves with dichoto-mously branched ribs from the Lias of Steierdorf. A similar but smaller type of leaf was afterwards described by Zigno[5] from Jurassic beds of Italy as *P. asarifolius*, and Nathorst[6] figured a closely allied form from Rhaetic rocks of Sweden. While some authors regarded *Hausmannia* and *Protorhipis* as ferns, others compared them with the leaves of *Baiera* (Gink-goales); Saporta suggested a dicotyledonous affinity for leaves of the *Protorhipis* type. The true nature of the fossils was recognised by Zeiller[7], who called attention to the very close resemblance in habit and in soral characters to the recent genus *Dipteris*. A comparison of the different species of *Dipteris*, including young leaves (fig. 231, p. 297), with those of the fossil species reveals a very striking agreement[8]. There can be no doubt, as Richter points out, that the names *Haus-mannia* and *Protorhipis* stand for one generic type.

[1] Nathorst (06²) p. 15.   [2] Richter (06).
[3] Dunker (46) A. p. 12.   [4] Andrae (53) A.
[5] Zigno (56) A. IX. fig. 2.   [6] Nathorst (78²) Pl. IX. fig. 2.
[7] Zeiller (97⁴) p. 51.   [8] Seward and Dale (01).

Hausmannia may be defined as follows :

Rhizome creeping, slender, dichotomously branched ; leaf-stalks slender (2—25 cm. long), bearing a leathery lamina (1—12 cm. long and broad), wedge-shaped below, occasionally cordate or reniform, entire or more or

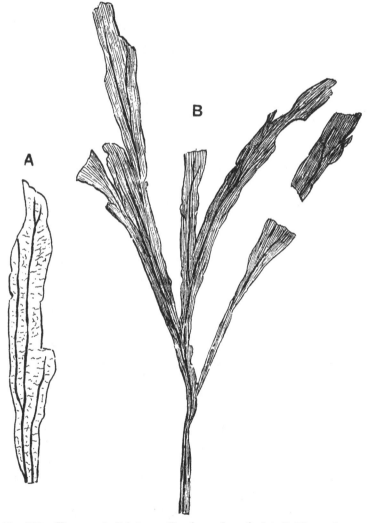

FIG. 288. *Hausmannia dichotoma.* (Specimens from the late Dr Marcus Gunn's Collection of Upper Jurassic plants, Sutherlandshire ; very slightly reduced.)

less deeply lobed into broad linear segments. The leaf is characterised by dichotomously branched main ribs which arise from the summit of the rachis as two divergent arms and radiate in a palmate manner, with repeated forking, through the lamina. Lateral veins are given off at a wide angle, and, by subdivision, form a fairly regular network similar to that in *Dictyophyllum*, *Clathropteris*, and *Dipteris*.

Hausmannia dichotoma, Dunker[1]. Fig. 288, A, B.

This Wealden species, represented in the North German flora and in beds of approximately the same age at Quedlinburg, has been discovered by Dr Marcus Gunn in Upper Jurassic rocks on the north-east coast of Scotland. The lamina (12 cm. or more in length) is divided into five to seven linear segments and bears a close superficial resemblance to leaves of *Baiera* and to recent species of *Schizaea* (fig. 222, p. 287). Each segment contains one or two main ribs (fig. 288, A). A similar form is described by Bartholin[2] and by Moeller[3] as *H. Forchammeri* from Jurassic rocks of Bornholm.

Hausmannia Kohlmanni, Richt. Fig. 278, F.

In this species, instituted by Richter from material obtained from the Lower Cretaceous beds of Strohberg[4], the comparatively slender rhizome bears fronds with petioles reaching a length in extreme cases of 25 cm. but usually of about 10 cm. The lamina (1—7 cm. long and 1—10 cm. broad) is described as leathery, obcordate, and divided into two symmetrical halves by a median sinus which, though occasionally extending more than half-way through the lamina, is usually shallow. The venation consists of two main branches which diverge from the summit of the petiole (fig. 278, F) and subdivide into dichotomously branched ribs; finer veins (not shown in the drawing) are given off from these at right angles and form more or less rectangular meshes as in other members of the Dipteridinae and in such recent ferns as *Polypodium quercifolium* (fig. 231, D, p. 297).

The imperfect lamina represented in fig. 289 may belong to

---

[1] Dunker (46) A., Pl. v. fig. 1.   [2] Bartholin (92) Pls. xi. xii.
[3] Moeller (02) Pls. iv.—vi.   [4] Richter (06) p. 21.

Hausmannia Richteri or may be a distinct species; it shows some of the finer veins connecting the shorter forked ribs, which formed part of the reticulate ramifying system in the mesophyll. This specimen was obtained from the plant-beds of Culgower on the Sutherlandshire coast, which have been placed by some geologists in the Kimmeridgian series.

The smaller type represented in fig. 278, E, is referred by Richter to a distinct species, *Hausmannia Sewardi*[1], founded on a few specimens from the Lower Cretaceous strata of Strohberg. This species is characterised by a stouter rhizome

Fig. 289. *Hausmannia sp.* Upper Jurassic, near Helmsdale, Scotland. From a specimen in the British Museum. (Nat. size.)

bearing smaller leaves consisting of a short petiole (3—4 cm. long) and an obovate lamina (1—2 cm. long and broad). There are usually two opposite leaflets on each leaf-stalk, and these may be equivalent to the two halves of a single deeply dissected lamina.

It is interesting to compare these different forms of *Hausmannia* with the fronds of recent species of *Dipteris* represented in fig. 231. The more deeply dissected type, such as *H. dichotoma*, closely resembles *D. Lobbiana* or *D. quinquefurcata*, while the more or less entire fossil leaves (fig. 278, E, F and fig. 289) are very like the somewhat unusual form of *Dipteris conjugata* shown in fig. 231, B, p. 297.

[1] Richter (06) p. 22.

Other species of the genus are recorded from Liassic rocks of Steierdorf[1] (Hungary) and of Bornholm[2]. Nathorst[3] has described a small Rhaetic species from Scania: a French Permian plant described by Zeiller[4] and compared by him with *H. dichotoma*, may be a Palaeozoic example of this Dipteris-like genus.

Some segments of leaves from the Eocene beds (Middle Bagshot) of Bournemouth, and now in the British Museum, described by Gardner and Ettingshausen[5] as *Podoloma polypodioides*, bear a close resemblance in the venation to the lamina of *Dipteris conjugata*.

---

[1] Andrae (53) A.   [2] Moeller (02) Pls. iv.–vi.
[3] Nathorst (78²) Pl. ix. fig. 2.   [4] Zeiller (79).
[5] Gardner and Ettingshausen (82) p. 29, Pl. iii. fig. 6.

CHAPTER XXII.

Marattiales (Fossil).

THE discovery of Pteridosperms has necessarily led to a considerable modification of the views formerly held that existing genera of Marattiaceae represent survivors of a group which occupied a dominant position in the forests of the Coal age. Mr Arber writes:—" The evidence, formerly regarded as beyond suspicion, that the eusporangiate ferns formed a dominant feature of the vegetation of the Palaeozoic period, has been undermined, more especially by the remarkable discovery of the male organs of *Lyginodendron* by Mr Kidston. At best we can only now regard them as a subsidiary group in that epoch in the past history of the vegetable kingdom[1]." Dr Scott expresses himself in terms slightly more favourable to the view that the Marattiaceae represent the aristocracy among the Filicales. He says:—" We now have to seek laboriously for evidence, which formerly seemed to lie open to us on all hands. I believe, however, that such careful investigation will result in the resuscitation of the Palaeozoic ferns as a considerable, though not as a dominant group[2]." Zeiller's faith[3] in the prospect of Marattiaceous ferns retaining their position as prominent members of Palaeozoic floras, though shaken, is not extinguished: he recognises that they played a subordinate part.

Reference has already been made to the impossibility of determining whether Palaeozoic fern-like fronds may be legitimately retained in the Filicales, or whether they must be removed into the ever widening territory of the Pteridosperms.

[1] Arber (06).p. 227.   [2] Scott (06) p. 189.   [3] Zeiller (05).

The difficulty is that the evidence of reproductive organs is very far from decisive. In the absence of the female reproductive organs, the seeds, we cannot in most cases be certain whether the small sporangium-like bodies on fertile pinnules are true fern sporangia or the microsporangia of a heterosporous pteridosperm. What is usually called an exannulate fern sporangium, such as we have in *Angiopteris* and in many Palaeozoic plants, has no distinguishing features which can be used as a decisive test. The microsporophylls of the Mesozoic Bennettitales produced their spores in sporangial compartments grouped in synangia like those of recent Marattiaceae; and in the case of *Crossotheca*, a type of frond always regarded as Marattiaceous until Kidston[1] proved it to be the microsporophyll of *Lyginodendron*, we have a striking instance of the futility of making dogmatic assertions as to the filicinean nature of what look like true fern sporangia. In all probability Dr Kidston's surmise that the supposed fern sporangia known as *Dactylotheca*, *Renaultia*, *Urnatopteris* are the microsporangia of Pteridosperms will be proved correct[2]. The question is how many of the supposed Marattiaceous sporangia must be assigned to Pteridosperms? There is, however, no reasonable doubt that true Marattiaceae formed a part of the Upper Carboniferous flora. All that can be attempted in the following pages is to describe briefly some of the numerous types of sporangia recognised on Palaeozoic fern-like foliage, leaving to the future the task of deciding how many of them can be accepted as those of ferns. It is impossible to avoid overlapping and some repetition in the sections dealing with true Ferns and with Pteridosperms. The filicinean nature of the stem known as *Psaronius* (see page 415) has not as yet been questioned.

The nomenclature of supposed Marattiaceous species from Carboniferous and Permian rocks is in a state of some confusion owing to a lack of satisfactory distinguishing features between certain types to which different generic names have been assigned. As we have already seen in the case of supposed

<hr />

[1] Kidston (06).   [2] Kidston (06) p. 429.

leptosporangiate sporangia, the interpretation of structural features in petrified or carbonised sporangia does not afford an example of unanimity among palaeobotanical experts.

Ptychocarpus.

This generic name, proposed by the late Professor Weiss[1], is applied to a type of fructification illustrated by the plant which Brongniart named *Pecopteris unita*, a species common in the Upper Coal-Measures of England[2]. It is adopted by Kidston for fertile specimens from Radstock which he describes as *Ptychocarpus oblongus*[3], but the precise nature of the fertile pinnules of this species cannot be determined.

Ptychocarpus unita (Brongn.[4]).   Fig. 291, A, B.   (= *Goniopteris unita*, Grand'Eury.)

This species has tripinnate fronds with linear pinnae bearing contiguous pinnules of the *Pecopteris* type (fig. 291, B), 4—5 mm. long, confluent at the base or for the greater part of their length. On the under surface of the fertile segments, which are identical with the sterile, occur circular synangia (fig. 291, A) consisting of seven sporangia embedded in a common parenchymatous tissue and radially disposed round a receptacle supplied with vascular tissue. The synangium is described as shortly stalked like those of *Marattia Kaulfussii* (fig. 245, B', p. 320). In shape, in the complete union of the sporangia, and presumably in the apical dehiscence, *Ptychocarpus* agrees very closely with *Kaulfussia* (fig. 245); but we cannot be certain that we have not a collection of microsporangia simulating a fern synangium.

A synangium closely resembling *Ptychocarpus* has been described by Mr Watson[5] from the Lower Coal-Measures of

---

[1] Weiss, C.E. (69) p. 94, Pl. xi. fig. 2. The specimens figured by Weiss bear a somewhat remote resemblance to that described by Renault (96) A, under the same generic name.

[2] Kidston (91²) p. 23.   [3] Kidston (88) p. 350.

[4] Renault (96) A. p. 9; Zeiller (88) A. p. 162; Grand'Eury (77) A. Pl. viii. fig. 13.

[5] Watson (06).

Lancashire as *Cyathotrachus altus,* but there is no convincing
evidence as to the nature of the plant on which it was borne.

Danaeites.

This generic name, instituted by Goeppert[1], has been used
by authors without due regard to the nature of the evidence
of affinity to *Danaea.* The type named by Stur *Danaeites
sarepontanus*[2] (fig. 291, E) bears small pecopteroid pinnules
with ovoid sporangia in groups of 8—16 in two contiguous
series on the lower face of the lamina. The sporangia dehisce
by an apical pore and are more or less embedded in the
mesophyll of the segments. No figures have been published
showing any detailed sporangial structure, and such evidence
as we have is insufficient to warrant the conclusion that the
resemblance to *Danaea* is more than an analogy.

Parapecopteris.

Parapecopteris neuropteroides, Grand'Eury.　Fig. 290, D.

The plant described by Grand'Eury[3] from the Coal-fields of
Gard and St Étienne, and made the type of a new genus, is
characterised by pinnules intermediate between those of *Pecopteris* and *Neuropteris*[4] and by the presence of two rows of united
sporangia along the lateral veins, as in *Danaea* and *Danaeites.*

Asterotheca.

Certain species of *Pecopteris* fronds from Carboniferous strata
are characterised by circular sori or synangia consisting of a
small number (3—8) of exannulate sporangia attached to a
central receptacle and free only at their apices. Strasburger[5]
suggested a Marattiaceous affinity for *Asterotheca* and Stur[6]
describes the species *Asterotheca Sternbergii* Goepp. (fig. 291,
C, D) as an example of a Marattiaceous fern. The latter
author retains Corda's genus *Hawlea*[7] for the fertile fronds

---

[1] Goeppert (36²) A. p. 380.
[2] Stur (85) A. p. 221, Pl. LXI.; Zeiller (88) A. p. 41.
[3] Grand'Eury (90) A. p. 288, Pl. VI. fig. 26.
[4] For an account of these genera, see Chap. XXVII.
[5] Strasburger (74).　　　　[6] Stur (85) A. p. 183.　　　　[7] Corda (45) A. Pl. LVII.

of the common Coal-Measures species *Pecopteris Miltoni,* while
on the other hand Kidston[1] includes this type in *Asterotheca.*

Pecopteris (Asterotheca) Miltoni (Artis).

1825.   *Filicites Miltoni,* Artis, Antedil. Phyt. Pl. XIV.

1828.   *Pecopteris Miltoni,* Brongniart, Prodrome, p. 58.

1828.   *Pecopteris abbreviata,* Brongniart, Hist. vég. foss. p. 337,
        Pl. CXV. figs. 1—4; Lindley and Hutton, Foss. Flor. Vol. III.
        Pl. 184.

1845.   *Hawlea pulcherrima,* Corda, Flor. Vorwelt, p. 90, Pl. LVII.
        figs. 7, 8.

1877—1888.   *Hawlea Miltoni,* Stur, Culm Flora, p. 293 ; Farne Carbon.
        Flora, p. 108, Pls. LIX. LX.

1888.   *Pecopteris (Asterotheca) abbreviata,* Zeiller, Flor. Valenc. p. 186,
        Pl. XXIV. figs. 1—4.

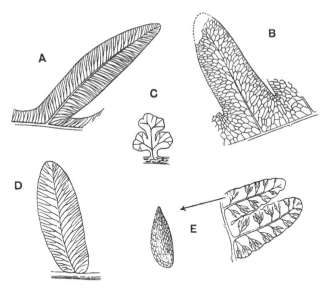

FIG. 290.   A.   *Alethopteris lonchitica.*   × 2½.
            B.   *Lonchopteris rugosa.*   × 2.
            C.   *Sphenopteris Hoeninghausi.*   × 4.
            D.   *Parapecopteris neuropteroides.*
            E.   *Pecopteris (Dactylotheca) plumosa* [= *P. (Dactylotheca) dentata*
                 Zeiller (88)]. × 4.
                 (A—C, E, after Zeiller; D, after Grand'Eury.)

For description
see Chap. XXVII.

[1] Kidston (91²) p. 20; Stur (85) A. Pl. LIX.

The fronds of this species reached a length of more than 3 metres and a breadth of 2 metres. They are characterised by the presence of aphlebiae[1] appressed to the rachis and by circular sori composed of a small number (3—6) of sporangia.

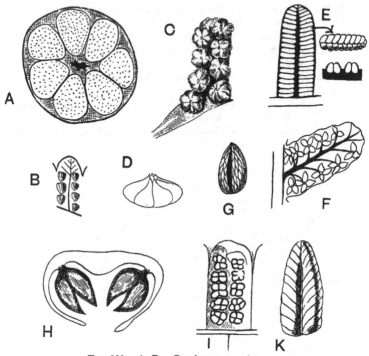

FIG. 291.   A, B.   *Ptychocarpus unita.*
C, D.   *Asterotheca Sternbergii.*
E.   *Danaeites sarepontanus.*
F.   *Hawlea Miltoni.*
G.   *Hawlea pulcherrima.*
H—K.   *Scolecopteris elegans.*
(A, B, after Renault; C—G, after Stur; H, I, after Strasburger;
K, after Sterzel.)

In habit and in the form of the pinnules this type is similar to *Dactylotheca plumosa.*

Hawlea.

Stur[2] retains this generic name for sori in which the

---

[1] See Chap. xxvii.   [2] Stur (85) A. p. 106.

sporangia are free and united only by the proximal end to
a central receptacle (fig. 291, F, G). He describes the individual sporangia as possessing a rudimentary annulus, a
comparatively strong wall, and terminating in a pointed distal
end. He emphasises the greater degree of cohesion between
the sporangia of *Asterotheca* as the distinguishing feature of
that genus; but this is a character difficult to recognise in some
cases, and from the analogy of recent ferns one is disposed to
attach little importance to the greater or less extent to which
sporangia are united, at least in such cases as *Asterotheca* and
Hawlea when the cohesion is never complete.

Scolecopteris.

Zenker[1] gave this name to detached fertile pinnules from
the Lower Permian of Saxony, which he described as *Scolecopteris elegans.* He recognised the fern nature of the sori
and suggested that the pinnules might belong to the fronds
of one of the "Staarsteinen" (*Psaronius*), a view which subsequent investigations render far from improbable. The sori,
which occur in two rows on the lower surface of the small
pecopteroid segments with strongly revolute margins (fig. 291,
H—K), contain 4—5 sporangia attached to a stalked receptacle
comparable with that of *Marattia Kaulfussii.* These pedicellate
synangia were fully described by Strasburger[2], who decided in
favour of a Marattiaceous alliance. The lower portions of the
distally tapered sporangia are concrescent, the distal ends being
free (fig. 291, H). Stur includes in *Scolecopteris* the common
species *Pecopteris arborescens* (fig. 376), but Kidston[3] states
that the British example of *Scolecopteris* is *S. polymorpha*
Brongn. from the Upper Coal-Measures.

Scolecopteris elegans Zenk. furnishes an example of a plant,
or plant fragment, which has been assigned to the animal
kingdom. Geinitz[4] described silicified pinnules as *Palaeojulus
dyadicus,* the generic name being chosen because of the re-

---

[1] Zenker (37).   [2] Strasburger (74).   [3] Kidston (91²) p. 20.
[4] Geinitz (72).   See Solms-Laubach (83), who gives in full the early history
of the genus *Scolecopteris.*

semblance to Millipedes such as the genus *Julus*. The mistake is not surprising to anyone who has seen a block of siliceous rock from Chemnitz crowded with the small pinnules with their concave surfaces formed by the infolding of the edges. Sterzel[1], who pointed out the confusion between Myriapods and Filices, has published figures which illustrate the deceptive resemblance of the pinnules, with their curved lamina divided by lateral veins into segments, to the body of a Millipede (fig. 291, K). He points out that Geinitz searched in vain for the head and legs of *Palaeojulus* and expressed the hope that further examination would lead to fresh discoveries : the examination of sections revealed the presence of sporangia and demonstrated the identity of *Palaeojulus* and *Scolecopteris*.

Discopteris.

Stur[2] instituted this genus for fertile fronds from the Upper Carboniferous Schatzlarer beds, including two species *Discopteris karwinensis* and *D. Schumanni*. He described the small Sphenopteroid pinnules as characterised by disc-shaped sori made up of 70—100 sporangia attached to a hemispherical receptacle : the absence of a true annulus led him to refer the genus to the Marattiaceae. In his memoir on the coal-basin of Heraclea (Asia Minor), Zeiller[3] instituted the species *Sphenopteris* (*Discopteris*) *Rallii* and figured sporangia resembling those described by Stur in the possession of a rudimentary " apical annulus." He compared the sporangia with those of recent Osmundaceae and Marattiaceae. In the later memoir on the Upper Carboniferous and Permian plants of Blanzy and Creusot, Zeiller[4] gives a very full and careful description of fertile specimens of *Sphenopteris* (*Discopteris*) *cristata*, a fern originally described by Brongniart as *Pecopteris cristata*[5]. Many of the Sphenopteroid pinnules of this quadripinnate fern frond show the form and structure of the sori with remarkable clearness in the admirable photographs reproduced in Plates I.—III. of Zeiller's Blanzy memoir. The lobed pinnules of this

---

[1] Sterzel (78); (80).   [2] Stur (85) p. 140.   [3] Zeiller (99) p. 17.
[4] Zeiller (06) p. 10.   [5] Brongniart (28) A. Pl. cxxv. fig. 4.

species are of oval-triangular form, 5—15 mm. long and 2·5—6 mm. broad[1]. An examination of the type-specimens of *Discopteris* from Vienna enabled Zeiller to correct Stur's original description of the sori: he found that the Austrian and French specimens, though specifically distinct, undoubtedly belong to one genus. The sori in *Discopteris cristata* are globular, as in the recent genera *Cyathea* and *Alsophila*, and frequently cover the whole face of the lamina. The individual sporangia are 0·4—0·5 mm. long and 0·15—0·2 mm. in diameter; they are exannulate, but for the annulus is substituted a group of thicker-walled and larger cells in the apical and dorsal region. The description by Stur of a hemispherical receptacle seemed to indicate an important difference between the Austrian and French species; but Zeiller found that this feature does not actually exist and that it was so described as the result of misinterpretation. Zeiller succeeded in isolating spores, 40—50 μ in diameter, from some of the sporangia of *D. cristata* and found that they exhibited the three-rayed pattern characteristic of fern-spores and which is indicative of their formation in tetrads. The conclusion arrived at is that the genus *Discopteris*, as represented by *D. karwinensis*, *D. cristata* etc., may be regarded as a true fern and included in the Marattiaceae. As Zeiller points out, the sori of *Discopteris* differ from those of recent Marattiaceae in their pluriseriate construction and agree in this respect with those of the Cyatheaceae. The comparison already made[2] between the sporangia of *D. Rallii* and those of recent Osmundaceae holds good: the genus affords another example of a generalised type, in this case probably a fern, combining features which are now distributed among the Marattiaceae, Osmundaceae and Cyatheaceae.

In addition to genera founded on true synangia or groups of free or partially united sporangia, the literature of Palaeozoic ferns contains several generic names applied to sporangia which occur singly on Sphenopteroid or Pecopteroid pinnules. The following may serve as examples; but it should be stated that these will probably be transferred eventually to the Pterido-

---

[1] Renault and Zeiller (88) A. Pl. xxiv.   [2] Page 325

sperms. It is, however, immaterial whether they are dealt with here or in the chapter devoted to the seed-bearing "ferns."

Dactylotheca.

Zeiller[1] created this genus for fertile fronds of *Pecopteris dentata* Brongn. (= *P. plumosa* Artis[2]), a common British species in the Upper and Middle Coal-Measures. Stur[3] included *P. dentata* in his list of species of *Senftenbergia*, the genus to which reference was made under the Schizaeaceae.

Pecopteris (Dactylotheca) plumosa (Artis). Figs. 290, E, 292, 293.

1825. *Filicites plumosus*, Artis, Antedil. Phyt. p. 17, Pl. xvii.
1828. *Pecopteris plumosa*, Brongniart, Hist. vég. foss. p. 348, Pls. cxxi. cxxii.
— *P. dentata*, Brongniart, *ibid*. Pls. cxxiii. cxxiv.
— *P. delicatulus*, Brongniart, *ibid*. Pl. cxvi. fig. 6.
1832. *Sphenopteris caudata*, Lindley and Hutton, Foss. Flor. Vol. i. Pl. xlviii.; Vol. ii. Pl. cxxxviii.
1834. *Pecopteris serra*, Lindley and Hutton, *ibid*. Vol. ii. Pl. cvii.
1834. *Schizopteris adnascens*, Lindley and Hutton, *ibid*. Vol. i. Pls. c. ci.
1836. *Aspidites caudatus*, Goeppert, Syst. fil. foss. p. 363.
1838. *Steffensia silesiaca*, Presl, in Sternberg, Flor. Vorwelt, Vers. ii. p. 122.
1869. *Pecopteris silesiacus*, Schimper, Trait. pal. vég. Vol. i. p. 517.
— *Cyathocarpus dentatus*, Weiss, Flora der jüngst. Stk. und Roth. p. 86.
1877. *Senftenbergia plumosa*, Stur, Culm Flora, ii. p. 187 (293).
— *S. dentata*, *ibid*.
1886. *Dactylotheca plumosa*, Kidston, Cat. Palaeozoic Plants, p. 128.
1888. *Dactylotheca dentata*, Zeiller, Flor. Valenc. Pls. xxvi.—xxviii.

For a fuller synonymy reference should be made to Kidston's account of this species[4], from which the above list is compiled. The large fronds of this species are tri- or quadripinnate. The pinnules vary much in shape and size and in degree of lobing, according to their position on the frond (fig. 293). The primary

[1] Zeiller (83) p. 184; (88) A. p. 30.
[2] Artis (25) A.
[3] Stur (75) A.
[4] Kidston (96) p. 205.

pinnae are subtended by two Aphlebiae (fig. 293, A) appressed
to the rachis, like the delicate leaves of the recent fern *Teratophyllum aculeatum* (see page 301). The sporangia (0·5—0·65)
are oval and exannulate and are attached parallel to the lateral
veins; they may occupy the whole of the space between the
midrib and the edge of the pinnules. This species occurs in the
Upper, Middle, and Lower Coal-Measures of Britain, reaching

FIG. 292. *Dactylotheca plumosa.* (After Kidston. Slightly reduced.)

its maximum in the Upper Coal-Measures. The aphlebiae
undoubtedly served to protect the young fronds, as shown by a
specimen figured by Kidston (fig. 293, B); they may also have
served other purposes, as suggested by the above comparison
with *Teratophyllum*, in the mature frond. Lindley and Hutton
regarded the aphlebiae as leaves of a fern climbing up the

rachis; which they named *Schizopteris adnascens*, a confusion
similar to that already mentioned in the description of *Hemitelia
capensis* (see p. 304).

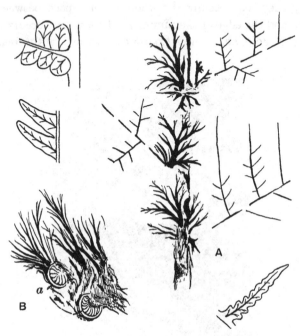

Fɪɢ. 293. *Dactylotheca plumosa*: A. Rachis with Aphlebiae. B, *a*, young pinnae
circinately folded. (After Kidston. A, B, ⅘ nat. size.)

Renaultia.

This name was proposed by Zeiller[1] for Upper Carboniferous
fertile pinnae of the Sphenopteroid type, bearing ovoid sporangia
either singly or in marginal groups of 2 to 5 at the ends of the
veins.   The appearance of the apical cells occasionally suggests
the presence of a rudimentary annulus.   Kidston has recorded
this type of fructification in Britain[2].   Stur describes fertile
pinnules of the same type under the generic name *Hapalopteris*[3].

[1] Zeiller (83) p. 185.   [2] Kidston (82).   [3] Stur (85).

Zeilleria.

This genus was founded by Kidston[1] for fertile pinnae of a very delicate fern, *Zeilleria delicatula* (Sternb.) characterised by filiform ultimate segments bearing an indusium-like body, spherical when immature and splitting at maturity into four small valves. Kidston, in his earlier paper, compared the species with recent Hymenophyllaceae. In the same genus he includes *Z. avoldensis*[2] (Stur) assigned by Stur to *Calymmatotheca*, a genus described by some authors as characterised by groups of radially elongated sporangia at the tips of the pinnules; these supposed sporangia are now known to be the valves of an indusium-like organ or cupule, as Stur asserted. There can be little doubt that the fertile fronds placed in *Calymmatotheca* and in *Zeilleria* were borne by Pteridosperms.

Urnatopteris.

The Upper Carboniferous fronds of a delicate Sphenopteris habit, to which this name was assigned by Kidston[3], were described by him as *Eusphenopteris tenella* (Brongn.)[4] and compared with Hymenophyllaceae; subsequently Kidston expressed the opinion that *Urnatopteris* may be a Marattiaceous fern, as Williamson[5] believed; he has since suggested that the sporangia are the microsporangia of a Pteridosperm[6]. The sterile and fertile pinnae differ in the absence of a lamina in the latter. The sporangia (or microsporangia) are characterised by a poricidal dehiscence.

The records from strata higher in the geological series than the Permian, disregarding many of doubtful value, afford ample testimony to the existence of Marattiaceae in Upper Triassic and Rhaetic floras.

Marattiopsis.

The generic name *Danaeopsis* was applied by Heer[7] to an Upper Triassic fern, previously described by Presl as *Taeni-*

---

[1] Kidston (84²).   [2] Kidston (87).
[3] Kidston (82) p. 32.   [4] Kidston (84²) p. 594.
[5] Williamson (83) A.   [6] Kidston (06).
[7] Heer (76) A. p. 71, Pl. xxiv. fig. 1.

opteris marantacea. A splendid specimen from the Keuper of Stuttgart is figured in Schimper's Atlas[1] showing the pinnate habit of the frond and the broadly linear segments, 25 cm. × 3·5 cm., bearing rows of contiguous sporangia. The large pinnules have a strong midrib giving off curved and forked lateral veins. Presl's species may most appropriately be included in the genus *Marattiopsis.* A specimen of *M. marantacea* described by Leuthardt[2] as *Danaeopsis marantacea* from the Upper Trias of Basel shows a peculiarity in the venation; the lateral veins often fork near their origin, as noticed by other authors, but each vein forks a second time near the edge of the lamina and the two arms converge, forming a series of intramarginal loops (fig. 265, B).

Marattiopsis Muensteri (Goepp.). Fig. 245, D, E.

This widely spread Rhaetic plant affords the best example of a post-Permian species which may be accepted as an authentic record of fossil Marattiaceae. Various generic names have been used for this species; Goeppert originally described the plant as *Taeniopteris Muensteri*[3]; Schimper[4] proposed the name *Marattiopsis,* and Schenk[5] substituted *Angiopteris* on the ground that the fertile pinnules resemble that genus rather than *Marattia.* *Marattiopsis,* if interpreted as indicating a *family* resemblance rather than special affinity to the genus *Marattia,* would seem to be the more appropriate designation.

This species has been figured by several authors and in many instances with fertile pinnules; the best illustrations are those published by Zeiller[6] in his monograph of Tonkin plants.

The pinnate fronds are characterised by a broad rachis bearing sessile broadly linear pinnules rounded at the base, obtusely pointed at the apex, reaching a length of 15—20 cm. and a breadth of 12—35 mm. From a

---

[1] Schimper (74) A. Pl. 38 ; see also Schenk (88) A. p. 31.
[2] Leuthardt (04) p. 29, Pl. xiii. figs. 1, 2.
[3] Goeppert (36²) A. Lief. i. and ii. Pl. iv.   [4] Schimper (69) A. p. 607.
[5] Schenk (83) A. p. 260.   [6] Zeiller (03) Pl. ix.

well-marked midrib are given off secondary veins dichotomously branched close to their origin. The linear synangia near the ends of the veins contain two rows of sporangial compartments and open as two valves as in *Marattia*. (Cf. fig. 245, A, p. 320.)

This species occurs in the Rhaetic beds of Scania, Franconia, and Tonkin. A similar type is figured by Fontaine from Jurassic beds in California as *Angiopteridium californicum*[1], and Bartholin[2] and Moeller[3] record *M. Muensteri* from the Lias of Bornholm. Schenk's species from China[4], *Angiopteris Richthofeni*, is a closely allied species, and a similar form is recorded from Jurassic and Caucasian strata[5]. The microscopical examination by Nathorst[6] of a group of spores from a synangium of *M. Muensteri* shows that they resemble those of recent Marattiaceae.

From the Upper Triassic plant beds of Lunz, Stur has included several species of ferns in the Marattiaceae, and of these Krasser[7] has recently published full diagnoses but unfortunately without illustrations. In addition to *Marattiopsis marantacea* (Presl) the list includes species referred to *Coniopteris*, to *Speirocarpus*, a genus founded by Stur, to *Oligocarpia*, *Asterotheca*, and *Bernouillia* (Heer).

As already pointed out, some at least of these Austrian ferns are more probably Osmundaceous than Marattiaceous.

Danaeopsis Hughesi, Feistmantel.

The pinnate fronds described by Feistmantel[8] from the Middle Gondwana rocks of India and recorded from Rhaetic strata in South Africa[9], China[10], and Tonkin[11], may belong to a member of the Marattiaceae, but no fertile specimens have been described. The close agreement between the sterile leaves from India and South Africa and the fertile fronds of *Marattiopsis marantacea* suggests generic identity.

[1] Fontaine, in Ward (00) Pl. LV. figs. 3—5.   [2] Bartholin (92) Pl. IX.
[3] Moeller (02).   [4] Schenk (83) A.
[5] Seward (07[4]) Pl. II. figs. 16—18.   [6] Nathorst (08).   [7] Krasser (09).
[8] Feistmantel (82) Pls. IV.—X.   [9] Seward (08) p. 95.
[10] Krasser (00) Pl. II.   [11] Zeiller (03) Pl. IX.

The Upper Triassic ferns described by Heer, Krasser[1], and Leuthardt[2] as *Bernouillia* have been referred to the Marattiaceae, but without trustworthy evidence in favour of this affinity.

The large leaves, 70 cm. long and 7 cm. broad, described by Zigno[3] from the Jurassic of Italy as *Danaeites Heeri*, are probably Cycadean. The Polish Jurassic species *Danaea microphylla*[4] is a more satisfactory record.

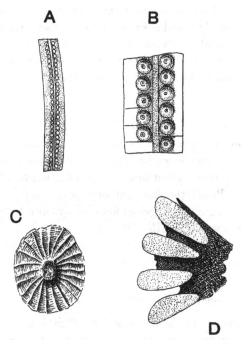

Fig. 294. A, B. *Nathorstia angustifolia*, Heer. (After Heer. A, nat. size.) C, D. Sorus of *N. latifolia*, Nath. (After Nathorst. C, × 12; D, × 45.)

Nathorstia.

This name was instituted by Heer[5] for pieces of pinnate fronds from Lower Cretaceous rocks of Greenland. The resemblance of the long pinnules to the fertile segments of

[1] Krasser (09) p. 21.   [2] Leuthardt (04) Pls. xix. xx.
[3] Zigno (56) A. Pl. xxv.   [4] Raciborski (94) A. Pl. vi.   [5] Heer (80).

Laccopteris is so close that generic identity might well be assumed, but it has recently been shown by Nathorst[1] that the soral characters justify Heer's use of a distinctive name for the Arctic fern. The circular sori arranged in two rows (fig. 294, A, B) are superficially identical with those of *Laccopteris*, but consist of concrescent sporangia forming a circular synangium (fig. 294, C, D) like those of *Kaulfussia* and *Ptychocarpus*. The lighter areas in fig. 294, D, represent the sporangia: fig. C shows the radial disposition of the numerous sporangial compartments round a central receptacle. From a stout midrib lateral veins arise at right angles, but their distal terminations are not preserved. It is probable, as Nathorst suggests, that Bayer's[2] species *Drynaria fascia* from the Lower Cretaceous rocks of Bohemia should be referred to Heer's genus. In the absence of well-preserved sori it would be exceedingly difficult, or even impossible, to distinguish between pinnules of *Laccopteris* and *Nathorstia*.

A Tertiary species, *Marattia Hookeri* (fig. 261, C, p. 350), described by Gardner and Ettingshausen[3] from the Eocene beds of the Isle of Wight is referred by them to the Marattiaceae because of a resemblance of the sterile pinnae to those of *M. Kaulfussii*; but this is insufficient evidence of relationship.

---

[1] Nathorst (08).
[2] Bayer (99).   [3] Gardner and Ettingshausen (82) Pl. XII. figs. 1—7.

CHAPTER XXIII.

Psaronieae.

THIS family name, first suggested by Unger, may be conveniently adopted for the numerous species of petrified tree-fern stems characteristic of the Lower Permian and Upper Carboniferous strata. In his monograph *Uber die Staarsteine* published in 1854, Stenzel[1] referred to the Psaronieae as a special sub-division of the Filices most nearly allied to the Polypodiaceae. There is now a consensus of opinion in favour of including *Psaronius* in the Marattiales, or at least of regarding the genus as more closely allied to the Marattiaceae than to any other family. While admitting that the balance of evidence is in favour of this view, it is probably wiser to retain the distinctive term Psaronieae on the ground that species of *Psaronius* differ in several respects from any recent ferns, and because of our comparative ignorance in regard to the nature of the fructification.

Psaronius.

This generic name was proposed by Cotta in his classic work *Die Dendrolithen*[2]. The stems so named, formerly included by Sprengel[3] in the genus *Endogenites*, had long been familiar as petrified fossils. Most of the specimens described by the earlier writers were obtained from Lower Permian rocks in the neighbourhood of Chemnitz, Saxony. The mottled appear-

---

[1] Stenzel (54) p. 803. [2] Cotta (32). [3] Sprengel (28).

ance presented by their polished surfaces is said to have given rise to the appellation *Staarsteine* (starling stones), a term expressing a resemblance, more or less remote, to a starling's breast. It has been suggested that this word is a corruption of *Stern Steine* or star stones[1], a descriptive term suggested by the stellate arrangement of the vascular strands in transverse sections of the roots. Parkinson[2], in his *Organic Remains of a former World,* speaks of these stems as starry stones. The history of our knowledge prior to 1854 is summarised by Stenzel. At first compared with corals or the stems of sea-lilies, *Psaronii* were recognised by Sprengel, who first used a lens in the examination of the fossils, as fern stems most nearly allied to those of recent Cyatheaceae. By other authors, e.g. Schlotheim and Sternberg, they were referred to Palms, and by Brongniart considered to be the lower portions of Lycopodiaceous (*Lepidodendron*) stems. Corda and many subsequent authors selected the Marattiaceae as the most closely allied family among existing plants.

Psaronius is represented by specimens obtained from the Lower Permian of Saxony and Upper Carboniferous rocks in Central France, also from Bohemia, Brazil and North America. As yet a few fragments only have been found in the English Coal-Measures. The genus was recognised by Williamson[3] who described the roots and a small piece of the vascular tissue of a stem which he called *P. Renaulti,* and this type has since been more fully described by Scott[4]. The roots of another species have been described by Butterworth[5] as *P. Cromptonensis.*

It was pointed out in the account of *Lepidodendron* that several generic names have been used for the same type of stem in different states of preservation; in *Psaronius* accidents of fossilisation have been responsible for a similar confusion in nomenclature. The name *Psaronius* is applied to petrified specimens which, as a rule, lack external features. Casts or impressions of Palaeozoic tree-fern stems provided with leaf-scars are described as species of *Caulopteris, Megaphyton,* and less commonly as *Ptychopteris* (figs. 297—299). The first name

[1] Stenzel (54) p. 753.   [2] Parkinson (11) A.
[3] Williamson (76).   [4] Scott (08).   [5] Butterworth (00).

is applied to stems exhibiting spirally disposed leaf-scars like those of recent tree-ferns; in *Megaphyton* the scars are distichously arranged, in two rows, while *Ptychopteris* is applied to decorticated stems. These terms are used for stems belonging to one generic type and possessing the structure of Psaronius stems.

The researches of Grand'Eury[1] led to the discovery that certain Psaronius stems bore fronds of the *Pecopteris* type some of which bore sori of the *Asterotheca* or *Scolecopteris* type. The same author[2] has also contributed many interesting facts,

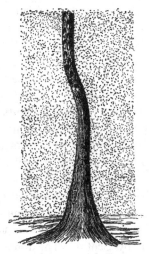

Fig. 295. *Psaronius* stem with roots. (Much reduced. After Grand'Eury.)

obtained by an examination of the relation of Psaronius stems to the sediments of French Coal-fields in which they occur, in regard to habitat and manner of growth. The specimen represented in fig. 295 shows a portion of a Psaronius stem, the upper part of which illustrates the Caulopteris state of preservation, while the lower part is covered by a mass of roots. It is probable, as Rudolph[3] suggests, that this rich development of roots, which gives to an old Psaronius stem the appearance of an

---

[1] Grand'Eury (77) A.   [2] Grand'Eury (77) A; (90) A.
[3] Rudolph (05).

elongated cone, may have served an important mechanical purpose analogous to the secondary thickening in a Dicotyledon or a Conifer. A specimen of *Psaronius Cottai* in the Hof Museum, Vienna, is cited in illustration of the enormous breadth of the root-system : the radii of the stem proper and of the encasing cylinder of roots bear the ratio 17 to 165. The comparatively frequent occurrence of a lacunar cortex in the roots points to the growth of the stems in swampy ground, a conclusion in harmony with the evidence afforded by the ana-tomical features of many other Palaeozoic genera.

Psaronius may be briefly defined as follows :—

Tree-fern stems, occasionally reaching a height of 50 feet or more, closely resembling in habit recent tree ferns, but exhibiting in the structure and arrangement of the vascular system a close agreement with recent Marattiaceae. Leaves, in such cases where a connexion between fronds and stems is known, large and highly compound and of the *Pecopteris* type, borne in more or less crowded spirals (*Psaronius polystichi*), in four rows (*P. tetrastichi*), or in two opposite rows (*P. distichi*). Leaves deciduous, leaving a clearly defined oval scar containing the impression of the leaf-trace in the form of an open U, or a closed oval with a small inverted V-shaped band a short distance below the upper end of the long axis of the oval (figs. 297, 298) ; in *Megaphyton* the alternate scars of the two opposite series are larger and characterised by a different form of meri-stele. The surface of the cortex below the leaf-scars occasionally shows impressions of pits similar to the lenticel-like organs on recent Tree-fern stems. The central region of the stem is occupied by a complex system of concentrically disposed steles (dictyosteles), which in transverse section present the appearance of flat or curved bands varying in extent and in degree of curvature. The vascular bands consist of xylem surrounded by a narrow zone of phloem ; the xylem is composed either exclusively of tracheae or of tracheae and parenchyma ; the protoxylem in the one instance in which it has been clearly recognised is endarch[1]. The steles are embedded in parenchymatous tissue and in some species are associated with mechanical tissue (e.g. *P. infarctus*, fig. 296, A, B). The central or vascular region of the stem may be surrounded externally by a cylinder of mechanical tissue interrupted by outgoing leaf-traces and adventitious roots. The leaf-traces arise as single bundles from an internal stelar band and pursue an obliquely radial course towards the outside, even-tually anastomosing with peripheral cauline steles, which in some species form with the leaf-traces the outermost zone of the vascular region. The leaf-traces have the form of loops which pass into the petioles as

---

[1] Scott (08) p. 302.

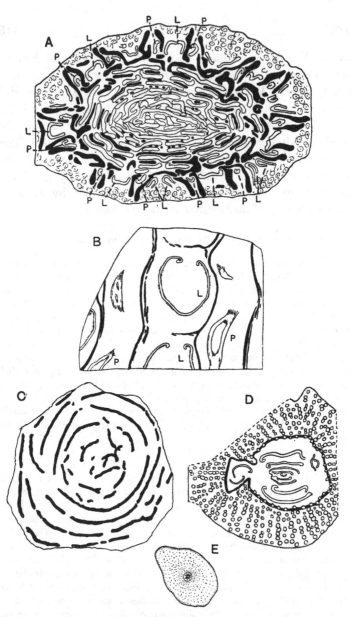

FIG. 296.   A.   *Psaronius infarctus* (P, peripheral steles ; L, leaf-traces).
          B.   *P. infarctus*, longitudinal tangential section through the
               peripheral region of the stem.
          C.   *P. coalescens.*
          D.   *P. musaeformis.*
          E.   *P. asterolithus* (root).
               (A—C, E, after Zeiller; D, after Stenzel.)

V-shaped meristeles or closed oval cylinders.   As a leaf-trace passes out compensating strands occupy the foliar gap.

The vascular region is surrounded by a parenchymatous cortex, which in younger plants, or in the apical region of an older plant, forms the surface of the stem to which the leaf-stalks are attached.   From the peripheral steles, or from the more external bands of the vascular network, roots are given off which pass in a sinuous vertical course through the cortex, appearing on the surface between the leaf-bases.   In older stems, after leaf-fall, the tissue immediately external to the vascular region produces secondary parenchyma with which the roots become intimately associated by their outermost cells.   As a result of the secondary cortical development and the gradual increase in the number of roots invading the cortical tissue from above, the stem is enclosed by a cylinder of roots and associated parenchymatous tissue of secondary origin.   In still older portions of a stem the more external roots are free from the stem-cortex and form a thick felted mantle, which increases in thickness towards the base of the tree.

The roots (fig. 296, E) are polyarch, 5—10 groups of xylem alternating with strands of phloem, and similar in structure to those of recent species of *Marattia* and *Angiopteris*; the stele is enclosed by an inner cortex of compact or lacunar tissue containing secretory sacs, and this is surrounded by a cylinder of mechanical tissue.   In one or two instances secondary xylem has been observed wholly or partially enclosing the root-stele[1].

Our knowledge of the anatomy of *Psaronius* is based largely on the investigations of Stenzel considerably extended by Zeiller's more intensive studies and, more recently, by the later work of Stenzel[2] and that of Rudolph.   A striking fact, which has led to various suggestions, is that in a transverse section of a *Psaronius* stem with its encasing cylinder of roots no signs of leaf-traces are met with in the root-region.   If the roots simply penetrated the cortex, as in some recent species of *Lycopodium* (fig. 125, A) or as in *Angiopteris*, we should expect to find leaf-traces in the outer region (root-cylinder) of Psaronius stems.   An explanation of the absence of leaf-traces which was suggested by Stenzel, is that the cortical zone formed a comparatively narrow band in the young leaf-covered stem; after leaf-fall it became the seat of active growth in its inner layers and so produced a constantly widening zone of secondary parenchyma, which pushed the superficial cortical tissue with

---

[1] Butterworth (00).   Pelourde (08[2]) has recently described the structure of the roots of several species of *Psaronius*.

[2] Stenzel (06).

the leaf-bases or leaf-scars farther out until it was exfoliated. Farmer and Hill[1] find it difficult to accept this explanation; but, as Rudolph shows, the radial arrangement of the cortical cells between the adventitious roots and their elongation in a radial direction are arguments in support of the secondary nature of the cortical zone.

In sections of the adventitious roots of *Psaronius Renaulti* figured by Williamson[2], the spaces between the cylindrical roots are partially occupied by cell-filaments which, at first sight, suggest root-hairs; it may well be, as Rudolph suggests, that these felted hairs represent the outermost and looser part of the growing secondary cortex which gradually passes into the covering mass of free extra-cortical roots.

As Stenzel[3] has shown, slender stems of *Zygopteris* (= *Ankyropteris*) are occasionally met with growing through the web of Psaronius roots.

Psaronius infarctus Unger. Fig. 296, A, B.

This species, which Zeiller[4] has investigated from sections of Unger's material, illustrates a type in which the vascular tissue is very richly developed and forms crowded concentric series of curved plates associated. in the more peripheral series, with bands of mechanical tissue. The outermost part of the vascular region consists of (i) a series of loops or variously curved bands of conducting tissue representing leaf-traces at different stages in their outward course, (ii) a series of similar vascular strands (peripheral steles of Zeiller) confined to the stem (cauline) and from which roots are given off, and (iii) bands of mechanical tissue associated with the leaf-traces and peripheral steles. The peripheral steles (fig. 296, A, B, P) form anastomoses with the leaf-traces and contribute to their formation.

The form of some of the vascular bands in the section of *Psaronius infarctus* shown in fig. 296, A, illustrates the occasional anastomosing of one dictyostele with another: the

[1] Farmer and Hill (02).   [2] Williamson (76) Pl. III.
[3] Stenzel (89) Pl. VI.
[4] Zeiller (90) p. 204, Pls. XVI. XVII.; see also Rudolph (05).

different degrees of looping of other bands represent stages in the giving off of leaf-traces which eventually pass out as **V**-shaped meristeles. Beyond the leaf-traces and sclerenchymatous bands the section consists of transverse sections of adventitious roots.

The surface-features of *Psaronius infarctus* are probably represented, as Zeiller points out, by the cast described by Lesquereux as *Caulopteris peltigera* (fig. 298, A).

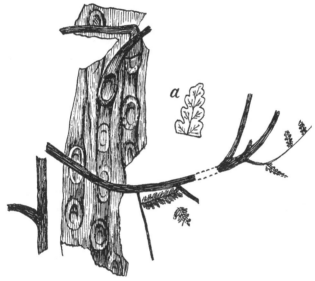

FIG. 297.   *Pecopteris Sterzeli*: *a*, pinnule. (After Renault and Zeiller. $\frac{1}{11}$ nat. size.)

The Psaronius shown in fig. 297 is one of the few examples illustrating the connexion between fronds and stem. The leaf (*Pecopteris Sterzeli* Zeill. and Ren.[1]) is quadripinnate and is described as reaching a length of at least 3 metres; the ultimate segments are entire or lobed. The stem is characterised by elliptical scars, 6—8 cm. × 3·5—4 cm., with leaf traces like those in *Caulopteris peltigera*. The fronds of *Pecopteris Pluckeneti*, a Pteridosperm, bear a very close

[1] Renault and Zeiller (88) A. Pls. v.—viii.

resemblance to those of *P. Sterzeli*, which are as yet known only in a sterile state.

Psaronius brasiliensis Unger, a species founded by Unger on a piece of silicified stem acquired by Martius in Brazil and now in the Rio Museum, is a good example of a tetrastichous species. Solms-Laubach[1] has recently told the history of this type, which is represented by sections, cut from the Rio stem, in several European collections. A well-preserved section in the British Museum is figured by Arber[2] in his catalogue of the Glossopteris flora and by other authors. Scott gives a concise description of the species in his *Studies in Fossil Botany*[3]. The roots of *P. brasiliensis* are stated by Pelourde[4] to have a lacunar cortex.

Psaronius musaeformis Corda[5].   Fig. 296, D.

This species from the Lower Permian of Chemnitz and the Coal-Measures of Bohemia affords an example of the distichous type in which the leaves are borne in two rows. The vascular bands, as seen in a section of the dictyosteles, occur in regular parallel series. The stelar region is separated from the cylinder of encasing roots by a sclerenchymatous sheath, broken at intervals where roots pass out from the vascular region.

Psaronius coalescens[6] (fig. 296, C) illustrates a somewhat different arrangement of vascular tissue which approaches more closely to the polycyclic structure characteristic of such recent ferns as *Matonia* and *Saccoloma*. A still closer resemblance to the solenostelic type is seen in *Psaronius Renaulti* from the Lower Coal-Measures of England which Scott[7] describes as characterised by a single annular stele, interrupted only by the exit of leaf-traces. As he points out, it is noteworthy that this species is distinguished by the simplest form of stele met with in the genus; it is the oldest species and may be regarded

---

[1] Solms-Laubach (04).   [2] Arber (05) Pl. vii.
[3] Scott (08) fig. 113; Zeiller (90) p. 246, Pl. xxi. fig. 1.
[4] Pelourde (08²).   [5] Stenzel (06) Pl. vi.; Goeppert (64) A.; Stenzel (06).
[6] Zeiller (90) Pl. xxiii.   [7] Scott (08) p. 301.

as the most primitive representative of the genus *Psaronius* so
far discovered.

Psaronius stems preserved as casts showing surface-features,
or in a decorticated state.

i. *Caulopteris.*

This generic name was instituted by Lindley and Hutton[1]
for tree-fern stems from the English Coal-Measures showing
circular or oval scars arranged quincuncially. The vascular tissue
of the petiole is represented by a U-shaped impression on the

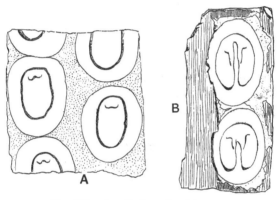

FIG. 298. A. *Caulopteris peltigera.*
B. *Megaphyton insigne.*
(After Grand'Eury.)   Much reduced.

scar, the ends of the U being incurved, or by a closed oval ring
with a wide-open and inverted V near its upper end. The
surface between the leaf-scars bears the impression of adventitious roots. *Caulopteris* is represented, in the Upper Coal-
Measures of England, by *C. anglica*[2] Kidst. The species
C. peltigera (fig. 298, A), originally described by Brongniart as
Sigillaria, illustrates the closed form of leaf-trace and, as Zeiller
suggests, it is the cast of a *Psaronius* stem which possessed
a vascular system on the same plan as that of *P. infarctus.*
C. Saportae[3] illustrates the open U-shaped type of petiole stele.

[1] Lindley and Hutton (33) A. Pl. XLII.   [2] Kidston (88) Pl. XXVI.
[3] Renault and Zeiller (88) A. Pl. XXXV. fig. 6.

Caulopteris peltigera has scars measuring 6—9 by 4—6 cm.; it occurs in the Commentry Coal-field of France in association with the fronds known as *Pecopteris cyathea*, a species which Kidston regards as identical with *P. arborescens*[1].

ii. *Megaphyton*.

The first use of this name was by Artis[2], who gave it to a long flattened cast, *Megaphyton frondosum*, found in Carboniferous strata in Yorkshire, characterised by two vertical rows of large scars and by impressions of sinuous roots. Kidston records the genus from the Middle and Upper Coal-Measures of Britain. A good example of this type of cast is afforded by *M. McLayi* Lesq.[3] from the Coal-Measures of North America, which has been recognised in European Carboniferous rocks. The leaf-scars are rounded or oval, broader than high; the vascular impression has the form of a closed ring (5—8 × 3—6 cm.), more or less circular and with a tendency to a rectangular outline, characterised by a deep inverted U-shaped sinus in the middle of the lower surface and by a W-shaped impression of an internal strand (fig. 298, B)[4].

iii. *Ptychopteris*.

This generic name, instituted by Corda[5], is applied to decorticated stems of *Psaronius*, the surface of which is that of the vascular region on which the form of the leaf-scars is more or less clearly defined. The scar-areas are limited by an impression of the sclerenchymatous sheath enclosing the leaf-meristele, and internal to this is the impression of the leaf-trace. In some specimens a layer of coaly material which represents the carbonised cortex and adventitious roots covers the *Ptychopteris* cast. The *Ptychopteris* cast represented in fig. 299 shows the decorticated surface of part of a long stem on which the leaf-scars are arranged as in *Megaphyton*. An example of

---

[1] Kidston (86) A. p. 113.   [2] Artis (25) A. Pl. xx.

[3] Lesquereux (66) A.

[4] Renault and Zeiller (88) A. Pl. xl.; Grand'Eury (90) A.

[5] Corda (45) A.; see also Grand'Eury (90) A.; Renault and Zeiller (88) A. Pls. xxxviii.—xl.

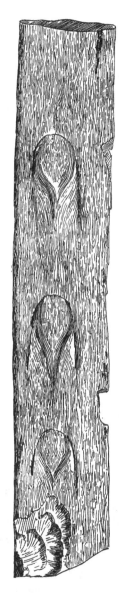

Fig. 299. *Ptychopteris.* ⅙ nat. size. From the Middle Coal-Measures of
Lancashire. (The Manchester Museum.)

Ptychopteris is figured by Fontaine and White[1] from Virginia as
Caulopteris gigantea.

Position of Psaronius.

A comparison of *Psaronius* with the Marattiaceae and other
recent ferns leads to the conclusion that, on the whole, the
evidence is in favour of the view usually held, namely that
this genus is more closely related to the Marattiaceae than to
any other recent ferns.   It is, however, important not to overlook the differences between *Psaronius* and recent genera of
Marattiaceae, or the resemblances between the extinct genus
and the Cyatheaceae.   In habit *Psaronius* agrees closely with
recent tree-ferns; in the vascular system and in the sequence

Fig. 300.   *Dicksonia antarctica* (half of stem in transverse section): *st*, stele;
s, sclerenchyma.

of events connected with the production of leaf-traces, there are
striking resemblances between *Psaronius* and the Cyatheaceous
fern *Saccoloma adiantoides* (= *Dicksonia Plumieri* Hook.) as
described by Mettenius[2].   The piece of stem of *Dicksonia
antarctica* represented in fig. 300 exhibits a fairly close agreement with species of *Psaronius*, e.g. *P. infarctus* (fig. 296, A, B).
Moreover, the peripheral steles, which Zeiller has shown are
confined to the stem and play an important part in the production of the roots and in the anastomoses with leaf-traces,

[1] Fontaine and White (80) Pl. xxxvi.; Zeiller (90) Pl. xiv.
[2] Mettenius (65); Tansley (08) p. 85.

are not represented in any Marattiaceous fern; on the other
hand, they are comparable with the accessory strands met with
in stems of recent Cyatheaceous tree-ferns[1] (cf. fig. 240). The
complex system of concentric dictyosteles is a feature more closely
matched in *Angiopteris* (Marattiaceae) than in any Cyatheaceous genus, the chief difference being in the more band-like
form of the steles in *Psaronius*, though in a stem of *Angiopteris*
figured by Mettenius we see a close approach to the extinct
type. The position of the protoxylem has unfortunately not
been clearly defined in *Psaronius* stems, but in *P. Renaulti* it
is stated by Scott[2] to be endarch, a position which some of the
protoxylem strands occupy in *Angiopteris*[3]. The occurrence of
large sieve-tubes described by Scott in *P. Renaultii* is another
feature shared by recent Marattiaceae. In many of the conti-
nental species of *Psaronius* the phloem has not been preserved,
and our knowledge of this tissue is comparatively meagre. In
the Marattiaceae the roots arise mainly from the inner portions
of the stele, while in *Psaronius* they are usually formed from
the external vascular bands. The formation of secondary cortical
tissue is a peculiarity of *Psaronius*; on the other hand, if
Butterworth[4] is correct in referring to that genus the roots
with secondary xylem, which he describes as *P. Cromptonensis*,
a comparison may be made with the occurrence of secondary
tracheae in the stem steles of *Angiopteris*[5].

The absence of mechanical tissue in the stem of *Angiopteris*
is in contrast with its occurrence in the fossil stems and in
recent tree-ferns; but this is a character of secondary importance and one which can be readily explained by the difference
in habit between *Angiopteris* and *Psaronius*.

The roots of *Psaronius*, more especially as regards the stelar
structure, are in close agreement with those of Marattiaceae.

The reference to Marattiaceae of the great majority of fertile
fern-like fronds from Permian and Carboniferous rocks constituted a strong *a priori* argument in favour of including
Psaronius stems in the same family, especially when it was
known that leaves with Marattiaceous synangia were borne by

---

[1] Rudolph (05).   [2] Scott, D. H. (08).   [3] Shove (00).
[4] Butterworth (00).   [5] Farmer and Hill (02).

species of this genus. It is, however, well to remember the
change in our views as to the dominance of Marattiaceae in
Palaeozoic floras consequent on the discovery of the Pteridosperms. The association of fronds bearing *Asterotheca* and
Scolecopteris types of fructification with *Psaronius* stems recorded
by Grand'Eury[1] is a point in favour of the Marattiaceous affinity
of this extinct genus, but it is not impossible that *Psaronius*
stems bore fronds which produced Pteridosperm organs of reproduction. In this connexion the specimen represented in
fig. 297 is of interest, as the fronds (*Pecopteris Sterzeli*) borne
on the *Psaronius* stems are hardly distinguishable from the
seed-bearing leaves known as *Pecopteris Pluckeneti*.

The position of *Psaronius* may be best expressed by assigning
it to a separate family, the Psaronieae, as advocated by Stenzel,
and by regarding it as one of the many instances of a generalised
type which in the sum of its characters approaches most nearly
to the Marattiaceae.

[1] Grand'Eury (77) A. p. 98.

CHAPTER XXIV.

Ophioglossales (Fossil).

THE fossils hitherto classed with the Ophioglossales are not such as afford any satisfactory evidence in regard to the past history or phylogeny of the group. In the generalised class of Palaeozoic ferns, the Botryopterideae, we find certain characters suggesting comparison with recent members of the Ophioglos-

FIG. 301. *Rhacopteris* sp., Ballycastle, Ireland. From a specimen in the Manchester Museum. [M.S.]

saceae, but no trustworthy records of these eusporangiate ferns are furnished by the older plant-bearing strata.

The genus *Rhacopteris* (fig. 301), characteristic of the Culm flora, has been compared with *Botrychium*, but on grounds which

are wholly inadequate. The species *R. paniculifera* Stur[1] is
characterised by a stout rachis bearing two rows of laterally
attached rhomboidal or subtriangular segments with a more or
less deeply lobed margin and spreading veins. The rachis
branches distally into two arms, and these are again symmetrically subdivided into fertile axes bearing clusters of small
spherical bodies 1 mm. broad, which Stur speaks of as exannulate sporangia similar to those of *Botrychium*. He includes
the species in the Ophioglossaceae. As Zeiller[2] pertinently
remarks, *Rhacopteris* differs essentially in habit from any recent
member of this family. *Rhacopteris* also includes species characterised by leaflets deeply dissected into linear segments; an
example of this form is represented by *Rhacopteris flabellata*
(Tate) recorded by Kidston[3] from rocks of Calciferous Sandstone age in Flintshire.

The specimen described by Renault[4] from the Carboniferous
rocks of Autun as *Ophioglossites antiqua* is equally unconvincing:
it consists of a carbonised fragment, 7 cm. × 1·5 cm., regarded as
part of a fertile lamina characterised by a vertical series of
transversely elongated slits, 7 mm. wide, some of which, on
slight magnification, are seen to contain a mass of small orange-
yellow granulations. The slits are compared with the surface-
openings of the sunken sporangia of *Ophioglossum*, and the
yellow bodies are identified as spores. The material is too
imperfect to justify the use of the name *Ophioglossites*.

Noeggerathia.

This genus of uncertain position may be briefly described
here, though it has little claim to recognition as a representative of the *Ophioglossales*. It is characteristic of Lower
Carboniferous rocks and is compared by Stur[5] with recent
Ophioglossaceae. *Noeggerathia foliosa* Sternb. (fig. 302) may
be cited as a typical example of the genus. It consists of an

[1] Stur (75) A. Pl. VIII.   [2] Zeiller (00) p. 55.
[3] Kidston (89³), Pls. I. II. For other figures of *Rhacopteris* see also Stur
(75) A.
[4] Renault (96) A. p. 30, Pl. LXXXII. figs. 7—9.
[5] Stur (75) A.

axis bearing ovate leaves with numerous spreading veins. The upper part of the axis forms a spike composed of fertile leaves in the form of transversely oval bracts 2 cm. broad with a serrate

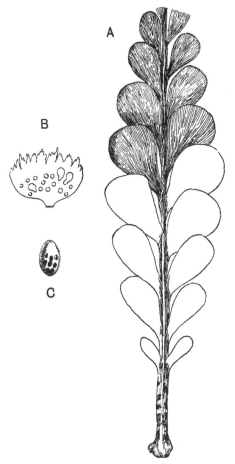

FIG. 302. *Noeggerathia foliosa.* (After Stur; A, reduced.)
B, Fertile leaf; C, Sporangium.

edge bearing on the upper face several sporangia (3 × 4 mm.) in some of which spores have been seen (fig. 302, B, C). In another form described by Weiss[1] the bracts bear a greater

[1] Weiss, C. E. (79).

number of sporangia characterised by the presence of an arillus-like basal ring.

Geinitz[1], who first described fertile specimens of *Noeggera-thia*, placed the genus in the Gymnosperms, and O. Feistmantel[2] was in favour of this view. C. Feistmantel[3], who described the small bodies in the sporangia, suggested comparison with Schizaeaceae, and Weiss[4] discussed various possibilities, asking but not answering the question, are the so-called sporangia

Fig. 303. *Chiropteris Zeilleri*, Sew. [From a specimen in the British Museum (v. 3268). Nat. size.]

rightly so named or are they fruits? Potonié[5] places the genus in the Cycadofilices. An important feature is the occurrence of the sporangia on the upper face of the bracts as in Lycopodiales and *Sphenophyllum*, but in other respects *Noeggerathia* bears no resemblance to these two groups. Sterile examples of the genus are similar in habit to *Rhacopteris*, but in the latter genus the leaves or leaflets are laterally attached and not obliquely inserted. Further, we may assume that in

---

[1] Solms-Laubach (91) A. p. 141.   [2] O. Feistmantel (75).
[3] C. Feistmantel (79).   [4] C. E. Weiss (79).   [5] Potonié (99) p. 167.

Rhacopteris the segments are leaflets of a compound leaf, whereas in *Noeggerathia* they are probably single leaves. We must leave the position of this Lower Carboniferous genus undecided, merely expressing the opinion that it is perhaps more nearly allied to the Cycads than to any other group.

The plant figured by Lindley and Hutton from the English Coal-Measures as *Noeggerathia flabellata*, which some authors quote as a species of *Noeggerathia*, is generally recognised as a *Psygmophyllum* and placed with some hesitation in the Ginkgoales.

Chiropteris.

This genus was founded by Kurr on a leaf characterised by anastomosing venation from Keuper beds near Stuttgart. A resemblance in form and venation to the leaves of recent species of *Ophioglossum* led authors to suggest the inclusion of Kurr's specimen in the Ophioglossaceae. We have, however, no justification for considering *Chiropteris* as a member of this family; it may be a fern, and that is all that can be said. The leaf represented in fig. 303 is the type-specimen of a South African Rhaetic species *Chiropteris Zeilleri*[1]. The genus is recorded also from Rhaetic rocks in Queensland.[2]

Newberry[3] describes some leaves from the Lower Cretaceous of Montana as species of *Chiropteris*: one of his types, *C. spatulata*, is almost certainly a *Sagenopteris*, similar to *S. Phillipsi* (figs. 327, 328) or *S. Mantelli*. A second species, *C. Williamsii*, is probably not generically identical with the specimen represented in fig. 303.

[1] Seward (03) p. 63.
[2] Carruthers (72²) Pl. xxvii. fig. 5; Seward (03) p. 62.
[3] Newberry (91) Pl. xiv.

CHAPTER XXV.

Coenopterideae. $\left\{ \begin{array}{ll} \text{I.} & \textbf{Botryoptereae.} \\ \text{II.} & \textbf{Zygoptereae.} \end{array} \right.$

THE term Botryopterideae, first used by Renault, has been
applied to a group of Palaeozoic ferns ranging from the Lower
Carboniferous to the Permian and containing several genera,
the distinguishing features of which are supplied by the
anatomical structure of the stems or, in many cases, by that
of the petiolar vascular strand. Scott[1] subdivides the Botryopterideae into the Botryopteris and the Zygopteris sections. In
an admirable monograph recently published by Paul Bertrand[2]
considerable changes are proposed in current nomenclature; he
substitutes the name Inversicatenales for Botryopterideae, a
designation, which as Scott remarks, is "probably too technical
to command general acceptance." A more serious criticism is
that the name Inversicatenales has reference to a character
(the inverse curvature of the leaf-trace in relation to the axis
of the stem) which is by no means universal in the group[3].

In the following account, necessarily incomplete, the generic
terminology of Bertrand is adopted, but this decision does not
carry with it any obligation to accept the name Inversicatenales.
We may speak of the types of Palaeozoic ferns dealt with in
the following pages as members of a group differing in many
respects from any existing genera of the Filicales, and exhibiting the characteristics associated with generalised plants.
Williamson, as early as 1883, spoke of Renault's Botryopte-

---

[1] D. H. Scott (08).   [2] P. Bertrand (09).   [3] Scott (09²).

rideae as comprising "altogether extinct and generalised" types[1]. For these generalised Palaeozoic ferns I propose to use the name Coenopterideae[2]. This term may be adopted in a wider sense than Renault's name Botryopterideae. The name Primofilices proposed by Arber[3] might be employed, but the implication which it carries is an argument against its adoption. We have not yet reached a stage in the investigation of extinct types at which we are able to recognise what are actually primary or primitive ferns. The search for origins will continue; as new discoveries are made our point of view shifts and the primitive type of to-day may to-morrow have to take a higher place. The epithet primitive or primary is in reality provisional: to adopt such a name as Primofilices suggests a finality which has not been, or is likely to be, achieved. The true ancestral type—the *Urform*—which we strive to discover eludes the pursuer like a will-o'-the-wisp.

Seeing that the number of true ferns of Palaeozoic age has been recently considerably reduced and is likely to suffer further reduction, the consideration of such undoubted Carboniferous and Permian examples of the Filicales as are left acquires a special importance. In the first place it is natural to ask whether the Palaeozoic ferns include any types which, if not themselves ancestral forms, may serve to indicate the probable lines of evolution of existing families. It is probable that in the near future our knowledge of the Coenopterideae will be considerably extended; as yet we possess meagre information in regard to those characters on which most stress has generally been laid in the classification of recent ferns, namely the structure of the spore-bearing organs. The sporangia of *Diplolabis* and *Stauropteris* (figs. 309, A; 322) are exannulate; in the former genus they occur in sori or synangia consisting of a small number of sporangia, while in the latter they are borne singly at the tips of ultimate ramifications of a highly

---

[1] Williamson (83²) A, p. 478.

[2] κοινόs = Lat. *communis*, common or general. I am indebted to my friend Mr L. H. G. Greenwood, Fellow of Emmanuel College, for supplying me with a name to express the idea of the generalized nature of these Palaeozoic ferns.

[3] Arber (06).

compound leaf. The resemblance of the synangium of *Diplolabis* to that of *Kaulfussia* (fig. 245, C) is not shared in an equal degree by the sporangia of *Stauropteris*, which are in some respects comparable with those of the Ophioglossaceae. In the Zygoptereae, or at least in the case of such fertile fronds as are known, and in *Botryopteris* (fig. 319), the sporangia occur in groups, and the pedicel of each sporangium is supplied with vascular tissue as in *Helminthostachys*. Another characteristic of the sporangia of the extinct types is the possession of an annulus several cells in breadth, a peculiarity which supplies a point of contact with the Osmundaceae. In the sporangia of *Kidstonia* we have a similar though not an identical type (fig. 256, E, p. 340). So far, then, as the evidence afforded by sporangial characters is concerned, it points to comparison with the Ophioglossaceae, the Osmundaceae, and the Marattiaceae. When we compare the steles of the stems we find a wide range of structure. All the genera agree in being monostelic; in *Tubicaulis* and *Grammatopteris* the protoxylem is exarch, in *Botryopteris* it is internal, while the foliar strand of *Stauropteris* and the stele of *Ankyropteris corrugata* are mesarch. The axillary branching of species of *Ankyropteris* suggests comparison with the Hymenophyllaceae.

The investigation of the vascular system of the petioles has afforded results which in the hands of P. Bertrand have led to conclusions in regard to inter-relationships. We must, however, not overlook the danger of attributing an excessive importance to this single criterion and of neglecting the facts of stem anatomy.

I. Botryoptereae.

Grammatopteris.

Renault instituted this genus for petrified stems from the Permo-Carboniferous beds of Autun. *Grammatopteris Rigolloti*[1], the type-species, is represented by a fragment, 12—15 mm. in diameter, surrounded by crowded petioles characterised by a vascular strand in the form of a short and comparatively broad

---

[1] Renault (96) A. p. 46, Pls. xxx. xxxi. See also Tansley (08) fig. 2, p. 13.

plate with the smallest tracheae at each end.   The solid xylem
of the stem stele (protostele) has peripheral groups of protoxylem.   Nothing is known as to the form of the leaves, but
sporangia similar to those of *Etapteris* (*Zygopteris*) were found
in association with the stem.   It is possible, as P. Bertrand
suggests, that Renault's species may be the stem of a
Tubicaulis.

Tubicaulis.

Tubicaulis solenites (Sprengel)[1].   Fig. 304.

This species from the Lower Permian of Saxony has been fully
described by Stenzel[2].   It is characterised by a very slender
erect stem bearing numerous spirally disposed leaves associated
with adventitious roots; the single stele (protostele) consists
exclusively of tracheae, described as intermediate between the
scalariform and reticulate type, surrounded by phloem.   Leaf-
traces are given off from the periphery of the stele where groups

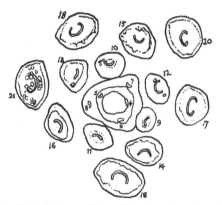

Fig. 304.   *Tubicaulis solenites*.   (From Tansley, after Stenzel.)   Stem and
    petioles: the latter numbered in the order of their age.

of smaller elements occur; these have the form of a wide-open
U-shaped strand with the base of the U facing the axis of the
stem.   As the trace passes out towards the leaves, the ends of
the U become more or less incurved.   The stem is said to reach

[1] Cotta (32) p. 15.   [2] Stenzel (89) Pls. I. II.

a metre in length and to bear compound fronds a metre long. The orientation of the leaf-trace with its concavity turned outwards is in striking contrast to the relation between leaf-trace and stem in recent ferns.

Tubicaulis Sutcliffii, Stopes[1].

In this species the vascular axis, 2 mm. in diameter, is almost cylindrical and of the protostelic type with the proto-xylem "near to or at the edge": the tracheae are scalariform or reticulate.   The leaf-traces, when first separated from the edge of the stele, are oval and gradually assume the curved form seen in *T. solenites* (fig. 304) with the convex side towards the axis of the stem.   The transition from the scalariform to the reticulate type of pitting on the tracheal walls referred to by Miss Stopes has also been noticed in some recent ferns (e.g. *Helminthostachys*) and in *Sigillaria* (fig. 200, C, p. 212). The fact that the scalariform type of pitting is practically universal in the xylem of recent ferns, would seem to show that this character has been acquired in the course of evolution and retained in preference to the reticulate form characteristic of several Palaeozoic species.   The distinction between the two methods of pitting is one of little phylogenetic importance.

Botryopteris.

This genus, founded by Renault on a specimen from Autun, is represented in the Lower Coal-Measures of England by *Botryopteris hirsuta* (= *Rachiopteris hirsuta* Will.), *B. ramosa* (= *R. ramosa* Will.[2]) (fig. 306) and *B. cylindrica* (fig. 305), also by *B. antiqua* (fig. 307) from the Culm of Pettycur, Scotland.

An important characteristic of the genus is the solid stele of the stem which agrees with that of *Tubicaulis* and *Grammatopteris,* except in the central or peripheral position of the smallest tracheae.

---

[1] Stopes (06).

[2] Williamson (89) A. p. 162.   The term *Rachiopteris* was adopted by Williamson for petrified petioles from the Coal-Measures which he believed to be filicinean.

Botryopteris forensis Renault[1]. Figs. 309, B; 319, D—G.

The stem of this species from the Upper Carboniferous of St Étienne is 1·7 cm. × 7·5 mm. in diameter. The solid stele consists of reticulate tracheae with the smallest elements on the outer edge. The comparatively broad cortex of the type-specimen is traversed by a leaf-trace in an almost vertical course and by vascular strands passing horizontally to roots. The petioles are circular in section and their vascular strand has the form of an ω in transverse section (fig. 319, G), the three projecting arms pointing to the axis of the stem. Both stem and leaves bore large multicellular hairs, spoken of by Renault as equisetiform because of the finely toothed sheaths of which they are composed. The compound fronds had fleshy lobed pinnules with dichotomously branched veins (fig. 309, B); stomata are said to be confined to the upper surface, an observation which leads Renault to describe the plant as aquatic on evidence which is hardly convincing.

The pyriform and pedicellate sporangia are borne in groups of two to six on the ultimate divisions of the frond; the wall is composed of two layers of cells and on one side of the sporangium is an annulus several cells in breadth (fig. 319, D, F). An interesting type of sporangium described by Oliver[2] from Grand'Croix in France may, as he suggests, belong to *Botryopteris forensis*; the differences between Renault's and Oliver's specimens being the result of the more perfect preservation of the tissues in the latter. The sporangium described by the English author is circular in section and measures 0·65 × 0·53 mm.; the wall is in part composed of a single layer of cells and in part of two to three layers, a character recalling the " annulate " sporangia of *Botryopteris*. Between the spore-mass and the wall is an interrupted ring of short tracheal elements similar to the xylem-mantle which occurs at the periphery of the nucellus of certain Palaeozoic gymnospermous seeds. In the absence of proof of a connexion between this sporangium and *Botryopteris* it is convenient to use the generic name *Tracheotheca* subsequently proposed by Oliver[3].

[1] Renault (75); (96) A. p. 47, Pl. XXXII.
[2] Oliver (02).   [3] Oliver (04) p. 395 (footnote).

In the recent ferns *Helminthostachys* and *Botrychium*, and, as
Oliver notices, in the microsporangia of the Australian Cycad
Bowenia spectabilis, vascular strands extend almost to the
sporogenous tissue, but the fossil sporangium is unique in
having a tracheal layer in immediate contact with the spores.
These xylem elements may, as Oliver suggests, have served the
purpose of conveying water to the ripening spores.

Botryopteris hirsuta (Will.)[1].

This English species has a slender axis bearing numerous
leaves with petioles equal in diameter to the stem. The surface
of the vegetative organs bears large multicellular hairs. The
leaf-traces resemble those of *B. forensis*, but the projecting teeth
which terminate in protoxylem elements are less prominent
than in the French species; the petioles were named by Felix
Rachiopteris tridentata[2]. As a leaf-trace passes into the stele
of the stem the three protoxylem strands unite and take up an
internal position in the solid stele. The stele may, therefore,
be described as endarch. The small tracheae at the edge of
the stele supply the xylem strands of adventitious roots.

Sporangia similar to those of *B. forensis* have been found in
association with the English species.

Botryopteris cylindrica (Will.). Fig. 305.

A plant originally described by Williamson[3] from the Lower
Coal-Measures of England as *Rachiopteris cylindrica* (fig. 305)·
and afterwards more fully dealt with by Hick[4], has a slender
stem with a cylindrical stele characterised by well-defined
central protoxylem elements in one or two groups. The leaf-
traces are semi-lunar in section with the protoxylem on the
flatter side. The stele of *Botryopteris cylindrica* (fig. 305, A)
is more cylindrical in section than that of *B. ramosa* (fig. 306)
and shows more clearly the differentiation into smaller central
and larger peripheral tracheae. In the section reproduced in
fig. 305, B the stele is giving off a branch almost identical in

¹ Scott, D. H. (08).   ² Felix (86) A.
³ Williamson (78) A. p. 351.   ⁴ Hick (96).

structure with the main vascular axis.   Scott[1], in referring to
the inclusion of this type in the genus *Botryopteris*, expresses
the opinion that its habit must have been very different from

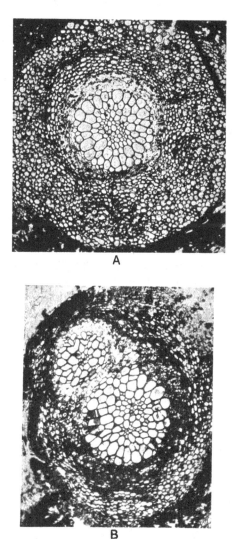

A

B

FIG. 305.   *Botryopteris cylindrica* (× 30).   From sections in the Cambridge
Botany School.

[1] Scott (08).

that of other species, and he suggests the institution of a new genus.

Botryopteris ramosa (Williamson).   Fig. 306.

This species, which bears a close resemblance to *Botryopteris hirsuta*, was originally described by Williamson from the Lower Coal-Measures of England as *Rachiopteris ramosa*[1], the specific name being chosen on account of the numerous and crowded branches given off from the main axis.   The section shown in fig. 306, A, illustrates Williamson's description of the stem as being "always surrounded [when seen in transverse sections] by a swarm of similar sections of the large and small branches, though of varying shapes and sizes."   The stele is composed of a solid and more or less cylindrical rod of xylem tracheae of the reticulate type surrounded by phloem (figs. A and D): one or more internal groups of smaller protoxylem elements occur in an approximately central position (fig. A, *px*).   The stele is in fact endarch like those of *Selaginella spinosa* and *Trichomanes reniforme*, a feature which, as Tansley[2] believes, probably entitles the vascular axis to be considered a primitive form of protostele.   In the specimens represented in fig. 306 the phloem and inner cortical tissues were almost completely destroyed before petrifaction.   The thick-walled outer cortex bears at its periphery numerous multicellular hairs.   Some of the xylem strands given off from the stele no doubt supplied adventitious roots, but in most cases the outgoing branches are leaf-traces and the numerous sections of axes of different sizes seen in fig. A point to a repeated subdivision of the crowded fronds. The structure of a petiole is shown in figs. C and D.   As seen in fig. C, the oval vascular strand has three protoxylem groups, *px*, on its flatter side; a well-defined epidermal layer is shown at *e* in fig. C.

Fig. B shows at *a* a section of a leaf-axis in the act of branching and the row of branchlets at *b* represents a further

[1] Williamson (91[2]) A. p. 261.   The two species described by Williamson as *Rachiopteris hirsuta* and *R. ramosa* were first identified as *Botryopteris* by Scott in 1898 (*British Assoc. Report*, Bristol Meeting, p. 1050).

[2] Tansley (08) p. 15.

stage in subdivision. At *sp* in fig. A the section has cut
through a single sporangium characterised by a group of larger
(" annulus ") cells on one side of the wall.

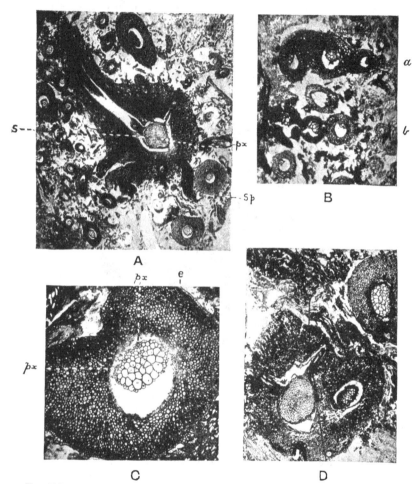

FIG. 306. A—D. *Botryopteris ramosa*; stem and frond axes. (A × 7; B × 15;
C × 26; D × 13. From sections in the Cambridge Botany School
Collection.) *px*, protoxylem; *sp*, sporangium ; *e*, epidermis.

This slender fern with its numerous repeatedly branched
leaves may perhaps have lived epiphytically on more robust
plants.

Botryopteris antiqua, Kidst.   Fig. 307.

This species, recently described by Kidston[1] from the Culm
of Pettycur near Burntisland, is represented by sections of a
small stem with a cylindrical stele 0·40 mm. in diameter composed entirely of scalariform tracheae without any recognisable
protoxylem.   The petioles are larger than the stem; the
meristele (fig. 307) is oval with protoxylem elements on the
slightly more rounded adaxial face.   As Kidston suggests, this

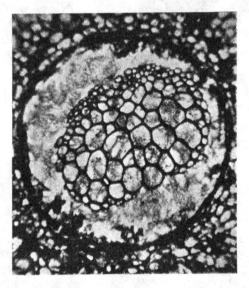

Fig. 307.   *Botryopteris antiqua*: Petiolar vascular strand.   (After Kidston: × 65.)

stem may belong to a scrambling plant which required support
to bear its relatively large leaves.   An interesting feature is
the absence of projecting teeth in the leaf-trace, a character in
marked contrast to the ω form assumed by the petioles of
Botryopteris forensis (fig. 319, G) and *B. hirsuta*.   This leads
Kidston to suggest that the vascular strand of the petiole
tends "to become more simple...as traced back in geological
time."   The greater similarity in this species between the stele

of the stem and that of the petiole is probably another mark of a more primitive type.

In these three types, *Grammatopteris*, *Tubicaulis*, and *Botryopteris*, we have monostelic plants, for the most part of very small size, with leaf-traces varying in shape from the oblong band-form in *Grammatopteris*, and the oval form of *Botryopteris antiqua*, to the ω type represented in its most pronounced form by *B. forensis*. In several species the stem stele is endarch. Our knowledge of the leaves is very meagre: in *B. forensis* they were repeatedly branched and apparently bore small fleshy pinnules; the sporangia, though differing from those of recent ferns, may be compared with the spore-capsules of Osmundaceae as regards the structure of the annulus. The abundance of hairs on the stems and leaves of some species, the tracheal sheath in the sporangium described by Oliver[1] as *Tracheotheca* (=*Botryopteris?*), and the apparent absence of a large well-developed lamina, may perhaps be regarded as evidence of xerophilous conditions.

II. Zygoptereae.

Corda[2] proposed the generic name *Zygopteris* for petrified petioles from the Permian of Saxony, included by Cotta in his genus *Tubicaulis*, which he named *T. primarius*. Corda's genus has been generally used for petioles of Palaeozoic ferns characterised by a vascular strand having the form of an **H** in transverse section (fig. 308, D). Since the generic name was instituted, information has been obtained in regard to the nature of the stems which bore some of the petioles of the *Zygopteris* type; and for other species of *Zygopteris*, the stems of which are still unknown, new generic names have been proposed. P. Bertrand[3] retains *Zygopteris* for one species only, *Z. primaria*. Fig. 308, D, shows the character of the petiolar vascular strand; the chief points are the comparatively long cross-pieces (antennae of P. Bertrand) inclined at an angle of 45° to the plane of symmetry of the petiole axis, and the groups

[1] Oliver (02).   [2] Corda (45) A.; see also Stenzel (89) p. 26.
[3] P. Bertrand (09) pp. 136, 212.

of protoxylem elements shown by the white patches in fig. D. In this as in other members of the Zygoptereae the main rachis of the leaf gives off four sets of branches in pairs alternately

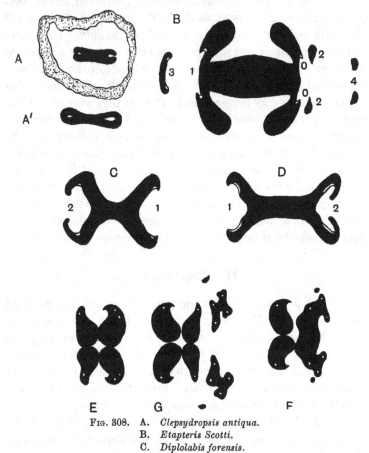

FIG. 308.   A.   *Clepsydropsis antiqua.*
          B.   *Etapteris Scotti.*
          C.   *Diplolabis forensis.*
          D.   *Zygopteris primaria.*
      E—G.   *Stauropteris oldhamia.*
The white patches in the xylem in figs. B—G mark the position of protoxylem elements.
(A, after Unger; B—G, after P. Bertrand.)

from the right and left side of the primary vascular axis. This method of branching of the stele in the primary rachis of several members of the Coenopterideae shows that the fronds bore pinnae laterally disposed, in some cases in one row and in others in two rows on each side of the rachis.   In a typical fern frond, as represented by recent and most fossil species, branching

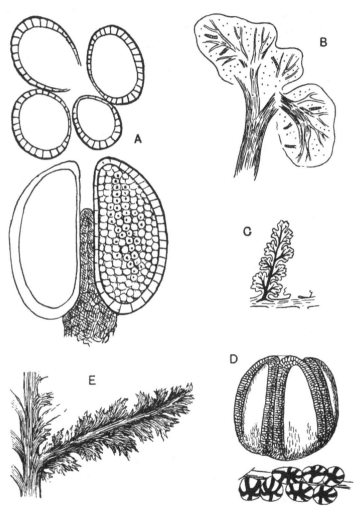

Fig. 309.   A.   *Diplolabis forensis.*
B.   *Botryopteris forensis.*
C, D.   *Corynepteris coralloides.*
E.   *Schizopteris (Etapteris) pinnata.*
(A, B, after Renault; C, D, after Zeiller; E, after Renault and Zeiller.)

of the rachis occurs in the plane of the frond, that is in the
plane represented by the horizontal arm of xylem in *Zygopteris
primaria* connecting the two antennae or cross-pieces.   In the
Zygoptereae the branches from the petiole vascular axis lie in
a plane at right angles to that of the frond; they lie in the
transverse and not in the horizontal plane.   The two strands
shown in fig. 308, B, 4, have been formed by the division of a
single strand, 3, in the transverse plane (i.e. in the plane of the
paper).   As Tansley[1] points out, a type of branching superficially similar to, though not identical with this, is seen in
some recent species of *Gleichenia* and *Lygodium*.   In this
connexion it is worthy of note that a fern figured by Unger
from Thuringia as *Sphenopteris petiolata* Goepp [2] bears pinnae
in two rows on the rachis which are characterised by repeated
branching and by a very narrow lamina or by slender naked
axes; the occurrence of this form of frond in rocks containing *Clepsydropsis antiqua* (fig. 308, A) suggests a possible
connexion between the petrified rachis and the impressions of
the leaves.

The vascular strand of the rachis of *Zygopteris primaria*
(fig. 308, D) is simpler than that of most of the Zygoptereae
and exhibits a close resemblance to the type of strand described
by Renault as *Diplolabis* (fig. 308, C).

Diplolabis.

Renault[3] instituted this genus for two species from the
Culm beds and Coal-Measures of France based on the structure
of the petioles.   The stems are unknown.   The main rachis
has a stele similar to that of *Zygopteris primaria*, but distinguished by its greater similarity, in transverse section, to
an X rather than to the letter H: the long transverse bar in
Zygopteris is here much reduced in size.   The petiole of
Diplolabis forensis Ren. (fig. 308, C) has a diameter of

---

[1] Tansley (08) p. 22.   [2] Unger and Richter (56) Pl. vi. fig. 19.
[3] Renault (96) A. p. 11.
[4] The *Diplolabis* type of strand is very similar in the form of the metaxylem
to the conducting strand of a lateral vein in *Scolopendrium officinarum* [cf.
Pelourde (09) fig. 3, p. 117].

1·5—2 cm. From the antennae a pair of small bundles is given off alternately from the right and left side, as in *Zygopteris*; the members of each pair coalesce after leaving the antennae and then separate to pass into the lateral branches of the frond. The position of the protoxylem and the formation of the lateral xylem strands previous to their separation are shown in fig. 308, C. On the side of the vascular strand shown in fig. C, 2, the two lateral extensions of the antennae are converging towards one another previous to their separation and subsequent union. The ovoid sporangia occur in groups of three to six and are coalescent below with a central receptacle; they have no annulus, but the cells on the side next the receptacle are smaller than those on the external wall (fig. 309, A). The synangial form of the sorus suggests comparison with Marattiaceae.

The species described by Renault from the Culm of Esnost is regarded by P. Bertrand as identical with that described by Solms, from the Culm of Falkenberg, as *Zygopteris Roemeri*[1]. *Diplolabis* is compared by P. Bertrand with *Metaclepsydropsis*, the generic name given to the Lower Carboniferous petiole described by Williamson as *Rachiopteris duplex*[2].

Mr Gordon has recently described in a preliminary note a new type of stem stele under the name *Zygopteris pettycurensis* from the Lower Carboniferous plant bed of Pettycur[3]: he regards the petioles attached to the stem as identical with *Zygopteris Roemeri* Solms-Laubach[4]. This species, founded by Solms-Laubach on petioles only, is placed by Bertrand[5] in the genus *Diplolabis* and regarded as identical with *D. esnostensis* Ren. The stele found by Mr Gordon may therefore be assigned to the genus *Diplolabis*: it includes two regions composed exclusively of tracheae and is cylindrical in transverse section. The inner xylem zone consists of short, square-ended, reticulately pitted elements and the outer zone is composed of long and pointed conducting tracheae. The scalariform protoxylem elements are

---

[1] Solms-Laubach (92). [2] Williamson (74) A. Pls. LIV. LV.
[3] Gordon (09). Mr Gordon's more complete account of this plant will shortly be published. I am indebted to him for furnishing me with the main facts in regard to the anatomical features.
[4] Solms-Laubach (92) Pl. II. fig. 13. [5] Bertrand, P. (09) p. 211.

situated between the two metaxylem zones. As Mr Gordon
says : this type of stem occupies a position " in the Zygopteroid
alliance " corresponding to that which *Thamnopteris Schlechtendalii* (p. 329) occupies in the Osmundaceous series. The
discovery of this stem supplies another link between the two
fern groups, Osmundaceae and Coenopterideae. Pelourde[1] has
described an imperfectly preserved vascular strand from a
locality near Autun as the type of a new genus *Flicheia esnostensis*.
Mr Gordon has pointed out to me that this is a partially rotted
petiole of *Diplolabis esnostensis* (= *Zygopteris Roemeri*).

In their recent account of fossil Osmundaceous genera, Kidston and Gwynne-Vaughan[2] speak of the central parenchyma of
the existing medullated stele as being derived from tracheal
tissue. They add that if the Zygopteroid line of descent is at
all close to the Osmundaceous, we must be prepared for the
existence of a *Zygopteris* with a solid xylem like that of
Thamnopteris: " such a discovery, in fact, we hopefully antici-
pate[3]." The new Pettycur stem amply justifies this prophecy.
It is noteworthy that Mr Gordon's stem affords an instance of
the occurrence of a type of stele, similar in its cylindrical form
and in the absence of parenchyma to that of *Botryopteris*, in a
plant bearing leaves characterised by the *Zygopteris* type of
vascular strand.

Metaclepsydropsis duplex (Will.) fig. 310, A[4]. [= *Rachiopteris
duplex*, Williamson 1874. *Asterochlaena* (*Clepsydropsis*)
duplex, Stenzel 1889. *Clepsydropsis*, Renault 1896.]

The vascular axis of the main axis of the frond is characterised by the hour-glass shape of the xylem which consists
entirely of tracheae, most of which are reticulately pitted. In
a transverse section (fig. 310, A) the two ends of the stele are
dissimilar; at one end of the long axis is a small bay of thin-
walled tissue (phloem) enclosed by a narrow band of xylem, and

[1] Pelourde (09).
[2] Kidston and Gwynne-Vaughan (08) p. 230.   [3] *Ibid.* (09) p. 664.
[4] A Culm species *Rachiopteris aphyllus* (Unger) is closely allied to *Metaclepsydropsis duplex*. [See Solms-Laubach (96) p. 30.]

at the other the bay is open and has two protoxylem groups. The latter represents the earliest stage in the production of secondary bundles: at a later stage the bay is closed by the elongation of the edges, the enclosed group of phloem is vertically extended, and the protoxylem strands are more widely separated. The curved band of xylem becomes detached as a curved arc and divides into two (fig. 310, A). In a single section of this species one often sees several strands of xylem enclosed in a common cortex with the main vascular axis; these are the xylem bundles of lateral pinnae. *Metaclepsydropsis duplex* shows the method of branching of the petiole vascular axis which has already been noticed in *Diplolabis* and *Zygopteris*. In reference to this feature, Williamson wrote in 1872—" I know of no recent fern in which the secondary branches of the petiole are thus given off in pairs, which pairs are distichously arranged on the primary axis, and each of which secondary petioles sustains ternary ones arranged distichously." By slightly altering the primary stele of this type of frond, by narrowing of the constricted portion of the hourglass and extending the lateral groups of xylem obliquely upwards, the form of stele shown in fig. 310, A, would be converted into the *Diplolabis* type (fig. 308, C).

Clepsydropsis.

Unger[1] instituted this genus as a subdivision of Corda's family Rhaciopterideae[2], the name having reference to the hour-glass form of the vascular axis[3]. The type-species *C. antiqua* (fig. 308, A) is spoken of as the commonest fossil plant in the Devonian rocks of Thuringia. In some sections the xylem has the form seen in fig. 308, A, in which an invagination of thin-walled tissue occurs at each end; in other sections (fig. 308, A') the bays become islands in the xylem. Solms-Laubach speaks of Unger's species as *Rachiopteris (Clepsydropsis) antiqua*. P. Bertrand[4], who has recently described Unger's plant, while recognising that *C. antiqua* and

---

[1] Unger and Richter (56) p. 165.   [2] Corda (45) A. p. 83.
[3] κλεψύδρα, water-clock.   [4] P. Bertrand (09) p. 127.

Metaclepsydropsis duplex closely resemble one another, draws
attention to certain differences in the structure of the xylem
which he regards as sufficient to justify a generic separation.
The leaf-traces of *Clepsydropsis* are described by Bertrand as
almost circular closed rings of xylem instead of an arc as in
Metaclepsydropsis.

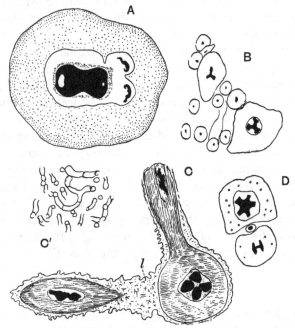

FIG. 310.   A.   *Metaclepsydropsis duplex*.
B, C.   *Stauropteris oldhamia*.
D.   *Ankyropteris scandens*.

[A, from a section in Dr Kidston's Collection (Lower Carboniferous); B, C, from
sections in the Cambridge Botany School; D, after Stenzel.]

Ankyropteris.

Stenzel adopted this name for a subdivision of Corda's genus
Zygopteris, applying it to a species described by Renault as
Z. Brongniarti, to a Permian species described by himself as
Z. (Ankyropteris) scandens, and to *Z. Lacattii* Ren.; *Rachio-
pteris Grayi* Will. and *Rachiopteris corrugata* Will. are also
included in this genus. The characters emphasised by Stenzel[1]

_____
[1] Stenzel (89) p. 25.

are (i) the double anchor-like form of the H-shaped petiole
strand in which the lateral arms (antennae) are curved like the
flukes of an anchor, and (ii) the emission of four rows of branches
instead of two. The latter distinguishing feature no longer
holds good, as *Z. primaria* also gives off four rows of bundles
and not two as Stenzel supposed. P. Bertrand has adopted
Stenzel's genus in a narrower sense[1].

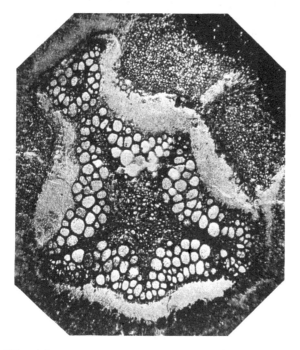

FIG. 311. *Ankyropteris Grayi.* Stele. (From a section in Dr Kidston's
Collection. × 18.)

Ankyropteris scandens Stenzel[2]. Fig. 310, D.

This Lower Permian species is very similar to or perhaps
identical with *Ankyropteris Grayi* (Williamson). The stem of

---

[1] Dr Scott points out to me that recent observations, which have not yet
been published, both by Dr Kidston and himself show that Bertrand's
terminology requires modification. There are many points to be cleared up
before we can hope to obtain a satisfactory classification of the Zygoptereae.

[2] Stenzel (89) p. 31, Pls. VI. VII.

A. scandens was found in association with the roots of a
Psaronius stem evidently petrified *in situ* as it burrowed,
like *Tmesipteris*, tropical aroids, and other recent plants, among
the living roots of the tree-fern. The stem, 10—11 mm.
in diameter, bore fronds with an H-shaped vascular strand,
small scale-leaves, and adventitious roots. The stele consists
of a five-angled cylinder of scalariform tracheae surrounding an
axial strand of parenchyma containing scattered tracheae of
smaller diameter. This axial tissue extends as a narrow strip
into each of the short and obtusely truncated arms (cf. fig. 311).
A striking feature is the production of a shoot in the axil of the
foliage-leaves (fig. 310, D), a manner of branching characteristic
of *Trichomanes* (see page 365).

Ankyropteris Grayi (Will.).   Fig. 311.

In describing this species, Williamson wrote—"That no
classification of these fossil ferns based solely upon transverse
sections of the petiolar bundles is or can be of much value, is
clearly shown when tested amongst those living ferns the
classification of which is chiefly based upon the sporangial
reproductive organs[1]." This is a view entirely opposed to that
which inspires P. Bertrand's recent monograph. Whether the
value attached to the vascular structure of petioles as a basis
of classification is upheld or not, it is noteworthy that since
Williamson expressed his opinion, our knowledge of the anatomy
of ferns and of the value of anatomical evidence has enormously
increased. The slender stem[2] of this Lower Coal-Measures
species agrees closely with that of *A. scandens*; it bore spirally
disposed fronds, scale-leaves, and roots. The stele has the form
of an irregular five-rayed star (fig. 311) in which the relative
length of the arms varies in different sections owing to the
separation of the distal ends to form leaf-traces. The axial
region is composed of parenchyma and associated narrow
tracheae, as in *A. scandens*. The xylem, with protoxylem
elements at the ends and especially at the angles of the arms,
is completely surrounded by phloem. The cortex consists

---

[1] Williamson (89) A. p. 158.   [2] *Ibid.*; see also Scott (08).

internally of parenchyma which becomes thicker-walled towards
the periphery and bears multicellular epidermal hairs. A leaf-
trace is detached in the form of a triangular strand and is
formed by the tangential extension of the distal end of an arm
of the stele. The trace, on its way through the cortex, divides
into two; the outer branch gradually changes from a slightly
curved band to an H-shaped meristele; the inner branch,
which supplied an axillary shoot, is similar to the stele of the

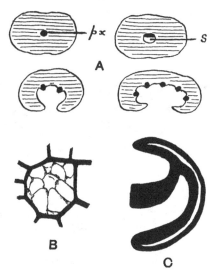

FIG. 312. A. *Thamnopteris Schlechtendalii.* Leaf-trace: *px*, protoxylem;
          s, island of parenchyma. (After Kidston and Gwynne-
          Vaughan.)
       B. *Ankyropteris corrugata.* Single trachea with tyloses.
       C. *A. bibractensis.* Part of foliar strand. (After P. Bertrand.)

stem, but smaller. Scott[1] has recently recorded the occurrence
of scale-leaves (aphlebiae) in this species like those described by
Stenzel in *A. scandens.*

Bertrand includes in *Ankyropteris* Renault's species *Zygopteris bibractensis*[2] and Williamson's species *Rachiopteris
corrugata*[3]: the former he names *A. bibractensis* var. *west-*

---

[1] Scott (07) p. 180.   [2] Renault (69); Williamson (74) A. p. 697.
[3] Williamson (77) Pls. v.—vii.

phaliensis. The fossil described by Williamson as *R. irregularis* or *inaequalis*[1] are the secondary branches of *A. bibractensis.*

Ankyropteris bibractensis, var. *westphaliensis.* Figs. 312, C; 313.

The rachis stele of this species, which is represented by portions of fronds only, has the form of a double anchor (fig. 313); the antennae are continued at the outer edge of their distal ends into a narrow band ("filament" of P. Bertrand) (fig. 312, C, and 313, *a*) composed of smaller tracheae and separated from the xylem of the antennae by a strip of thin-walled tissue (phloem ?). A group of protoxylem occurs at the

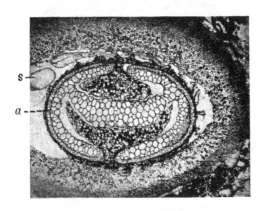

Fɪɢ. 313.　*Ankyropteris bibractensis*: *s*, stigmarian rootlet; *a*, narrow loop of xylem.　(Cambridge Botany School; × 6).

junction of the filament and antennae. The whole of the xylem is surrounded by phloem.

The section reproduced in fig. 313 shows the characteristic form of the petiolar vascular axis, consisting of a horizontal band of metaxylem with groups of much smaller tracheae on both the upper and lower margins. At the junction between the antennae, curved like the flukes of an anchor, and the horizontal band of xylem, the latter is only one trachea in breadth. The narrow loops of smaller xylem elements are

[1] Williamson (89) A. Pl. vɪɪɪ. fig. 28.

shown on the outer edge, a (fig. 313), of the antennae separated from the arcs of larger tracheae by a dark line which represents a crushed band of delicate tissue. The spaces enclosed by the incurved antennae are largely occupied by parenchymatous ground-tissue. The cylinder of outer cortex consists internally of comparatively thin-walled parenchyma succeeded externally by a zone of dark and thicker-walled cells characterised by a fairly regular arrangement in radial series, as if formed by a secondary meristem; there is, however, no indication of a meristematic layer. Below the small-celled epidermis are a few layers of thinner-walled cells which are not arranged in radial series. The structure of the outer part of the cortex is similar to that in the petiole of recent species of *Angiopteris* (fig. 243, p. 319) and *Marattia*, in which a more delicate hypoderm is succeeded by a band of mechanical tissue.

The rachis of this type of frond gives off two rows of lateral branches from the vascular axis, the plane of symmetry being at right angles to the primary rachis. Each pinna bore at its base two aphlebiae supplied with vascular strands from the leaf-traces.

We have no certain information in regard to the sporangia of this species, but Scott points out that "pear-shaped sporangia, with a very broad and extensive annulus, are commonly found associated with *Zygopteris bibractensis* and *Z. corrugata* in the petrifactions of the English Lower Coal-Measures[1]."

Ankyropteris corrugata (Will.). Figs. 312, B; 314—317.

The stem of this type of Zygoptereae was described by Williamson from the Lower Coal-Measures of Lancashire as *Rachiopteris corrugata* and included by him in the sub-group Anachoropteroides. The stele (fig. 314, B) is oval in transverse section; it consists of a cylinder of xylem tracheae enclosing a central region occupied by parenchymatous tissue and scattered narrow scalariform tracheae. The central tissue extends radially in the form of narrow arms which reach almost to the outer edge of the tracheal tissue and divide it up into 5—7 groups.

[1] Scott (08) p. 322.

A cylinder of thin-walled tissue encloses the xylem and in this occur groups of large sieve-tubes (fig. 314, D, *Sv*).

In a section of this species in the Williamson Collection[1] the long axis of the stele has a length of 5 mm. and the diameter of the stem as a whole is 2·5 cm. The greater part of the extra-stelar tissue consists of large parenchymatous cells passing near the periphery into a band of darker and thicker-walled tissue.

Reniform vascular strands traverse the cortex in an obliquely ascending course on their way to the leaves, also smaller bundles, some of which are given off directly from the stele, while others are branches of the petiole vascular strands. The petioles described by Williamson as *Rachiopteris insignis*[2] were afterwards recognised by him as those of *Ankyropteris corrugata*, though this conclusion was not published[3]. Williamson's species *R. insignis* must not be confused with Unger's Culm species *Arctopodium insigne*, which Solms-Laubach[4] refers to as *Rachiopteris insignis*. The leaf-bundle of *Ankyropteris corrugata* is at first reniform in contour (fig. 314, C, *P*), but as it becomes free from the stem it gradually assumes the H-shaped form (figs. 315—317). This petiolar strand differs from that of *Ankyropteris bibractensis* (fig. 313) in the shorter and less strongly curved antennae; and, as Williamson first noticed, the tracheae are frequently filled with thin-walled parenchyma (fig. 312, B). The existence of scale-leaves or aphlebiae like those of *Ankyropteris scandens* and *A. Grayi* has been recorded by Scott in *A. corrugata*[5].

The section represented in fig. 314, C, shows the relatively small size of the stele *S* in the stem of *Ankyropteris corrugata*. The main mass of the cortex consists of uniform parenchyma passing near the surface into darker and stronger tissue: two vascular bundles are shown in the cortex, one of which forms the conducting strand of a petiole, *P*, which has nearly freed itself from the stem: the other bundle, as shown by the examination of a series of sections, eventually passes into

[1] No. 245.
[2] Williamson (80) A. p. 507.   [3] Scott (06).
[4] Solms-Laubach (96).   [5] Scott, D. H. (06) p. 519.

XXV]   ANKYROPTERIS   457

another leaf-stalk.   A root of another plant has invaded the
cortex at *R*, fig. 314, C.

The form and structure of the stele is diagrammatically
represented in fig. 314, B.   The outer portion (black) consists

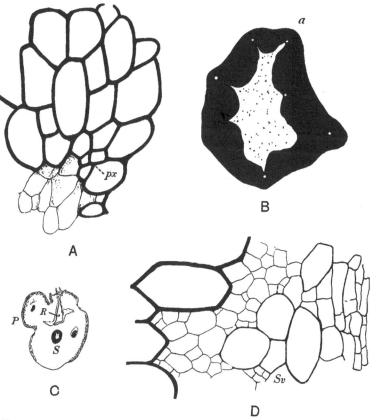

FIG. 314.   *Ankyropteris corrugata.*   *R*, intruded root; *P*, petiole; *S*, stele, *Sv*
        sieve-tubes.
        A, B.   From a section in the University College Collection.
        C.   After Williamson.
        D.   From a section in the Williamson Collection (British Museum).

of a cylinder of scalariform tracheae in which the position o
groups of smaller elements (protoxylem) is shown by the white
patches. The xylem is thus seen to be mesarch. The prominent group of xylem on the lower right-hand side of the section

consists of tracheae, cut across in an oblique direction, which are about to pass out as a separate strand. The centre of the stele is occupied by parenchymatous tissue in which are included scattered tracheae, either singly or in small groups. These medullary tracheae are rather narrower than those of the main xylem cylinder. A characteristic feature is the radial outward extension into the xylem of the medullary parenchyma, which tends partially to divide the tracheal cylinder into broad groups.

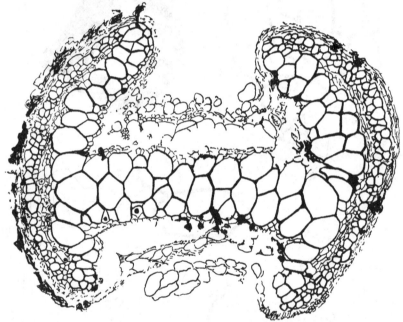

Fig. 315. *Ankyropteris corrugata* (Will.). Petiolar vascular strand. [From a section in the University College (London) Collection; after Tansley × 35.]

Fig. 314, A, enlarged from fig. B, *a*, shows the mesarch position of a protoxylem group, and a few of the parenchymatous cells of one of the narrow arms of the axial tissue. At *Sv* in fig. D a group of large sieve-tubes is seen separated from the xylem by a few parenchymatous cells, and beyond the sieve-tubes are some tangentially elongated cells. Both the

sieve-tubes, *Sv,* and the flattened cells resemble tissues in a
corresponding position in the steles of modern Osmundaceae.

In a section of *Ankyropteris corrugata* in the Williamson
Collection the radial arrangement of the more external metaxylem elements suggests the addition of secondary tracheae[1].
This suggestion of secondary thickening, a point which requires
much more thorough investigation, is interesting in relation to
a new type of stem named by Scott *Botrychioxylon*[2], but not
yet fully described. This generic name has been given to

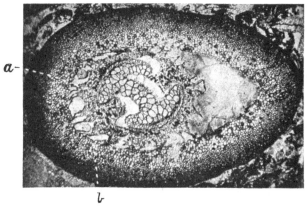

FIG. 316. *Ankyropteris corrugata.* Petiole. *a,* narrow xylem loop; *b,* spaces
        in cortex. From a section in the Cambridge Botany School Collec-
        tion. (× 10.)

a stem stele which closely resembles that of *Ankyropteris corrugata* except in the regular radial arrangement of the peripheral
xylem elements. The name *Botrychioxylon* was chosen by Scott
because of the secondary xylem characteristic of the recent
genus *Botrychium* (fig. 247, p. 322).

In the petiolar vascular strand represented in fig. 315 the
narrow band of tracheae which forms a loop external to the
antennae is clearly seen, also the small-celled parenchyma
between the loops and the larger metaxylem elements of the

[1] British Museum, section No. 245. Cf. figures by Williamson and Bertrand:
Williamson (77) Pl. v. fig. 19; Bertrand, P. (09) Pl. xii. fig. 87.
[2] Scott (07) p. 182; (08) p. 318.

antennae. The crushed tissue lying on the outer face of each
of the loops probably represents the phloem and pericycle; the
thin-walled elements above and below the horizontal band of
metaxylem are probably sieve-tubes.

Fig. 316 shows a transverse section of a petiole of this
species: the loops, *a*, of small tracheae are seen bending round
the outer edge of the antennae. The inner and more delicate
cortical tissue is partially preserved and spaces, *b*, have been
formed in it as the result of contraction previous to petrifaction.

FIG. 317. *Ankyropteris corrugata.* From a section in the Cambridge Botany
      School Collection. (× 9.)

In the petiole represented in fig. 317 the tracheae of the
horizontal band are considerably crushed; the section is, however, of interest because of the presence of *Lyginodendron*
roots, *l*, in the space originally occupied by the inner cortex.

In a paper on the tyloses of *Rachiopteris corrugata*, Weiss[1]
draws attention to the fact that similar inclusions have not
been found in the tracheae of recent ferns. The occurrence of
thin-walled parenchymatous cells in the large tracheae of
Ankyropteris corrugata petioles and of other species is a
striking feature. Williamson[2] compared these cells with the
tyloses in the vessels of recent flowering plants, and in a later
paper[3] he suggested that the included cells may belong to
saprophytic or parasitic fungi. It is, as Weiss points out,

[1] Weiss, F. E. (06).   [2] Williamson (77).   [3] Williamson (88) A.

difficult to explain the occurrence of tyloses in tracheae not immediately in contact with living parenchyma. It may be that the pits in the tracheae of *Ankyropteris* were open spaces as in the xylem of recent ferns described by Gwynne-Vaughan, and if so this would facilitate the invasion of the conducting elements by growing cells. A comparison is made by Weiss between certain cell-groups found by him in the tracheae of *Ankyropteris* and by Miss Jordan[1] in the vessels of the recent dicotyledon *Cucumis sativa*. In a more recent paper on tyloses Miss McNicol[2] expresses the opinion that pseudoparenchyma in the tracheae of the fossil petioles owes its origin to fungal hyphae.

Williamson compared the petiole bundles of *Ankyropteris corrugata* with those of recent Osmundaceae, a comparison based on the structure of the leaf-trace before its separation from the stem and its assumption of the H-form. It is noteworthy, however, that this comparison has acquired a greater significance as the result of recent work. The stele of *Ankyropteris* bears a fairly close resemblance to that of *Zalesskya* described by Kidston and Gwynne-Vaughan; in both types the xylem is represented by two kinds of tracheal tissue. In the Permian Osmundaceous genus the centre of the stele consists of short storage tracheids, while in *Ankyropteris* we may regard the central parenchyma and scattered tracheae as derivatives of the solid xylem core of some ancestral type. Moreover, the appearance and arrangement of the phloem and the tangentially elongated elements external to it (fig. 314) remind one of the extra-xylem zone in recent Osmundaceae. That the Osmundaceae and Zygoptereae are closely related groups there can be little doubt; of this affinity and common origin[3] *Ankyropteris corrugata* affords striking evidence.

The difference between the steles of *Ankyropteris Grayi* and *A. scandens* (figs. 310, D; 311) and that of *Ankyropteris corrugata* is comparatively small. In the two former species the cylindrical form has become stellate owing to the radial extension of the xylem arms. It may be that this more elaborate style of vascular construction is connected with the

[1] Jordan (03).   [2] McNicol (08).   [3] Gwynne-Vaughan (09).

climbing habit of *A. scandens* and possibly *A. Grayi*. The
radial extension of the xylem and the consequent alternation of
the yielding parenchymatous cortex and the more rigid tracheal
arms would probably render the water-conducting elements less
liable to injury in a twisting axis[1]. In *Anachoropteris De-
caisnii*[2], described by Renault, and, more especially in *Asterochlaena laxa*[3] Stenzel, a Lower Permian type from Saxony
(fig. 324), the xylem of the stele is much more deeply lobed
than in *Ankyropteris Grayi* or *A. scandens*.

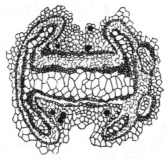

FIG. 318. *Etapteris Scotti*, P. Bert. (From Tansley, after Renault.)

Etapteris Scotti. Figs. 308, B; 309, E; 318.

P. Bertrand has proposed this name for a species of petiole
from the Lower Coal-Measures of England referred by Binney[4]
to *Zygopteris Lacattii* Ren., and included by Williamson[5] in
his comprehensive genus *Rachiopteris*. Bertrand[6] regards the
English species, which is recorded also from Germany[7], as
distinct from Renault's type[8] and therefore proposes a new
name. The petiole stele has the H-form, but its structure
is simpler than that of the *Ankyropteris* petiole.

The horizontal band of xylem has at each end two oval
groups of tracheae connected with it by a single row of xylem

---

[1] Compare figures of the vascular cylinders of climbing Dicotyledons given
by Schenck (93).

[2] For a figure of the stele see Tansley (08) p. 25, fig. 20.

[3] Stenzel (89) Pls. III. and IV.

[4] Binney (72).   [5] Williamson (74) A.   [6] Bertrand, P. (09).

[7] Felix (86) A.   [8] Renault (69).

elements (fig. 318). From the lower part of each oval group
a small strand is detached; the two strands from one side of
the stele coalesce and then separate to pass into two pinnae.
Fig. 308, B, shows four stages in the giving-off of the secondary
branches. This species, therefore, produces four rows of branches
in alternate pairs from the right and left sides of the petiole.

The first stage is shown at 0, 0, fig. 308, B; the two
projecting groups of protoxylem mark the points of departure
of a pair of small strands. At 1, the projections are more
prominent, and at 2 a pair of strands has become detached: at
a later stage, 3, these two strands unite to divide later (4) into
two slightly curved bundles.

Our knowledge of the fructification of *Etapteris* is based on
Renault's account of sporangia, which he regarded as belonging
to *Zygopteris* (*Etapteris*) *Lacattii*. They have the form of

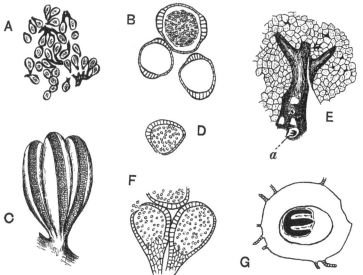

FIG. 319.   A—C.   Sporangia of *Etapteris* (?).
D—G.   *Botryopteris forensis*.   (After Renault.)

elongated slightly curved sacs (2·5 × 1·3 mm.) borne in clusters
(fig. 319, A—C) on slender ramifications of the fertile frond,
which is characterised by the absence of a lamina.   Each

sporangium has a pedicel, and three to eight sporangia are attached to a common peduncle; the walls of the sporangia are at least two cell-layers in thickness and the annulus consists of a band of thick-walled cells passing from the crest down each side (figs. B and C), thus differing from the sporangia of *Botryopteris* (fig. 319, D, F) in which the broad annulus is confined to one side.

It is practically certain that the fronds described by Grand'Eury[1] as *Schizopteris pinnata* (fig. 309, E) and *Schizostachys frondosus* represent respectively the sterile and fertile

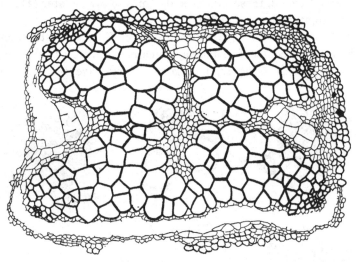

Fig. 320. *Stauropteris oldhamia.* (After Tansley. From a section in Dr Scott's Collection. × 60.)

leaves of *Etapteris*. Zeiller[2] gives expression to this by substituting the generic name *Zygopteris* for *Schizopteris*, and we may now speak of the leaves as *Etapteris*. Dr White[3] has referred to a new genus, *Brittsia*, some impressions of pinnate fronds from the Coal-Measures of Missouri which, as he points out, bear a close resemblance to *Schizopteris pinnata* Grand'Eury (fig. 309, E). No sporangia have been found; it is, however, probable that *Brittsia problematica* represents fragments of a

---

[1] Grand'Eury (77) A. Pl. xvii.   [2] Renault and Zeiller (88) A.
[3] White (99) p. 97.

leaf borne by a plant closely allied to *Etapteris* (*Zygopteris*). The broad rachis bears crowded pinnae given off at a wide angle; the small pinnules are rather deeply lobed or pinnatifid (3—10 mm. long by 1·5—3 mm. broad). The lamina is traversed by irregularly lobed and occasionally anastomosing veins. In the fertile pinnae the segments have no lamina but bear bundles of pedicellate sporangia.

It should be noticed that the sporangia described by Renault and by other authors as those of *Zygopteris* (fig. 319, A—C) have not been found in organic continuity with a frond showing a well-preserved vascular strand. It is, however, certain that this characteristic annulate sporangium, borne on branched and slender pedicels, was produced on fronds with a much reduced lamina belonging to some species of the Zygoptereae, *Etapteris* and probably also *Ankyropteris*.

Stauropteris.

This genus was instituted by Binney for petioles from the Lower Coal-Measures of Oldham (Lancashire).

Stauropteris oldhamia Binney[1] is characterised by a stele (figs. 308, E—G; 310, C; 320; 321) composed of four groups of xylem which Bertrand regards as homologous with the antennae of *Diplolabis*, *Ankyropteris*, and *Etapteris*, the horizontal cross-piece of these genera being absent in *Stauropteris*. Williamson spoke of this species as "one of the most beautiful and also one of the most perplexing of the plants of the Coal-Measures"; he discussed its possible affinity with both Lycopods and ferns, deciding in favour of the latter group[2]. In transverse section the petiolar vascular axis is approximately square, the xylem groups forming the ends of the diagonals; the tracheal groups are separated by phloem and the centre of the stele in the primary rachis is also occupied by that tissue, which is connected by four narrow strips with the external phloem. The structure of the petiolar vascular axis is very clearly shown in the drawing by Mrs Tansley reproduced in fig. 320. Proto-

---

[1] Binney (72); Williamson (74) A. p. 685.
[2] Williamson (74) A. p. 685.

xylem elements occur close to the surface of each of the four arms of the xylem; the bays between the two lateral and the two lower xylem groups contain large sieve-tubes. Portions of the inner cortex are seen in places abutting on the small-celled pericyclic tissue.

The right and left halves of the stele are not absolutely identical (fig. 320; fig. 308, E); this is due to the fact that secondary branches are given off in four rows, two alternately from the right and left sides. The preparation for the departure of the lateral strands alters the configuration of the stelar xylem groups. The protoxylem groups are not external but separated from the surface by one or two layers of metaxylem. In fig. 308, E, the occurrence of two protoxylem strands in the right-hand groups of metaxylem marks an early stage in the detachment of branches. These two protoxylems are the result of division of single protoxylem strands like those in the left-hand half of the stele. At a later stage the petiolar stele assumes the form shown in fig. 308 F, and two small bundles are detached to supply aphlebiae: this is followed by the stage shown in fig. G, where two four-armed strands are passing out to a pair of branches of the leaf axis. The separation of these two meristeles leaves the right-hand half of the stele in the condition seen on the left-hand side of fig. E. The diagrammatic sketch represented in fig. 310, C, shows one pair of branches in organic connexion with the rachis, and each of these arms contains an obliquely cut vascular strand like those in fig. 308 G.

The cortex consists for the most part of fairly thick-walled parenchyma (fig. 321) which in the hypodermal region is replaced by a zone of thin-walled lacunar tissue. A few stomata have been recognised in the epidermis[1]. The lower left-hand branch seen in fig. 310, C, has been shaved by the cutting wheel so that the aerenchymatous tissue, l, is shown in surface-view: a portion of this tissue is enlarged in fig. C'. The same delicate chlorophyllous tissue forms a folded and shrivelled layer with an uneven margin on the surface of the rachis and lateral branches. This hypha-like tissue, which was discovered by

---

[1] Bertrand, P. (09) Pl. vii. fig. 48.

Scott[1] and figured by Bertrand[2], doubtless represents the much reduced lamina of the highly compound leaves; it may be compared with the green outer cortex of *Psilotum* shoots and with the lacunar tissue in the capsule of the common moss, *Funaria hygrometrica*.

The rachis reproduced in fig. 321 is surrounded by an enormous number of sections, some transverse, others more or less vertical, of branchlets of various sizes. Fig. 310, B, shows the three-rayed vascular axis of a branch of a lower order than

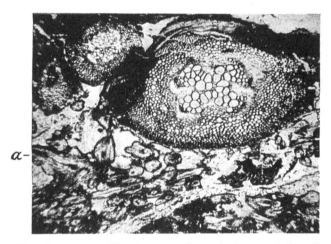

Fig. 321. *Stauropteris oldhamia*: *a*, sections of pinnae. (×10. From a section in the Cambridge Botany School Coll.)

those seen in fig. C, and the single vascular strands of still finer ramifications of the leaf. The extraordinary abundance of axes of different sizes, many of which are cut in the plane of branching, in close association with the rachises of *Stauropteris* affords a striking demonstration of the extent to which the subdivision of the frond was carried in a small space. The leaves must have presented the appearance of a feathery plexus of delicate green branchlets devoid of a lamina, some of which bore terminal sporangia. It may be that the delicate fronds were borne on a slender rhizome which lived epiphytically in a moist atmosphere on the stouter stems of a supporting plant.

[1] Scott (05³) p. 115.   [2] Bertrand, P. (09) Pl. VII.

The sporangia[1] of *Stauropteris oldhamia* are exannulate and nearly spherical, with a wall of more than a single row of cells; they occur at the tips of slender and doubtless pendulous branchlets. The discovery by Scott[2] of germinating spores (fig. 323) in a sporangium of this type supplies an interesting piece of evidence in favour of the fern nature of these reproductive organs. Similar germinating spores have been described by Boodle[3] in sporangia of *Todea*.

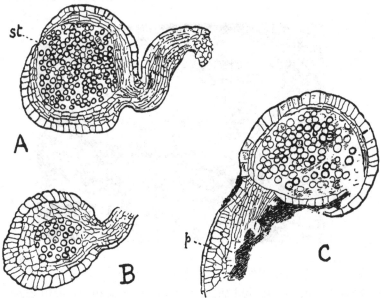

FIG. 322. Sporangia of *Stauropteris oldhamia*. *St*, stomium : *p*, palisade tissue. (From Tansley, after D. H. Scott, from a drawing by Mrs D. H. Scott.)

Stauropteris burntislandica.

This Lower Carboniferous plant identified by Williamson with the Oldham plant from the Lower Coal-Measures is referred by Bertrand to a distinct species. In the structure of the rachis stele it agrees closely with *Stauropteris oldhamia*; the main vascular strand gives off four rows of branches, two from each side, and aphlebiae were present at the common

[1] Scott (04); (05³).   [2] Scott (06²).   [3] Boodle (08).

base of each pair of pinnae. Mrs Scott[1], who has recently
described the sporangia of this species, speaks of one specimen
in which germinating spores were found. The same author
gives an account of some curious spindle-shaped bodies which
she found in association with *S. burntislandica*. The nature
of these organs is uncertain; Mrs Scott inclines to regard
them as glands borne in pairs on lateral pedicels of the
frond: she adopts for these the name *Bensonites fusiformis*
proposed by Dr Scott. If there is a reasonable probability, as
there certainly seems to be, in favour of connecting these organs
with *Stauropteris*, it is legitimate to question the desirability
of adding to the long list of names included in the group
Coenopterideae.

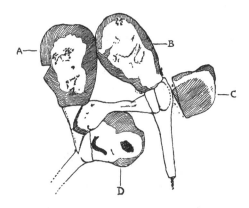

Fig. 323.   Germinating spores from a sporangium of *Stauropteris*.   (From
Tansley, after D. H. Scott.)

Corynepteris.   Fig. 309, C, D.

This genus was founded by Baily[2] on fragments of a fern
from Carboniferous rocks in County Limerick, Ireland, characterised by a peculiar type of fructification which he named
Corynepteris stellata. More complete examples of the same
genus have been described by Zeiller[3] from the Coal-field of

[1] Scott, R. (08) Pl. xxxiv. figs. 1, 2.
[2] Baily (60) Pl. xxi. κορύνη, a club or mace.   [3] Zeiller (83).

Valenciennes. The sporangia are large, ovoid, and sessile; the annulus (fig. 309, D) has the form of a complete vertical band several cells in breadth: five to ten sporangia are grouped round a receptacle. Zeiller describes two species as *Sphenopteris* (*Corynepteris*) *coralloides* Gutb. and *S.* (*Corynepteris*) *Essinghii* And.; in both the fronds are quadripinnate and bear aphlebiae at the base of the pinnae. The former species is recorded by Kidston[1] from the South Wales Coal-field. A single pinnule of *C. coralloides* is shown in fig. 309, C. Potonié[2] refers this frond to his genus *Alloiopteris*: the portion of a pinna represented in fig. 354 G shows the characteristic modified pinnule next the rachis. Zeiller draws attention to the occurrence of two parallel lines on the rachis of a specimen of *Corynepteris coralloides* which he figures[3], and suggests that these may indicate the existence of an H-shaped form of vascular strand like that of *Etapteris* and *Ankyropteris*. The sorus of *Corynepteris* is comparable with that of the Marattiaceae, but the broad annulus is a difference which suggests affinity to *Etapteris*. The sorus is similar to that in *Diplolabis* (fig. 309, A), but in that genus the sporangia are exannulate.

The vascular axis in the stems of different members of the Coenopterideae assumes a variety of types. In *Botryopteris antiqua* the xylem forms a solid protostele in which no protoxylem strands have been recognised; in other species, e.g. *B. ramosa*, the cylindrical stele is similar to that of *Trichomanes radicans* (Hymenophyllaceae) in the more or less central position of the protoxylem. In *Botryopteris forensis* the protostele is said to be exarch. The probability is that the central Botryopteris type is the endarch protostele, a form of vascular axis which may be regarded as primitive. The leaf-traces of the Lower Carboniferous *Botryopteris antiqua* are simple oval strands differing but slightly from the cylindrical stele of the stem. In the Upper Carboniferous British species the petiolar vascular strand has become more specialised and farther removed from that of the stem; in *B. forensis* the

[1] Kidston (94).   [2] Potonié (02) p. 492.   [3] Zeiller (88) A. Pl. x.

distinction between leaf and stem steles is still more pronounced. It is perhaps legitimate to regard these types as representing an ascending series, the more primitive of which are distinguished by the greater similarity between leaf and stem, organs differentiated from a primitive thallus[1], that is from a vegetative body. Portions of this ultimately became specialised as lateral members or leaves, while a portion acquired the character of a radially constructed supporting axis or stem.

The vascular strand characteristic of the Zygoptereae is represented by the H-shaped form as seen in *Ankyropteris corrugata* or in a more complex form in *A. bibractensis*. This style of strand may be regarded as a development from the simple strands of *Grammatopteris* and *Tubicaulis* or *Botryopteris antiqua* along other lines than those followed by *B. forensis*. The extension of the xylem in two symmetrically placed arms at the ends of the cross-piece of the H is correlated with the habit of branching of the leaf-system which forms one of the striking peculiarities of many of the Zygoptereae. The solid type of stele characteristic of the Botryoptereae is closely matched by that in the Lower Carboniferous stem discovered by Mr Gordon[2]. By the partial transformation of the central xylem region into parenchymatous tissue and the concentration of water-conducting elements in the peripheral region the style of *Ankyropteris corrugata* was developed. The vascular strand of the older plant, which is of the *Diplolabis* type, may be regarded as a more primitive style than that of the H-form of petiole strand represented by *Ankyropteris corrugata*. A further stage in evolution is seen in the stem stele of *Ankyropteris Grayi* and *A. scandens*, both of which have the H-form of meristele. This step in increasing complexity of stem stele, though probably connected with the increasing specialisation of the leaf-traces, as held by Mr Tansley, may also be associated with the development of a climbing habit. In *Asterochlaena laxa* Stenzel (fig. 324) and *A. ramosa* (Cotta)[3] the tendency towards a stellate expansion of the originally cylindrical form

---

[1] Tansley (08).   [2] See p. 447.
[3] Stenzel (89) p. 15, Pls. III. IV.

of stele reaches a higher degree, with the result that a style is
evolved which agrees closely with that of the conducting tissue
of some existing Dicotyledonous Lianes.

Attention has already been drawn to the generalised features
exhibited by the Coenopterideae both in the anatomy of the
steles and in the structure of the sporangia. The conclusion
arrived at is that while the Coenopterideae foreshadow in some

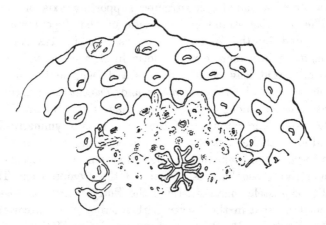

FIG. 324. *Asterochlaena laxa*: part of stem with petiole and a few roots.
From Tansley, after Stenzel.

of their characters more than one group of more recent ferns,
some at least of their members afford convincing evidence of
the correctness of the view—which is also that of Dr Kidston
and Mr Gwynne-Vaughan—that the Osmundaceae and the
Coenopterideae are offshoots of a common stock.

CHAPTER XXVI.

HYDROPTERIDEAE { Marsiliaceae.
 Salviniaceae.

THE unsatisfactory and meagre records in regard to the past history of these heterosporous Filicales render superfluous more than a brief reference to the recent species.

Marsiliaceae.

This family is usually spoken of as including the two genera *Marsilia* and *Pilularia*. Lindman[1] has however founded a third genus, *Regnellidium*, on a Brazilian plant which is distinguished by some well-defined characters from all species of *Marsilia*. The members of the Marsiliaceae live for the most part in swampy situations. *Marsilia* is represented in Europe by *M. quadrifoliata* L. which occurs in Portugal, France, Germany and other parts of the Continent, extending also to Kashmir, Northern China, and Japan. Of the other 53 species, 17 are recorded from different regions in Africa, while others occur in South America, Asia[2], Australia, and elsewhere.

Pilularia globulifera L. is the only British representative of the Hydropterideae. The remaining four species of the genus occur in South America, California, New Zealand, Australia, and *P. minuta* Dur. is met with in the South of France, Algeria, and Asia Minor in subtropical or warm temperate regions.

The Marsiliaceae are regarded as more nearly related to the

---

[1] Lindman (04).   [2] Baker (87) A.; Sadebeck, in Engler and Prantl (02).

Schizaeaceae than to any other family of homosporous ferns[1]. Their heterospory, the production of sporangia in closed fruit-like sporocarps, and the anatomical features associated with existence in marshy habitats, tend to obscure the resemblances to the true ferns.

The genus *Marsilidium* proposed by Schenk[2] for a piece of an axis, bearing apparently a whorl of six leaflets, from the Wealden of Osterwald, cannot be regarded as satisfactory evidence of the existence of the Marsiliaceae in the Wealden flora of North Germany.

The six leaflets of *Marsilidium speciosum*, having a length of 5 cm., are similar in shape to the four leaflets of recent species of *Marsilia*, but they differ in the repeated dichotomy of the veins from the reticulate venation of the recent forms. It is worthy of note, however, that in Lindman's Brazilian type *Regnellidium diphyllum* (fig. 326, A), the leaflets are characterised by dichotomous and not by anastomosing veins.

Hollick[3] has described some impressions of imperfect orbicular leaves with a "finely flabellate obscurely reticulated(?) venation" from Cretaceous rocks of Long Island as *Marsilia Andersoni*, but these are too fragmentary to be accorded this generic designation. My friend Dr Krasser informs me that he is describing some well-preserved leaves from Cretaceous beds of Grünbach in Lower Austria as *Marsilia Nathorsti*[4]. He compares these with the recent form *Marsilia elata*, a variety of *M. Drummondi*.

Another Lower Cretaceous species *Marsilia perucensis* has been figured by Frič and Bayer[5] as a stalked fruit-like body from Bohemia. This was originally described by Velenovský as *M. cretacea*, but under this name Heer[6] had previously recorded a supposed sporocarp from Greenland. These fossils have little claim to recognition as examples of Marsiliaceous plants.

The fragment figured by Heer[7] from Tertiary rocks of

---

[1] Campbell (04); Bower (08) p. 551.   [2] Schenk (71) p. 225.
[3] Hollick (94) Pl. LXXI.   [4] Mentioned by Krasser (06) in a preliminary note.
[5] Frič and Bayer (01) p. 86, fig. 34.   [6] Heer (82) Pl. XVI.
[7] Heer (55) A. Vol. III. p. 156, Pl. CXLV. fig. 35.

Oeningen as *Pilularia pedunculata* is too small to determine with reasonable accuracy. Other supposed representatives of the family mentioned in palaeobotanical literature are not of sufficient importance to describe.

Salviniaceae.

The two genera of Salviniaceae, *Salvinia* and *Azolla*, are water plants, and are usually described as annuals which survive the less favourable season in the form of detached sporocarps. Goebel[1] states that all the tropical species of *Salvinia* known to him have an unlimited existence.

Salvinia natans, Hoffm., the only European species, extends from the South of France to Northern China and the plains of India: the other twelve species are mostly tropical. *Azolla*, represented by four species, occurs in Western and Southern North America, South America, Madagascar, Australia, New Zealand, and is widely spread in tropical Asia and Africa.

Species of *Azolla* frequently form a considerable proportion of the floating carpet of vegetation on inland waters[2] growing under conditions which might be supposed favourable for preservation in a fossil state.

The Salviniaceae, though probably rather farther removed than the Marsiliaceae from the homosporous Filicineae, are considered by Bower[3] to be related to the Gradatae, but modified in consequence of their aquatic habit and the assumption of heterospory.

No undoubted examples of fossil species of *Azolla* have been described. *Salvinia*, on the other hand, is represented by several Tertiary species, for the most part founded on leaves only, and Hollick[4], who published a list of fossil Salvinias, has described detached leaves as *Salvinia elliptica* Newb. from what may be Upper Cretaceous rocks from Carbonado, Washington.

---

[1] Goebel (05).
[2] See Seward (94) p. 441, for a description of the floating plants on the lagunas of Gran Chaco (S. America) by Prof. Graham Kerr.
[3] Bower (08) p. 611.   [4] Hollick (94).

Some of the leaves figured as Tertiary Salvinias are of no value as evidence of the former distribution of the genus[1].

From the Coal-beds of Yen-Bäi (Tonkin), probably of Miocene age, Zeiller[2] has figured some well-preserved impressions of oval or orbicular leaves, 15 mm. long and 10—20 mm. broad, characterised by reticulate venation and by cordate bases, which he refers to Heer's Swiss species *Salvinia formosa*[3].

Dr Zeiller[4] in the most recently published part of his series of valuable résumés of palaeobotanical literature refers to a description by Brabenec of specimens of this species from Bohemian Tertiary beds showing both microspores and megaspores.

One of the most complete specimens so far discovered has recently been described by Fritel[5] from Eocene beds of the Paris Basin as *Salvinia Zeilleri*. This species, founded on portions of stems bearing floating leaves, submerged root-like leaves, and sporocarps, is compared with a recent tropical American species *S. auriculata*.

It is noteworthy that no authentic records of Hydropterideae have been discovered in Palaeozoic rocks[6]. Comparisons have been made in the case of the genera *Traquairia* Carr. and *Sporocarpon* Will. with the reproductive organs of *Azolla*[7], but these rest on a wholly insufficient basis.

Dawson[8] proposed the generic name *Protosalvinia* for some spores of Devonian age, which he regarded on inadequate grounds as evidence of Palaeozoic Hydropterideae.

Zeiller[9], in discussing the possible relationships of the problematical type *Chorionopteris gleichenioides* Cord., suggests a possible alliance with the Hydropterideae. Corda founded the genus *Chorionopteris*[10] on some small fragments of pinnules, 6—7 mm. long, found in the Carboniferous rocks of Radnitz in Bohemia.

---

[1] E.g. Lesquereux (78) Pl. LXIV. fig. 14; Pl. v. fig. 10. Staub (87) Pl. XIX. fig. 2.

[2] Zeiller (03) Pl. LI. figs. 2, 3.

[3] Heer (55) A. Vol. III. p. 156, Pl. CXLV. figs. 1—315.

[4] Zeiller (09²) p. 95.   [5] Fritel (08).

[6] See also Arber (06) p. 228.   [7] Solms-Laubach (91) A. p. 183.

[8] Dawson (86).   [9] Zeiller (88) A. p. 58.   [10] Corda (45) A. Pl. LIV.

The lobes of the pinnules are incurved distally to form a capsule, containing four sporangia, which apparently opened on dehiscence into four valves; the spores are of one size. The material is however insufficient for accurate determination.

There is no evidence contributed by fossil records which indicates a high antiquity for the Hydropterideae. It is unsafe to base any conclusion on the absence of undoubted Palaeozoic representatives of this group; but the almost complete absence of records in pre-Tertiary strata is a fact which may be allowed some weight in regard to the possible evolution of the hetero-sporous filicales at a comparatively late period in the earth's history.

A description of the Mesozoic genus *Sagenopteris* may be conveniently included in this chapter, though as in many other instances the inclusion of a genus under the heading of a recent family name does not by any means imply that the position of the extinct type is regarded as settled.

Sagenopteris.

This generic name was applied by Presl[1] to small fronds composed of four or rarely two palmately disposed leaflets with a more or less distinct midrib and anastomosing secondary veins. Schimper[2] compared *Sagenopteris* with *Marsilia*, but did not regard the resemblance as evidence of relationship. Nathorst[3] expressed the opinion that certain fruit-like bodies obtained from the Rhaetic beds of Scania are of the nature of sporocarps and were borne by *Sagenopteris*, with the leaves of which they were associated. He published a drawing of part of a fruit showing on its partially flattened surface some raised oval bodies which are considered to be spores. Dr Nathorst kindly placed at my disposal the drawings reproduced in fig. 325 made from some of his specimens found at Bjuf in Scania.

In contour and superficial features, e.g. the veining on the wall, these bodies bear a fairly close resemblance to the

---

[1] Presl, in Sternberg (38) A.   [2] Schimper (69) A.   [3] Nathorst (78) p. 17.

sporocarps of recent species of *Marsilia*. They were found in
association with the leaves of *Sagenopteris undulata* Nath.,
an abundant Scania type similar in form to the English Jurassic
species *S. Phillipsi* (figs. 327, 328). Heer was independently
led by an examination of some examples of the Swedish "fruits"
to compare them with the sporocarps of *Marsilia*. A small
spherical body is figured by Zigno[1] close to a leaf of his
species *S. angustifolia*, which may be a sporocarp. In a recent
paper, Salfeld[2] says that he found fructification on the lower
face of the leaflets of *S. Nilssoniana* Brongn. from German
Jurassic rocks, but he brings forward no evidence in support of

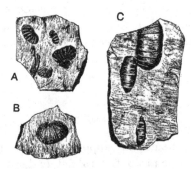

Fig. 325.   Sporocarp-like bodies found in association with the leaves of *Sagenopteris*. (Nat. size.   From drawings supplied by Dr Nathorst.)

this statement. The systematic position of *Sagenopteris* is by
no means settled. In a previous account of the genus I
expressed the view that it is probably a member of the true
ferns[3], but the resemblance of Dr Nathorst's drawings to the
Marsilian sporocarps influences me in favour of his opinion that
Sagenopteris may belong to the Hydropterideae. The evidence,

---

[1] Zigno (56) A. Pl. xx.   [2] Salfeld (09) p. 17.

[3] In a footnote to Fontaine's description of Jurassic plants of Oregon, Lester
Ward writes:—"Seward treats *Sagenopteris* as a fern, classing it now (*Jur. Fl.
Yorkshire Coast*, 1900, p. 161) in the family Polypodiaceae, although in his
Wealden Flora, 1894, p. 129, he placed it in the Schizaeaceae." [Ward (05)
p. 83, note *b*.] My words are "I am disposed to regard *Sagenopteris* as probably
a genus of ferns" (*loc. cit.* 1900, p. 161). I have never referred this plant to
the Polypodiaceae or Schizaeaceae or to any other family.

as Solms-Laubach[1] states, is not wholly satisfactory: Schenk points out that the frequent occurrence of detached *Sagenopteris* leaflets suggests that they easily fell off the petiole, whereas in *Marsilia* the leaflets do not fall off independently. The discovery of a new type of Marsiliaceae in Brazil, which Lindman has described as *Regnellidium diphyllum*[2] (fig. 326, A), affords an additional piece of evidence bearing on the comparison

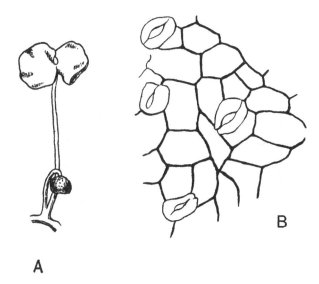

A

FIG. 326. A. *Regnellidium diphyllum* Lind. Single leaf and stalked sporocarp. (⅞ nat. size. After Lindman.)
B. Cuticle of *Sagenopteris rhoifolia*. (After Schenk.)

of *Sagenopteris* with members of this family. In *Regnellidium* the leaves differ from those of *Marsilia* in bearing two instead of four leaflets, and in the former the veins are repeatedly forked, and do not anastomose as in *Marsilia*. In the possession of only two leaflets *Regnellidium* agrees with some forms of *Sagenopteris* (fig. 328).

[1] Solms-Laubach (91) A. p. 182.
[2] I am indebted to my friend Dr Nathorst for calling my attention to Lindman's paper.

Sagenopteris Phillipsi (Brongniart)[1].   Figs. 327, 328.

   1828. *Glossopteris Phillipsi*, Brongniart, Hist. vég. foss. p. 225,
         Pls. LXI. *bis*, LXIII.
   1838. *Sagenopteris Phillipsi*, Presl, in Sternberg's Flor. Vorwelt, vii.
         p. 69.

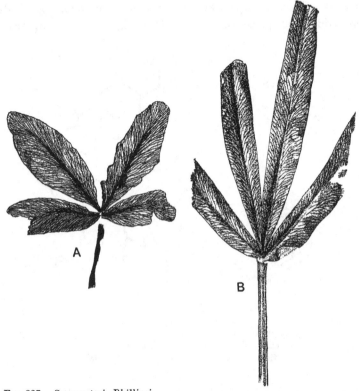

FIG. 327.   *Sagenopteris Phillipsi*.
         A. From the type-specimens of Lindley and Hutton (*Glossopteris
             Phillipsi*).   Gristhorpe Bay, Yorkshire.   British Museum, No.
             39221.   Slightly reduced.   M.S.
         B. From a specimen in the British Museum (39222).   Nat. size.
             Figured by Lindley and Hutton as *Glossopteris Phillipsi*.

   The fronds of this common Jurassic species, which is
recorded from many European localities, from North America,
Australia, the Antarctic regions[2], and elsewhere, are very variable
as regards the form, size, and number of the leaflets.

   [1] For a fuller synonymy, see Seward (00) p. 162.   [2] Nathorst (04[2]).

Frond petiolate, in some forms the petiole bears four linear or oval-lanceolate leaflets having a distinct midrib and oblique anastomosing veins. In others a shorter winged petiole bears one or two shorter and broader, somewhat obcuneate, leaflets without a midrib.

It is probable that Bunbury[1] was correct in his opinion that the specimen figured by Lindley and Hutton[2] as *Otopteris cuneata*, characterised by two leaflets (fig. 328), is not specifically distinct from the normal form with four leaflets (fig. 327).

Similarly, such specimens as that represented in Pl. XVIII., fig. 3 of the first part of my *Jurassic Flora*, in which a short stalk bears only one leaflet may, provisionally at least, be included in Brongniart's species. Yabe[3] describes a form with two leaflets from Jurassic rocks of Korea as *Sagenopteris*

FIG. 328.　*Sagenopteris Phillipsi.*　From a specimen in the Manchester University Museum.　Nat. size.

bilobata which resembles *S. Phillipsi*; and Moeller[4] records a specimen similar to that represented in fig. 328 from Bornholm as *S. cuneata* (Lind. and Hutt.).

The leaf shown in fig. 327, A, in which the longest segments are 4·5 cm. in length, represents the most abundant form and illustrates the very close agreement between *S. Phillipsi* and the Rhaetic species *S. rhoifolia.* Fig. 327, B, which is drawn from a specimen figured by Lindley and Hutton[5], shows a leaf with longer (6·5 cm.) and much narrower segments. Broader leaflets are occasionally met with in which the lamina reaches a length of 11 cm.[6]

Leaves with leaflets narrower (3 mm. broad) than those

[1] Bunbury (51) A.　　　　　[2] Lindley and Hutton (35) A. Pl. CLV.
[3] Yabe (05) Pl. III. fig. 16.　　　[4] Moeller (02) Pl. VI. fig. 10.
[5] Lindley and Hutton (33) A. Pl. LXIII. fig. 2.　　[6] Seward (00) p. 169, fig. 26.

represented in fig. 327, B, are described by Zigno[1] from Jurassic
beds of Italy as *S. angustifolia* and by Moeller[2] from the Jurassic
of Bornholm as *S. Phillipsi* f. *pusilla*. A coarser type of
venation than that of *S. Phillipsi* is occasionally found in
Jurassic examples, as in *S. grandifolia* Font.[3] from Oregon and
S. Nathorsti Barth. from Bornholm[4].

Sagenopteris is recorded also from several Rhaetic floras.
The best known species, *S. rhoifolia* Presl[5], is hardly distinguishable from some forms of *S. Phillipsi* or from the Italian
Jurassic species described by Zigno as *S. Goeppertiana*[6], though
the leaflets are usually rather larger. This species was first
described by Brongniart as *Filicites Nilssoniana*[7], and a few
authors[8] have adopted this specific name because of its priority
over Presl's designation. As Nathorst remarks, to give up
the well-known name *S. rhoifolia* for *S. Nilssoniana* is "mere
pedantry." The epidermis of *S. rhoifolia* as figured by Schenk[9]
consists of cells with straight and not undulating walls:
stomata occur on the lower surface (fig. 326, B).

Rhaetic leaves of the type represented by *S. rhoifolia* have
a wide geographical distribution.

The specimens described by Feistmantel from the Damuda
series of India as *Sagenopteris longifolia* are no doubt fronds
of *Glossopteris longifolia*[10].

The Wealden species *Sagenopteris Mantelli* (Dunk.)[11] agrees
closely in habit and in the form of the leaflets with *S. Phillipsi*
and *S. rhoifolia*. It is probable that some of the leaves
described by Velenovský[12] from Lower Cretaceous rocks in
Bohemia as *Thinnfeldia variabilis* are portions of *Sagenopteris*
fronds. *S. Mantelli* is recorded from several European localities,
from California[13], and elsewhere.

[1] Zigno (56) A. Pl. xxi.   [2] Moeller (02) Pl. vi. figs. 8, 9.
[3] Ward (05) Pl. xv. fig. 5.
[4] Bartholin (92) Pl. v. fig. 9.   [5] Presl, in Sternberg (38).
[6] Zigno (56) A. Pls. xxi. xxii.; Raciborski (94) A. Pl. xx. figs. 13—18.
[7] Brongniart (25) Pl. xii. fig. 1.
[8] Fontaine, in Ward (05); Salfeld (09) Pl. i.   [9] Schenk (67) A. Pl. xiii.
[10] Arber (05) p. 75.   [11] Seward (94²) A. p. 130.   [12] Velenovský (85) Pl. ii.
[13] Fontaine, in Ward (05) Pl. lxv. Newberry's *Chiropteris spatulata* from
Montana may be founded on leaflets of *Sagenopteris Mantelli*. Newberry (91).

Sagenopteris appears to have been widely distributed during the Rhaetic, Jurassic and Lower Cretaceous floras. The very great similarity between the specimens recorded from these three formations renders the genus an uncertain guide in regard to geological age. Decisive evidence as to its position in the plant kingdom is at present lacking: the inclusion of the genus as a possible member of the Hydropterideae has still to be justified.

CHAPTER XXVII.

GENERA OF PTERIDOSPERMS, FERNS, AND *PLANTAE INCERTAE SEDIS*.

THE genera and species described in this Chapter are founded on sterile leaves or portions of leaves, and in the great majority of cases the reproductive organs are either imperfectly known or have still to be discovered. Some of the genera, the smaller number, are no doubt true ferns, while most of them may safely be regarded as plants which will ultimately be shown to belong to some other group, in most cases that of the Pteridosperms. It is possible that a few of the types may be members of the Cycadophyta rather than of the Pteridospermeae, but evidence as to systematic position is for the most part of a negative kind or too incomplete to lead to any definite expression of opinion as to the cycadean or pteridosperm nature of the imperfectly known Palaeozoic or Mesozoic species. Many of the genera are of little botanical interest, though even the most problematical are of importance as criteria of geological age. Genera which there is good reason for including in the Pteridosperms are dealt with in this section, in order that the Chapter in Volume III. devoted to this important group may be limited to more completely known types.

In most text-books it is customary to employ family names for sterile fern-like fronds which possess similar venation features or have in common certain vegetative characters, the value of which it is impossible to estimate. In the following account family or group names are not adopted, on the ground that such slight utility as they may have is more than counterbalanced by the risk attending a grouping under one name of plants which may agree only in unessential characters. The practice of classifying fossil plants has been carried to excess. Grouping together genera as a matter of convenience unavoidably creates a prejudice in favour of actual relationship, which may or may not exist.

Taeniopteris.

This generic name was instituted by Brongniart[1] for simple linear or broadly linear leaves with a prominent midrib from which secondary veins, simple or dichotomously branched, are given off at right angles or obliquely. The frond of the type-species *Taeniopteris vittata* (fig. 332), characteristic of Jurassic floras, was compared by Brongniart with the pinnules of *Danaea* and *Angiopteris*. Among recent ferns the Taeniopteris form of frond and venation is represented by *Oleandra neriiformis*, *Asplenium nidus*, and many other species. Though usually applied to fronds which there is good reason for regarding as simple leaves, the generic designation *Taeniopteris* has been extended to include pinnate fronds, e.g. the Upper Palaeozoic species *T. jejunata* Grand'Eury, and *T. Carnoti* Ren. and Zeill. (fig. 330, A). The compound fronds from the Lower Coal-Measures of Missouri described by Dr White[2] as *T. missouriensis* are characterised by decurrent and confluent Taeniopteroid pinnules. In a later reference[3] to this plant White pertinently adds, "perhaps it belongs more properly in *Alethopteris*."

Leaves of the *Taeniopteris* type are described by several authors as species of *Oleandridium*, *Angiopteridium*, *Danaeites*, *Marattia*, and other genera. In such species of Taeniopteroid leaves as have been dealt with in a former Chapter, the occurrence of sori justifies the substitution of a name denoting a close relationship to existing members of the Marattiaceae, but in the absence of fertile specimens the provisional designation *Taeniopteris* should be retained. It is often difficult to decide between *Taeniopteris* and *Nilssonia* as the more suitable

---

[1] Brongniart (28) A. p. 61.   [2] White (93).
[3] White (99) p. 143.

name to apply to fragments of fossil leaves of Mesozoic age. *Taeniopteris* is, however, distinguished from the Cycadean genus by the greater prominence of the rachis, also by the dichotomous branching of the secondary veins, usually close to their origin and at varying distances between the axis of the frond and the edge of the lamina. The genus *Taeniopteris*, though most abundant in Rhaetic and Jurassic strata, occurs also in Upper Carboniferous and Lower Permian rocks. The generic name *Macrotaeniopteris* instituted by Schimper[1] has been used for leaves differing only in size from the usual type of *Taeniopteris*, but there is no adequate reason for its retention.

The species included in *Taeniopteris* afford no satisfactory evidence as to their systematic position. It is obviously unwise to adopt such generic titles as *Oleandridium, Marattiopsis*, etc., merely because of resemblance in the venation of sterile fragments to *Oleandra* or Marattiaceous ferns.

Some specimens of *Taeniopteris* fronds described by Mr Sellards[2] from Permian rocks of Kansas, which are referred to later, have furnished unconvincing evidence of reproductive organs.

Taeniopteris multinervis, Weiss.   Fig. 329, A, B.

The late Dr Weiss[3] instituted this species (which he designated *Taeniopteris multinervia*, though the specific name *multinervis* is constantly used) for a fragment of a leaf from the Lower Permian of Lebach characterised by numerous forked veins given off at right angles from a prominent rachis (fig. 329, B). This type of frond is recorded from the Permian of Trienbach (Alsace) by Zeiller[4], by Renault[5] and Zeiller[6] from the Upper Carboniferous of Autun, and from other localities. The lamina of the simple leaf reaches a breadth of 6 cm. and a length of 40 cm. (fig. 329, A); the numerous secondary veins (25—36 per cm. of lamina) are either at right angles to the rachis or given off at an acute angle. The mesophyll consists of

[1] Schimper (69) A. p. 610.   [2] Sellards (01).
[3] Weiss, C. E. (69) p. 98, Pl. vi. fig. 13.   [4] Zeiller (94) p. 169.
[5] Renault (96) A. p. 1.   [6] Zeiller (90) Pls. xii. xiii.

polygonal cells some of which are elongated at right angles to the surface of the lamina. A very similar form is described by Fontaine and White from the Permian of Virginia as *T. Lescuriana*[1].

It is futile to expect to be able to separate the numerous

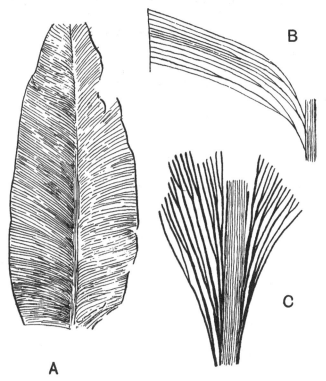

FIG. 329.   A.   *Taeniopteris multinervis*, Weiss.   (⅔ nat. size.   After Zeiller.)
               B. . *T. multinervis*.   (Enlarged.   After Zeiller.)
               C.   *Lesleya Delafondi*.   (× 2.   After Zeiller.)

Taeniopteris leaves into well-defined species: all we can do is to group the specimens under different names, using as artificial distinctions such characters as the shape of the leaf, the number of veins per centimetre, and the prominence of the rachis. Another Virginian species of Permian age described by Fontaine

[1] Fontaine and White (80) Pl. xxxiv.

and White[1], *T. Newberriana,* is said to bear sori, but no
satisfactory information is given as to the nature of these
organs. Specimens referred with some hesitation to this
species and to a similar species, *T. coriacea,* have been described
by Sellards[2] from material obtained from Permian beds in
Kansas. The lamina of the simple linear fronds is characterised
by the occurrence of small oval bodies half immersed in the
substance of the leaf between the secondary veins (figs. 330, D, E).
One of these bodies is represented in an apparently dehisced
condition in fig. 330, D. Sellards suggests the possibility that
these bodies are sporangia, but, as he points out, they afford no
indication of cellular structure nor are they in direct connexion
with the veins.

Taeniopteris jejunata, Grand'Eury[3].

This species differs from *T. multinervis* in its bipinnate fronds;
the linear or oval-linear pinnae are attached by a short stalk to
the primary rachis and reach a length of 25 cm.; the veins are
less crowded, 12—15 per centimetre.

T. jejunata is recorded from the Coal-fields of the Loire and
Commentry[4] in France, from the Lower Permian of Thuringia[5],
and elsewhere.

Taeniopteris Carnoti, Ren. and Zeiller[6]. Fig. 330, A.

This species, founded on portions of pinnate fronds from the
Coal-field of Commentry, is characterised by rather broader
(25—30 mm.) pinnules, with short pedicels and a cordate base,
reaching a length of 25—30 cm. The secondary forked veins are
more numerous than in *T. jejunata.* In *T. multinervis* the
pinnules are still broader and have a stronger midrib.

Several species of *Taeniopteris* have been described from
Triasso-Rhaetic rocks in Europe, India, Tonkin and elsewhere.

---

[1] Fontaine and White (80) Pl. xxxiv. figs. 1—8.   [2] Sellards (01).
[3] Grand'Eury (77) A. p. 171.   [4] Renault and Zeiller (88) A.
[5] Potonié (93) A. p. 145, Pl. xvii. fig. 3.
[6] Renault and Zeiller (88) A. p. 282, Pl. xxii. fig. 10.

In some cases it is practically impossible to recognise clear specific distinctions between Rhaetic and Jurassic types.

From the Damuda and Panchet series of India (Triasso-Rhaetic) Feistmantel has described large sterile fronds as *Macrotaeniopteris Feddeni*[1] which reach a breadth of 20 cm.: these may be compared with the Indian species *Taeniopteris lata* Oldham[2], and to *T. gigantea* from the Rhaetic of Franconia[3] and Scania. A specimen of this species figured by Nathorst[4] from Scania has a lamina 33 cm. broad. Other examples are afforded by *M. Wianamattae* Feist.[5] from rocks of the same age in Australia and by *Taeniopteris superba* Sap.[6] from Lower Rhaetic rocks near Autun.

From the Rhaetic of Tonkin, Zeiller records several species, among which may be mentioned *T. Jourdyi* Zeill.[7] and *T. spatulata* MacClelland (fig. 330, B, C). Both have simple fronds. Those of *T. Jourdyi* reach a length of 10—40 cm. and a breadth of 10—70 mm.; the rachis is characterised by crowded and discontinuous transverse folds, and the secondary veins (35—50 per cm.) are usually at right angles to the rachis. This Tonkin species is compared by Zeiller with the European Rhaetic species *T. tenuinervis* Brauns.

The polymorphism of the fronds is a striking feature: in one case described by Zeiller the lamina appears to be divided into segments like those characteristic of the leaf of the Cycadean genus *Anomozamites*. It is obviously difficult in many instances to distinguish between detached Taeniopteroid pinnae of a compound frond and complete simple leaves. In some compound fern fronds, as in the recent Polypodiaceous genus *Didymochlaena*, the pinnules are deciduous, and the same feature undoubtedly characterised the fronds of many extinct species. A specimen figured by Zeiller which shows several petioles of *T. Jourdyi* attached to a thick stem[8] demonstrates the simple nature of the leaves. In other cases, e.g. *T. vittata*, specimens occur in which the slightly enlarged petiole-base has a clean-cut surface indicating abscission from a rhizome (fig. 332).

[1] Feistmantel (81) A. Pls. xxi. A. xxii. A.   [2] Oldham and Morris (63) p. 41.
[3] Schenk (67) A. Pl. xxviii. fig. 12.   [4] Nathorst (78) Pl. ix.
[5] Feistmantel (90) A. Pl. xxvii.   [6] Saporta (73) A. Pls. lxi. lxii.
[7] Zeiller (02) Pls. x.—xiv. p. 66.   [8] Zeiller (02) Pl. xi. fig. 4.

The fronds described by Zeiller as *T. spatulata*[1] (fig. 330, B, C) closely resemble Jurassic leaves from Victoria referred to *Taeniopteris Daintreei* McCoy[2].

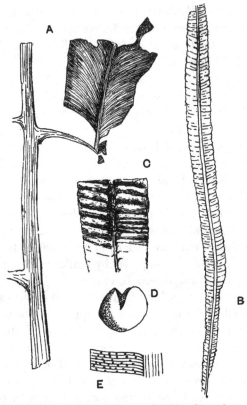

Fig. 330. A. *Taeniopteris Carnoti*, Ren. and Zeill. (Nat. size. After Renault and Zeiller.)
            B. *T. spatulata*, McClell. (Nat. size. After Zeiller.)
            C. *T. spatulata.* (× 3. After Zeiller.)
            D. Supposed sporangium of *T. coriacea.* (× 15. After Sellards.)
            E. *T. coriacea.* (× 2. After Sellards.)

Whether specifically identical or not, these leaves represent a type distinguished from the other species of the genus by the small breadth of the linear-lanceolate or linear-spathulate

[1] Zeiller (02) Pl. XIII. For synonymy, see also Arber (05) p. 124.
[2] Seward (04) figs. 18—22.

lamina, which may be 6—15 cm. in length and 3—12 mm. broad. The lamina is often characterised by transverse folds (fig. 330, C).

Taeniopteris Carruthersi. **Fig. 331.**

  1872. *Taeniopteris Daintreei*, Carruthers, Quart. Journ. Geol. Soc. Vol. XXVIII. Pl. XXVII. fig. 6.
  1883. *T. Carruthersi*, Tenison-Woods, Proc. Linn. Soc. N. S. Wales, Vol. VIII. p. 117.

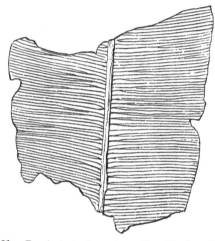

Fig. 331. *Taeniopteris Carruthersi*, Ten.-Woods. Nat. size.

The simple fronds included under this specific name are characterised by a strong midrib from which numerous simple or forked secondary veins are given off at a right angle or slightly inclined. The breadth of the lamina decreases gradually towards the petiole. The Australian species named by McCoy *Taeniopteris Daintreei*, to which Carruthers referred the Queensland fossils, has a much narrower and more linear form of frond, and for this reason Tenison-Woods instituted a new specific name. *T. Carruthersi* represents a form of leaf met with in Rhaetic, or possibly Upper Triassic, rocks in S. Africa[1] and Australia. A very similar, perhaps an identical type, was described from Argentina by Geinitz[2] as *T. mareyiaca*: among

[1] Seward (08) p. 98.   [2] Geinitz (76) Pl. II. figs. 1—3.

many other examples of this form of frond may be mentioned
T. immersa[1] Nath. from the Rhaetic rocks of Scania and *T.
virgulata* from the Rhaetic of Tonkin[2].

A comparison of *Taeniopteris Carruthersi* or various other
" species " of Rhaetic fronds with the Jurassic species *T. vittata*
illustrates the slight and unimportant differences on which
specific separation is based. It is hopeless to attempt to draw
a satisfactory distinction between the numerous Taeniopteris
fronds from Upper Triassic and Jurassic rocks.

Taeniopteris vittata, Brongniart.   Fig. 332.

The simple leaves to which Brongniart applied this name
are characteristic of the Inferior Oolite flora of England, and
examples of the same type are recorded from Jurassic rocks of
India, Poland, the Arctic regions, Japan, China, Australia and
other countries[3].

Leaf linear-lanceolate, reaching a length of more than 20 cm. and a
breadth of 3 cm. The lamina increases gradually in breadth from the
base and tapers towards the apex. Numerous secondary veins are given
off at right angles from a broad midrib: the lateral veins may be simple
or forked close to their origin, near the margin, or in the intermediate
portion, of the lamina.

It is exceedingly difficult to use *Taeniopteris* leaves of this
form as evidence in regard to the Jurassic or Rhaetic age of
plant-bearing strata.   The species *T. tenuinervis* Brauns, as
figured by Schenk[4] from the Rhaetic rocks of Germany and
Persia, and recorded from several other regions, presents a
close agreement with *T. vittata*. *Oleandridium lentriculi-
forme* Etheridge[5] from the Hawkesbury series of Australia is
another similar leaf.   The species *T. vittata* from the Yorkshire
coast, represented in fig. 332, shows a well-preserved petiole
with a clean-cut base like that of the petioles of *Oleandra
neriiformis* and other recent ferns which are detached from the
rhizome by the action of an absciss-layer.

1 Nathorst (78) Pl. xix.   2 Zeiller (02) Pl. xiv.
3 For synonymy and distribution, see Seward (00) pp. 159, 304.
4 Schenk (67) A. Pl. xxv. See also Bartholin (92) Pl. ix. fig. 7.
5 Etheridge (94²).

Fig. 332. *Taeniopteris vittata.* (British Museum No. 39217. ⅔ nat. size.)

A broader form of frond with similar venation was described by Lindley and Hutton[1] as *Taeniopteris major*. An examination of the type-specimen from the Inferior Oolite of Yorkshire, now in the Manchester Museum, led me to doubt the necessity of specific separation from *T. vittata*[2].

A smaller frond of the same general type as *T. vittata* is recorded from Wealden strata of North Germany and England under the name *T. Beyrichii*[3].

Weichselia.

This generic name was instituted by Stiehler[4] for impressions of bipinnate sterile fronds, presumably ferns, from Lower Cretaceous rocks near Quedlinburg. The same type of leaf from English Wealden beds had previously been referred by Mantell and other authors to *Pecopteris*, and by Brongniart to his genus *Lonchopteris*[5]. It is, however, advisable to follow Nathorst's example[6] and restrict the latter name to Palaeozoic species. As already suggested, it would obviate confusion to substitute a new generic designation for *Lonchopteris* in the case of Triassic species which are probably members of the Osmundaceae. The type-species of Stiehler, *Weichselia Ludowicae*[7], does not differ in any important character from *Weichselia Mantelli*, the species originally described by Stokes and Webb from the Wealden of England as *Pecopteris reticulata*.

Weichselia Mantelli (Brongn.)[8].   Fig. 333.

1824. *Pecopteris reticulata*, Stokes and Webb, Trans. Geol. Soc. [2]. Vol. I. p. 423, Pls. XLVI. XLVII.
1828. *Lonchopteris Mantelli*, Brongniart, Prod. p. 6 ; Hist. vég. foss. p. 369, Pl. CXXXI.
1894. *Weichselia Mantelli*, Seward, Wealden Flora, Vol. I. p. 114. Pl. X. fig. 3.
1899. *Weichselia reticulata*, Fontaine, in Ward, Ann. Rep. U. S. Geol. Surv. p. 651.

[1] Lindley and Hutton (33) A. Pl. XCII.
[2] Seward (00) p. 14.   [3] Schenk (71) Pl. XXIX.; Seward (94²) A. p. 125.
[4] Stiehler (58) Pls. XII. XIII.   [5] See p. 576.   [6] Nathorst (90).
[7] For figures, see Stiehler *loc. cit.* and Hosius and Von der Marck (80) Pls. XLIII. XLIV.
[8] For synonymy, see Fontaine, in Ward (99) p. 651; Seward (94²) A. p. 114; Seward (00) p. 20.

Frond bipinnate, rachis broad; pinnae very long, of uniform breadth and with prominent axes; pinnules crowded, entire, with obtuse apex, usually oblong but more or less triangular or rounded towards the distal

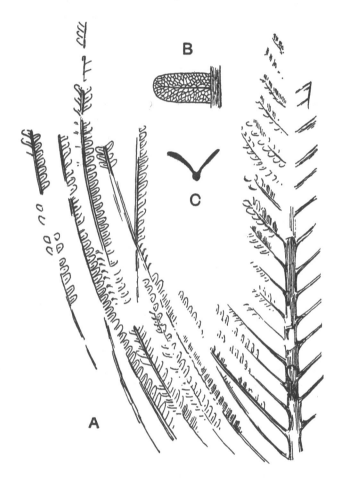

Fig. 333. *Weichselia Mantelli.*

    A. Part of a frond from the Wealden of Sussex, England. (British
          Museum; v. 2630. ¾ nat. size.)

    B. Pinnule from Bernissart, Belgium (× 3).

    C. *Weichselia erratica,* Nath. Section of pinna. (After Nathorst.)

ends of the pinnae. The pinnules, which may reach a length of 9 cm., are characterised by a fleshy lamina attached by the whole breadth of the base; the two rows of segments on each secondary rachis are usually inclined towards one another so that they form with the axis of the pinna a wide-open V instead of lying in one plane (fig. 333, C). From a median rib are given off numerous anastomosing branches (fig. 333, B).

This characteristic Wealden species is recorded from England, Germany, France, Belgium, Austria, Russia, Bornholm, North America, and Japan. It is by no means certain that *Weichselia Mantelli* is a true fern: no satisfactory evidence of fructification has been adduced.

The broad and strong rachis is comparable with that of a Cycadean leaf and the thick lamina suggests a plant of xerophilous habit. I have retained the specific name *Mantelli* on the ground of long established usage instead of following Fontaine in his adherence to strict priority.

Glossopteris.

The name *Glossopteris* was proposed by Brongniart in 1822[1] for an imperfect leaf-impression which he called *Filicites* (*Glossopteris*) *dubius,* but the specimen so named has since been identified as part of a sporophyll of a *Lepidostrobus.* The author of the genus afterwards published[2] a diagnosis, based on well-preserved leaves from Permo-Carboniferous rocks in Australia and India, of the type-species *Glossopteris Browniana,* the Indian examples being distinguished as *G. Browniana* var. *indica* while the Australian form was named *G. Browniana* var. *australasica.* Schimper[3] afterwards raised the Indian fossils to specific rank as *G. indica* though some authors[4] have continued to consider the two forms as insufficiently distinct to be regarded as different species.

The genus *Glossopteris* may be defined as follows:

Leaves simple, varying considerably in size, shape, and venation characters, but almost without exception characterised by repeatedly anastomosing lateral veins. The leaves are of two kinds: (i) *foliage leaves;* apparently always sterile, usually spathulate, with an obtuse apex, a well-marked midrib which may persist to the apex or die out in the upper half of the lamina, characterised by its slight prominence and comparatively great breadth especially in the basal part of the frond. In

---

[1] Brongniart (22) A. Pl. II. fig. 4.   [2] Brongniart (28²) A. Pls. LXII. LXIII.
[3] Schimper (69) A. p. 645.   [4] Seward (97²) A. p. 317.

most cases the lamina extends as a narrow margin to the leaf-base, but in a few forms there is a short petiole (fig. 334). Though usually spathulate, the frond may be linear-lanceolate, or ovate; the apex is sometimes acute. Leaves vary in length from 3 to 40 cm. and may in larger forms have a breadth of 10 cm. Numerous lateral veins curve upwards and outwards to the margin of the lamina or pursue a straight course almost at right-angles to the midrib. (ii) *Scale-leaves*[1] which differ from the foliage-leaves in their much smaller size and in the absence of a midrib; they are deltoid, oval or cordate in shape and generally terminate in an acute apex; the edge of the lamina may be slightly incurved so that the leaf presents a convex upper surface supplied with anastomosing veins. The scale-leaves, which vary in length from about 1 to 6 cm., probably acted as sporophylls. The only evidence as to the nature of the fructification so far obtained is represented by empty sporangium-like organs (1·2—1·5 mm. long by 0·6—·0·8 mm. broad) frequently associated with the scale-leaves[2].

The leaves, in some cases at least, were borne near together on a cylindrical stem or rhizome which produced branched adventitious roots[3]. The fossils long known as *Vertebraria* were recognised by Zeiller[4] and by Oldham[5] as the stems of *Glossopteris*.

The systematic position of *Glossopteris* must for the present be left an open question. Though usually spoken of as a fern, it is noteworthy that despite the enormous abundance of its foliage leaves in the Permo-Carboniferous strata of India, Australia, South Africa, and South America, no single example has been discovered which shows undoubted remains of sori or sporangia. Many authors have described fertile leaves of *Glossopteris*; but it was not until Arber's discovery of sporangia in close association with the scale-leaves that any light was thrown on the nature of the reproductive organs.

The probability is that *Glossopteris* was not a true fern but a member of that large and ever-increasing class, the Pteridosperms. This opinion is based largely on negative evidence. Such sporangia as have been described may have contained microspores and the plant may have been heterosporous. The occurrence of seeds in association with Glossopteris fronds recorded by more than one writer[6], though by no means decisive and possibly the result of chance association, is favour-

[1] For figures see Zeiller (96) A.; Zeiller (02), (03); Arber (05); Seward (97) A.
[2] Arber (05²); Seward (07²).   [3] Bunbury (61) Pl. xi.
[4] Zeiller (96) A.   [5] Oldham (97); Zeiller (97²).
[6] Seward (97) A. (07²); Arber (02²) p. 20; Zeiller (96) A. p. 374.

able to this view.   Dr White[1] has suggested that the small
leaves described by Zeiller[2] as *Ottokaria bengalensis* from Lower
Gondwana (Permo-Carboniferous) rocks of India, and similar
fossils recorded by himself from Brazil as *O. ovalis*, may represent "sporangiferous" organs of *Glossopteris* or *Gangamopteris*,
"both of which are probably pteridospermic."   There is, however, no conclusive evidence in support of this suggestion.

The genus, whatever its position may be, has a special interest
for the geologist and for the student of plant distribution; it is
a characteristic member of a Permo-Carboniferous flora which
flourished over an enormous area, including India, South Africa,
—extending from Cape Colony to Rhodesia and German East
Africa[3],—Australia, and South America[4].   This flora, known as
the Glossopteris flora, differed considerably in its component
genera from that which overspread Europe and North America
and some more southern regions in the Upper Carboniferous
and Permian periods.

The discovery by Amalitzky[5] of *Glossopteris*, and other genera
characteristic of the Glossopteris flora, in the Upper Permian
rocks in Vologda (Russia) demonstrates the existence of a
northern outpost of the southern botanical province, and
Zeiller's discovery of the genus in the Rhaetic flora of Tonkin[6]
shows that *Glossopteris* persisted beyond the limits of the
Palaeozoic epoch.   Dr David White[7] has recently proposed to
re-christen the Glossopteris flora the Gangamopteris flora on the
ground that *Gangamopteris* is strictly Palaeozoic in its range,
whereas *Glossopteris* persisted into the Mesozoic era; this is
perhaps hardly a sufficient reason for giving up so well
established a title as the Glossopteris flora.   A fuller account
of this southern flora must be reserved for another volume.

[1] White (08) p. 535.   [2] Zeiller (02).
[3] Potonié (00).
[4] Seward (04³) ; Zeiller (97³); Arber (05) p. 17; D. White (07).
[5] Amalitzky (01) ; Zeiller (98²).   [6] Zeiller (02).
[7] D. White (07) p. 617 (footnote 2).

Glossopteris Browniana, Brongniart[1].　Figs. 334—36.

The specific name *Browniana* is now applied to obtusely
pointed leaves which sometimes reach a length of 15 cm., but
are usually rather shorter.　In form and venation they closely
resemble the leaves of the recent genus *Antrophyum* and
species of *Acrostichum*.　The comparatively broad midrib may

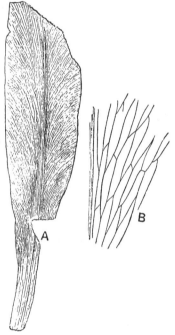

Fɪɢ. 334.　*Glossopteris Browniana,* Brongn.　A.　Nat. size : B × 3½.

be replaced in its proximal portion by several parallel veins :
from it are given off numerous lateral veins which form a
reticulum characterised by meshes approximately equal in size
and elongated in a direction parallel to the general course of
the secondary veins (fig. 334).

The drawings, originally published by Zeiller[2], reproduced
in fig. 335 illustrate the venation and its range of variation ;

[1] For synonymy, see Arber (05) p. 48.　　　　[2] Zeiller (96) A.

the meshes are usually hexagonal and arranged as shown in
figs. A and B, but occasionally (fig. 335, C) they follow a more
steeply inclined course.

Small leaves with a more or less distinct midrib, 2—3 cm.
in length, supply transitional stages between foliage- and scale-
leaves. In the true scale-leaves spreading and occasionally
anastomosing veins take the place of the midrib and lateral
veins of the ordinary frond. McCoy[1] in describing some
Australian specimens of *Glossopteris* in 1847 spoke of scale-like
appendages of the rhizome which he compared with the large
ramenta of *Acrostichum* and other ferns. It was, however, Zeiller[2]

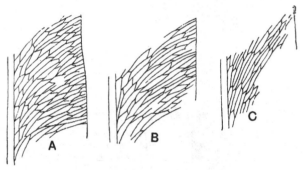

FIG. 335.   *Glossopteris Browniana*, Brongn.   (After Zeiller.   × 2.)

who first recognised the leaf-nature of these scales and adequately described them; additional figures of scale-leaves have
been published by Mr Arber[3] and by myself[4]. The importance of these small leaves has been considerably increased by
Mr Arber's discovery of associated sporangia which, as he
suggests, were probably borne on their lower concave surface.

The sporangia (fig. 336) are compared by Arber with the
micro-sporangia of recent Cycads and with the Palaeozoic
sporangia described by Zeiller as *Discopteris Rallii* (fig. 256, D);
the latter are distinguished by the well-defined group of thicker
walled cells representing the annulus of true fern sporangia.
We know nothing as to the contents of the Glossopteris

[1] McCoy (47).   [2] Zeiller (96) A.
[3] Arber (05); (05²).   [4] Seward (97) A; (07).

sporangia, whether they contained microspores or whether they are the spore-capsules of a homosporous plant.

The rhizome of *Glossopteris Browniana* has been described in detail by Zeiller, who first demonstrated that the fossils originally assigned by Royle[1] to the genus *Vertebraria* represent the stem of this and, as we now know, of some other species of *Glossopteris*. *Vertebraria* occurs in abundance in Permo-Carboniferous strata in association with *Glossopteris*; the differences between Australian, Indian, and South forms, though expressed

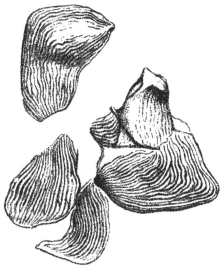

Fig. 336.   *Glossopteris Browniana*, Brongn.   Sporangia.   (× 30).   After Arber.

by specific names, are insignificant.   The stems are usually preserved in the form of flattened, single or branched, axes sometimes bearing slender branched roots and characterised by one or two, or less frequently three, longitudinal grooves or ridges (fig. 337) from which lateral grooves or ridges are given off at right angles, dividing the surface into more or less rectangular areas 1 cm. or more in length.   The surface of these areas is often slightly convex and in some specimens the outlines of cells may be detected.   Mr Oldham has described some interesting examples

[1] Royle (33).

of *Vertebraria* from India in which the longitudinal and transverse grooves are occupied by a dark brown ferruginous substance
or by the carbonised remains of plant-tissues (fig. 338, C, D).
In transverse section, a *Vertebraria* cast appears to be divided
into a number of wedge-shaped segments radiating from a
common centre.   Prof. Zeiller[1] has figured specimens of *Vertebraria* with portions of Glossopteris fronds still attached.

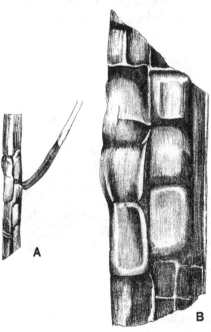

Fɪɢ. 337.   *Vertebraria indica*, Royle.   Nat. size.   (After Feistmantel.)

The rhizome of *Glossopteris*, as represented by the Vertebraria
casts, is aptly compared by Zeiller[2] with that of the recent
Polypodiaceous fern *Onoclea struthiopteris*. Sections of the
recent stem (fig. 338, E, F) show that the form is irregularly
stellate owing to the presence of prominent wings which
anastomose laterally at intervals as shown by the examination
of a series of sections. The leaf-traces are derived from the
steles of adjacent wings. Fig. 338 (B and A) represents some-

---

[1] Zeiller (96) A.   [2] Zeiller (96) A.

what diagrammatically a longitudinal and transverse view of a
Vertebraria; the radiating arms represented in the transverse
section (fig. A) are the stem ribs or wings and the segments
between them are intrusions of sedimentary material. The
rectangular areas characteristic of the surface of a *Vertebraria*
are the intruded segments of rock: these are separated at
intervals by transverse grooves, which mark the course of

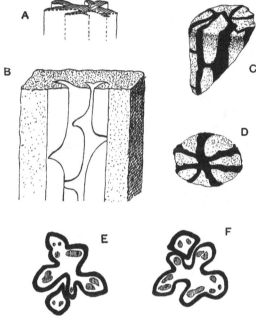

FIG. 338.　A, B.　*Vertebraria indica.*　(After Zeiller.)
　　　　　　C, D.　*V. indica.*　(Nat. size.　After Oldham.)
　　　　　　E, F.　*Onoclea struthiopteris.*　(× 2.　After Zeiller.)

vascular strands given off at each anastomosis of the longitudinal
wings to supply the leaves.

Mr Oldham, who discovered the connexion between *Glossopteris* and *Vertebraria* independently of Dr Zeiller, does not
agree with the interpretation of the structural features of the
rhizome which Zeiller bases on a comparison between *Vertebraria*
and *Onoclea struthiopteris.* Oldham[1] describes *Vertebraria* as

[1] Oldham (97).

consisting of a central axis "joined to an outer rind by a series of radial septa," the spaces between the septa being divided

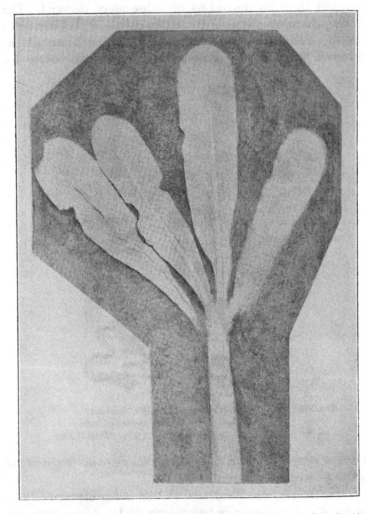

Fig. 339.   *Glossopteris* fronds attached to rhizome.   (From a specimen lent by Dr Mohlengraaff.   Considerably reduced.)

into chambers by transverse partitions.   His view is that the rhizome of *Glossopteris* was a cylindrical organ and not an irregularly winged axis like the stem of *Onoclea*. Zeiller[1] has replied in detail to Oldham's interpretation and adheres to his original view, that the rhizome consisted of a solid axis with radial wings or flanges which at intervals anastomosed transversely in pairs at the nodes. It may, however, be possible that the spaces between the longitudinal and transverse grooves on a Vertebraria axis, which have been filled with the surrounding rock, were originally occupied in part at least by secondary wood, and the transverse strips of carbonaceous material[2] lying in the grooves may represent medullary-ray tissue and accompanying leaf-traces. The longitudinal striations seen in some specimens of *Vertebraria* on the areas between the grooves may be the impressions of woody tissue. It is impossible without the aid of more perfectly preserved material to arrive at a satisfactory conception of the structural features of a complete Glossopteris rhizome.

In the specimen of *Glossopteris Browniana* shown in fig. 339 several leaves are attached to an axis which shows none of the surface-features of *Vertebraria*. I am indebted to the kindness of Dr Mohlengraaff of Delft for the loan of this specimen which was obtained from Permo-Carboniferous rocks in the Transvaal. An axis figured by Etheridge[3] from an Australian locality bears a tuft of *Glossopteris* leaves, possibly *G. Browniana*; in place of the rectangular areas characteristic of *Vertebraria* it shows transversely elongated leaf-scars or, on the internal cast, imbricate rod-like projections which Etheridge suggests represent vascular bundles.

Glossopteris indica, Schimper.   Figs. 340, A, 341.

It is a question of secondary importance whether or not the fronds which Brongniart spoke of as a variety of *Glossopteris Browniana* should be recognised as specifically distinct. The careful examination by Zeiller of the venation characters has, however, afforded justification for separating *G. Browniana* and *G. indica*. We must admit that the slight and not very constant

---

[1] Zeiller (02).   [2] Zeiller (02) Pl. v. fig. 7.   [3] Etheridge (94).

differences in the size and form of the meshes produced by the anastomosing of the lateral veins are characters which cannot

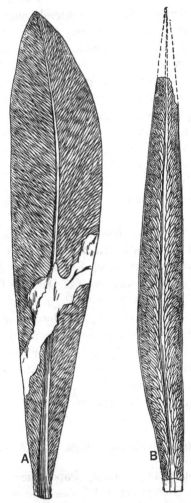

Fig. 340. A. *Glossopteris indica*, Schimper. (½ nat. size.), B. *Glossopteris angustifolia*, Brongniart. (Nat. size.) From Arber, after Feistmantel.

be recognised as having more than a secondary value, though, as a matter of convenience, we employ them as aids to determination. The arbitrary separation of sterile leaves, which differ by small degrees from one another in form and in the details of venation, by the application of specific names is a thankless task necessitated by custom and convenience; it is, however, idle to ignore the artificial basis of such separation. Mr Arber has recently published, in his valuable *Glossopteris Flora*, an analytical key which serves to facilitate the description and determination of different types of frond[1].

The large leaves of *Glossopteris indica*, reaching a length in extreme cases of 40 cm. and a breadth of 10 cm , are characterised by a rather greater regularity in the arrangement of the

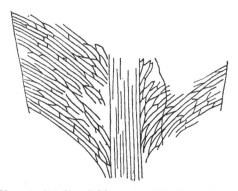

Fig. 341. *Glossopteris indica*, Schimp. (× 1½.) From Arber, after Zeiller.

meshes and by the greater parallelism of the upper and lower sides of each mesh (fig. 341) and by less difference in size between the venation meshes than in *G. Browniana*, the leaves of which are usually smaller. The relatively thick epidermis consists of rectangular cells with stomata in depressions[2]. The scale-leaves[3], rather larger than those of *G. Browniana*, are more or less rhomboidal with rounded angles and reach a length of 1·5—6 cm. and a breadth of 1·5—2·5 cm. The rhizome is practically identical with that of *G. Browniana*[4].

This species occurs in great abundance in the Permo-Carboniferous rocks of India, Australia, and in various parts of

[1] Arber (05) p. 47.   [2] Zeiller (96) A. p. 368, fig. 13.
[3] Zeiller (02) ; (03).   [4] Oldham (97).

South Africa, and elsewhere. It has been recognised also by Amalitzky[1] in Upper Permian beds in Russia and by Zeiller in the Rhaetic series of Tonkin[2].

Glossopteris angustifolia, Brongniart.   Figs. 340, B; 342.

It is convenient to retain this designation for linear fronds with an acute or obtuse apex and a venation-reticulum composed of long and narrow meshes (fig. 340, B). It is by no means unlikely, as Arber suggests, that the same plant may have produced leaves of the *G. indica* type and narrower fronds which conform to *G. angustifolia*. In his description of some

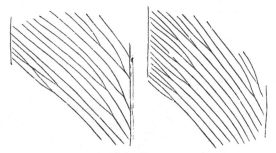

Fig. 342.   *Glossopteris angustifolia* var. *taeniopteroides*.   (× 3½.)

Indian specimens of *G. indica*, Zeiller draws attention to the variation exhibited in regard to the extent of anastomosing between the secondary veins: some examples with very few cross-connexions agree more closely with *Taeniopteris* than with *Glossopteris* as usually defined[3]. The venation shown in fig. 342 illustrates an extreme case of what is almost certainly a Glossopteris leaf of the *G. angustifolia* type. This specimen, which was discovered by Mr Leslie in the Permo-Carboniferous sandstone of Vereeniging (Transvaal), has been referred to a variety of Brongniart's species as *G. angustifolia* var. *taeniopteroides*[4] on account of the almost complete absence of any cross-connexions. The reference to *Glossopteris*, which my friend

[1] Amalitzky (01).   [2] Zeiller (03) Pl. xvi.
[3] Zeiller (02).   [4] Seward and Leslie (08) p. 113.

Dr Zeiller suggested, is amply justified by the form of the leaf
as a whole, by the angle at which the lateral veins leave the
midrib, a feature in contrast to the wider angle at which the
lateral veins are usually given off in *Taeniopteris* (figs. 329, 332),
and by the similarity to the Indian specimens already mentioned.
Several authors have described leaves or leaflets under the generic

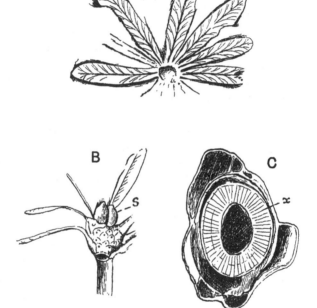

Fig. 343. *Blechnoxylon talbragarense*, Eth.: *s*, scale-leaves; *x*, secondary
xylem. (After Etheridge. A × 2; B × 3; C much enlarged.)

name *Megalopteris*[1] from Carboniferous and Permian rocks
which bear a close resemblance to the South African variety,
but in some cases at least *Megalopteris* is known to be a pinnate
and not a simple leaf.   The leaf figured by Jack and Etheridge
as *Taeniopteris* sp.[2] from Queensland may also be an example of

[1] Dawson (71) A. Pl. XVII.; Fontaine and White (80) p. 11; White (95)
p. 315; Arber (05³) p. 307, Pl. XX.
[2] Jack and Etheridge (92).

Glossopteris. Comparison may be made also with the Palaeozoic leaves described in the first instance by Lesquereux and more recently by Renault and Zeiller as species of *Lesleya*[1] (fig. 347).

Blechnoxylon talbragarense, Etheridge.   Fig. 343.

Under this name Etheridge[2] described some specimens from the Permo-Carboniferous Coal-Measures of New South Wales, which he regards as a fern, comparable, in the possession of a cylinder of secondary xylem, with the recent genus *Botrychium* and with *Lyginodendron* and other members of the Cycadofilices. The slender axis (1—3 mm. in diameter) appears to consist of a zone of radially disposed tissue (fig. 343, C, x), which is probably of the nature of secondary xylem, enclosing a pith and surrounded externally by imperfectly preserved remnants of cortex. Unfortunately no anatomical details could be made out, but the general appearance, if not due to inorganic structure, certainly supports Etheridge's determination. The stem bore at intervals clusters of linear-lanceolate leaves (reaching 12 mm. in length) in close spirals (fig. 343, A and B); the leaves are characterised by a strong midrib and forked secondary veins. Small "pyriform" bodies of the nature of scale-leaves occur in association with the fronds (fig. 343, B, s).

In his description of this interesting plant, Etheridge quotes an opinion which I expressed in regard to the comparison of the stem with those of *Botrychium, Lyginodendron,* and other genera. No satisfactory evidence has been found as to the nature of the fructification. Although the leaves of *Blechnoxylon* are much smaller than those of *Glossopteris,* I am now disposed to regard the genus as closely allied or even generically referable to *Glossopteris.* The crowded disposition of the leaves is like that in *Glossopteris,* shown in fig. 339 and in the figures published by Etheridge and by Oldham; the association of scale-leaves and foliage-leaves is another feature in common. The absence

---

[1] Lesquereux (79) A. Pl. xxv.; Renault and Zeiller (88) A. Pl. xxiii. See p. 517.

[2] Etheridge (99).

of a reticulum of anastomosing veins can no longer be considered a fatal objection to the suggestion that the Australian type may be a species of *Glossopteris*. If the view that *Blechnoxylon* is not a distinct genus is correct, the occurrence of secondary

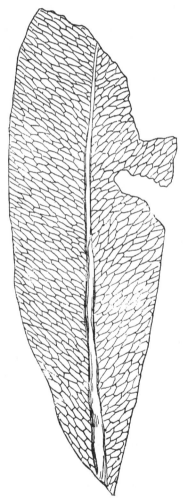

Fig. 344.   *Glossopteris retifera*.   (Nat. size.   From Arber, after Feistmantel.)

xylem is favourable to the opinion already expressed that *Glossopteris* is more likely to be a Pteridosperm than a true fern.

The data at present available render it advisable to retain Mr Etheridge's name: the comparison with *Glossopteris* lacks confirmation.

Glossopteris retifera, Feist.   Fig. 344.

In some *Glossopteris* leaves the anastomosing secondary veins form a coarser reticulum, as in the example represented in fig. 344. The name *G. retifera* was given by Feistmantel[1] to Indian fronds of this type; similar forms have been described as *G. conspicua* and *G. Tatei*. The type illustrated by *G. retifera* is recorded also from Permo-Carboniferous rocks in Zululand[2], Natal, the Transvaal, Cape Colony, and the Argentine.

Gangamopteris.

In 1847 McCoy[3] described a leaf-fragment from Permo-Carboniferous rocks in New South Wales as *Cyclopteris angustifolia*. The type-specimen of this species, which is now in the Sedgwick Museum, Cambridge, has been re-described by Mr Arber[4]. Subsequently[5] McCoy instituted the generic name *Gangamopteris* for leaves, like that previously referred by him to *Cyclopteris*, from the Bacchus Marsh Sandstone of New South Wales, but he did not publish a diagnosis of the genus until several years later[6]. Feistmantel[7], who has described many species of *Gangamopteris* from the Lower Gondwana strata of India, slightly modified the original diagnosis. The genus is represented by sterile fronds only. We know nothing of the stem, and such evidence as is available in regard to the form of the fertile leaves is of a circumstantial kind. It is, however, highly probable that *Gangamopteris* is not a true fern but a Pteridosperm.

Leaves simple, sessile, varying in shape; obovate or spathulate, broadly lanceolate or rarely linear; the apex is usually blunt (fig. 345) but occasionally gradually tapered. In general appearance a Gangamopteris leaf is similar to that of *Glossopteris indica*, the chief distinction being the

---

[1] Feistmantel (80) Pls. xxviii. A., xli. A.   [2] Seward (07).
[3] McCoy (47).   [4] Arber (02²).   [5] McCoy (60) p. 107 (footnote).
[6] McCoy (75).   [7] Feistmantel (79).

absence of a midrib. Gangamopteris leaves are on the whole larger than those of *Glossopteris*; many of them reach a length of 20 cm. and some of the large Indian fronds are nearly 40 cm. long. The venation of *Gangamopteris* shows a greater uniformity in the size and shape of the meshes than that of *Glossopteris*. The middle of the lamina, especially in the lower part, is occupied by a few vertical veins from which branches curve upwards and outwards towards the edge of the lamina. The secondary veins are connected by frequent anastomoses and agree very closely with those of *Glossopteris*. The lamina becomes narrower towards the base, which is either cuneate or in some cases slightly auriculate (fig. 345).

As I have elsewhere pointed out[1], the presence or absence of a midrib is not in itself a character of real taxonomic importance. In the recent fern *Scolopendrium vulgare* the frond has a prominent midrib, while in *S. nigripes* there is no median rib. Mr Arber has expressed the opinion that "it is extremely doubtful whether the genus *Gangamopteris* should not be merged in *Glossopteris*[2]." The retention of the two names is, however, convenient, and it would tend to confusion were we to carry to its logical conclusion the view that the recognised distinction between the two genera may not be a mark of generic difference.

Gangamopteris is confined to Palaeozoic strata, a fact which leads White[3] to speak of the Gangamopteris rather than of the Glossopteris Flora. It occurs in South America, South Africa, Australia, and India, extending as far north as Kashmir; it has been discovered by Amalitzky in Permian rocks of Russia[4]. The Russian rocks in which *Glossopteris* and *Gangamopteris* were found are no doubt of Permian age. In Australia, South Africa, Brazil and Argentina, and in the Indian Coal-fields, *Gangamopteris* is a characteristic genus of Lower Gondwana rocks. These strata are usually spoken of as Permo-Carboniferous in order to avoid the danger of attempting on insufficient data a precise correlation with European formations.

Feistmantel speaks of *Gangamopteris* as most abundant in the Talchir-Karharbári beds, though it is represented also in the overlying Damuda series. In Australia the genus occurs in rocks which correspond in position and in their plant fossils

[1] Seward and Woodward (05) p. 2.
[2] Arber (02[2]) p. 14.   [3] White, D. (07).   [4] Arber (05).

with the Talchir-Karharbári beds of India; similarly, in South
Africa and South America the Gangamopteris beds are homo-
taxial with those of India and Australia. The leaf described
by Carruthers[1] from Brazil as *Noeggerathia obovata* (the type-
specimen is in the British Museum) is no doubt specifically
identical with *Gangamopteris cyclopteroides* Feist.[2] In a paper
by Mr Hayden on Gangamopteris beds in the Vihi Valley,
Kashmir, evidence is adduced in support of the conclusion that
the rocks are "not younger than Upper Carboniferous and may
belong to the base of that subdivision or even to the Middle
Carboniferous[3]." It would seem that *Gangamopteris* was a
very widely spread genus during the latter part of the Carboniferous period in the vast Southern Continent to which the
name Gondwana Land is often applied, and that it flourished in
the Southern Flora during at least part of the Permian period:
with other members of the Glossopteris Flora it migrated to the
North where it has been preserved in Permian rocks of Northern
Russia. The Glossopteris Flora must have had its birth in the
Southern hemisphere. The conclusion seems inevitable that
the leaves of *Glossopteris* and *Gangamopteris* in the shales and
sandstones of India, South Africa, South America, and Australia
are relics of the vegetation of a continent of which these regions
are the *disjuncta membra*. Darwin wrote to his friend Hooker
in 1881, "I have sometimes speculated whether there did not
exist somewhere during long ages an extremely isolated continent, perhaps near the South Pole[4]." It is probable that
Gangamopteris is one of the genera which flourished on this
continent.

Gangamopteris cyclopteroides, Feistmantel[5].   Fig. 345.

1876.   Feistmantel, Records Geol. Surv. India, Vol. IX. Pt iii. p. 73.

The specimen represented in fig. 345 illustrates the characters
of this commonest representative of the genus.

[1] Carruthers (69²) p. 9, Pl. VI. fig. 1.   [2] Seward (03) p. 83.
[3] Hayden (07); Seward (07⁵).
[4] Darwin (87) A. Vol. III. p. 248.
[5] For synonymy, see Arber (05) p. 104.

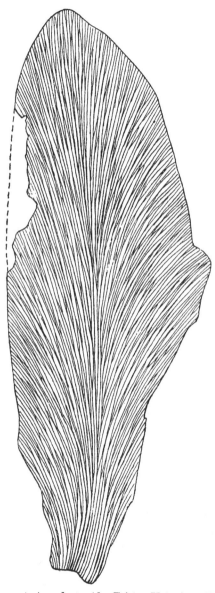

FIG. 345.　*Gangamopteris cyclopteroides*, Feist.　(Nat. size.　From Arber, after
Feistmantel.)

Gangamopteris kashmirensis, Seward.

1905. Seward, Mem. Geol. Surv. India, Vol. II. Mem. ii.

This type agrees closely with *G. cyclopteroides* in size and
in the form of the leaf, but it is distinguished by the flatter
form of the arch formed by the lateral veins, by their greater
inclination to the margin of the lamina, and by the more
acutely pointed apex of the lamina.   This species, though not
very sharply distinguished from *G. cyclopteroides,* is important
as coming from beds which have been assigned on other than
palaeobotanical evidence to an Upper or possibly a Middle
Carboniferous horizon[1].

We have no definite information in regard to the nature of
the reproductive organs of *Gangamopteris,* but such evidence as
there is supports the view expressed by Dr White[2] and shared
by some other authors that *Gangamopteris* and *Glossopteris*
should be assigned to the Pteridosperms.   Despite the abundance of *Gangamopteris* leaves, no fertile specimen has been
discovered.   This negative evidence may prove to be as correct
as that which led Stur[3] to exclude *Neuropteris, Alethopteris* and
Odontopteris from the ferns.   The only evidence of a positive
kind is that furnished by Dr David White in his recent Report
on the Palaeozoic Flora of South Brazil.   This author describes
some small Aphlebia-like leaves under two new generic names
Arberia[4] and *Derbyella*[5]   The differences between the two sets
of specimens, so far as can be determined from the reproductions of imperfect impressions, are slight, and it is by no means
clear that a distinction of generic rank exists.   These scale-
leaves are on the average about 2 cm. in length; the lamina is
oval or rounded and has more or less prominent lobes.   In
Derbyella there are indications of anastomosing veins.   The
specimens referred to *Arberia minasica* are, as White points
out, very similar to the fossil described by Feistmantel from
Lower Gondwana rocks of India as probably a portion of an
inflorescence of *Noeggerathiopsis*[6].   Feistmantel's specimen is

[1] Seward and Smith Woodward (05) ; (07[5]).
[2] White (08) pp. 473, 483.   [3] Stur (84) p. 638.
[4] White (08) p. 537, Pl. VIII. figs. 8–10.   [5] *Ibid.* p. 543, Pl. IX. figs. 1–3.
[6] Feistmantel (80) Pl. XXVII. fig. 5.

represented in fig. 346: the curled lobes may have originally borne seeds. In the Brazilian examples the abruptly truncated lobes "bear evidence of separation from reproductive bodies." An important point is the association of these scale-leaves with *Gangamopteris* fronds and with gymnospermous seeds of the *Samaropsis* type. On the leaves assigned to *Derbyella aurita* circular depressions occur at the base of the lobes which are described as probably due to sporangia.

Dr White's discovery gives us increased confidence in expressing the view that *Gangamopteris* bore its reproductive organs on specialised leaves very different from the sterile fronds; it also strengthens the suspicion that the genus is a member of the class of seed-bearing fern-like plants.

Fɪɢ. 346.   *Arberia* sp. (=*Noeggerathiopsis* of Feistmantel). (Nat. size. After Feistmantel.)

Lesleya.

This generic designation was instituted by Lesquereux[1] for simple oval-linear leaves from the Coal-Measures of Pennsylvania. The leaves so named are probably generically identical with the specimen doubtfully assigned by Brongniart[2] to the Coal-Measures, and made by him the type of the genus *Cannophyllites* on the ground of a resemblance to the leaves of the recent flowering plant *Canna*. Fig. 347 illustrates the form of a *Lesleya* leaf from the Coal-basin of Gard, named by

[1] Lesquereux (30) A. p. 142; Pl. xxv.   [2] Brongniart (28) A. p. 129.

Grand'Eury *L. simplicinervis*[1], a type in which the veins are frequently unbranched and not repeatedly forked as in most examples of the genus (fig. 329, C). The features of the genus are, the oval-linear or lanceolate shape of the presumably simple frond, its entire or, in one species at least (*L. Delafondi*, Zeill.), finely dentate margin, the stout rachis giving off at a very acute angle numerous dichotomously branched secondary veins.

Fig. 347. *Leslya simplicinervis*, Grand'Eury. (Reduced: after Grand'Eury.)

In *L. Delafondi* (fig. 329, C), described by Zeiller[2] from the Lower Permian of Autun, the frond may reach a length of more than 20 cm. and a breadth of 8 cm. Similar species are represented by *L. ensis*[3] from the coal-field of Commentry, and *L. grandis*[4] from Upper Carboniferous rocks of North America. The genus is characteristic of Upper Carboniferous and Lower Permian strata: the form of the leaf and the direction

---

[1] Grand'Eury (90) A. Pl. VIII. fig. 5.   [2] Zeiller (90) p. 166, Pl. XIII. fig. 2.
[3] Renault and Zeiller (88) A. Pl. XXIII. fig. 6.   [4] Lesquereux, *loc. cit.*

of the secondary veins suggest comparison with *Glossopteris*, but in *Lesleya* there are no cross-connexions between the veins. Nothing is known as to the fructification, a fact which naturally evokes the opinion that the genus is a Pteridosperm[1] and not a true fern. Some years before the discovery of Pteridosperms, Grand'Eury[2] suggested that *Lesleya* might be a Gymnosperm ; his opinion being based on the woody nature of the rachis and on the simple venation of *Lesleya simplicinervis*.

Neuropteridium.

In their monograph of fossil plants from the Bunter Series of the Vosges, Schimper and Mougeot[3] described some pinnate leaves of ferns as species of the genus *Neuropteris*. In 1869 Schimper[4] placed these in a new sub-genus *Neuropteridium*, in order to draw attention to the fact that their fronds appear to be simply pinnate and not bipinnate or tripinnate as in *Neuropteris*. The type-species of *Neuropteridium* is *N. grandifolia* Sch. and Moug. from the Bunter Sandstones of the Vosges. The genus includes Triassic European species and the widely distributed Permo-Carboniferous species from Brazil[5] originally described by Carruthers as *Odontopteris Plantiana*. It is probable that some Carboniferous plants, particularly species from the lower members of the formation, referred to the genus *Cardiopteris*, are not generically distinct from the Indian and southern hemisphere type *Neuropteridium validum* (= *Odontopteris Plantiana*).

Fronds pinnate, linear ; a broad rachis bears pinnules which may be either semicircular or broadly linear with an entire or lobed margin. The longer pinnules may exceed 6 cm. in length. The pinnules agree with those of *Neuropteris* in being attached by the median portion of the lamina and not by the whole base, which is more or less auriculate. In some cases the repeatedly forked veins diverge from the centre of the pinnule base ; in others there is a midrib which persists for a short distance only, and in some species the more persistent median vein gives the segments a closer resemblance to those of *Neuropteris*. Fructification unknown, with the exception of obscure indications of sporangia (?) on the fertile leaves of a Triassic species.

[1] White (05²) p. 381.   [2] Grand'Eury (90) A. p. 305.
[3] Schimper and Mougeot (44) A.   [4] Schimper (69) A. p. 447.
[5] Carruthers (69²).

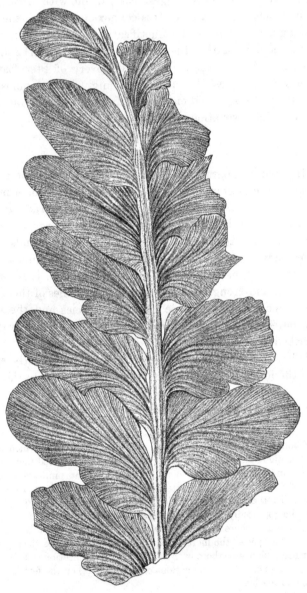

FIG. 348. *Neuropteridium validum*, Feist. Nat. size. From the Karharbári Coal-field, India. From Arber, after Feistmantel.

Neuropteridium validum. (Feistmantel[1]).　Fig. 348.

> 1869.　*Odontopteris Plantiana,* Carruthers, Geol. Mag.　Vol. vi. p. 9,
> Pl. vi. figs. 2, 3.
> 1878.　*Neuropteris valida,* Feistmantel, Mem. Geol. Surv. India, Foss.
> Flor. Gondwana Syst., Vol. iii. p. 10, pl. ii.—vi.
> 1880.　*Neuropteridium validum,* Feistmantel, *Ibid.* 2, p. 84.

The specimen represented in fig. 348 illustrates the main features of *Neuropteridium validum.* This species is referred to by Dr White[2] as *N. Plantianum* on the ground of priority, and with a view to perpetuate the name of the English engineer Nathaniel Plant who discovered the species in a Brazilian Coalfield in the province of Rio Grande do Sul.　Feistmantel's specific name is however retained as being much better known.　An examination of Mr Plant's specimen in the British Museum led me[3] to speak of the Brazilian species as identical with *N. validum* described by Feistmantel from Lower Gondwana rocks of India.　Zeiller[4] had previously drawn attention to the resemblance between the two sets of specimens.　The frond of *N. validum* may exceed 50 cm. in length.　The lower pinnules may be entire and semicircular in form while the upper and larger segments, which may reach a length of 5 or 6 cm., are characterised by broad lobes (fig. 348).

This type is represented in the flora of the Talchir-Karharbári series (Lower Gondwana) of India[5], in Permo-Carboniferous rocks of Brazil and Argentine[6], and in the sandstones of Vereeniging on the borders of the Transvaal and Cape Colony.　It is a characteristic member of the Glossopteris Flora and occurs in association with *Glossopteris* and *Gangamopteris.*

Neuropteridium intermedium (Schimper).　Fig. 349.

This species has been figured by Schimper and Mougeot[7] from the Bunter of the Vosges and more fully described by

[1] See Arber (05) p. 116 ; Seward (03) p. 85.
[2] White (08) p. 483.　　　[3] Seward (03) p. 83.　　　[4] Zeiller (95) p. 616.
[5] Feistmantel (79).　　　　　　[6] Kurtz (94).
[7] Schimper and Mougeot (44) A. Pl. xxxviii.

Blanckenhorn[1] from the Bunter beds of Commern. The pinnate
leaves reach a length of 65 cm.; the lower semicircular pinnules
pass gradually into broadly linear segments characterised by an
auriculate base and a Neuropteris type of venation (fig. 354,
D', E). In the example reproduced in fig. 349 from one of

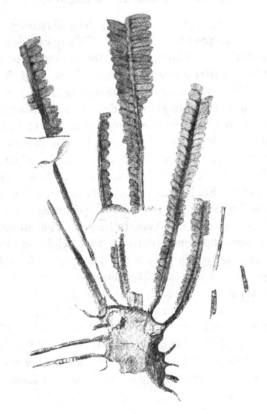

Fig. 349. *Neuropteridium intermedium* (Schimp.). (After Blanckenhorn.
¼ nat. size.)

Blanckenhorn's figures, the fronds are attached to a short and
thick rhizome bearing roots and portions of old petioles.

An example of another Triassic species is afforded by *Neuropteridium grandifolium* Schimp. and Moug., which agrees very
closely with *N. validum* in the size and shape of the pinnules.

[1] Blanckenhorn (85) p. 127, Pls. XVII.—XIX.

The occurrence in Lower Mesozoic European rocks of fronds hardly distinguishable from the older southern species may be regarded as favourable to the view already expressed, that some at least of the Permo-Carboniferous plants migrated north of the Equator. The resemblance between the Vosges Triassic species of *Schizoneura*[1] and the examples of this genus recorded from the Lower Gondwana rocks of India affords additional evidence of a northern migration.

Our knowledge of the reproductive organs of *Neuropteridium* is practically *nil.* There is no doubt that Zeiller[2] and Blanckenhorn[3] are correct in regarding the Bunter fronds assigned by Schimper and Mougeot to the genus *Crematopteris* as the fertile leaves of *Neuropteridium intermedium* or some other species from the same horizon. These fronds bear crowded pinnules similar to those of *Neuropteridium intermedium*, *N. Voltzii*[4], and other species, exhibiting on the exposed surface numerous carbonaceous spots which may be the remains of sporangia.

Cardiopteris.

Schimper[5] applied this generic name to Lower Carboniferous fronds of a simple-pinnate habit which had previously been described as species of *Cyclopteris.* *Cardiopteris frondosa* may serve as a typical example. This species, originally described by Goeppert as *Cyclopteris frondosa* (fig. 350), is recorded from Lower Carboniferous rocks in the Vosges district[6] in Silesia, Moravia[7], and Thuringia[8]. The pinnules, which are attached in opposite pairs to a broad rachis, vary in length from 2 to 10 cm. and have a breadth of 2 to 8 cm.; in manner of attachment and venation they agree with those of *Neuropteridium validum.* The venation is very clearly shown in a drawing of some large pinnules figured by Stur[9].

The specimen of *Cardiopteris frondosa*, a portion of which

---

[1] Vol. I. p. 292.   [2] Zeiller (00²).   [3] Blanckenhorn (85) p. 129, Pl. XXI.
[4] Blanckenhorn *loc. cit.* The specimens figured by this author are in the Strassburg Museum, as are also some of those figured by Schimper and Mougeot.
[5] Schimper (69) A. p. 452. [6] Schimper and Koechlin-Schlumberger (62) A.
[7] *Ibid.*   [8] Fritsch, K. (97).   [9] Stur (75) A. Pl. XIV. fig. 1.

is shown in fig. 350 on a slightly reduced scale, was originally figured by Schimper from an unusually good example in the Strassburg Museum. Schimper's drawing hardly does justice to the original specimen.

A frond bearing rather narrower pinnules, alternately placed on the rachis, which Fritsch has described as *Cardiopteris Hochstetterii* var. *franconica* from the Culm of Thuringia, bears a close resemblance to *Neuropteridium validum* but differs in the entire margin of the pinnules. An Upper Carboniferous

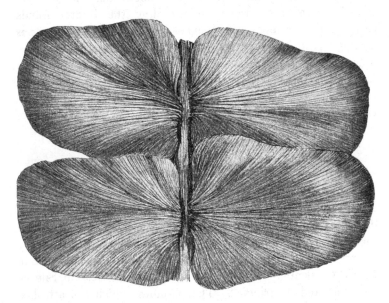

Fig. 350.   *Cardiopteris frondosa* (Goepp.).   (¾ nat. size.   After Schimper.)

species from Russia described by Grigoriew[1] as *Neuropteris*, cf. *cordata* var. *densineura*, represents another form of similar habit.

Schuster[2] has recently proposed a new generic name *Ulvopteris* for a fragment of a pinna from the Coal-Measures of Dudweiler in Germany bearing large pinnules, which he compares with those of *Cardiopteris* and species of *Rhacopteris*. The specimen appears to be indistinguishable from some of

¹ Grigoriew (98) Pl. IV.   ² Schuster (08) p. 184.

those already referred to as conforming to *Neuropteridium*, and it is difficult to recognise any reason for the creation of a new generic name.

We cannot hope to arrive at any satisfactory decision in regard to the precise affinity between *Neuropteridium validum* and species referred to *Cardiopteris* and other genera so long as portions of sterile fronds are the only tests at our disposal. It is difficult to determine whether a specimen consisting of an axis bearing pinnules represents a large pinna of a bipinnate frond or if it is a complete pinnate leaf. There is, however, no adequate reason for supposing that the presumably pinnate fronds from the Gondwana Land rocks are generically distinct from the Lower Carboniferous European species *Cardiopteris frondosa*. Granting the probability that both genera are Pteridosperms and closely allied to one another, the two generic names may be retained on the ground of long usage and in default of satisfactory evidence confirmatory of generic identity. *Cardiopteris* would thus stand for a type of frond characteristic of the Lower Carboniferous strata of Europe, while *Neuropteridium* is retained for the Southern species *N. validum*, and for others from the Trias of the Vosges.

Aphlebia.

This name was proposed by Presl[1] for large leaf-like impressions having a pinnate or pinnatifid form and characterised by a confused irregular type of venation, or by a fine superficial striation or wrinkling which simulates veins. Gutbier had previously described similar fossils as *Fucoides*, and other authors have described Aphlebiae as species of *Rhacophyllum*, *Schizopteris*, and other genera[2]. The term *Aphlebia* is retained, not as denoting a distinct genus but (i) as a descriptive name for detached leafy structures similar to those figured by Presl, which are now recognised as laminar appendages of the petioles of ferns or fern-like fronds, and (ii) as an epithet for highly modified pinnules which frequently occur at the base of the

---

[1] Presl, in Sternberg (38) A.
[2] For synonymy, see Zeiller (88) A. p. 301.

primary pinnae of Pecopteroid and Sphenopteroid fronds
(e.g. *Dactylotheca plumosa*, fig. 293)[1].

Modified pinnules, similar in their reduced and deeply
dissected lamina to those represented in fig. 293, are frequently
found at the base of the primary pinnae of Palaeozoic species of
Sphenopteris and other genera of Pteridosperms or ferns, including
members of the Coenopterideae. Potonié[2] gives a list of various
types of Aphlebiae in his paper on these organs. A striking
case has recently been described by Zeiller in a French Upper
Carboniferous species, *Sphenopteris Matheti*[3]. It would seem
that the larger examples of Aphlebiae are more frequently
associated with the compound leaves of Pteridosperms than
with those of Ferns[4].

As examples of the larger types of Aphlebiae reference may
be made to *Aphlebia crispa* (Gutb.)[5], which reaches a length of
nearly 60 cm. and has the form of a more or less triangular
pinnate leaf divided into decurrent deeply lobed segments, to
a similar species represented by *A. Germari* (= *Schizopteris
lactuca* Germ.)[6] which simulates the leaves of endive (*Cichorium
endivia L.*), and to some large forms figured by Grand'Eury[7] as
species of *Schizopteris*.

Aphlebiae such as that figured by Kidston[8] as *Rhacophyllum
crispum*, with narrow ultimate segments, might easily be
mistaken for the impressions of an alga.

The term *Aphlebia* may be applied also to the Cyclopteroid
pinnules on the petioles of some species of *Neuropteris*, *Odontopteris* and *Archaeopteris*. Goebel[9] has referred to the application
by Potonié and other authors of the term Aphlebioid to the
pinnules which serve as bud-protecting organs in recent fronds
of *Gleichenia* (fig. 226, p. 290); he expresses the opinion that it
is superfluous and misleading to make use of a special designation for structures which are undoubtedly modified pinnules.
In the case of fossils it is, however, convenient to employ
the term *Aphlebia* as a descriptive name for modified pinnules

[1] Page 406.   [2] Potonié (03) p. 162.   [3] Zeiller (06) Pls. VI. VII.
[4] Arber (06).   [5] Renault and Zeiller (88) A. ; Zeiller (88) A. Pl. LI.
[6] Renault and Zeiller (88) A. Pl. XXIV.   [7] Grand'Eury (90) A. Pl. XIX.
[8] Kidston (91) Pl. XXXV.   [9] Goebel (05) p. 318.

or stipular structures which cannot be connected with definite species of fronds. It is clear that some Aphlebiod leaflets, such as those of *Dactylotheca*, served as protective organs for the

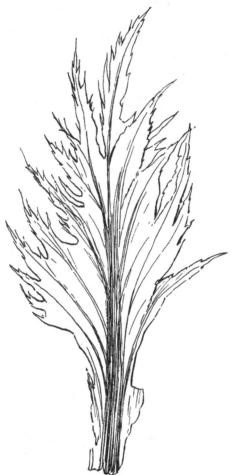

Fig. 351. Scale-leaf of *Gunnera manicata*. (Slightly reduced. M.S.)

unexpanded pinnae[1], and in all probability the large Aphebiae served the same purpose as the fleshy stipules of *Angiopteris* and *Marattia* which cover the uncoiled fronds. The pinnatifid

[1] See p. 406, fig. 293 ; Potonié (03) also figures a young frond of *Dactylotheca plumosa* partially covered by Aphlebiae.

scale-leaves of considerable size (fig. 351) which occur in the
leaf-axils or as ochrea-like stipules on the fronds of *Gunnera*
(a tropical and subtropical Dicotyledonous genus) bear a very
close resemblance to some Palaeozoic Aphlebiae, e.g. *Aphlebia
crispa* (Gutb.).   The recent and fossil scale-leaves may be
regarded as similar in function as in form; moreover the delicate
coiled fronds of Palaeozoic Pteridosperms or ferns, like those of
some recent flowering plants, may have been kept moist by a
secretion of mucilage.   The pinnatifid stipules of *Marattia
fraxinea* (fig. 241, B, p. 317) resemble certain fossil Aphlebiae,
and the wrinkled surface of the recent stipules presents an
appearance similar to that which in some fossil forms has been
erroneously described as veining.   It is not improbable that
mantle-leaves of such recent ferns as *Polypodium quercifolium*
(fig. 234, M, p. 303) are comparable with some fossil Aphlebiae
which may have served as humus-collectors for Palaeozoic
epiphytes.

The filiform appendages on the petioles of the recent fern
Hemitelia capensis (fig. 235, p. 304) have often been compared
with the aphlebioid leaflets of fossil fronds.

Potonié who has discussed the nature of Aphlebiae regards
them as vestiges of a once continuous lamina, which formed a
winged border to the branched axes of more primitive forms of
fronds.   It is possible that the pinnules between the pinnae on
the rachis of *Archaeopteris* and the Cyclopteroid leaflets of
Neuropteris and *Odontopteris* may have the morphological
significance attributed to them by Potonié.   In some cases
it is probable that the Aphlebiae, whether vestiges or not,
served the purpose of protecting either the whole frond or
individual pinnae.   Aphlebiae, though especially characteristic
of Palaeozoic leaves, are occasionally met with in the form of
modified pinnules at the base of the primary pinnae on
Mesozoic ferns, e.g. in *Coniopteris hymenophylloides*[1].

In some fern fronds the lowest pinnule of each pinna
differs in shape or size from the normal ultimate segments,
but it would be almost affectation to extend the use of the
term *Aphlebia* to such pinnules.   The Jurassic species

[1] Seward (00) Pl. xxi. fig. 1.

Cladophlebis lobifolia (Phill.) is a case in point[1]. In this fern, which some authors speak of, without sufficient reason, as *Dicksonia lobifolia*[2], the lowest pinnule is large and different in shape from the others.

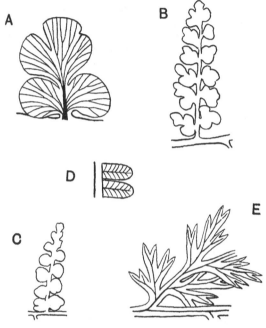

Fɪɢ. 352.   A. *Sphenopteris obtusiloba.* Pinnule. (Enlarged. After Zeiller.)
B, C. *S. obtusiloba.* (⅞ nat. size. After Zeiller.)
D. *Pecopteris arborescens.* (Slightly enlarged. After Zeiller.)
E. *Sphenopteris furcata* (= *Diplotmema furcatum*). (Slightly en-
larged. After Zeiller.)

Sphenopteris.

Sphenopteris is one of the many generic names which we owe to Brongniart[3]. It is the generic designation used for a great number of Palaeozoic and later fronds, most of which are those of true ferns while some Palaeozoic species are undoubted Pteridosperms. The genus, which is purely provisional, includes

---

[1] *Ibid.* p. 145.   [2] Raciborski (94) A. Pl. xɪ.   [3] Brongniart (22) A.

members of widely different families possessing pinnules of the
same general type, such as is represented in some recent species
of *Davallia, Asplenium*, and other ferns.

The fronds of *Sphenopteris* may be bipinnate, tripinnate, or quadripinnate; the rachis may be dichotomously branched or the branching may
be of the pinnate type characteristic of most recent ferns.   The pinnules
are small; they vary considerably in shape even in a single frond, but the
chief characteristics are: the lobed lamina, contracted and often wedge-
shaped at the base (fig. 352), the dichotomously branched veins radiating
from the base or given off from a median rib at an acute angle.   The
lamina may be divided into a few bluntly rounded lobes (fig. 352, C) or
deeply dissected into linear or cuneate segments (fig. 352, A, B, E).

Examples of Sphenopteroid leaves have already been
described under the genera *Coniopteris, Onychiopsis, Ruffordia*,
etc.   Among the numerous examples of *Sphenopteris* species
from the Carboniferous rocks mention may be made of *Sphenopteris obtusiloba* Brogn.[1] (fig. 352, A—C), which occurs in the
Middle and Lower Coal-Measures of Britain[2].   This type is
characterised by the almost orbicular, oval or triangular pinnules
which may reach a length of 15 mm.; they are occasionally
entire, but more usually divided into 3 to 5 rounded lobes.   The
forked veins radiate from the base of the pinnule.   The rachis
may be dichotomously branched.   Fructification unknown.

The species *S. furcata* Brogn.[3], characteristic of the Middle
and Lower Coal-Measures of Britain (fig. 352, E), is referred
to under Stur's genus *Diplotmema*[4] in which it is included
by some authors solely because of the dichotomous habit of
branching of the pinnae.

The pinna represented in fig. 353 illustrates a similar type
of pinnule.   This species, which is very common in the Calciferous
Sandstone of Scotland, was described by Lindley and Hutton
as *Sphenopteris affinis*[5].

The fronds of *Sphenopteris affinis* were discovered by
Mr Peach[6] in a fertile condition, but he regarded the reproductive
organs as those of a plant parasitic on the *Sphenopteris* fronds.

---

[1] For synonymy, see Kidston (86) p. 68.   [2] Kidston (94) p. 298.
[3] Zeiller (88) A. p. 147, Pls. iv. v. ; Kidston (86) p. 80.   [4] See p. 535.
[5] Lindley and Hutton (31) A. Pl. xlv.   [6] Peach (78).

Kidston[1] substituted Stur's genus *Calymmatotheca* for *Sphenopteris* on the ground that the sporangia figured by Peach under the name *Staphylopteris Peachii* bear a close resemblance

Fig. 353.　*Sphenopteris affinis*, Lind. and Hutt. From the Calciferous Sandstone of Burdiehouse (Scotland). (Sedgwick Museum, Cambridge.) M.S.

to the organs which Stur described as valves of an indusium in his species *Calymmatotheca Stangeri*[2]. An examination of Stur's

[1] Kidston (87) p. 145.　　　　[2] This species will be described in Vol. III.

specimens by Miss Benson[1] and by Prof. Oliver and Dr Scott
has confirmed Stur's interpretation of the appendages at the
tips of the fertile pinnae as valves of an indusial or cupular
structure. The superficially similar bodies on the fertile pinnae
of *S. affinis* are however true sporangia, and cannot legitimately
be included in the genus *Calymmatotheca* as described by Stur.
For this reason Miss Benson institutes a new genus *Telangium*,
the type-species of which, *T. Scotti* from the Lower Coal-
Measures of Lancashire, is based on petrified material. The
Scotch species *Sphenopteris affinis* (= *Calymmatotheca affinis* of
Kidston) is also transferred to *Telangium*; the sporangia are
considered by Miss Benson to be microsporangia. This with other
species is no doubt correctly included in the Pteridosperms.
A complete frond of *Sphenopteris affinis*, showing a regular
dichotomy of the main axes, is represented by an admirable
drawing in Hugh Miller's *Testimony of the Rocks*[2].

Some of the Palaeozoic species of *Sphenopteris* probably
represent the fronds of true ferns, but others are known to have
been borne by Pteridosperms. *S. Hoeninghausi* (fig. 290, C,
p. 399) is the foliage of *Lyginodendron*, and Scott[3] speaks of
three species, *S. dissecta*, *S. elegans*, and *S. Linkii* as the leaves
of *Heterangium*. Grand'Eury[4] has recorded the occurrence in
French Coal-Measures of seeds in association with other
Sphenopteroid fronds.

Mariopteris, Diplotmema, Palmatopteris.

The discovery of sporangia on the fronds of several Palaeozoic species of *Sphenopteris* and *Pecopteris* has led to the
institution of new generic names, which indicate an advance in
knowledge beyond the stage implied by the use of those
provisional designations based solely on the form and venation
of the pinnules. Other names have been created by authors in
place of *Sphenopteris* and *Pecopteris* on the ground that a
striking feature in the mode of branching of fronds is sufficiently
important to justify generic recognition even in the absence

[1] Benson (04).   [2] Miller (57), Frontispiece.
[3] Scott (05²) p. 144.   [4] Grand'Eury (05²).

of fertile specimens. As examples of designations based primarily on the branch-system of compound leaves, the genera *Mariopteris, Diplotmema,* and *Palmatopteris* may be briefly considered (fig. 354 A—C). Dr Kidston[1] is of opinion that the creation of new genera for purely vegetative characters of fronds is of no real advantage, and he prefers to retain the older provisional names for species known only in the sterile condition. On the other hand, if we are sufficiently familiar with specimens large enough to enable us to recognise a well-defined morphological character, it may serve a useful purpose to employ a generic designation for features which may have a phylogenetic value. A comparative examination of Palaeozoic, Mesozoic, and recent compound fronds, including both Pteridosperms and true ferns, brings to light certain distinguishing features characteristic of the older types which, as Potonié maintains[2], point to the derivation of the pinnate habit from a primitive dichotomous system of branching. For a more complete discussion of this question reference should be made to Potonié's suggestive papers. Among recent ferns *Matonia* and *Dipteris,* two survivals from the past, afford instances of fronds with a branching system of the dichotomous type.

Similarly, in *Gleichenia, Lygodium,* and more rarely in species of Polypodiaceae (*e.g. Davallia aculeata,* fig. 232) dichotomy is a striking feature of the fronds. In the great majority of recent ferns the fronds have assumed a pinnate habit. Among Palaeozoic fern-like fronds dichotomous branching of the main rachis and of the pinnae is much more common. Potonié draws attention to several other features which distinguish Palaeozoic fronds from the majority of later species: the frequent occurrence of pinnules borne directly on the main rachis (fig 354, D), and of modified pinnules or Aphlebiae on the rachis and petiole, are characters to which he attributes an evolutionary significance. The main point is that a comparative examination of leaf-form affords evidence in favour of the view that the modern type of frond, with its naked rachis bearing two rows of pinnae, has been derived from a less specialised type in which the distinction between the

---

[1] Kidston (01²) p. 191.　　　　[2] Potonié (95); (99).

parts of the leaf is much less evident. The primitive leaf was probably a dichotomously branched axis provided with a continuous lamina which eventually became broken up into separate lobes or pinnules.

As the dichotomy of the frond became less regular, a pinnate habit was acquired, as is clearly seen in many Palaeozoic types which constitute connecting links between forked and pinnate fronds (fig. 354, D). The Aphlebiae may be remnants of the once-continuous lamina on the petiole, and the normal pinnules borne on the rachis may be regarded as the attributes of fronds in which the division of physiological labour had not reached the stage which characterises the leaves of recent ferns.

Mariopteris.

This name, which is due to Zeiller[1], is applied by him to Palaeozoic fronds characterised by a double bifurcation of the rachis of the primary pinnae. *Mariopteris muricata* (= *Pecopteris muricata* Schloth.) may be taken as the type of the genus. This species is common in the Lower and Middle Coal-Measures of Britain and rare in the Upper Coal-Measures[2]. It is described by Kidston[3] as one of the most polymorphic and widely distributed Coal-Measure species. The pinnules as seen in fig. 364, B, are of the Sphenopteroid type. No fertile specimens are known, but it is significant that Grand'Eury[4] has recorded the association of *Mariopteris muricata* and seeds.

The main rachis gives off alternate naked branches, each of which bifurcates at its apex into two short naked axes, and these are again forked, the ultimate branches having the form of bipinnate pinnae provided with large Sphenopteroid pinnules (fig. 354, B). Zeiller includes in *Mariopteris* some species which Stur[5] referred to his genus *Diplotmema*. Possibly some of the Palaeozoic fronds with a zigzag rachis may have been climbers like *Lygodium*.

[1] Zeiller (79²).   [2] Kidston (01²) p. 195.
[3] For synonymy, see Kidston (03) p. 771.
[4] Grand'Eury (08).   [5] Stur (75) A. p. 120.

Diplotmema.

This generic name is employed by Zeiller[1] and other authors in a more restricted sense than that in which it was originally

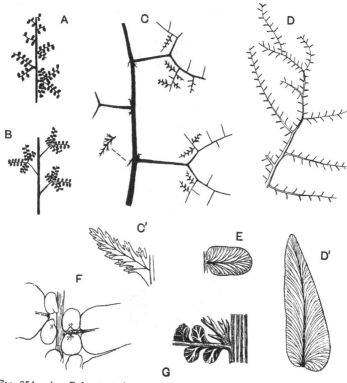

FIG. 354. A. *Palmatopteris.*
B. *Mariopteris.*   (A, B, after Potonié.)
C. *Diplotmema Zeilleri,* Stur.   (After Zeiller.)
C'. *D. Zeilleri.*   Pinnule.   (× 3.   After Zeiller.)
D. *Neuropteris macrophylla.*   (British Museum   .)
D'. *N. macrophylla.*   Pinnule.   (Slightly enlarged. After Kidston.)
E. *N. heterophylla.*   Pinnule.   (Slightly enlarged. After Zeiller.)
F. *N. Scheuchzeri.*   (Slightly reduced. After Kidston.)
G. *Alloiopteris Essinghii.*   (Enlarged. After Potonié.)

used by Stur.   The Upper Carboniferous species *Sphenopteris furcata* Brongn. (fig. 352, E) may serve as the type.   This species occurs in the Middle and Lower Coal-Measures of Britain[2].

[1] Zeiller (79²); (88) A. p. 142.   [2] Kidston (94) p. 240.

The main rachis gives off branches as in *Mariopteris,* but in *Diplotmema* each naked lateral branch is forked at its apex into two opposite pinnae bearing deeply dissected Sphenopteroid pinnules. Zeiller[1] and Stur have recorded fertile specimens of *Diplotmema,* but in no case have actual sporangia been

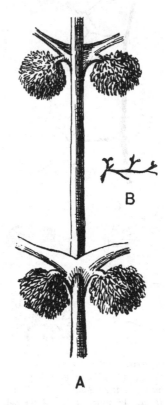

FIG. 355. A. *Cephalotheca mirabilis,* Nath. Fertile pinnae. (Partially re-
                     stored   After Nathorst.)
           B. *C. mirabilis.* Sterile pinnule. Nat. size. (After Nathorst.)

discovered. In the species *Diplotmema Zeilleri* Stur (fig. 354, C, C′) two Aphlebiae occur at the base of each secondary axis[2]. It has been pointed out by Potonié that in *Diplotmema furcatum* the equal dichotomy of the lateral branches is not characteristic

---

[1] Zeiller (88) A, p. 147.   [2] *Ibid.* Pl. XVI.

of the frond as a whole. In the case of branches higher on the rachis the dichotomy becomes unequal and the forked axis is gradually replaced by a simple pinna (fig. 354, A). For this type of frond, Potonié proposed the generic name *Palmatopteris* in place of *Diplotmema,* which he discards. The long comparatively slender rachis of *P. furcata* suggests comparison with the liane species of *Lygodium*[1].

Cephalotheca.

This genus was proposed by Nathorst[2] for some peculiar bipinnate fertile fronds from the Upper Devonian rocks of Bear Island. The pinnae bear slender forked ultimate segments represented by a few detached fragments (fig. 355, B), associated with the rachises. The fertile pinnae are given off in opposite pairs from the main axis over which they are concrescent (fig. 355, A). A mop-like cluster of sporangia is borne on the lower surface and close to the base of a fertile pinna : the exannulate sporangia are compared with those of *Scolecopteris.* Nathorst compares *Cephalotheca* with a Belgian species of Upper Devonian age described by Crépin[3] as *Rhacophyton condrusorum* and by Gilkinet[4] as *Sphenopteris condrusorum.* A similar fossil is also described by Baily[5] as *Filicites lineatus* from the Kitorkan Grits of Ireland.

The position of *Cephalotheca* cannot be definitely determined from the available data, but it is more probable that it was a seed-bearing Pteridosperm and not a true fern. Zeiller[6] has recently expressed the same opinion.

Thinnfeldia.

The genus *Thinnfeldia,* founded by Ettingshausen in 1852[7] on some Hungarian Liassic specimens, though frequently included in the Filicales, cannot be said to occupy that position by virtue of any well-authenticated filicinean features. It is by

---

[1] Potonié (92).   [2] Nathorst (02) p. 15.
[3] Crépin (75).   Previously described by Crépin (74) as *Psilophyton.*
[4] Gilkinet (75).   [5] Nathorst (02).
[6] Zeiller (09²) p. 20.   [7] Ettingshausen (52).

no means improbable that many of the species referred to this genus are closely allied to Palaeozoic Pteridosperms.

Thinnfeldia may be briefly defined as follows :

Fronds simple and pinnatifid, pinnate or bipinnate : rachis broad and occasionally dichotomously branched. Pinnules often fleshy or coriaceous ; broadly linear, entire or lobed, provided with a midrib from which simple or forked secondary veins are given off at an acute angle : or the laminae may be short and broad without a midrib and traversed by several slightly divergent and forked veins.

No satisfactory evidence of reproductive organs has so far been adduced.

The genus is chiefly characteristic of Upper Triassic, Rhaetic, and Jurassic floras, though it was in all probability represented in Permian floras. Several species, many of which are value-less, are recorded also from Cretaceous and Tertiary formations. Search should be made for fertile specimens or for evidence as to the association of seeds with *Thinnfeldia* fronds.

Some Permian fossils from Kansas which Sellards[1] has made the type of a new genus, *Glenopteris*, appear to be indistinguishable generically from leaves of Lower Mesozoic age universally recognised as typical examples of *Thinnfeldia*.

Thinnfeldia odontopteroides (Morris)[2].   Figs. 356—358.

This is a very variable species as regards the shape and size of the ultimate segments and their venation. It is a type of extended geographical range characteristic of Rhaetic or Upper Triassic rocks in Australia, South Africa, India, South America, and various European localities.

Frond bipinnate ; the broad rachis may be dichotomously branched. Pinnules with a thick lamina which may be almost semi-circular in form, deltoid, broadly oval or broadly linear, and often confluent at the base. Short and broad pinnules occur on some fronds directly attached to the main rachis between the pinnae. The longer and narrower pinnules (fig. 356, C), resembling those of the Palaeozoic genus *Alethopteris*, have a well-defined midrib, while the smaller segments are characterised by several slightly divergent veins which spring directly from the rachis (fig. 356, A). Epidermal cells polygonal or, above the veins, rectangular in shape ; stomata, which are slightly sunk, occur on both the upper and lower epidermis. Fertile specimens unknown.

---

[1] Sellards (00).   [2] For synonymy, see Seward (03) p. 52.

The portion of a lobed pinnule shown in fig. 356, B, illustrates a form of segment intermediate between the linear type with a midrib and a row of shorter pinnules without a median vein. Fig. 356, D, represents another instance of variation in the arrangement of the veins in segments of different sizes. Various specific and generic names have been assigned to Thinnfeldia fronds of Rhaetic age on the ground of the occurrence of pinnules longer and narrower than those

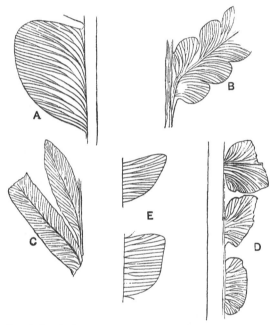

FIG. 356.　A—D.　*Thinnfeldia odontopteroides* (Morris).
E.　*Ptilozamites.*　(E, after Nathorst.)

usually associated with *T. odontopteroides*; but in view of the range of variation met with in a single leaf it is advisable to extend rather than to restrict the boundary of what we are pleased to regard as a specific type.

The name *Thinnfeldia lancifolia* has been applied by Morris to fossils from Australia which may be identified with *T. odontopteroides*, and the same designation is employed by

Szajnocha and by Solms-Laubach [1] for Rhaetic specimens from
South America. Similar fronds are described by Geinitz [2] as
Thinnfeldia tenuinervis from Argentine Rhaetic strata.
Odontopteris macrophylla Curran, *T. falcata* Ten.-Woods, *Glei-
chenia lineata* Ten.-Woods, and *Cardiopteris Zuberi* Szaj.
afford other examples of what are probably closely allied forms [3].

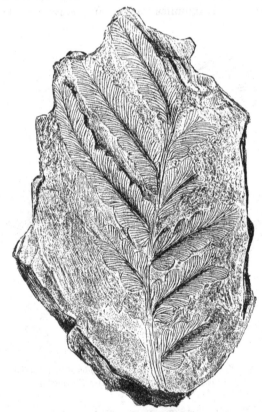

FIG. 357.   *Thinnfeldia odontopteroides* (Morris).   ⅖ nat. size.

Some exceptionally large examples of *T. odontopteroides* are
figured by Feistmantel [4] from the Hawkesbury series of New
South Wales in which the bipinnate frond has a breadth of

---

[1] Solms-Laubach and Steinmann (99) Pl. xiv. fig. 2 ; Szajnocha (88).
[2] Geinitz (76) Pl. i.   [3] Seward (08) p. 95.   [4] Feistmantel (90) A. Pl. xxiv.

25—30 cm. A specimen from the Molteno beds of South Africa, probably of Rhaetic age, represented in fig. 357, illustrates a smaller leaf with pinnules of the linear type, some of which are partially divided into shorter pinnules with forked veins. The example represented in fig. 358, from Cyphergat (S. Africa), shows two equal branches of a rachis with small contiguous segments.

Fig. 358. *Thinnfeldia odontopteroides.* From a specimen in the British Museum (v. 2490). 1½ nat. size.

Some specimens figured by Zeiller[1] from the Rhaetic strata of Tonkin as *Pecopteris* (*Bernouillia* ?) sp. may be portions of *Thinnfeldia* fronds, and the large leaves which he refers to *Ctenopteris Sarreni* differ but slightly from the Australian specimens described by Feistmantel as *T. odontopteroides.*

[1] Zeiller (03).

Thinnfeldia rhomboidalis, Ettingshausen.   Figs. 359, 360, C.

Under this name Ettingshausen described the type-specimen of the genus from Lower Lias strata at Steierdorf in Hungary. He assigned the plant to the Coniferae on the ground of a resemblance of the pinnules to the phylloclades of *Phyllocladus*. *Thinnfeldia rhomboidalis* bears a close resemblance to *T. odon-topteroides*, but the pinnules are usually longer and narrower, as shown in the English specimen from the Lower Lias of

Fig. 359.   *Thinnfeldia rhomboidalis*, Ettings.   Slightly reduced.   From an English Liassic specimen in the British Museum.   [M.S.]

Dorsetshire represented in fig. 359.   The darker margin of the pinnules shown in fig. 360, C, gives the impression of a revolute lamina, but a microscopical examination points to a thicker cuticle at the edge of the segments.

The species is recorded from Jurassic rocks of France, Germany, Italy, India, Australia, and elsewhere[1].

Palaeobotanical literature contains numerous records of

[1] For references, see Seward (04) p. 31.

Jurassic, Cretaceous and some Tertiary species referred to *Thinnfeldia*, but many of these are probably not generically identical with *T. odontopteroides* or *T. rhomboidalis*. Mr Berry[1] in a paper on *The American species referred to Thinnfeldia* concludes that the genus is "a rather indefinite one...and badly in need of revision." He regards the Middle and Upper Cretaceous American species as Conifers related to *Phyllocladus* and probably forming a link between the Podocarpeae and Taxeae: for these forms he proposes the generic name *Protophyllocladus*. The opinion has been expressed elsewhere[2] that this "problematical[3]" genus rests on an unsatisfactory basis; the available data do not justify the use of a name which implies the existence in North American Cretaceous floras of a type related to the New Zealand and Tasmanian Conifer *Phyllocladus*. We are not in a position to assign a single species of *Thinnfeldia* to the Filicales or the Gymnosperms.

A leaflet from Jurassic rocks of Poland figured by Raciborski[4] shows what this author regards as the impression of a circular sorus: no sporangia have been found. A specimen in the British Museum[5], which is said to come from Rhaetic beds in Queensland, shows a row of contiguous polygonal prominences on each side of the midrib which resemble the sori of a fern; but until sporangia are discovered we cannot determine the precise nature of this apparently fertile frond.

A species described by Fontaine[6] from the Potomac beds (Wealden-Jurassic) of North America as *Thinnfeldia variabilis* affords a good example of a plant which cannot be identified with any degree of confidence either as a fern or a seed-bearing type. Mr Berry draws attention to the former application of this name by Velenovský to a distinct Lower Cretaceous Bohemian species and proposes for Fontaine's plant the name *T. Fontainei*; he maintains that no one has doubted the fern-nature of the Potomac plant. *T. variabilis* may indeed be a fern, but the evidence is not such as to preclude legitimate doubts as to the correctness of this suggestion. Solms-Laubach[7],

---

[1] Berry (03). [2] Seward (04) p. 31. [3] Hollick and Jeffrey (09) p. 24.
[4] Raciborski (94) A. Pl. xx. figs. 1, 2; Zeiller (00²) p. 98.
[5] v. 5950. [6] Fontaine (89) Pls. xvii. xviii.
[7] Solms-Laubach (91) A. p. 141.

in referring to Schenk's view that *Thinnfeldia* and its allies may represent a group intermediate between Ferns and Gymnosperms, admits that it is a possible supposition; he is, however, inclined to consider *Lomatopteris* and *Cycadopteris*, "genera especially comparable with *Thinnfeldia*," as more probably ferns.

At this point we may conveniently consider a series of genera which occupy an equally uncertain position and bear a very close resemblance to *Thinnfeldia*.

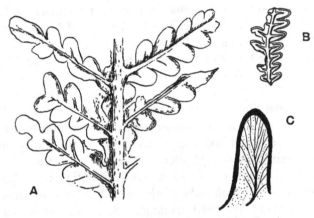

FIG. 360. A. *Lomatopteris jurensis.* (⅔ nat. size. After Kurr.)
          B. *L. Schimperi.* (⅔ nat. size. After Salfeld.)
          C. *Thinnfeldia rhomboidalis*, Ett. (Slightly enlarged. British Museum. No. 52672.)

Lomatopteris.

The generic name *Lomatopteris* was proposed by Schimper[1] for some bipinnate fronds originally described by Kurr[2] from Jurassic rocks of Würtemberg as *Odontopteris* (?) *jurensis* (fig. 360, A). I have elsewhere expressed the opinion[3] that this German species may be identical with *Thinnfeldia rhomboidalis* Ett. Kurr's type-specimen, a portion of which is reproduced in fig. 360, A, consists of a frond or large pinna

[1] Schimper (69) A. p. 472.   [2] Kurr (45) Pl. II. fig. 1.
[3] Seward (04) p. 30.

characterised by a prominent and broad rachis giving off alternate linear pinnae bearing bluntly rounded, contiguous and basally concrescent pinnules having a thick or revolute border and a central rib. The lateral veins are visible in the ultimate segments of Kurr's fossil. Saporta[1] has described several species, which he refers to Schimper's genus, from French Jurassic strata: it is, however, difficult to recognise some of the examples represented in his illustrations as specifically distinct forms. This author notices the resemblance of *Lomatopteris* to *Thinnfeldia*, not only in habit but in the structure of the epidermal cells[2]. In *Lomatopteris* and in *Thinnfeldia* the cells have straight and not sinuous walls and the slightly sunken stomata are surrounded by a ring of epidermal cells. Salfeld[3] has recently described portions of fronds from Jurassic rocks of South-West Germany, which he identifies as *Lomatopteris jurensis*. He disagrees with my view that *Lomatopteris* does not differ sufficiently from *Thinnfeldia* to be accorded generic autonomy, chiefly on the ground that the folded-over edge of the pinnules is a distinguishing feature of *Lomatopteris*. There is, however, no difference, in appearance at least, between the leaflets of some species of *Thinnfeldia*, e.g. *T. rhomboidalis* from Liassic rocks of England[4], and those referred to *Lomatopteris*. In a later paper, Salfeld[5] describes some Portlandian fragments from North Germany as *Lomatopteris Schimperi*, identifying them with a Wealden fossil of somewhat doubtful affinity, which Schenk[6] makes the type of his species. The Portlandian specimens are described as tripinnate, with thick decurrent obtusely terminated pinnules with a revolute edge. The general form of the frond is very similar to that of *L. jurensis*. Salfeld publishes a photograph of a large specimen which he describes as fertile and a drawing of a piece of a pinna: the latter is reproduced in fig. 360, B. He speaks of sori occurring in two rows, probably attached to lateral veins, in the groove between the midrib and the revolute edge of the lamina. The sporangia are described as " nicht

---

[1] Saporta (73) A.   [2] Schenk (67) A.   [3] Salfeld (07) p. 192.
[4] Seward (04) p. 34, fig. 2, Pl. IV.   [5] Salfeld (09).
[6] Schenk (76) Pl. XXVI. fig. 7.

näher bekannt[1]." An examination of the figures reveals
nothing as to the nature of the "sori." The specimens are
considered by Salfeld to afford decisive evidence against the
view that *Lomatopteris* and *Thinnfeldia* are generically identical.
Nothing has so far been published which constitutes a valid
argument in favour of retaining Schimper's generic name.

Cycadopteris.

Zigno[2] founded the genus *Cycadopteris* on Italian Jurassic
impressions regarded by Schimper as indistinguishable from
Lomatopteris. As Solms-Laubach[3] points out, the supposed
sori of *Cycadopteris* described by Zigno are not convincing.
There appear to be no satisfactory reasons for separating
Cycadopteris from *Lomatopteris*, nor do the fronds described
under these names exhibit any important differences from
Thinnfeldia.

Ptilozamites.

Nathorst[4] founded this genus on a remarkable series of
specimens from the Rhaetic Coal-beds of Scania and assigned
it to the Cycadophyta. The species *Ptilozamites Heeri* may be
taken as a representative type. The leaves are linear and
simply pinnate. In the example shown on a much reduced
scale in fig. 361 the frond is 53 cm. long and 2·1 cm. broad.
The upper edge of each pinnule is straight or slightly concave;
the lower edge is rounded; the veins are slightly divergent
and dichotomously branched (fig. 356, E, p. 539). In some of
Nathorst's specimens the broad rachis is forked as in many
Thinnfeldias.

As a comparison of fig. 356, A and E, shows, the pinnules
of some specimens of *Thinnfeldia odontopteroides* are identical
with those of *Ptilozamites*. In the latter genus the rachis is
either unbranched or occasionally forked, while in *Thinnfeldia*
the branching may be of the dichotomous or pinnate type. In
Ptilozamites the segments appear to be always without a mid-

[1] Salfeld (09) p. 34.   [2] Zigno (56) A.
[3] Solms-Laubach (91) A. p. 114.   [4] Nathorst (78).

rib, while a median vein frequently occurs in those of *Thinn-feldia*. There can be little doubt as to the very close alliance between the Rhaetic species referred to these two genera. The

FIG. 361. *Ptilozamites Heeri*, Nath.   (⅓ nat. size. After Nathorst.)

name *Ptilozamites* should perhaps be retained for such long
and narrow fronds as that shown in fig. 361: no species
included in *Thinnfeldia* is known in which the rachis reached
so great a length without branching. The habit of *Ptilozamites Heeri* predisposes one in favour of Nathorst's opinion
that the fronds are Cycadean: we have no information in regard
to the nature of the reproductive organs.

Ctenopteris.

This name was instituted by Saporta[1], at Brongniart's
suggestion, for Liassic species characterised by pinnules like
those of *Thinnfeldia*, but distinguished by the bipinnate habit
of the frond. Saporta compares the genus with the Palaeozoic
leaves known as *Odontopteris*, and with Italian Jurassic plants
referred by Zigno to his genus *Dichopteris*.

The name *Ctenozamites* is applied by Nathorst[2] to the type
of frond which Saporta, Zeiller, and other authors refer to
Ctenopteris. Nathorst instituted *Ctenozamites* for fossils agreeing in the form and venation of the pinnules with his genus
Ptilizamites but differing in being bipinnate and not pinnate.

Fronds of *Ctenopteris* are characteristic of the Jurassic and
Rhaetic series; they are known only in the sterile condition.
As Zeiller[3] says, *Ctenopteris* may be a member of the Cycadofilices, an extinct group founded on Palaeozoic plants combining Cycadean and Filicinean characters, and some of which
are now known to be Pteridosperms. It is probable that the
genus is not a true fern: it is more likely to be a member
of the Cycadophyta or of some generalised extinct group.

Ctenopteris cycadea (Brongniart).   Fig. 362.

1828. *Filicites cycadea*, Brongniart, Hist. Vég. foss. p. 387, Pl. CXXIX.
1832. *Odontopteris cycadea*, Berger, Verstein. Coburg Geg. p. 23,
Pl. III.
1873. *Ctenopteris cycadea*, Saporta, Pal. Franç. Vol. I. p. 355
Pls. XL. XLI.

[1] Saporta (73) A. p. 352.   [2] Nathorst (78) p. 122.   [3] Zeiller (03) p. 52.

Frond bipinnate, broad rachis giving off branches at an acute angle; pinnules broadly linear, slightly falcate, with several slightly divergent forked veins.

A frond very similar to the Lower Lias specimen from Dorsetshire represented in fig. 362 was described by Leckenby as *Ctenis Leckenbyi* (Bean MS.) from the Inferior Oolite of Yorkshire[1]. Leckenby recognised the possibility of a Cycadean

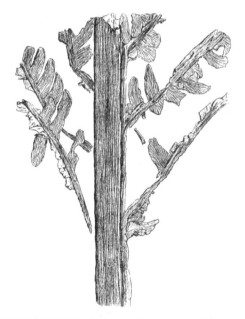

FIG. 362. *Ctenopteris cycadea*, Brongn. (½ nat. size.) From a specimen in the British Museum. [M.S.]

affinity, but regarded the bipinnate habit as an objection. The branched fronds of the Australian Cycad *Bowenia* supply an answer to this objection. Several good examples of *Ctenopteris cycadea* are figured by Schenk[2] from Rhaetic rocks of Persia. Zeiller's Tonkin Rhaetic species, *C. Sarrani*[3], affords a striking illustration of the difficulty of drawing a clear line of separation between *Ctenopteris* and some species of *Thinnfeldia*.

[1] Leckenby (64) A. Pl. x. fig. 1; Seward (04) p. 36.
[2] Schenk (87).   [3] Zeiller (03) Pls. VI.—VIII.

Ctenopteris is in all probability very closely related to *Thinnfeldia* and *Ptilozamites*.

Dichopteris.

This genus was proposed by Zigno[1] for some large specimens from the Jurassic plant-beds of Northern Italy.

The bipinnate leaves are characterised by the great breadth of the rachis which is dichotomously branched in the distal region (fig. 363); the linear pinnae reach a considerable length. Pinnules relatively small, oblong and slightly contracted at the base; the decurrent and confluent lamina forms a narrow wing to the main axis. Veins slightly divergent and forked, as in *Ptilozamites*.

Dichopteris visianica, Zigno. Fig. 363.

A specimen of this species in the Padua Museum has a total length of 83 cm. It has been elsewhere suggested[2] that a fragment figured by Zigno as a fertile example of this type is probably part of a frond of the Osmundaceous fern *Todites*. Since this opinion was expressed I have had an opportunity of examining the actual specimen at Padua: the circular patches described by Zigno as sori appear to be irregularities in the matrix and not an original feature.

Brongniart[3] instituted the genus *Pachypteris* for some imperfectly preserved English Jurassic fossils from Whitby, which he described as *P. lanceolata*. Specimens have since been described[4] from the Inferior Oolite rocks of the Yorkshire coast. Brongniart described the pinnules as being without veins or as possessing only a midrib. It is almost certain that the apparent absence of veins in most specimens[5] is due to the fleshy nature of the segments and that the species *P. lanceolata* should be transferred to *Dichopteris*.

Krasser[6] has described a species from Cretaceous rocks of the island of Lesina, off the Dalmatian coast, as *Pachypteris dal-*

---

[1] Zigno (56) A. Pls. XII. XIII. [2] Seward (00) p. 170.
[3] Brongniart (28) A. p. 49. [4] Seward (00) p. 171.
[5] Saporta (73) A. p. 368. [6] Krasser (95).

matica which is very similar in habit to the English specimens and to Zigno's *Dichopteris visianica.*   One of Krasser's specimens is practically identical with *Dichopteris lanceolata* (Brongn.),

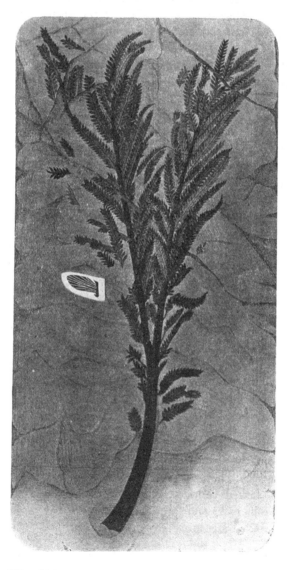

Fig. 363.   *Dichopteris visianica*, Zigno.   (⅓ nat. size.   After Zigno.)

while in others the small pinnules are replaced in some of the
pinnae by a continuous lamina with a few distal serrations.
The latter form a link between the *Dichopteris* and *Thinnfeldia*
type of segment. Krasser gives a full résumé of opinions
expressed by other authors in regard to the position of *Pachypteris* (= *Dichopteris*) and decides in favour of a Cycadean
alliance.

A French Jurassic plant which Saporta[1] made the type of a
new genus *Scleropteris*, and described as *S. Pomelii*, appears
to be indistinguishable from *Dichopteris*.

Dichopteris, though conveniently retained as a distinct
genus, agrees so closely, in the broad and forked rachis and in
the fleshy pinnules, with *Thinnfeldia* that it would seem reasonable to regard the two genera as members of the same group.

Several authors have drawn attention to the striking
resemblance in form and venation between the fronds of the
Palaeozoic genus *Odontopteris* and those of *Ctenopteris* and
Thinnfeldia. In *Odontopteris*, as in *Neuropteris*, another Palae-
ozoic genus, the rachis occasionally bifurcates as in *Thinnfeldia*
and *Dichopteris*, and the ultimate segments of some species of
Odontopteris (fig. 366, A) are practically identical with those of
Thinnfeldia and *Ptilozamites*.

Odontopteris is probably a Pteridosperm. There is no
adequate reason for supposing that this group of plants which
played a prominent part in the Permo-Carboniferous floras was
no longer in existence during the Mesozoic era.

Odontopteris.

Brongniart[2] instituted the genus *Odontopteris* for compound
fronds from the Coal-Measures characterised by pinnules attached by the whole breadth of the base and traversed by
numerous forked veins. *Odontopteris* is very rare in British
Carboniferous rocks and "appears to be restricted to the Middle
and Upper Coal-Measures[3]."

[1] Saporta (73) A. Pl. XLVII.   [2] Brongniart (28) A. p. 60.
[3] Kidston (01²) p. 196.

Fronds large, bipinnate or tripinnate, the main rachis, which may be dichotomously branched, bears long linear pinnae with broadly linear or deltoid pinnules, acute or blunt, attached by the whole of the base; the lower margin of the lamina, which is usually entire and rarely lobed (e.g.

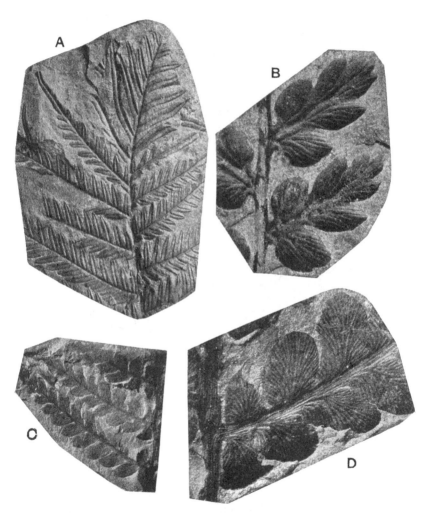

Fig. 364. A. *Alethopteris lonchitica* (Schloth.). $\frac{1}{2}$ nat. size.
            B. *Mariopteris muricata* (Schloth.). × 2.
            C. *Odontopteris* cf. *alpina* (Presl). $\frac{3}{5}$ nat. size.
            D. *O.* cf. *alpina*. Portion of pinna enlarged.
                  (A—D. From photographs by Dr Kidston.)

Odontopteris osmundaeformis)[1], is often decurrent on the axis of the pinna. The basal pinnule of each pinna is frequently attached by a contracted base, and the lamina may differ in form from that of the normal segments.

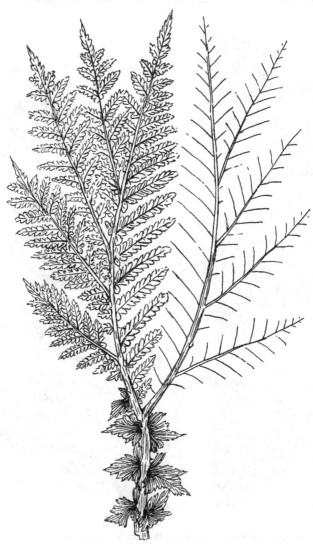

FIG. 365.   *Odontopteris minor*, Brongn.   (Rather less than ⅓ nat. size.   After Zeiller.)   [The pinnules are omitted in the right-hand branch.]

[1] Potonié (93) A. Pl. xv.

Pinnules often occur on the main rachis, and in some species the petiole bears modified pinnules which are larger than the ultimate segments of the pinnae and in some cases Cyclopteroid in shape. The pinnules are traversed by numerous dichotomously branched veins; if a midrib is present it dies out in the basal part of the lamina. In some species (genus *Mixoneura*) pinnules of the Neuropteroid type, characterised by a well-defined midrib, occur in association with typical Odontopteroid pinnules on the same pinna.

The species represented in fig. 364, C, D, from the Middle Coal-Measures of Barnsley, Yorkshire, illustrates the form and venation of the *Odontopteris* type of pinnule. Another species, *O. Reichiana* Gutb.[1], is also recorded by Kidston from the Lower Coal-Measures of Lancashire. Some unusually good specimens of the type-species of the genus *Odontopteris minor*, Brongn., have been figured by Zeiller[2] from the Coal-Measures of Blanzy (fig. 365) which show the dichotomy of the main axis and the occurrence of Aphlebiae on the petiole. The late Dr Weiss[3] divided *Odontopteris* into two sections, *Xenopteris* and *Mixo-neura*, the pinnules of the former having the form shown in fig. 364, D; while in species of the latter sub-genus some of the pinnules are identical in form and venation with those of *Neuropteris* except that they are attached by the whole breadth of the base. Zeiller[4] employs *Mixoneura* as a generic designation. In an American species *O. Wortheni* Lesq.[5] the pinnules bear numerous hairs like those on some species of *Neuropteris* (fig. 373, p. 570). The large size of the fronds of *Odontopteris* suggested to Weiss[6] that they were borne on the stems of tree-ferns, but Grand'Eury's[7] examination of specimens in the Coal-beds of central France led him to picture the plant as bearing a tuft of leaves on a short subterranean stem. Renault and Zeiller[8], on the other hand, obtained evidence in the Commentry Coalfield of fronds borne on elongated stems which grew on the ground and were supported by stronger plants. Stur[9] was the first to suggest that *Odontopteris* should be excluded from the

---

[1] Kidston (89) p. 409.
[2] Zeiller (06) Pls. xix.–xxii.; (00²) p. 100, fig. 73.   [3] Weiss, C. E. (70).
[4] Zeiller (06) p. 90.   [5] Lesquereux (80) A. p. 131; Weiss (70).
[6] Weiss (69) p. 37.   [7] Grand'Eury (77) A. Pl. A.
[8] Renault and Zeiller (88) A. p. 219.   [9] Stur (84).

ferns. Grand'Eury's[1] supposed fertile pinnules of *Odontopteris* do not afford any satisfactory evidence of the sporangial nature of the small swellings which he figures at the ends of the veins. This author pointed out several years ago that the petioles of some species of *Odontopteris* possess the anatomical features of *Myeloxylon*, a type of leaf-stalk which is now known to belong to Pteridosperms. In a recent paper Grand'Eury[2] records the association of *Odontopteris* fronds with small seeds (*Odontoptero-carpus*), a discovery which leaves little or no doubt as to the Pteridospermic nature of the genus. The fronds of *Odontopteris* are very similar in habit to those of *Neuropteris*, another Pteridospermic genus.

The similarity between some *Odontopteris* and *Thinnfeldia* leaves, to which attention has already been called, is well illustrated by *O. genuina* Grand'Eury[3], a pinnule of which is represented in fig. 366, A. *Odontopteris* is a fairly widespread genus in Upper Carboniferous and Lower Permian rocks, and is recorded also from Triassic strata: it is represented in the Coal-fields of North America and in several parts of Europe[4].

In some fronds included in *Odontopteris* the pinnae are characterised by a broad irregularly lobed lamina which also forms a winged border to the rachis. Examples of this form are afforded by *Odontopteris Browni* Sew.[5] from the Burghersdorp Series (Triassic?) of Cape Colony, and *O. Fischeri* described by Brongniart[6] from the Permian of Russia. The Russian species would perhaps be more appropriately placed in the genus *Callipteris*, as Weiss[7] suggests; the absence of venation in *O. Browni* renders generic identification unsatisfactory.

---

[1] Grand'Eury (77) A. Pl. xiii.   [2] Grand'Eury (08).
[3] Renault and Zeiller (88) A. Pl. xxiv.
[4] Weiss, C. E. (69); Goeppert (64) A.; Potonié (93) A, (04); Lesquereux (80) A., p. 124; White (99) p. 125.
[5] Seward (08) p. 97, Pl. viii.
[6] Brongniart, in Murchison, Verneuil, and Keyserling (45) Pl. A.
[7] Weiss, C. E. (70) p. 871.

Callipteris.

Brongniart[1] instituted this genus for certain species of supposed ferns previously referred to the genera *Pecopteris, Alethopteris,* and *Neuropteris. Callipteris* is a characteristic Permian plant which is almost certainly a Pteridosperm.

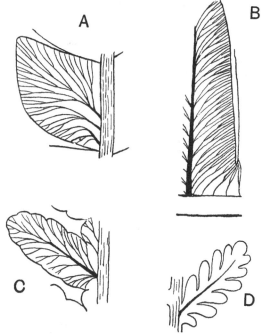

FIG. 366. A.   *Odontopteris genuina* (Grand'Eury). (× 2⅔. After Renault and Zeiller.)
B.   *Callipteridium gigas* (Gutb.). (× 2⅔. After Zeiller.)
C.   *Callipteris Pellati* (Zeill.). (× 1¾. After Zeiller.)
D.   *C. lyratifolia* (Goepp.). (× 1¾. After Zeiller.)

Zeiller has pointed out that such descriptions of fertile specimens as have been written are unsatisfactory. A few years ago, however, Grand'Eury[2] recorded the occurrence of seeds in association with Callipteris fronds in the Autun district, and in some cases they were found attached to the pinnae and rachis. The seeds are ovoid or spherical (5—10 mm. broad) and smaller than

[1] Brongniart (49) A. p. 24.   [2] Grand'Eury (06).

those of *Neuropteris*. The drawings of fertile segments published by Weiss[1] afford no indication of reproductive organs. Potonié[2] figures some pinnules of *Callipteris conferta* in which the thick lamina is covered with sinuous grooves probably made by some insect larvae: as he suggests, similar markings may have been mistaken for the remains of sori. The occurrence of *Callipteris* fronds recorded by Weber and Sterzel[3] in association with *Medullosa* stems in the Lower Permian of Saxony is in accordance with Grand'Eury's conclusion.

Fronds reaching 1 metre in length, bipinnate or tripinnate, main rachis frequently exhibiting a combination of dichotomous and pinnate branching. Pinnae linear, usually crowded, decurrent on the rachis ; the pinnules on the lower side of the pinnae are continued on to the rachis. Pinnules of the Pecopteroid type, entire or slightly lobed, or of the Sphenopteroid type and more or less deeply dissected (fig. 366 C, D), the lamina of adjacent pinnules concrescent ; on the lower pinnae the lamina may be continuous as in an Alethopteris pinnule. A midrib may extend almost to the bluntly rounded apex of the ultimate segments, giving off oblique, simple, or forked veins, the lowest of which arise directly from the rachis ; in the Sphenopteroid forms the lateral veins are given off at a more acute angle.

A striking feature of the genus is the occurrence of pinnules on the main rachis, as in *Odontopteris*. Zeiller has wisely extended the application of *Callipteris* to fronds possessing this character irrespective of the entire or lobed form of the ultimate segments. He found among the numerous examples of the genus obtained from Autun[4] and Lodève[5] transitional forms connecting such species as *C. conferta* (fig. 367) and *C. Pellati* Zeill. (fig. 366, C) in which the Pecopteroid pinnules are slightly lobed, with *C. lyratifolia* (Goepp.) (fig. 366, D), *C. flabellifera*[6] (Weiss), and *C. Bergeroni* Zeill. characterised by deeply lobed Sphenopteroid segments.

Callipteris conferta (Sternberg)[7].   Fig. 367.

1723.   Scheuchzer, Herb. Diluv. Pl. II., fig. 3.
1826.   *Neuropteris conferta*, Sternberg, Flor. Vorwelt p. 17.
1849.   *Callipteris conferta*, Brongniart, Tableau, p. 24.

[1] Weiss (69) Pls. VI. VII.   [2] Potonié (93) A. Pl. I. figs. 1, 2.
[3] Weber and Sterzel (96) p. 99.   [4] Zeiller (90) p. 84.   [5] Zeiller (98³).
[6] For figures of this and other species, see Potonié (07).
[7] For synonymy, see Zeiller (90) p. 87 and Potonié (07) p. 2.

This polymorphic species (fig. 367) is one of the most characteristic Permian plants. The oval-linear pinnules, attached by the whole base, occur on both pinnae and rachis; this feature, the thick texture of the lamina, and the linear, obliquely set, pinnae render the fronds easily recognisable. The fronds bore seeds.

FIG. 367. *Callipteris conferta.* From the Permian of Aschbach, Rhenish Prussia (British Museum, No. 39052).

In a recent account of some Permian plants from Germany, Schuster[1] refers a portion of a frond to *Callipteris conferta* (Sternberg) var. *polymorpha* Sterzel, which is characterised by unusually large and polymorphic pinnules. In size and shape the pinnules recall those of *Neuropteridium validum* Feist.

[1] Schuster (08) Pl. VIII. fig. 7.

Callipteridium.

The name *Callipteridium*, created by Weiss[1] as a sub-genus of *Odontopteris*, is applied by Zeiller and other authors to a few Upper Carboniferous and Permian species characterised by the occurrence of simply pinnate pinnae on the main rachis between the bipinnate primary pinnae. Single pinnules are borne directly on the rachis of the primary pinnae between the pinnate branches. The form and venation of a typical pinnule are shown in fig. 366, B. *Callipteridium pteridium*, originally recorded by Schlotheim as *Filicites pteridius*[2], has been fully described by Renault and Zeiller from unusually large specimens found in the Commentry Coal-field[3]. This species illustrates the peculiar morphological features of the genus. The main rachis of the tripinnate fronds, several metres long, shows a combination of dichotomous and pinnate branching; from the zigzag and forked axis are given off bipinnate pinnae and, between these, shorter pinnate branches. The pinnules closely resemble those of *Callipteris conferta* but reach a greater length; the pinnules borne on the rachises of the lateral branches differ from the others in their broader base and more triangular lamina.

No fertile specimens have been found. It is probable that *Callipteridium* was not a true fern, and that White[4] is correct in including it among the Pteridosperms.

Archaeopteris.

In 1852 Forbes[5] published a brief description of some supposed fern fronds, found by the Geological Surveyors of Ireland in Upper Devonian rocks of Kilkenny, under the name *Cyclopteris hibernica*. The Irish specimens were more fully described by Baily[6] in 1858. Fronds of the same type were referred by other authors to *Cyclopteris*, *Adiantites* or *Noeggerathia*, until Schimper[7] proposed the generic name *Palaeopteris* on the ground that the fronds described by Forbes and

[1] Weiss, C. E. (70).   [2] Schlotheim (20) A. p. 406.
[3] Renault and Zeiller (88) A. Pl. xix.   [4] White (05²) p. 388.
[5] Forbes (53) p. 43.   [6] Baily (59) p. 75.   [7] Schimper (69) A. p. 473.

Baily are distinguished by the nature of their fertile pinnae
from the sterile leaves included in Brongniart's provisional
genus *Cyclopteris*. The earlier use of *Palaeopteris* by Geinitz

Fig. 368. *Archaeopteris hibernica*. (From a specimen in the Science and Art
Museum, Dublin. Rather less than ⅙ nat. size.)

for an entirely different plant led Dawson[1] to institute the genus *Archaeopteris*. The genus *Archaeopteris* may be defined as follows:

Fronds bipinnate, reaching a considerable length (90 cm.); the stout rachis bears long linear pinnae; sterile pinnules obovate or cuneate with an entire, lobed, fimbriate, or laciniate lamina traversed by divergent dichotomously branched veins. The fertile pinnae usually occur on the lower part of the rachis; pinnules with a much reduced lamina bear numerous fusiform or oval exannulate sporangia (fig. 369, A, E, H), sessile or shortly stalked, singly, or in groups of two or three. The base of the petiole is characterised by a pair of partially adnate stipules (fig. 369, C, D), and single pinnules or scales occur in some species on the rachis between the pinnae and on the petiole.

Archaeopteris hibernica (Forbes).   Figs. 368, 369, A—C.

The specimen from Kilkenny represented in fig. 368 has a length of over 80 cm. The upper pinnae bear numerous imbricate obovate pinnules (fig. 369, A, B) with an entire or very slightly fimbriate margin, while on the shorter lower pinnae the ultimate segments are reduced to a slender axis bearing numerous fusiform sporangia, 2—3 mm. in length. Kidston[2] has pointed out that sporangia occasionally occur on the edge of ordinary pinnules, and he first recognised the stipular nature of the scale-like appendages which Baily noticed on the swollen petiole base (5 cm. broad) of the Irish species (fig. 369, C). Restorations of *Archaeopteris hibernica* have been figured by Baily[3] and by Carruthers[4], but the description of the fertile pinnae by the latter author requires modification in the light of Kidston's description of the Dublin specimens.

Archaeopteris is recorded from Upper Devonian rocks of the South of Ireland, Belgium, Germany, Southern Russia, Bear Island, and Ellesmere Land in the Arctic regions, Canada, Pennsylvania, and elsewhere. Many of the specimens described under different names bear a close resemblance, which in some cases probably amounts to specific identity, to *A. hibernica*.

---

[1] Dawson (71) A. p. 48; (82).
[2] Kidston (91[2]) p. 30, Pl. III.; (06) p. 434.
[3] Baily (75) Pl. XXVIII.   [4] Carruthers (72[2]) Pl. II.

A. Jacksoni originally described by Dawson[1] and more recently by Smith and White[2] from Devonian rocks of Maine, the Canadian type *A. gaspiensis* Daws., and some species figured by Lesquereux[3] from Pennsylvania, are examples of forms which present a striking similarity in habit to the Irish species. The Belgian Devonian fossils named by Crépin[4] *Palaeopteris hibernica* var. *minor* are regarded by him as probably identical with Goeppert's species *Cyclopteris Roemeriana* from the neighbourhood of Aachen. Heer recorded *Archaeopteris Roemeriana* from Upper Devonian beds in Bear Island, and Nathorst[5], who has published a more complete account of the Arctic forms, draws attention to the resemblance of some of them to *A. hibernica*. A species described by Schmalhausen[6] from the Upper Devonian of Southern Russia as *A. archetypus* (fig. 369, D) appears to differ from *A. hibernica* in the slightly less reduced lamina of the fertile segments. This species has been more adequately illustrated by Nathorst[7] from material collected in Ellesmere Land: he is unable to confirm Schmalhausen's statement that the pinnae are spirally disposed.

The species *A. fimbriata* (fig. 369, G) described by Nathorst from Bear Island is characterised by the more deeply dissected lamina of the sterile pinnules. In *A. fissilis* Schmal. from Russia and Ellesmere Land the lamina (fig. 369, E, F) is cut up into filiform segments: a fertile pinnule of this species is represented in fig. 369, E.

Some sterile impressions figured by Krasser[8] from Palaeozoic strata (Lower Carboniferous or Upper Devonian ?) in the province of Nanshan in China as *Noeggerathia acuminifissa* are considered by Zeiller[9] to be portions of an *Archaeopteris* or *Rhacopteris* frond. The resemblance to the former genus is however by no means close enough to warrant a reference to *Archaeopteris*. The sterile specimens described by Stur[10] from the Culm of Altendorf as species of *Archaeopteris* are

---

[1] Dawson (71) A. [2] Smith and White (05) p. 39.
[3] Lesquereux (80) A. [4] Crépin (74).
[5] Nathorst (02). [6] Schmalhausen (94). [7] Nathorst (04).
[8] Krasser (00) Pl. I. figs. 3—7. [9] Zeiller (03²) p. 27.
[10] Stur (75) A. Pls. VIII. XII. XVI.

probably not generically identical with the Irish and Arctic
species.   The dichotomous branching of the rachis in *A. Tscher-
maki* and *A. Dawsoni* is a feature unknown in *Archaeopteris*.

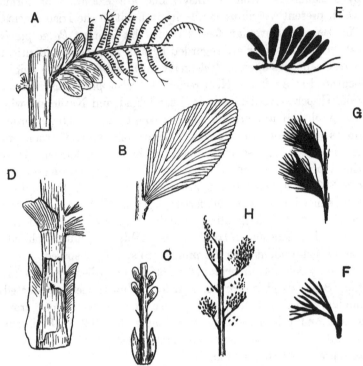

FIG. 369. A. *Archaeopteris hibernica*. Fertile pinna. Dublin Geological
Survey Museum. (Reduced. After Kidston.)
B. *A. hibernica*. Pinnule. (Slightly enlarged. After Carruthers.)
C. *A. hibernica*. Base of petiole. (Dublin Museum. After Kidston.)
D. *A. archetypus*. Base of petiole : Ellesmere Land. (After
Nathorst. ⅔ nat. size.)
E. *A. fissilis*. Sporangia. (Slightly enlarged. After Schmal-
hausen.)
F. *A. fissilis*. Sterile pinnule. Ellesmere Land. (Slightly en-
larged. After Nathorst.)
G. *A. fimbriata*. Bear Island. (After Nathorst. ⅔ nat. size.)
H. *Archaeopteris sp*. Ellesmere Land. (After Nathorst. ⅔ nat.
size.)

In the absence of fertile pinnae the separation of *Archaeopteris*
from *Rhacopteris* is by no means easy.

Archaeopteris was regarded by Carruthers as a fern closely allied to recent species of Hymenophyllaceae, but this conclusion was based upon an interpretation of the fertile segments which Kidston[1] has shown to be incorrect. The latter author regarded the presence of stipules and the structure of the exannulate sporangia as evidence of a Marattiaceous alliance. In a later reference to *Archaeopteris*, Kidston expresses the opinion that the genus is not a true fern but a member of the Cycadofilices or Pteridosperms, a view shared by Grand'Eury[2] and doubtless by many other palaeobotanists. The sporangia of *Archaeopteris* appear to be of the same type as those of *Dactylotheca* (fig. 290, E, p. 399). Schmalhausen gave expression to his disagreement with Nathorst and other authors who referred *Archaeopteris* to the Marattiaceae by proposing the distinctive group-name Archaeopterideae.

There can be little doubt that the reproductive organs of *Archaeopteris* so far discovered are microsporangia, and that the plant bore seeds. The sporangia are larger than those of any known fern and, as Kidston points out, they are similar to those of *Crossotheca* which he has shown to be microsporangia of the Pteridosperm *Lyginodendron*. The presence of stipules in *Archaeopteris hibernica*, *A. fimbriata*, *A. archetypus* (fig. 369, D) and probably throughout the genus does not materially affect the question of taxonomic position. Stipules are a characteristic feature of Marattiaceae and, in a reduced form, of Osmundaceae, but similar appendages are borne at the base of the petiole of the Cycad *Ceratozamia*. The occurrence of Aphlebiae on the rachis of *Archaeopteris* is a feature shared by the fronds of *Neuropteris* and other Pteridosperms.

Neuropteris.

The fronds for which Brongniart[3] created this genus, though suspected by Stur in 1883 as wrongly classed among the ferns, have only recently been shown to be the leaves of Pteridosperms. As yet only one case is recorded in which

---

[1] Kidston (88²).   [2] Grand'Eury (08).   [3] Brongniart (22) A.

Neuropteris pinnae occur in organic connexion with seeds[1], but it is almost certain that the genus as a whole must be

FIG. 370.   *Neuropteris* frond with *Cyclopteris* leaflets.   English Coal-Measures.
(From a block given to me by Mr Carruthers.   A.C.S.)

[1] Kidston (05[2]).

placed in this generalised group. Renault[1] pointed out that the petioles of Neuropteris fronds from Autun had the anatomical features of *Myeloxylon* (petiole of *Medullosa*). Since Kidston's important discovery of seed-bearing pinnae of *N. heterophylla*, Grand'Eury[2] has recorded the association of Neuropteris fronds with seeds in French Coal-fields. By some of the older authors *Neuropteris* was compared with *Osmunda* because of a similarity in venation. In the frequent dichotomy of the frond and in the occurrence of pinnules on the rachis, *Neuropteris* closely resembles *Odontopteris*[3]: there can be little doubt as to the close relationship of the Pteridosperms possessing these two types of foliage. *Neuropteris* may be defined as follows:

Fronds reaching a considerable size, probably in some cases a length of 10 metres[4]; bi- or tri-pinnate; the rachis may be dichotomously branched (figs. 354, D; 370); both rachis and petiole bear single pinnules, those on the latter frequently differ from the normal leaflets in their larger Cyclopteroid laminae (fig. 370). Pinnules entire, rarely slightly lobed, broadly linear, attached by a small portion of the base, which is usually more or less cordate. In *N. Grangeri* Brongn. the pinnules are attached by a short pedicel[5]. The midrib always dies out before reaching the blunt or pointed apex of the lamina and gives off at an acute angle numerous secondary veins characterised by their arched course and repeated forking.

Potonié describes the secondary veins of the pinnules of *Neuropteris pseudogigantea*[6] as occasionally anastomosing, a feature which may be regarded as a step towards the reticulate venation of the closely allied genus *Linopteris*.

Renault[7] described some petrified pinnules of *Neuropteris* in which the mesophyll shows a differentiation into upper palisade tissue and lacunar tissue below; the lower epidermis is infolded at intervals where grooves (probably stomatal) occur like those on the leaves of an Oleander (*Nerium oleander*).

The rachises of Neuropteris fronds are described by Grand'-Eury under the generic name *Aulacopteris*[8].

---

[1] Renault (76). [2] Grand'Eury (08). [3] White (99) p. 128.
[4] Grand'Eury (77) A. p. 122. [5] Zeiller (90) Pl. xi. fig. 6.
[6] Potonié (99) p. 113. [7] Renault (82) A. Vol. iii.; Zeiller (90) p. 139.
[8] Grand'Eury (77) A. p. 105.

Neuropteris heterophylla, Brongniart[1].   Figs. 354, E; 371.

This species is characteristic of the Lower Coal-Measures of Britain; it occurs also in the Middle Coal-Measures and is a common type in Upper Carboniferous rocks in various parts of the world.   The fronds are large and tripinnate, the rachis

Fig. 371.   *Neuropteris heterophylla*. From a specimen in the Manchester Museum. ½ nat. size. M.S.

is often dichotomously branched and Cyclopteroid pinnules may occur on the petiole. The pinnules, 5—20 mm. in length and 3—8 mm. broad, have a rounded apex (fig. 354, E, p. 535).

As shown in fig. 371 which represents a primary pinna, the

[1] Brongniart (22) A. Pl. II. fig. 6. For synonymy, see Kidston (03) p. 773; Zeiller (88) A. p. 261.

small pinnules on the lower branches are gradually replaced in the upper portion of the specimen by falcate segments.

Neuropteris macrophylla, Brongniart[1]. Figs. 354, D, D′ ; 372.

The rachis of the large fronds of this species illustrates the dichotomous habit of many Neuropteris fronds, also the occurrence on the petiole of large Cyclopteroid pinnules (cf. fig. 370). The small piece of a pinna reproduced in fig. 372 shows the slender attachment of the segments, the blunt apex, and the

Fig. 372. *Neuropteris macrophylla*, Brongn.
From a photograph by Mr Hemingway.

Neuropteroid venation. Single pinnules of this species may be distinguished from those of *N Scheuchzeri* by the blunter apex, the absence of the pair of small Cyclopteroid pinnules on the same branch and by the absence of hairs. *N. macrophylla* is characteristic of the Upper Coal-Measures of Britain.

[1] For synonymy, see Kidston (88) p. 354.

Neuropteris Scheuchzeri, Hoffmann.   Figs. 354, F ; 373.

Fragments of this well-known Coal-Measure species were
figured by Scheuchzer in his *Herbarium Diluvianum*[1] as *Lithos-
munda minor,* and by Lhywd (Luidius[2]) as *Phyllites mineralis* as
early as 1760. *Neuropteris Scheuchzeri,* so named by Hoffmann
in 1826, is a type which many authors have described under
different names. Lesquereux[3] figured it as *N. hirsuta* from the
Coal-fields of Pennsylvania, and under the same name it is
recorded by Fontaine and White[4] from Permian rocks of
Virginia. The oval patches on the surface of a pinnule described by these authors as sori are certainly not of that nature.
The same species is described by Bunbury[5] from Nova Scotia

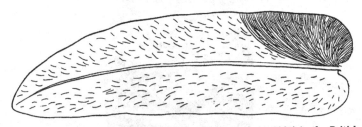

FIG. 373. *Neuropteris Scheuchzeri.* From a specimen (v. 2009) in the British
        Museum. ¾ nat. size.

as *N. cordata* Brongn. var. *angustifolia.* For a full synonymy
of the species reference should be made to lists published by
Kidston[6], White[7], and Zeiller[8].

The large tripinnate fronds are characterised by the long
linear- or oval-lanceolate pinnules (fig. 373)[9] with a pointed
apex and numerous bristle-like hairs on the lamina ; two much
smaller Cyclopteroid segments occur at the base of the pinnae
which are terminated by the linear leaflets (fig. 354, F, p. 535).
Neuropteris Scheuchzeri is characteristic of the Upper and
Middle Coal-Measures of Britain and is recorded from several

[1] Scheuchzer (1723) A. p. 129, Pl. x. fig. 3.
[2] Lhywd (1760) A. Pl. v. fig. 190.
[3] Lesquereux (79) A. Pl. VIII.   [4] Fontaine and White (80) p. 47.
[5] Bunbury (47) Pl. XXI.   [6] Kidston (94) p. 357; (03) p. 806.
[7] White (99) p. 132.   [8] Zeiller (88) A. p. 251.   [9] See Vol. I. p. 45.

localities in North America and the Continent. Zalessky[1] has recently recorded the species from the Coal-Measures of Donetz. The frequent occurrence of detached pinnules points to a caducous habit. Even single leaflets can, however, be identified by their large size, the pointed apex, and hairy lamina. The hairs are preserved as fine oblique lines simulating veins; they were so described by Roemer[2] who took them for cross-connexions between the secondary veins and referred the pinnules to Gutbier's genus *Dictyopteris*.

Another example of *Neuropteris* with hairy pinnules is described from the Commentry Coal-field by Renault and Zeiller as *N. horrida*[3]. The oval-linear, bluntly rounded, pinnules are characterised by a median band of hairs on each surface and a narrower strip at the edge of the lamina.

Cyclopteris.

This generic name was created by Brongniart in 1828[4] for specimens which he believed to be complete single leaves of orbicular or reniform shape similar to those of *Trichomanes reniforme*. The lamina is traversed by numerous dichotomously branched veins which spread from the centre of the base.

It was suspected by Lindley and Hutton[5] that certain Cyclopteris leaves belonged to the frond of a species of *Neuropteris*, and some years later Lesquereux[6] concluded that Brongniart's genus was founded on orbicular leaflets of *Neuropteris*. In 1869 Roehl[7] figured a specimen of *Neuropteris* bearing Cyclopteroid pinnules on its rachis. It is now universally admitted that *Cyclopteris* is not a distinct genus and that the specimens so named were borne as modified pinnules on the main rachis of *Neuropteris* and *Odontopteris*. It is, however, convenient to retain the name for detached leaflets which cannot be referred to the fronds on which they were borne. A specimen found by Mr Hemingway in the Upper Coal-Measures

---

[1] Zalessky (07) Pl. xxiv. fig. 5.   [2] Zeiller (88) A. p. 251.
[3] Renault and Zeiller (88) A. p. 251, Pl. xxxii. [4] Brongniart (28) A. p. 51.
[5] Lindley and Hutton (33) A. p. 28.
[6] Lesquereux (66) A.   [7] Roehl (69).

of Yorkshire and described in 1888[1] affords a striking example
of the large size attained by what was probably a frond of
Neuropteris. The piece of main rachis reached a length of
over 120 cm. and bore five pairs of Cyclopteris pinnules, some
of which were 7 cm. long and 5 cm. broad. The complete
frond must have reached a length of at least 4 metres. Fig. 370
shows some typical Cyclopteroid leaflets on the petiole of a
Neuropteris frond.

Linopteris.

The Upper Palaeozoic fronds included in this genus are
more familiar as species of *Dictyopteris*. Potonié[2] has, however,
pointed out that the creation of this name by Lamouroux in
1809 for a genus of Brown Algae which is still retained, makes
it advisable to fall back upon the designation *Linopteris*.
Gutbier[3] proposed the genus *Dictyopteris* in 1835: *Linopteris*
was first used by Presl[4] in 1838. The fronds so named are
identical with species of *Neuropteris* except in the anastomosis
of the secondary veins; *Linopteris* bears to *Neuropteris* the
same relation as *Lonchopteris* bears to *Alethopteris*. As in
Neuropteris, Cyclopteroid pinnules occur on the petioles of *Linopteris*, but the veins form a fine reticulum. Grand'Eury[5] records
the association of *Linopteris Brongniarti* with seeds belonging
to the genus *Hexagonocarpon*, a fact which points to the
Pteridosperm nature of the foliage.

Some fertile pinnules of *Linopteris Schutzei* (Roemer) are
described by Zeiller[6] from Autun as bearing on the under
surface of the lamina two rows of long and pointed sporangia,
probably united in groups. The presumption is that these are
microsporangia.

Fig. 374 is a reproduction of a careful drawing, originally
published by Zeiller[7], of a pinnule of the type-specimen of
Gutbier's species *Linopteris neuropteroides*. This species differs
from *Linopteris obliqua*, instituted by Bunbury[8] for specimens

[1] Seward (88).   [2] Potonié (99) p. 153 (note).   [3] Gutbier (35).
[4] Presl, in Sternberg (38) A.   [5] Grand'Eury (04).
[6] Zeiller (90) Pl. xi. fig. 9.
[7] Zeiller (99) p. 46.   [8] Bunbury (47) A. p. 427.

obtained by Lyell[1] from the Coal-Measures of Nova Scotia, in the smaller size of the meshes. *Linopteris obliqua* occurs in the Upper and Middle Coal-Measures of Britain; it is recorded by Zeiller from Asia Minor, by Lesquereux[2] from Pennsylvania, and by other authors from several European localities. The pinnules frequently occur detached from the frond and like those of some species of *Neuropteris* were caducous. *Linopteris* is rare in British strata.

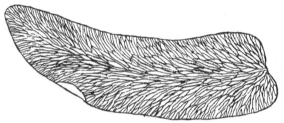

FIG. 374. *Linopteris neuropteroides*, Gutb. (Pinnule of type-specimen. Enlarged. After Zeiller.)

Alethopteris.

The name *Alethopteris*, instituted by Sternberg[3], is applied to compound fronds often reaching a considerable size, exhibiting the following features:

The linear pinnules are attached by the whole breadth of the base, with the lower edge of the lamina decurrent and usually continuous with that of the next pinnule (figs. 290, A, p. 399; 375). The ultimate segments are entire, with an acute or rounded apex and often characterised by a fairly thick lamina convex on the upper surface. From a prominent midrib, continued to the apex of the pinnule, numerous simple and forked secondary veins are given off at a wide angle, the decurrent portion of the lamina being supplied by veins direct from the axis of the pinna. In the upper part of a frond or primary pinna the pinnules may be replaced by a continuous, lobed, or entire simple lamina. The main rachis occasionally exhibits dichotomous branching, but the fronds are for the most part constructed on the pinnate plan. Single Cyclopteroid pinnules[4] occur on the petiole of some species of the genus.

[1] Lyell (45) A. Vol. II. p. 202.   [2] Lesquereux (80) A. p. 146.
[3] Sternberg (26) A.   [4] Grand'Eury (04).

In certain species of *Alethopteris* the pinnules appear to have been deciduous as in *Didymochlaena* among recent ferns[1]. A piece of cuticle from the upper surface of a pinnule of *Alethopteris Grandini* (Brongn.) figured by Zeiller[2] shows very clearly the polygonal form and straight walls of most of the epidermal cells, those above the veins being almost rectangular. The position of the sunken stomata is revealed by small circular spaces surrounded by a circle of cells.

The absence of fertile specimens of this common genus of Upper Carboniferous plants led Stur[3] to exclude it from the ferns. Although no seeds have so far been found in organic connexion with an Alethopteris frond, it is certain that some species, probably all, represent the foliage of Pteridosperms. Renault was the first to describe petrified specimens of *Aletho-pteris* fronds exhibiting the anatomical structure of *Myeloxylon* (leaf-axis of *Medullosa*). The calcareous nodules from English Coal-seams contain numerous fragments of the Myeloxylon type of rachis bearing Alethopteroid pinnules.

The constant association of the fronds of *Alethopteris lonchitica* and *Trigonocarpon* seeds noticed by Mr Hemingway in the Coal-Measures of Yorkshire led him to regard the species as seed-bearing: it has since been recognised as the foliage of the Pteridosperm *Medullosa anglica*[4].

Grand'Eury[5] has recorded the association in French Coal-fields of species of *Alethopteris* with *Trigonocarpon* and *Pachy-testa* seeds.

Alethopteris lonchitica (Schlotheim)[6]. Figs. 364, A; 290, A.

This species, described by Schlotheim in 1820 as *Filicites lonchiticus* and previously figured by Scheuchzer[7], is abundant in the Middle and Lower Coal-Measures of Britain[8]. It is characterised by large tripinnate fronds, probably quadripinnate in the lower part, bearing primary pinnae of a more or less

[1] Grand'Eury (90) A.   [2] Zeiller (90) Pl. ix. fig. 6, A.   [3] Stur (83).
[4] Scott (07) p. 206 ; Scott and Maslen (06) p. 112.
[5] Grand'Eury (04).
[6] For synonymy, see Kidston (03) p. 772: Zeiller (88) A.
[7] Scheuchzer (1723) A. Pl. i. fig. 4.   [8] Kidston (94) p. 245.

triangular form divided into pinnate branches replaced in the apical region by linear segments. The pinnules, 8—30 mm. long and 3—5 broad, are linear- or oval-lanceolate with an obtuse apex; the upper margin of the lamina is slightly contracted at the base, while the lower edge is decurrent.

FIG. 375. *Alethopteris Serlii* (Brongn.). From a specimen in the York Museum. ¾ nat. size.

Alethopteris Serlii (Brongniart)[1].　　Fig. 375.

This species, figured by Parkinson in 1811, closely resembles *A. lonchitica*, but is distinguished by the more crowded and relatively longer pinnules which are joined to one another by a narrow connecting lamina (Fig. 375). The secondary veins in *A. Serlii* are rather finer and more numerous. Grand'-Eury[2] records the association of the seed *Pachytesta* with fronds of this species in the Coal-Measures of St Étienne.

---

[1] For synonymy, see Kidston (94) p. 596; (03) p. 806; White (99) p. 117.
[2] Grand'Eury (04).

A. Serlii is very abundant in the Upper Coal-Measures but rare in the Middle Coal-Measures of Britain[1].

Lonchopteris.

This name was proposed by Brongniart[2] for sterile fronds from Upper Carboniferous rocks which are practically identical with species of *Alethopteris*, but differ in the reticulate venation of the pinnules. It has been pointed out in a previous chapter[3] that *Lonchopteris* is usually used for Palaeozoic species, the Wealden leaves, which were placed in this genus by Brongniart, being transferred to *Weichselia*.

There can be little doubt as to the close relationship of *Lonchopteris* with *Alethopteris*: both may be referred to the Pteridosperms. *Lonchopteris rugosa* Brongn.[4] (fig. 290, B, p. 399) and *L. Bricei* Brongn., both British species, are fairly common in Upper Carboniferous strata. In *L. rugosa*, a Middle Coal-Measures species, the anastomosing secondary veins form polygonal meshes (fig. 290, B, p. 399) smaller than those of *L. Bricei*.

Pecopteris.

Reference has already been made to this genus in the chapter on Marratiales, so far as regards certain species of fertile fronds the sporangia of which resemble those of recent Marattiaceae. It is, however, by no means safe to assume that such Pecopteris fronds were borne on stems having the anatomical characters of ferns. The sporangia in some at least of the species may have contained microspores. In one Upper Carboniferous species usually referred to Pecopteris, *P. Pluckeneti,* Schlot., Grand'Eury[5] has recorded the occurrence of seeds on the pinnules of the ordinary fronds. This species will be referred to in Volume III. The substitution of such generic names as *Ptychocarpus, Asterotheca, Hawlea, Dactylotheca* and others for the purely provisional designation *Pecopteris* indicates a step towards a conclusion as to natural affinity The probability is that *Pecopteris,* as applied to Palaeozoic

---

[1] Kidston (94) p. 245.   [2] Brongniart (28) A. p. 59.
[3] Page 494.   [4] Kidston (94) p. 596.   [5] Grand'Eury (05).

species, in many cases stands for the compound fronds of true ferns, but the possibility of the inclusion of those of Pteridosperms in the same category is by no means excluded. The designation *Pecopteris* may conveniently be retained for sterile bipinnate, tripinnate, or quadripinnate fronds bearing pinnules having the following characteristics:

Lamina short, attached to the rachis by the whole of the base and at a wide angle, with the edges parallel or slightly converging towards the usually blunt apex; adjacent pinnules may be continuous basally by a narrow lamina. A well-marked midrib extends to the apex and gives off simple or forked lateral veins almost at right angles (fig. 352, D, p. 529).

Hydathodes like those on the leaflets of *Polypodium vulgare* and other recent ferns[1] are occasionally seen at the ends of the lateral veins of Pecopteris pinnules.

In addition to the examples of Palaeozoic fronds with the *Pecopteris* form of pinnule referred to in chapter XXII., the species *Pecopteris arborescens* may be briefly described.

Pecopteris arborescens (Schlotheim)[2]. Figs. 352, D : 376.

The species named by Schlotheim *Filicites arborescens* in 1804 is characteristic of the Upper Coal-Measures and is recorded also from Permian strata[3].

Fronds large; the rachis, which may reach a breadth of 3 cm.[4], gives off long ovoid-lanceolate pinnae in two alternate rows (fig. 376); pinnules small, 1·5—4 mm. long and 1—2 mm. broad, contiguous, with rounded apex, attached approximately at right angles; the upper surface of the lamina is slightly convex and may be hairy[5]. The fertile pinnules, identical in shape with the sterile, bear groups of ovoid exannulate sporangia (synangia). The midrib extends to the apex of the pinnule and gives off simple veins at a wide angle (fig. 352, D).

Our knowledge of the reproductive organs is very meagre. Grand'Eury described the synangia as consisting of 3—5 sporangia borne on a central receptacle; sporangia have been described also by Stur[6], Renault and Zeiller[7], and Potonié[8],

---

[1] Potonié (92²); (93) p. 54.   [2] For synonymy, see Kidston (88) p. 366.
[3] Zeiller (90) p. 45; Potonié (93) A. p. 57.
[4] Germar (44) Pls. xxxv. xxxvi.   [5] Kidston (88) p. 366.
[6] Stur (83).   [7] Renault and Zeiller (88) A. p. 196.
[8] Potonié (93) A. p. 48.

but no fertile British specimens are recorded.   Stur places this
species in the genus *Scolecopteris*, and Potonié regards the
sporangia found by him on Permian fronds, which may be
identical with *Pecopteris arborescens*, as conforming to those of
the *Asterotheca* type.   It is impossible to decide on the evidence available whether this species is a Pteridosperm or a
fern, but there is a natural inclination in doubtful cases to give
preference to the first of these two choices.

Fig. 376.   *Pecopteris arborescens* (Schloth.).   From the Upper Coal-Measures of
Radstock.   From a photograph by Dr Kidston.   Reduced.

The numerous fronds from Carboniferous and Permian rocks
described as species of *Pecopteris* exhibit a considerable range
of variation in the form of the pinnules.   In many species the
pinnules are of the type represented in fig. 352, D; in others
the lamina of the ultimate segments is slightly contracted at
the base and the secondary veins are given off at a more acute

angle, as in *Pecopteris polymorpha*, Brongn.[1] In *Pecopteris unita*, Brongn., already described as *Ptychocarpus unita*[2], the pinnules are joined together except in the apical region. Some fronds included in *Pecopteris* possess pinnules in which Pecopteroid and Sphenopteroid features are combined; *P. Sterzeli*, Zeill.[3] and *P. Pluckeneti*, Schlot. are examples of fronds in which the pinnules are lobed as in *Sphenopteris*, but the base of the lamina is only slightly contracted and the venation is not that of typical Sphenopteris species.

The species to which Potonié has applied the generic name *Alloiopteris*[4] also illustrates the impossibility of drawing a sharp line between *Pecopteris* and *Sphenopteris*. The fronds already described in chapter XXV. under the designation *Corynepteris* bear pinnules with a contracted base; in some species the lamina is lobed, but in others (fig. 354, G) it is entire with a midrib nearer one edge than the other. The species which Potonié assigns to *Alloiopteris*, like many other Sphenopteroid and Pecopteroid fronds, are characterised by the occurrence of an abnormal pinnule (aphlebia) at the base of each pinna (fig. 354, G, p. 535). Young fronds of *Pecopteris* are occasionally met with showing very clearly the circinate vernation of the pinnae as in the leaves of *Cycas* and *Angiopteris* represented in fig. 220, p. 283. The genus *Spiropteris* was created by Schimper[5] for coiled unexpanded fronds of fossil ferns; it is however superfluous to apply a distinctive term to specimens of this kind.

The designation *Pecopteris* is employed chiefly for leaves of Palaeozoic age which are unknown in the fertile state, or do not afford sufficient evidence as to the nature of the sporangia to justify the substitution of a special generic name. Many Mesozoic species have also been referred to *Pecopteris*, but most of these are more appropriately included in Brongniart's later genus *Cladophlebis*. The pinnules of *Cladophlebis*, as Brongniart pointed out, are intermediate between *Pecopteris* and *Neuropteris*; they are usually attached by the whole breadth of

---

[1] Zeiller (00[2]) p. 88.   [2] Page 397.
[3] Renault and Zeiller (88) A. p. 178, Pls. v.—viii. *Ante*, p. 419.
[4] Potonié (02).   [5] Schimper (69) A. p. 688.

the base, as in *Pecopteris*, but the more acute origin, more
arched form, and more frequent dichotomy of the lateral veins
are features shared by *Neuropteris*. As a rule, Mesozoic sterile
fronds with straight or folded, entire or dentate pinnules are
of the Cladophlebis type: this genus is especially characteristic
of Rhaetic and Jurassic floras. Examples of *Cladophlebis*
pinnules are shown in figs. 256, 257 (pp. 340, 342). It is to
be regretted that authors do not make more use of the generic
name *Cladophlebis* in describing sterile fronds, instead of
following the misleading and unscientific practice of employing
such genera as *Pteris*, *Asplenites*, and others on wholly insufficient grounds.

LIST OF WORKS REFERRED TO IN THE TEXT.

Amalitzky, W. (01) Sur la découverte, dans les dépôts permiens supérieurs du Nord de la Russie, d'une flore Glossoptérienne &c. *Compt. Rend.* vol. CXXXII. p. 591.

Arber, E. A. N. (02) On the distribution of the Glossopteris Flora. *Geol. Mag.* vol. IX. p. 346.

—— (02²) The Clarke collection of fossil plants from New South Wales. *Quart. Journ. Geol. Soc.* vol. LVIII. p. 1.

—— (03) Notes on some fossil plants collected by Mr Molyneux in Rhodesia. *Quart. Journ. Geol. Soc.* vol. LIX. p. 288.

—— (05) Catalogue of the fossil plants of the Glossopteris Flora in the British Museum. A monograph of the Permo-Carboniferous Flora of India and the Southern Hemisphere. *London.*

—— (05²) The Sporangium-like organs of *Glossopteris Browniana*, Brongn. *Quart. Journ. Geol. Soc.* vol. LXI. p. 324.

—— (05³) The fossil flora of the Culm-measures of N.W. Devon &c. *Phil. Trans. R. Soc.* vol. CXCVII. p. 291.

—— (06) On the past history of Ferns. *Annals Bot.* vol. XX. p. 215.

Arber, E. A. N. and H. H. Thomas. (08) On the structure of *Sigillaria scutellata*, Brongn. and other Eusigillarian stems, in comparison with those of other Palaeozoic Lycopods. *Phil. Trans. R. Soc.* vol. CC. p. 133.

Armour, H. M. (07) On the Sorus of *Dipteris. New Phytologist*, vol. VI. p. 238.

Atkinson, G. F. (94) The study of the biology of Ferns, by the collodion method. *New York.*

Auerbach, J. and H. Trautschold. (60) Ueber die Kohlen von Central Russland. *Nouv. Mém., Soc. Imp. Naturalistes Moscou,* vol. XIII.

Bäsecke, P. (08) Beiträge zur Kenntniss der physiologischen Schneiden der Achsen und Wedel der Filicinen &c. *Bot. Zeit.* Heft II–IV. Abt. I.

Baily, W. H. (59) On the fructification of *Cyclopteris hibernica* (Forbes), from the Upper Devonian or Lower Carboniferous strata at Kiltorkan Hill, County Kilkenny. *Brit. Assoc. Rep. London,* 1859, p. 75. (*Leeds meeting*, 1858.)

—— (60) On *Corynepteris*, a new generic form of fossil fern. *Journ. Geol. Soc. Dublin,* vol. VIII. p. 237.

—— (75) Figures of characteristic British fossils. *London.*

Baker, J. G. (67) Descriptions of six new species of simple fronded Hymenophyllaceae. *Journ. Linn. Soc.* vol. IX. p. 335.

—— (68) On the geographical distribution of Ferns. *Trans. Linn. Soc.* vol. XXVI. p. 305.

—— (68) See Hooker, Sir W. J. and J. G. Baker.

—— (88) On a further collection of ferns from West Borneo. *Journ. Linn. Soc.* vol. XXIV. p. 256.

Balfour, J. H. (57) On the occurrence in coal of peculiar vegetable organisms resembling the Sporangia of *Lycopodium*. *Trans. R. Soc. Edin.* vol. XXI. p. 187.

Barber, C. A. (89) On the structure and development of the bulb in *Laminaria bulbosa*, Lamour. *Annals Bot.* vol. III. p. 41.

Barrois, C. (04) Sur les Spirorbes du Terrain Houiller de Bruay. (Pas de Calais.) *Ann. Soc. Geol. Nord*, vol. XXXIII. p. 50.

Bartholin, C. T. (92) Nogle i den bornholmske Juraformation forekommende Planteforsteninger. *Bot. Tidsk. Kjövenhaven*, vol. XVIII. Heft 1.

Bather, F. A. (07) Nathorst's use of Collodion imprints in the study of fossil plants. *Geol. Mag.* vol. IV. p. 437.

—— (08) Nathorst's method of studying cutinised portions of fossil plants. *Geol. Mag.* vol. V. p. 454.

Bayer, G. (99) Einige neue Pflanzen der Perucer Kreideschichten in Böhmen. *Sitz. K. böhm. Ges. Wiss. Prag.*

Bennie, J. and R. Kidston. (88) On the occurrence of Spores in the Carboniferous formation of Scotland. *Proc. R. Physc. Soc. Edin.* vol. IX. p. 82, 1885–88.

Benson, Margaret. (04) *Telangium Scotti*, a new species of Telangium (Calymmatotheca) showing structure. *Ann. Bot.* vol. XVIII. p. 161.

—— (08) *Miadesmia membranacea*, Bertrand; a new Palaeozoic Lycopod with a seed-like structure. *Phil. Trans. R. Soc.* vol. CXCIX. p. 409.

—— (08²) The Sporangiophore—a unit of structure in the Pteridophyta. *New Phytologist*, vol. VII. p. 143.

Bernard, C. (04) Le bois centripète dans les feuilles de Conifères. *Bot. Cent.* Beiheft, vol. XVII. p. 241.

Berridge, E. M. (05) On two new specimens of *Spencerites insignis*. *Ann. Bot.* vol. XIX. p. 273.

Berry, E. W. (03) The American species referred to *Thinnfeldia*. *Bull. Torrey Bot. Club*, vol. XXX. p. 438.

Bertrand, C. E. (81) Recherches sur les Tmésiptéridées. *Arch. Bot. Nord de la France*, vol. I. p. 252.

—— (82) Phylloglossum. *Arch. Bot. Nord France*, vol. II. p. 70.

—— (91) Remarques sur le *Lepidodendron Harcourtii* de Witham. *Trav. Mém. Fac. Lille*, vol. II. Mém. 6.

—— (94) Sur une nouvelle Centradesmide. *Assoc. Franç. pour l'Avancement des Sciences.*

—— (99) On the structure of the stem of a ribbed *Sigillaria*. *Ann. Bot.* vol. XIII.

Bertrand, C. E. and **F. Cornaille.** (02) Études sur quelques caractéristiques de la structure des Filicinées actuelles. *Trav. Mem. L'Univ. Lille,* tome x. Mém. 29.

Bertrand, P. (09) Études sur la Fronde des Zygoptéridées. *Lille.*

Binney, E. W. (44) On the remarkable fossil trees lately discovered near St Helen's. *Phil. Mag.* vol. xxiv. p. 165.

—— (46) Description of the Dukinfield *Sigillaria. Quart. Journ. Geol. Soc.* vol. ii. p. 390.

—— (48) On the origin of coal. *Manchester Phil. Soc. Mem.* vol. viii. p. 148.

—— (62) On some plants showing structure from the Lower Coal-Measures of Lancashire. *Proc. Geol. Soc.* vol. xxviii. p. 106.

—— (65) A description of some fossil plants showing structure, found in the Lower Coal-Seams of Lancashire and Yorkshire. *Phil. Trans. R. Soc.* vol. clv. pt. ii. p. 579.

—— (71) Observations on the structure of fossil plants found in the Carboniferous Strata. *Palaeont. Soc. London.*

—— (72) Observations &c. Pt. iii. *Lepidodendron. Palaeont. Soc.*

Bischof, —— (53) Sigillarien aus dem bunten Sandstein Bernburgs. *Zeit. Gesammt. Naturwiss.* Bd. i.

Blanckenhorn, M. (85) Die Fossile Flora des Buntsandsteins und des Muschelkalkes der Umgegend von Commern. *Palaeontographica,* vol. xxxii. p. 117.

Bommer, C. (03) Les causes d'erreur dans l'étude des empreintes végétales. *Nouv. Mém. Soc. Belg. Geol. &c.* Fasc. i.

Boodle, L. A. (00) Comparative anatomy of the Hymenophyllaceae, Schizaeaceae, and Gleicheniaceae. *Annals Bot.* vol. xiv. p. 455.

—— (01) Comparative anatomy of the Hymenophyllaceae, Schizaeaceae, and Gleicheniaceae. *Annals Bot.* vol. xiv. p. 359.

—— (04) On the occurrence of Secondary Xylem in *Psilotum. Ann. Bot.* vol. xviii. p. 505.

—— (04²) The structure of the leaves of the bracken (*Pteris aquilina,* L.) in relation to environment. *Journ. Linn. Soc. London,* vol. xxxv. p. 659.

—— (08) On the production of dwarf male prothalli in sporangia of *Todea. Ann. Bot.* vol. xxii. p. 231.

Boulay, N. (76) Le Terrain Houiller du Nord de la France et les Végétaux Fossiles. *Lille.*

Bower, F. O. (91) Is the Eusporangiate or the Leptosporangiate the more primitive type in the ferns? *Annals Bot.* vol. v. p. 109.

—— (93) On the structure of *Lepidostrobus Brownii,* Schmpr. *Annals Bot.* vol. vii. p. 329.

—— (94) Studies in the Morphology of Spore-producing Members. Equisetaceae and Lycopodineae. *Phil. Trans. R. Soc.* vol. clxxxv. p. 473.

—— (96) Studies &c. II. Ophioglossaceae. *London.*

—— (97) Studies &c. III. Marattiaceae. *Phil. Trans. R. Soc.* vol. clxxxix. p. 35.

Bower, F. O. (00) Studies &c. IV. The Leptosporangiate Ferns. *Phil. Trans. R. Soc.* vol. CXCII. p. 29.

—— (04) Studies &c. V. General Comparisons and Conclusions. *Ibid.* vol. CXCVI. p. 191.

—— (08) The origin of a Land Flora. *London.*

Bowman, J. E. (41) Observations on the characters of the fossil trees lately discovered on the line of the Bolton railway. *Trans. Manchester Geol. Soc.* vol. I. p. 112.

Braun, A. (63) Ueber die Isoetes Arten der Insel Sardinien. *Monatsber. k. Preuss. Akad. Wiss. Berlin.*

Braun, F. (75) Die Frage nach der Gymnospermie der Cycadeen. *Monatsber. k. Preuss. Akad. Wiss.* p. 241.

Brebner, G. (02) The anatomy of Danaea and other Marattiaceae. *Ann. Bot.* vol. XVI.

Brodie, P. B. (42) On the discovery of insects in the Lower Beds of the Lias of Gloucestershire. *Brit. Assoc. Rep.* 1842, p. 58.

—— (45) A history of fossil insects in the Secondary Rocks of England. *London.*

Brongniart, A. (25) Note sur les végétaux fossiles de l'Oolite à fougères de Mamers. *Ann. Sci. Nat.* tome IV. p. 417.

—— (37) Histoire des Végétaux Fossiles, vol. II. *Paris.*

—— (45) See Murchison, R., de Verneuil, and Count Keyserling.

—— (68) Notice sur un Fruit de Lycopodiacées Fossiles. *Compt. Rend.* vol. LXVII. [*Aug.* 17*th*].

Brown, R. (45) On the geology of Cape Breton. *Quart. Journ. Geol. Soc.* vol. I. p. 207.

—— (46) On a group of erect trees in the Sydney coalfield of Cape Breton. *Quart. Journ. Geol. Soc.* vol. II. p. 393.

—— (47) Upright *Lepidodendron* with *Stigmaria* roots, in the roof of Sydney Main Coal. *Quart. Journ. Geol. Soc.* vol. IV. p. 46.

—— (49) Description of erect Sigillariae with conical tap-roots, found in the roof of the Sydney Main Coal, in the island of Cape Breton. *Proc. Geol. Soc.* (*March* 21), p. 354.

—— (51) Some account of an undescribed fossil fruit (*Triplosporites*) *Trans. Linn. Soc.* vol. XX. p. 469.

Browne, Lady Isabel. (09) The phylogeny and inter-relationships of the Pteridophyta. *New Phytologist* (*Reprint*, No. 3). *Cambridge.*

Bruchmann, H. (74) Ueber Anlage und Wachsthum der Wurzeln von *Lycopodium* und *Isoetes.* *Jen. Zeit. Naturwiss.* vol. VIII. p. 522.

—— (97) Untersuchungen über *Selaginella spinulosa*, A. Br. *Gotha.*

—— (98) Ueber die Prothallien und die Keimpflanzen mehreren Europäischer Lycopodien. *Gotha.*

Buckman, J. (45) See Murchison, R. I.

—— (50) On some fossil plants from the Lower Lias. *Quart. Journ. Geol. Soc.* vol. VI. p. 413.

Bunbury, C. J. F. (47) On fossil plants from the coal formation of Cape Breton. *Quart. Journ. Geol. Soc.* vol. III. p. 423.

Bunbury, C. J. F. (61) Notes on a collection of fossil plants from Nagpur, Central India. *Qvart. Journ. Geol. Soc.* vol. XVII. p. 325.

Butterworth, J. (00) Further research on the structure of *Psaronius*, a tree-fern of the Coal-Measures. *Mem. Proc. Manchester Lit. Phil. Soc.* vol. XLIII.

Campbell, D. H. (04) The affinities of the Ophioglossaceae and Marsiliaceae. *Amer. Naturalist*, vol. XXVIII. no. 454.

—— (05) The structure and development of Mosses and Ferns. *New York.*

—— (07) On the distribution of the Hepaticae and its significance. *New Phytol.* vol. VI. p. 203.

Carruthers, W. (65) On *Caulopteris punctata*, Göpp, a tree-fern from the Upper Greensand of Shaftesbury in Dorsetshire. *Geol. Mag.* vol. II. p. 484.

—— (69) On the structure of the stems of arborescent Lycopodiaceae of the Coal-Measures. *Month. Micr. Journ.* p. 177.

—— (69²) On the plant-remains from the Brazilian coal-beds, with remarks on the genus *Flemingites*. *Geol. Mag.* vol. VI. p. 5.

—— (70) On the nature of the scars in the stems of *Ulodendron*, *Bothrodendron*, and *Megaphytum*. *Monthly Micr. Journ.* p. 144.

—— (72) On the structure of the stems of the arborescent Lycopodiaceae of the Coal-Measures. *Monthly Micr. Journ.* p. 50.

—— (72²) Notes on fossil plants from Queensland, Australia. *Quart. Journ. Geol. Soc.* vol. XXVIII.

—— (72³) Notes on some fossil plants. *Geol. Mag.* vol. IX.

—— (73) On some Lycopodiaceous plants from the Old Red Sandstone of the North of Scotland. *Journ. Bot.*, November, 1873.

—— (73²) On *Halonia* of Lindley and Hutton and *Cyclocladia* of Goldenberg. *Geol. Mag.* vol. X. p. 145.

Cash, W. and **J. Lomax.** (90) On *Lepidophloios* and *Lepidodendron*. *Brit. Ass. Rep. (Leeds)*, p. 810.

Challenger Reports. (85) Reports on the scientific results of the voyage of H. M. S. Challenger during the years 1873–76. *Narrative*, vol. I. pts. 1 and 2. *London.*

—— (85²) *Ibid.* Botany, vol. I.

Chick, Edith. (01) See Tansley, A. G. and Edith Chick.

Chodat, R. (08) Les Ptéridopsides des temps Paléozoiques. *Arch. Soc. Phys. Nat.* [4], vol. XXVI. *Genoa.*

Christ, H. (96) Zur Farn-Flora der Sunda Inseln. *Ann. Jard. Buitenzorg*, vol. XIII. p. 90.

—— (97) Die Farnkräuter der Erde. *Jena.*

—— (04) *Loxsomopsis costaricensis* nov. gen. et spec. *Bull. Herb. Boissier* [2] vol. IV. No. 5.

—— and **K. Giesenhagen.** (99) Pteridographische Notizen, I. *Archangiopteris* nov. gen. *Flora*, Bd. LXXXVI. p. 72.

Chrysler, M. A. (08) Tyloses in tracheids of Conifers. *New Phytologist*, vol. VII.

Compton, R. H. (09) The anatomy of *Matonia sarmentosa*. *New Phyt.* vol. VIII. p. 299.

Copeland, E. B. (08) New genera and species of Bornean ferns. *The Philippine Journ. Sci.* vol. III. (C. Botany) p. 343.

—— (09) The ferns of the Malay Asiatic region. *Ibid.* vol. IV. No. 1.

Corda, A. J. (46) See Reuss, A. E.

—— (52) See Germar, E. F.

Cornaille, F. (02) See Bertrand, C. E. and F. Cornaille.

Cotta, C. B. (32) Die Dendrolithen in Beziehung auf ihren inneren Bau. *Dresden.*

Coward, K. H. (07) On the structure of *Syringodendron*, the bark of *Sigillaria*. *Mem. Proc. Manchester Lit. Phil. Soc.* vol. LI., pt. ii.

Crépin, F. (74) Description de quelques plantes fossiles de l'étage des Psammites de Condroz. *Bull. Acad. Roy. Belg.* [2] vol. XXXVIII.

—— (75) Observations sur quelques plantes fossiles des dépôts Dévoniens. *Soc. Roy. bot. Belg.* vol. XIV. p. 214.

Dale, E. (01) See Seward, A. C. and E. Dale.

Dangeard, P. A. (91) Mémoire sur la morphologie et l'anatomie des *Tmesipteris*. *Le Botaniste* [1] Fasc. VI. p. 163.

Darwin, C. (03) More letters of Charles Darwin, edited by F. Darwin and A. C. Seward (vol. II.). *London.*

David, W. E. and **E. F. Pittman.** (93) On the occurrence of *Lepidodendron australe* (?) in the Devonian rocks of N. S. Wales. *Rec. Geol. Surv. N. S. Wales*, vol. III. pt. iv. p. 194.

Davy, E. W. (07) Summer in British Central Africa. *Gard. Chron.* vol. XLI. [3] p. 263.

Dawes, J. S. (48) Remarks upon the internal structure of *Halonia*. *Quart. Journ. Geol. Soc.* vol. IV. p. 289.

Dawson, Sir J. W. (61) On an undescribed fossil fern from the Lower Coal-Measures of Nova Scotia. *Quart. Journ. Geol. Soc.* vol. XVII. p. 5.

—— (66) Acadian Geology. *London.*

—— (82) The fossil plants of the Erian and Upper Silurian Formations of Canada. Pt. II. *Geol. Surv. Canada.*

—— (86) On Rhizocarps in the Erian (Devonian) period in America. *Bull. Chicago Acad. Sci.* vol. I. no. 9.

Debey, M. H. and **C. von Ettingshausen.** (59) Die urweltlichen Acrobryen des Kreidegebirges von Aachen und Maestricht. *Denksch. k. Akad. Wiss. Wien*, vol. XVII.

Diels, L. (02) Filicales. *Die Natürlichen Pflanzenfamilien.* Engler and Prantl, Teil I. Abt. 4.

Eichwald, E. (60) Lethaea Rossica. Vol. I. *Stuttgart.*

Engler, A. (09) Syllabus der Pflanzenfamilien. *Berlin.*

—— and **K. Prantl.** (02) Die Natürlichen Pflanzenfamilien. *Leipzig.*

Ernst, A. (08) The new flora of the volcanic island of Krakatau. (Trans. by A. C. Seward.) *Cambridge.*

Etheridge, R. (80) A contribution to the study of British Carboniferous Tubicolar Annelida. *Geol. Mag.* vol. VII. p. 109.

Etheridge, R. (90) *Lepidodendron australe*, McCoy.—Its synonyms and range in Eastern Australia. *Rec. Geol. Surv. N. S. Wales*, vol. II. pt. 3, p. 119.

—— (92) See Jack, R. L. and R. Etheridge.

—— (94) On the mode of attachment of the leaves or fronds to the caudex in *Glossopteris*. *Proc. Linn. Soc. N. S. Wales*, vol. IX. p. 228.

—— (94²) On the occurrence of *Oleandridium* in the Hawkesbury Sandstone series. *Rec. Geol. Surv. N. S. Wales*, vol. IV. pt. 2, p. 49.

—— (99) On a fern (*Blechnoxylon talbragarense*) with secondary wood, forming a new genus, from the Coal-Measures of the Talbragar District, N. S. Wales. *Rec. Australian Mus.* vol. III. pt. 6.

Ettingshausen, C. von. (52) Begründung einiger neuen oder nicht genau bekannten Arten der Lias- und der Oolithflora. *Abt. K. K. Geol. Reichs*, vol. I. Abth. 3.

—— (59) See Debey, M. H. and C. von Ettingshausen.

—— (66) Die Fossile Flora des Mährischschlesischen Dachschiefers. *Denksch. K. K. Akad. Wien*, vol. XXV. p. 77.

—— (82) See Gardner, J. S. and C. von Ettingshausen.

Farmer, J. B. (90) On *Isoetes lacustris*, L. *Annals Bot.* vol. V. p. 37.

Farmer, J. B. and W. G. Freeman. (99) On the structure and affinities of *Helminthostachys zeylanica*. *Annals Bot.* vol. XIII. p. 421.

Farmer, J. B. and T. G. Hill. (02) On the arrangement and structure of the vascular strands in *Angiopteris evecta* &c. *Annals Bot.* vol. XVI. p. 371.

Faull, J. H. (01) The anatomy of the Osmundaceae. *Bot. Gaz.*, vol. XXXII. p. 381.

Feistmantel, C. (79) Ueber die Nöggerathien. *Sitz. K. böhm. Ges. Wiss.*

Feistmantel, O. (72) Ueber Baumfarrenreste der böhmischen Steinkohlen. Perm. und Kriedeformation. *Abt. k. böhm. Ges. Wiss.* vol. VI. [5].

—— (75) Die Versteinerungen der böhmischen Kohlenablagerungen. *Palaeontograph.* vol. XXIII. 1875-76.

—— (75²) Ueber das Vorkommen von *Noggerathia foliosa*, Stbg., in dem Steinkohlengebirge von Oberschlesien. *Zeit. Deutsch. Geol. Ges.*

—— (77) Jurassic (Liassic) Flora of the Rajmahal group from Golapili (near Ellore), S. Godaveri District. *Mem. Geol. Surv. India*, vol. I. pt. iii.

—— (79) The Flora of the Talchir-Karharbari Beds. *Mem. Geol. Surv. India. Foss. Flor. Lower Gondwanas*, I.

—— (80) The Flora of the Damuda and Panchet Divisions. *Mem. Geol. Surv. India. Ibid.* vol. III.

—— (82) The Fossil Flora of the South Rewah Gondwana Basin. *Ibid.* vol. IV. pt. i.

Fischer, F. (04) Zur Nomenclatur von *Lepidodendron* und zur Art-kritik dieser Gattung. *Abh. k. Preuss. Geol. Landes.* [N.F.] Heft XXXIX.

Fitting, H. (07) Sporen im Buntsandstein. Die Makrosporen von *Pleuromeia. Ber. Deutsch. Bot. Ges.* vol. XXV. p. 434.

Fliche, P. (03) Sur les Lycopodinées des Trias en Lorraine. *Compt. Rend.* 1903 (*April 6th*).

—— (09) Sur une fructification de Lycopodinée trouvée dans le Trias. *Compt. Rend.* tome CXLVIII. p. 259.

Fontaine, W. M. (83) Contributions to the knowledge of the Older Mesozoic Flora of Virginia. *Monographs U. S. Geol. Surv.* Vol. VI.

—— (89) The Potomac or younger Mesozoic Flora of Virginia. *U. S. Geol. Surv. Monographs,* VI.

—— (00) See Ward, L. F.

—— (05) See Ward, L. F.

Fontaine, W. M. and I. C. White. (80) The Permian or Upper-Carboniferous Flora of W. Virginia and Pennsylvania. *Second Geol. Surv. Report of Progress.*

Forbes, E. (51) Note on fossil leaves represented in Plates II.—IV. in the Duke of Argyll's paper on Tertiary Leaf-beds in the Isle of Mull. *Quart. Journ. Geol. Soc.* Vol. VII. p. 103.

—— (53) On the fossils of the Yellow Sandstone of the South of Ireland. *British Assoc. Rep. London,* 1853, p. 43 (*Belfast meeting,* 1852).

Ford, S. O. (02) On the anatomy of *Ceratopteris thalictroides* L., *Annals Bot.* vol. XVI. p. 95.

—— (03) See Seward, A. C. and S. O. Ford.

—— (04) The anatomy of *Psilotum triquetrum. Annals Bot.* vol. XVIII. p. 589.

—— (06) See Seward, A. C. and S. O. Ford.

Freeman, W. G. (99) See Farmer, J. B. and W. G. Freeman.

Fric, A. and E. Bayer. (01) Studien im Gebiete der Böhmischen Kreideformation. *Arch. Naturwiss. Landes. Böhmen,* Bd. XI.

Fritel, P. H. (08) Sur une espèce fossile nouvelle du genre Salvinia. *Journ. Bot. Ann.* XXI. No. 8, p. 190.

Fritsch, F. E. (05) See Tansley, A. G. and F. E. Fritsch.

Fritsch, K. von. (97) Pflanzenreste aus Thüringer Culm Dachschiefer. *Zeitsch. Naturwiss.* Bd. LXX.

Gardeners' Chronicle. (82) [*Selaginella grandis* n. sp.] Vol. XVII. p. 40 (July 8).

Gardner, J. S. and C. von Ettingshausen. (82) A monograph of the British Eocene Flora. Vol. I. Filices. *Palaeont. Soc. London.*

Geikie, Sir Archibald. (03) Text-book of Geology. 2 vols. *London.*

Geinitz, E. (73) Versteinerungen aus dem Brandschiefer der unteren Dyas von Weissing bei Pillaritz in Sachsen. *Neues Jahrb. Min.* p. 691.

Geinitz, H. B. (72) Fossile Myriopoden in dem Rothliegenden bei Chemnitz. *Sitzb. Naturwiss. Ges. Isis,* p. 128.

—— (76) Ueber rhaetische Pflanzen und Thierreste in den Argentinischen Provinzen La Rioja &c. *Palaeontograph.* Suppl. iii.

Germar, E. F. (52) *Sigillaria Sternbergi* Münster, aus dem bunten Sandstein. *Zeit. deutsch. Geol. Ges.* p. 183.

—— (44–53) Die Versteinerungen des Steinkohlengebirges von Wettin und Löbejün im Saalkreise. *Halle.*

Geyler, H. T. (77) Ueber Fossile Pflanzen aus der Juraformation Japans. *Palaeont.* vol. xxiv. p. 221.

Giesenhagen, C. (90) Die Hymenophyllaceen. *Flora*, p. 411.

—— (92) Ueber Hygrophile Farne. *Flora*, Bd. 76 (Ergänz. Band).

Gilkinet, A. (75) Sur quelques plantes de l'étage du poudingue de Burnot. *Acad. Roy. Belg.* [2] vol. xl. No. 8.

Goebel, K. (91) Pflanzenbiologische Schilderungen. Teil 2. *Marburg.*

—— (05) Organography of Plants. Vol. ii. (Trans. by J. B. Balfour.) *Oxford.*

Goeppert, H. R. (41) Die Gattungen der Fossilen Pflanzen. *Bonn.*

—— (32) Fossile Flora des Uebergangsgebirges. *Nov. Act. Caes. Leop.-Carol.* Vol. xiv. (*Supplt.*).

—— (54) See Roemer, F. A.

Goldenberg, F. (55–62) Flora Saraepontana Fossilis. *Saarbrücken.*

Gordon, W. T. (08) On the Prothallus of *Lepidodendron Veltheimianum. Trans. Bot. Soc. Edinburgh*, vol. xxiii. p. 330.

—— (09) On the structure of a new *Zygopteris. Nature.* (Botany at the British Association.) Vol. lxxxi. p. 537.

Grand'Eury, C. (75) See Renault, B. and C. Grand'Eury.

—— (04) Sur les graines des Neuroptéridées. *Compt. Rend.* cxxxix. p. 782.

—— (05) Sur les graines trouvées attachées au *Pecopteris Pluckeneti,* Schlot. *Compt. Rend.* vol. cxl. p. 920.

—— (05²) Sur les graines de Sphenopteris &c. *Compt. Rend.* cxli. p. 812.

—— (06) Sur les graines et inflorescences des *Callipteris. Compt. Rend.* cxliii. p. 664.

—— (08) Sur les organes et la mode de végétation des Neuroptéridées et autres Ptéridospermes. *Compt. Rend.* clxvi. p. 1241.

Gresley, W. S. (89) Note on further discoveries of *Stigmaria* (? *ficoides*) and their bearing upon the question of the formation of Coal-Beds. *Midland Naturalist*, vol. xii. p. 25.

Greville, R. K. (31) See Hooker, W. J. and R. K. Greville.

Grigoriew, N. (98) Sur la flore paléozoique supérieure recueillie aux environs des villages Troitskoie et Longanskoie dans le bassin du Donetz. *Bull. Com. Géol. St. Pétersbourg*, tome xvii.

Gwynne-Vaughan, D. T. (01) Observations on the anatomy of solenostelic ferns. *Annals Bot.* vol. xiv. p. 71.

—— (03) Observations &c. *Ibid.* vol. xvii. p. 689.

—— (05) On the anatomy of *Archangiopteris Henryi* and other Marattiaceae. *Annals Bot.* vol. xix. p. 259.

—— (07—09) See Kidston, R. and D. T. Gwynne-Vaughan.

—— (08) On the real nature of the Tracheae in the ferns. *Annals Bot.* vol. xxii. p. 517.

Gwynne-Vaughan, D. T. and **R. Kidston.** (08) On the origin of the adaxially curved leaf-trace in the Filicales. *Proc. R. Soc. Edinb.* vol. XXVIII. pt. vi. p. 433.

Hall, Kate M. (91) See Jennings, A. Vaughan, and Kate M. Hall.

Halle, T. G. (07) Einige krautartige Lycopodiaceen Paläozoischen und Mesozoischen Alters. *Arkiv Bot.* Bd. VII. No. 5.

Hannig, E. (98) Ueber die Staubgrübchen auf den Stämmen und Blattstielen der Cyathaeaceen und Marattiaceen. *Bot. Zeit.* p. 9.

Harvey-Gibson, R. J. (94) Contributions towards a knowledge of the anatomy of the genus *Selaginella*, Spr. *Annals Bot.* vol. VIII. p. 133.

—— (96) Contributions &c. *Ibid.* vol. X. p. 77.

—— (97) Contributions &c. *Ibid.* vol. XI. p. 123.

—— (02) Contributions &c. *Ibid.* vol. XVI. p. 449.

Haughton, S. (59) On *Cyclostigma*, a new genus of fossil plants from the Old Red Sandstone of Kiltorkan, W. Kilkenny. *Journ. Roy. Dublin Soc.* vol. II. p. 407.

Hawkshaw, J. (42) Description of the fossil trees found in the excavations for the Manchester and Bolton railway. *Trans. Geol Soc.* [2] vol. VI. p. 173.

Hayden, H. H. (07) The stratigraphical position of the Gangamopteris beds of Kashmir. *Rec. Geol. Surv. India*, vol. XXXVI. pt. i.

Heer, O. (71) Fossile Flora der Bären Inseln. *Flor. Foss. Arct.* vol. II.

—— (74) Die Kreideflora der Arctischen Zone. *K. Svensk. Vetens-kaps-Akad. Hand.* Bd. XII. [*Flor. Foss. Arct.* vol. III. 1875.]

—— (75) *Flora Fossilis Arctica*, vol. III.

—— (76) Jura-Flora Östsibiriens und des Amurlandes. *Ibid.* vol. IV. [ii].

—— (80) Nachträge zur Fossilen Flora Grönlands. *K. Svensk. Vetenskaps-Akad. Hand.* Bd. XVIII. [*Flor. Foss. Arct.* vol. VI. 1882.]

—— (82) Die Fossile-Flora Grönlands. *Flor. Foss. Arct.* vol. VI.

Hegelmaier, F. (72) Zur Morphologie der Gattung *Lycopodium*. *Bot. Zeit.* 1872, p. 773.

Hick, T. (93) On a new fossil plant from the Lower Coal-Measures. *Journ. Linn. Soc.* vol. XXIX. p. 86.

—— (93²) Supplementary note on a new fossil plant. *Ibid.* p. 216.

—— (96) On *Rachiopteris cylindrica*. *Mem. Proc. Manchester Lit. Phil. Soc.* vol. XLI. pt. i.

Hieronymus, G. (02) Selaginellaceae. *Die Natürlichen Pflanzenfamilien.* Engler and Prantl, Teil I. Abt. 4, p. 621.

Hill, T. G. (00) See Scott, D. H. and T. G. Hill.

—— (02) See Farmer, J. B. and T. G. Hill.

—— (04) On the presence of a parichnos in recent plants. *Annals Bot.* vol. XVIII. p. 654.

—— (06) On the presence of a parichnos in recent plants. *Ibid.* vol. XX. p. 267.

Hofmeister, W. (62) On the germination, development, and fructification of the Higher Cryptogamia, and on the fructification of the Coniferae. *Ray Soc.* 1862.

Hollick, A. (94) Fossil Salvinias, including description of a new species *Torrey Bot. Club*, vol. XXI. No. 6, p. 253.

—— (04) Additions to the Palaeobotany of the Cretaceous Formation on Long Island. *Bull. New York Bot. Gard.* vol. III. p. 403.

—— and **E. C. Jeffrey.** (09) Studies of Cretaceous coniferous remains from Kreischerville, New York. *Mem. New York Bot. Gard.* vol. III.

Hooker, Sir J. D. (48) On the vegetation of the Carboniferous Period, as compared with that of the present day. *Mem. Geol. Soc. Great Britain*, vol. II. pt. ii. p. 387.

—— (48²) Remarks on the structure and affinities of some *Lepidostrobi*. *Ibid.* p. 440.

—— (59) Stangeria paradoxa. *Bot. Mag.* Tab. 5121.

Hooker, Sir W. J. and **J. G. Baker.** (68) Synopsis Filicum. *London.*

Hooker, Sir W. J. and **R. K. Greville.** (31) Icones Filicum. Vol. II. *London.*

Hosius and **von der Marck.** (80) Die Flora der Westfälischen Kreideformation. *Palaeont.* Bd. XXVI. p. 127.

Hovelacque, M. (92) Recherches sur le *Lepidodendron selaginoides*, Sternb. *Mém. Soc. Linn. Normandie*, vol. XVII.

Hudson, W. H. (92) The Naturalist in La Plata. *London.*

Jack, R. L. and **R. Etheridge.** (92) The geology and palæontology of Queensland and New Guinea. *Brisbane.*

Jahn, J. J. (03) Ueber die Étage H. im Mittel-Böhmischen Devon. *Verh. Reichsanst. Wien*, No. 4, p. 73.

Jeffrey, E. C. (98) The morphology of the central cylinder in vascular plants. *Brit. Assoc. Rep. (Toronto Meeting)*, p. 869.

—— (98²) The Gametophyte of *Botrychium virginianum*. *Trans. Canad. Inst.* vol. V. p. 265.

—— (00) The morphology of the central cylinder in the Angiosperms. *Ibid.* vol. VI. p. 599.

—— (03) The structure and development of the stem in the Pteridophyta and Gymnosperms. *Phil. Trans. R. Soc.* vol. CXCV. p. 119.

—— (09) See Hollick, A. and E. C. Jeffrey.

Jennings, A. Vaughan and **Kate M. Hall.** (91) Notes on the structure of *Tmesipteris*. *Proc. R. Irish Acad.* [3] vol. II. p. 1.

Jones, C. E. (05) The morphology and anatomy of the stem of the genus *Lycopodium*. *Trans. Linn. Soc.* vol. VII. p. 15.

Jordan, Rose. (03) On some peculiar tyloses in *Cucumis sativus*. *New Phytologist*, vol. II. p. 208.

Karsten, G. (95) Morphologische und biologische Untersuchungen über einige Epiphytenformen der Molukken. *Ann. Jard. Buitenzorg*, vol. XII. p. 117.

Kidston, R. (82) On the fructification of *Eusphenopteris tenella* and *Sphenopteris microcarpa*. *R. Physc. Soc. Edinb.* vol. VII.

—— (83) Report on the fossil plants collected by the Geological Survey of Scotland in Eskdale and Liddesdale. *Trans. R. Soc. Edinburgh*, vol. XXX. p. 531.

592 LIST OF WORKS

Kidston, R. (84) On a new species of *Lycopodites*, Goldenberg (*L. Stockii*), from the Calciferous Sandstone Series of Scotland. *Ann. Mag. Nat. Hist.* [5] vol. XIV. p. 111.

—— (84²) On the fructification of *Zeilleria* (*Sphenopteris*) *delicatula*, Sternb. sp. ; with remarks on *Urnatopteris* (*Sphenopteris*) *tenella*, Brongnt., and *Hymenophyllites* (*Sphenopteris*) *quadridactylites*, Gutbier sp. *Quart. Journ. Geol. Soc.* vol. XL. p. 590.

—— (85) On the relationship of *Ulodendron*, Lindley and Hutton, to *Lepidodendron*, Sternberg ; *Bothrodendron*, Lindley and Hutton; *Sigillaria*, Brongniart ; and *Rhytidodendron*, Boulay. *Ann. Mag. Nat. Hist.* vol. XVI. p. 123.

—— (86²) On a new species of *Psilotites* from the Lanarkshire Coal-field. *Ann. Mag. Nat. Hist.* 1886, p. 494.

—— (86³) On the occurrence of *Lycopodites Vanuxemi*, Göppert, in Britain, with remarks on its affinities. *Journ. Linn. Soc.* vol. XXI. p. 560.

—— (86⁴) Notes on some fossil plants collected by Mr R. Dunlop, Airdrie, from the Lanarkshire Coal-field. *Trans. Geol. Soc. Glasgow*, vol. VIII. p. 47.

—— (87) On the fructification of some ferns from the Carboniferous formation. *Trans. R. Soc. Edinb.* vol. XXXIII. pt. i.

—— (88) See Bennie, J. and R. Kidston.

—— (88) On the fossil flora of the Radstock series of the Somerset and Bristol Coal-field (Upper Coal-Measures). Part I. *Trans. R. Soc. Edinb.* vol. XXXIII. pt. 2.

—— (88²) On the fructification and affinities of *Archaeopteris hibernica*, Forbes sp. *Ann. Mag. Nat. Hist.* [6] vol. II.

—— (89) On the fossil plants in the Ravenhead Collection in the Free Library and Museum, Liverpool. *Trans. R. Soc. Edinb.* vol. XXXV. pt. ii.

—— (89²) Additional notes on some British Carboniferous Lycopods. *Ann. Mag. Nat. Hist.* vol. IV. p. 60.

—— (89³) On some fossil plants from Teilia quarry, Gwaenysgor, near Prestatyn, Flintshire. *Trans. R. Soc. Edinb.* vol. XXXV. pt. ii.

—— (91) On the fructification of *Sphenophyllum trichomatosum*, Stern. from the Yorkshire Coal-field. *Proc. R. Soc. Edinb.* vol. XI. p. 56.

—— (91²) On the fructification and internal structure of Carboniferous ferns in their relation to those of existing genera. *Trans. Geol. Soc. Glasgow*, vol. IX. pt. i.

—— (91³) On the Fossil Flora of the Staffordshire Coal-fields. II. *Trans. R. Soc. Edinb.* vol. XXXVI. p. 63.

—— (93) On *Lepidophloios*, and on the British species of the genus. *Trans. R. Soc. Edinb.* vol. XXXVII. pt. iii. p. 529.

—— (94) On the various divisions of British Carboniferous rocks as determined by their fossil flora. *R. Physc. Soc. Edinb.* vol. XII. p. 183.

—— (96) On the Fossil Flora of the Yorkshire Coal-fields. I. *Trans. R. Soc. Edinb.* vol. XXXVIII. p. 203.

Kidston, R. (97) On the Fossil Flora of the Yorkshire Coal-fields. II. *Trans. R. Soc. Edinb.* vol. xxxix. pt. i. p. 33.

—— (01) Carboniferous Lycopods and Sphenophylls. *Trans. Nat. Hist. Soc. Glasgow,* vol. vi. pt. i. p. 25.

—— (01²) The flora of the Carboniferous Period. *Proc. Yorks. Geol. Polyt. Soc.* vol. xiv. pt. ii.

—— (02) The flora &c. Second Paper. *Ibid.* vol. xiv. pt. iii. p. 344.

—— (03) The fossil plants of the Carboniferous rocks of Canonbie, Dumfriesshire, and of parts of Cumberland and Northumberland. *Trans. R. Soc. Edinb.* vol. xl. pt. iv. p. 741.

—— (05) On the internal structure of *Sigillaria elegans* of Brongniart's " Histoire des Végétaux Fossiles." *Ibid.* vol. xli. pt. iii. p. 533.

—— (05²) On the fructification of *Neuropteris heterophylla. Trans R. Soc. London,* vol. cxcvii. p. 1.

—— (06) On the microsporangia of the Pteridospermeae, with remarks on their relationship to existing Ferns. *Phil. Trans. R. Soc.* vol. cxcviii. p. 413.

—— (07) Note on a new species of *Lepidodendron* from Pettycur (*L. Pettycurense*). *Proc. R. Soc. Edinb.* 1906–07, p. 207.

—— (07²) Preliminary note on the internal structure of *Sigillaria mammillaris* Brongniart and *S. scutellata* Brongniart. *Ibid.* vol. xxvii. p. 203.

—— (08) On a new species of *Dineuron* and of *Botryopteris* from Pettycur, Fife. *Trans. R. Soc. Edinb.* vol. xlvi. pt. ii. p. 361.

—— (08) See Gwynne-Vaughan, D. T. and R. Kidston.

Kidston, R. and **D. T. Gwynne-Vaughan.** (07) On the fossil *Osmundaceae.* Pt. I. *Trans. R. Soc. Edinb.* vol. xlv. pt. iii. p. 759.

—— (08) *Ibid.* pt. II. *loc. cit.* vol. xlvi. pt. ii. p. 213.

—— (09) *Ibid.* pt. III. *loc. cit.* vol. xlvi. pt. iii. p. 651.

Kitchin, F. L. (08) The invertebrate fauna and palaeontological relations of the Uitenhage series. *Ann. S. African Mus.* vol. vii. pt. ii.

Knowlton, F. H. (98) A catalogue of the Cretaceous and Tertiary plants of North America. *Bull. U. S. Geol. Surv.* No. 152.

—— (99) Fossil flora of the Yellowstone National Park. *U. S. Geol. Surv. Mem.* xxxii. pt. ii.

—— (02) Report on a small collection of fossil plants from the vicinity of Porcupine Butte, Montana. *Bull. Torrey Bot. Club,* vol. xxix.

Kny, L. (75) Die Entwickelung der Parkeriaceen dargestellt an *Ceratopteris thalictroides,* Brongniart. *Nova Acta K. Leop.-Car. Deutsch. Akad. Naturf.* vol. xxxvii.

Koehne, W. (04) Sigillarienstämme, Unterscheidungsmerkmale, Arten, Geologische Verbreitung &c. *Abh. K. Preuss. Geol. Landes.* [N. F.], Heft xliii.

594 LIST OF WORKS

Krasser, F. (95) Kreidepflanzen von Lesina. *Jahrb. K.-K. Geol. Reichs.* Bd. xLV. p. 37.

—— (96) Beiträge zur Kenntniss der Fossilen Kreideflora von Kunstadt in Mähren. *Beit. Paläont. Geol. Österreich.-Ung. und des Orients,* Bd. x. Heft 3.

—— (00) Die von W. A. Obrutschew in China und Centralasien 1893-94 gesammelten fossilen Pflanzen. *Denksch. K. Akad. Wiss. Wien,* Bd. LXX.

—— (06) Ueber die fossile Kreideflora von Grünbach in Niederösterreich. *Sitzb. K. Akad. Wiss. Wien* (Anz. iii.).

—— (09) Die Diagnosen der von D. Stur in der obertriadischen Flora der Lunzerschichten als Marattiaceenarten unterschiedenen Farne. *Sitz. Kais. Akad. Wiss. Wien,* Bd. cxviii. Abt. i.

Kubart, B. (09) Untersuchungen über die Flora des Ostrau-Karwiner Kohlenbeckens. I. Die Spore von *Spencerites membranaceus* n. sp. *Denksch. K. Akad. Wiss. Wien,* Bd. LXXXV.

Kühn, R. (90) Untersuchungen über die Anatomie der Marattiaceen. *Flora.*

Kurr, J. G. (45) Beiträge zur fossilen Flora der Juraformation. *Stuttgart.*

Kurtz, F. (94) Contribuciones á la Palaeophytologia Argentina. *Revist. Mus. de la Plata,* vol. vi.

Lang, W. H. (99) The prothallus of *Lycopodium clavatum* L. *Annals Bot.* vol. xiii. p. 279.

—— (08) Preliminary statement on the morphology of the cone of *Lycopodium cernuum* and its bearing on the affinities of *Spencerites. Proc. Roy. Soc. Edinb.* vol. xxviii. pt. v. p. 356.

Leclerc du Schlon. (85) Recherches sur la dissemination des Spores dans les Cryptogames vasculaires. *Ann. Sci. nat.* [7], vol. ii. p. 5.

Leslie, T. N. (06) See Mellor, E. T. and T. N. Leslie.

Lesquereux, L. (78) Contributions to the fossil flora of the Western Territories. Pt. ii. The Tertiary floras. *U. S. Geol. Surv. Report.*

Leuthardt, F. (04) Die Keuperflora von Neuewelt bei Basel. Teil ii. *Abh. Schweiz. Pal. Ges.* Bd. xxxi. p. 25.

Lignier, O. (03) Equisétales et Sphénophyllales. Leur origine filicinéenne commune. *Bull. Soc. Linn. Normandie* [5], vol. vii. p. 93.

—— (08) Sur l'origine des Sphénophyllées. *Bull. Soc. Bot. France* [4], vol. viii. p. 278.

Lindman, C. A. M. (04) *Regnellidium* novum genus Marsiliacearum. *Arkiv Bot. K. Svensk. Vetenskaps-Akad. Stockholm.*

Lloyd, E. and **L. M. Underwood.** (00) A review of the species of *Lycopodium* of North America. *Bull. Torrey Bot. Club,* vol. xxvii. p. 147.

Logan, W. E. (42) On the character of the beds of clay immediately below the coal-seams of S. Wales. *Trans. Geol. Soc. London,* vol. vi. p. 491.

Lomax, J. (90) See Cash, W. and J. Lomax.

Lomax, J. (00) See Wild, G. and J. Lomax.

—— (05) See Weiss, F. E. and J. Lomax.

Luerssen, C. (89) Die Farnpflanzen. *Rabenhorst's Kryptogamen Flora,* Bd. III. *Leipzig.*

Lulham, R. B. J. (05) See Tansley, A. G. and R. B. J. Lulham.

Lyon, F. M. (01) A study of the sporangia and gametophytes of *Selaginella apus* and *S. rupestris. Bot. Gaz.* vol. XXXII. p. 125.

McCoy, Sir F. (47) On the fossil botany and zoology of the rocks associated with the coal of Australia. *Ann. Mag. Nat. Hist.* [1], vol. XX. p. 151.

—— (60) A commentary on "A communication made by the Rev. W. B. Clarke" &c. *Trans. R. Soc. Victoria,* vol. v. p. 96.

—— (74) Prodromus of the Palaeontology of Victoria. *Geol. Surv. Vict., Decades* I—V.

McNicol, Mary. (08) On Cavity Parenchyma and Tyloses in Ferns. *Ann. Bot.* vol. XXII. p. 401.

Marion, A. F. (90) Sur le *Gomphostrobus heterophylla. Compt. Rend.* p. 892.

Maslen, A. J. (99) The structure of *Lepidostrobus. Trans. Linn. Soc.* vol. v. p. 357.

—— (06) See Scott, D. H. and A. J. Maslen.

Mellor, E. T. and **T. N. Leslie.** (06) On a fossil forest recently exposed in the bed of the Vaal River at Vereeniging. *Trans. Geol. Soc. S. Africa,* vol. IX. p. 125.

Mettenius, G. (60) Beiträge zur Anatomie der Cycadeen. *Abh. K. Sächs. Ges. Wiss.* Bd. VII. p. 567.

—— (63) Ueber den Bau von *Angiopteris. Abh. K. Sächs. Ges. Wiss.* Bd. IX.

Miller, H. (57) The testimony of the rocks. *Edinburgh.*

Möller, H. (02) Bidrag till Barnholms Fossila Flora. Pteridofyter. *Lunds Univ. Årsskrift,* Bd. XXXVIII. No. 5.

Mohl, H. von. (40) Ueber den Bau des Stammes von *Isoetes lacustris. Linnaea,* vol. XIV.

Morris, J. (40) *Ex* Prestwich's, J., Memoir on the Geology of Coalbrook Dale. *Trans. Geol. Soc.* vol. v. p. 413.

—— (45) See Strzelecki, Count.

—— (63) See Oldham, T. and J. Morris.

Motelay, L. and **Vendryès.** (82) Monographie der Isoetaceae. *Actes Soc. Linn. Bordeaux,* Tom. XXXVI. p. 309.

Münster, G. Graf zu. (42) Beiträge zur Petrefacten-Kunde. Heft 5. *Bayreuth.*

Murchison, R. I., J. Buchman, and **H. E. Strickland.** (45) Outline of the Geology of the neighbourhood of Cheltenham. *London.*

Murchison, R., E. de Verneuil, and **Count A. Keyserling.** (45) Géologie de la Russie d'Europe. Vol. II. *London and Paris.*

Nathorst, A. G. (78) Beiträge zur fossilen Flora Schwedens. Ueber einige Rhätische Pflanzen von Palsjö in Schonen. *Stuttgart.*

Nathorst, A. G. (78²) Om Floran i Skanes kolförande Bildningar. *Sver. Geol. Unders.* Ser. C.

—— (90) Ueber das angebliche Vorkommen von Geschieben des Hörsandsteins in den norddeutschen Diluvialablagerungen. *Arch. Ver. Freund. Nat. Mecklenb.*, Jahr. XLIV.

—— (02) Zur fossilen Flora der Polarländer. I. Zur Oberdevonischen Flora der Bären-Insel. *K. Svensk. Vetenskaps-Akad. Hand.* Bd. XXXVI. No. 3.

—— (02²) Beiträge zur Kenntniss Mesozoischen Cycadophyten. *Ibid.* Bd. XXXVI. No. 4.

—— (04) Die Oberdevonische Flora des Ellesmere Landes. *Rep. Second Norwegian Arctic Exp. in the Fram* (1898–02), No. 1.

—— (04²) Sur la flore fossile antarctique. *Compt. Rend.* (June 6.)

—— (06) Om Några Ginkgoväxter från Kolgrufvorna vid Stabbarp i Skåne. *Lunds Univ. Årsskrift*, N F. Bd. II. No. 8.

—— (06²) Bemerkungen über *Clathropteris meniscoides*, Brongn., und *Rhizomopteris cruciata*, Nath. *K. Svensk. Vetenskaps-Akad. Hand.* Bd. XLI. No. 2.

—— (06³) Ueber *Dictyophyllum* und *Camptopteris spiralis*. *Ibid.* No. 5.

—— (07) Ueber die Anwendung von Kollodium-Abdrücken bei der Untersuchung fossilen Pflanzen. *Arkiv Bot., Stockholm*, Bd. VII. No. 4.

—— (07²) Ueber *Thaumatopteris Schenki*. *K. Svensk. Vetenskaps-Akad. Hand.* Bd. XLII. No. 3.

—— (08) Paläobotanisch. Mitteilungen, III. *Ibid.* Bd. XLIII. No. 3.

Newberry, J. S. (91) The Flora of the Great Falls Coal-Field, Montana. *Amer. Journ. Sci.* vol. XLI. p. 191.

Newton, R. Bullen. (09) Fossils from the Nubian Sandstone of Egypt. *Geol. Mag.* vol. VI. p. 352.

Oldham, R. D. (97) On a plant of *Glossopteris* with part of the rhizome attached, and on the structure of *Vertebraria*. Rec. *Geol. Surv. India*, vol. XXX. pt. i. p. 45.

Oldham, T. and J. Morris. (63) Fossil Flora of the Gondwana system. Vol. I. pt. i. Fossil Flora of the Rajmahal series in the Rajmahal Hills. Mem. *Geol. Surv. India* [2], *Calcutta*.

Oliver, F. W. (02) A vascular Sporangium. *New Phytologist*, vol. I. p. 60.

—— (04) On the structure and affinities of *Stephanospermum*, Brongn., a genus of fossil Gymnosperm seeds. *Trans. Linn. Soc.* vol. VI. pt. 8.

Peach, C. W. (78) On the circinate vernation, fructification, and varieties of *Sphenopteris affinis* and on *Staphyllopteris* (?) *Peachii* of Etheridge and Balfour, a genus of plants new to British rocks. *Quart. Journ. Geol. Soc.* vol. XXXIV. p. 131.

Pelourde, F. (08) Recherches sur la position systématique des plantes fossiles dont les tiges ont été appelées *Psaronius, Psaroniocaulon, Caulopteris*. *Bull. Soc. bot. France* [4], tome VIII.

Pelourde, F. (08[2]) Recherches comparatives sur la structure de la racine chez un certain nombre de *Psaronius. Ibid.* p. 352.

—— (09) Recherches comparatives sur la structure des fougeres fossiles et vivants. *Ann. Sci. nat.* vol. x. p. 115.

—— (09[2]) Observations sur un nouveau type de pétiole fossile, le *Flicheia esnostensis. Mém. Soc. d'hist. nat. d'Autun,* tome xxi.

Penhallow, D. P. (92) Additional notes on Devonian plants from Scotland. *Canadian Rec. Sci.* vol. v. no. 1.

—— (02) *Osmundites skidegatensis. Trans. R. Soc. Canada* [2], vol. viii. sect. 4.

Phillips, J. (75) Illustrations of the Geology of Yorkshire, pt. i. The Yorkshire Coast (edit. 3). *London.*

Pittman, E. F. (93) See David, E. and E. F. Pittman.

Potonié, H. (89) Der im Lichthof der Königlichen Geologischen Landes-anstalt und Bergakademie aufgestellte Baumstumpf mit Wurzel aus dem Carbon des Piesberges. *Jahrb. K. Preuss. Geol. Landes.* p. 246.

—— (91) *Bericht. Deutsch. bot. Ges.* vol. ix. p. 256.

—— (92) Ueber einige Carbonfarne. *Jahrb. K. Preuss. Geol. Landes.* 1891.

—— (92[2]) Die den wasserspalten physiologisch-entsprechenden Organe bei fossilen und recenten Farn Arten. *Sitz.-Ber. Ges. naturforsch. Freunde zu Berlin. July* 19, 1892.

—— (93[2]) Anatomie der beiden "Male" auf dem unteren Wangenpaar und der beiden Seitennärbchen der Blattnarbe des Lepidodendron-Blattpolsters. *Ber. deutsch. Bot. Ges.* Bd. xi. Heft 5, p. 319.

—— (93[3]) Eine gewöhnliche Art der Erhaltung von *Stigmaria* als Beweis für die Autochthonie von Carbon-Pflanzen. *Zeits. Deutsch. Geol. Ges.*

—— (95) Die Beziehung zwischen dem echt-gabeligen und dem fiederigen Wedel-Aufbau der Farne. *Ber. Deutsch. Bot. Ges.* Bd. xiii. Heft 6.

—— (99) Lehrbuch der Pflanzenpalaeontologie. *Berlin.*

—— (00) Fossile Pflanzen aus Deutsch- und Portugiesisch-Ost-Afrika. *Deutsch-Ost-Afrika,* Bd. vii.

—— (01) Fossile Lycopodiaceae und Selaginellaceae. Engler and Prantl : *Die Natürlichen Pflanzenfamilien,* Teil i. Abt. iv. p. 715.

—— (01[2]) Die Silur- und die Culm-Flora. *Abh. K. Preuss. Geol. Landes.* Heft xxxvi.

—— (02) Ueber die fossilen Filicales &c. Engler and Prantl : *Die Natürlichen Pflanzenfamilien,* Teil i. Abt. iv. p. 473.

—— (03) Zur Physiologie und Morphologie der fossilen Farn-Aphle-bien. *Ber. Deutsch. Bot. Ges.* Bd. xxi. Heft 3.

—— (04) Abbildungen und Beschreibungen fossilen Pflanzen-Reste der Palaeozoischen und Mesozoischen Formationen. Lief. ii. *K. Preuss. Geol. Landes- und Bergakad.*

—— (05) *Ibid.* Lief. iii.

—— (06) *Ibid.* Lief. iii.

Potonié, H.   (07)   Abbildungen und Beschreibungen &c.   Lief. v.

Prantl, K.   (81)   Untersuchungen zur Morphologie der Gefässkrypto-gamen. Heft II. *Leipzig.*

——   (02)   See Engler, A. and K. Prantl.

Pritzel, E.   (02)   Lycopodiales.   Engler and Prantl: *Die Natürlichen Pflanzenfamilien,* Teil I. Abt. IV. p. 563.

Raciborski, M.   (91)   Ueber die Osmundaceen und Schizaeaceen der Juraformation. *Engler's Bot. Jahrb.* vol. XIII. p. 1.

Reid, C.   (99)   The Origin of the British Flora. *London.*

Reinecke, F.   (97)   Die Flora der Samoa-Inseln. *Engler's Bot. Jahrb.* Bd. XXIII. p. 237.

Renault, B.   (69)   Étude sur quelques végétaux silicifiés des environs d'Autun. *Ann. Sci. nat.* [5], vol. XII. p. 161.

——   (75)   Étude du genre *Botryopteris. Ann. Sci. nat.* [6], vol. I. p. 220.

——   (76)   Étude du genre *Myelopteris. Mém. Acad. Sci. l'Instit. France,* Tome XXII.

——   (79)   Structure comparée de quelques tiges de la Flore Carbonifère. *Nouv. Arch. Mus. Paris.*

——   (81)   Étude sur les *Stigmaria. Ann. Sci. Géol.* tome XII.

——   (90)   Sur une nouvelle Lycopodiacée houillère (*Lycopodiopsis Derbyi*). *Compt. Rend.,* April 14.

Renault, B. and C. Grand'Eury.   (75)   Recherches sur les végétaux silicifiés d'Autun. *Mém. Acad. Paris,* vol. XXII.

Renault, B. and A. Roche.   (97)   Sur une nouveau Diploxylée. *Bull. Soc. d'hist. nat. d'Autun.*

Renier, A.   (08)   Origine raméale des cicatrices Ulodendroides du *Bothrodendron punctatum,* Lind. et Hutt. *Compt. Rend., June* 29.

Reuss, A. E.   (46)   Die Versteinerungen der Böhmischen Kreideformation. *Stuttgart,* 1845–46.

Rhode, J. G.   (20)   Beiträge zur Pflanzenkunde der Vorwelt. *Breslau.*

Richter, P. B.   (06)   Beiträge zur Flora der unteren Kreide Quedlinburgs. Teil I. Die Gattung *Hausmannia,* Dunker, und einige seltenere Pflanzenarte. *Leipzig.*

Richter, R.   (56)   See Unger, F. and R. Richter.

Roche, A.   (97)   See Renault, B. and A. Roche.

Roehl, Major von.   (69)   Fossile Flora der Steinkohlen Formation Westphalens. *Palaeont.* Bd. XVIII. p. 1.

Roemer, F. A.   (54)   Beiträge zur geologischen Kenntniss des nord-westlichen Harzgebirges. *Palaeont.* Bd. III.

Royle, J. F.   (33)   Illustrations of the Botany and other branches of natural history of the Himalayan Mountains. *London,* 1833–39.

Rudolph, K.   (05)   Psaronien und Marattiaceen. *Denksch. Kais. Akad. Wiss. Wien,* Bd. LXXVIII.

Sadebeck, H.   (02)   See Engler, A. and K. Prantl.

Salfeld, H.   (07)   Fossile Land-pflanzen der Rät.- und Juraformation Südwestdeutschlands. *Palaeont.* Bd. LIV.

Salfeld, H. (09) Beiträge zur Kenntniss jurassische Pflanzenreste aus Norddeutschland. *Palaeont.* Bd. LVI.

Salter, J. W. (58) On some remains of terrestrial plants in the Old Red Sandstone of Caithness. *Quart. Journ. Geol. Soc. Proc.* vol. XIV. p. 77.

Saporta, le Marquis de. (88) Dernières adjonctions à la flore fossile d'Aix-en-Provence. *Ann. Sci. nat.* [7], vol. VII.

—— (94) Flore fossile du Portugal. *Direct. Trav. Géol. Portugal. Lisbon.*

Saxelby, E. M. (08) The origin of the roots in *Lycopodium Selago. Annals Bot.* vol. XXII. p. 21.

Schenck, H. (93) Beiträge zur Biologie und Anatomie der Lianen. Th. II. *Jena.*

Schenk, A. (71) Beiträge zur Flora der Vorwelt. Die Flora der Nordwestdeutschen Wealdenformation. *Palaeont.* Bd. XIX. p. 203.

—— (76) Zur Flora der Nordwestdeutschen Wealdenformation. *Palaeont.* Bd. XXVIII. p. 157.

—— (85) Die während der Reise des Grafen Bela Széchenyi in China gesammelten fossilen Pflanzen. *Palaeont.* Bd. XXXI. p. 165.

—— (87) Fossile Pflanzen aus der Albourskette. *Bibl. bot.* (*Uhlworm und Haenlein*), Heft VI.

Schmalhausen, J. (77) Die Pflanzenreste aus der Ursa-Stufe im Fluss-Geschiebe des Ogur in Ost-Sibirien. *Bull. Acad. Imp. Sci. St Petersburg.* Tome XXII. p. 278.

—— (94) Ueber Devonische Pflanzen aus dem Donetz-Becken. *Mém. Com. Géol.* vol. VIII. *St Petersburg.*

Schuster, J. (08) Zur Kenntniss der Flora der Saarbrücken Schichten und des pfälzischen Oberrotliegenden. *Geognost. Jahresheft,* XX. 1907.

Scott, D. H. (96) An introduction to structural botany. Pt. II. *London.*

—— (97) On the structure and affinities of fossil plants from the Palaeozoic rocks. On *Cheirostrobus,* a new type of fossil cone from the Lower Carboniferous strata (Calciferous Sandstone series). *Phil. Trans. R. Soc.* vol. CLXXXIX. p. 1.

—— (98) On the structure and affinities &c. II. On *Spencerites,* a new genus of Lycopodiaceous cones from the Coal-Measures founded on the *Lepidodendron Spenceri* of Williamson. *Ibid.* vol. CLXXXIX. p. 83.

—— (00) Studies in fossil botany. *London.*

—— (01) On the structure and affinities &c. IV. The seed-like fructification of *Lepidocarpon,* a genus of Lycopodiaceous cones from the Carboniferous formation. *Phil. Trans. R. Soc.* vol. CXCIV. p. 291.

—— (02) The Old Wood and the New. *New Phytologist,* vol. I. p. 25.

—— (04) Germinating spores in a fossil fern Sporangium. *Ibid.* vol. III. p. 18.

Scott, D. H. (04²) On the occurrence of *Sigillariopsis* in the Lower Coal-Measures of Britain. *Annals Bot.* vol. XVIII. p. 519.

—— (05) On the structure and affinities &c. v. On a new type of Sphenophyllaceous cone (*Sphenophyllum fertile*) from the Lower Coal-Measures. *Phil. Trans. R. Soc.* vol. CXCVIII. p. 17.

—— (05²) What were the Carboniferous ferns? *Journ. R. Micr. Soc.* p. 137.

—— (05³) The Sporangia of *Stauropteris oldhamia*. *New Phyt.* vol. IV. p. 114.

—— (06) On the structure of some Carboniferous ferns. *Journ. R. Micr. Soc.* p. 518.

—— (06²) The occurrence of germinating spores in *Stauropteris oldhamia*. *New Phyt.* vol. V. p. 170.

—— (06³) The structure of *Lepidodendron obovatum*, Sternberg. *Annals Bot.* vol. XX. p. 317.

—— (07) The present position of Palaeozoic botany. *Progressus Rei Botanicae*, Bd. I. p. 139.

—— (08) Studies in fossil botany (edit. II). Vol. I. *London.*

—— (09) *Ibid.* Vol. II.

—— (09²) Dr Paul Bertrand on the Zygopterideae. *New Phyt.* vol. VIII. p. 266.

Scott, D. H. and T. G. Hill. (00) The structure of *Isoetes Hystrix*. *Annals Bot.* vol. XIV. p. 413.

Scott, D. H. and A. J. Maslen. (06) On the structure of *Trigonocarpon olivaeforme*. *Ann. Bot.* vol. XX. p. 109.

Scott, J. (74) Notes on the tree ferns of British Sikkim. *Trans. Linn. Soc.* vol. XXX. p. 1.

Scott, Rina. (06) On the megaspore of *Lepidostrobus foliaceus*. *New Phyt.* vol. V. p. 116.

—— (08) On *Bensonites fusiformis*, sp. nov., a fossil associated with *Stauropteris burntislandica*, P. Bertrand, and on the sporangia of the latter. *Annals Bot.* vol. XXII. p. 683.

Sellards, E. H. (00) A new genus of ferns from the Permian of Kansas. *Kansas Univ. Quart.* vol. IX.

—— (01) Permian plants. *Taeniopteris* of the Permian of Kansas. *Ibid.* vol. X. no. 1.

Seward, A. C. (88) On a specimen of *Cyclopteris* (Brongniart). *Geol. Mag.* vol. V. p. 344.

—— (90) Notes on *Lomatophloios macrolepidotus* (Gold.). *Proc. Camb. Phil. Soc.* vol. VII. pt. ii.

—— (90²) Specific variation in Sigillarieae. *Geol. Mag.* vol. VII. p. 213.

—— (91) On an erect tree stump with roots, from the coal of Piesberg near Osnabrück. *Ibid.* vol. VIII.

—— (92) Fossil plants as tests of climate. *London.*

—— (94) Coal: its structure and formation. *Sci. Progr.* vol. II. pp. 355—431.

Seward, A. C. (99) Notes on the Binney collection of Coal-Measure plants. *Proc. Phil. Soc. Camb.* vol. X. p. 137.

—— (99²) On the structure and affinities of *Matonia pectinata*, R. Br. with notes on the geological history of the Matonineae. *Phil. Trans. R. Soc.* vol. CXCI. p. 171.

—— (00) Catalogue of the Mesozoic plants in the Department of Geology, British Museum. The Jurassic Flora. I. The Yorkshire Coast. *London.*

—— (03) Fossil floras of Cape Colony. *Ann. S. African Museum,* vol. IV. pt. i.

—— (04) The Jurassic Flora. II. Liassic and Oolitic floras of England. *Cat. Mesoz. Plants, British Museum. London.*

—— (04²) On a collection of Jurassic plants from Victoria. *Rec. Geol. Surv. Victoria,* vol. I. pt. iii.

—— (04³) Presidential address. *Report of the 73rd meeting of the Brit. Assoc.* (Southport) p. 824.

—— (06) The anatomy of *Lepidodendron aculeatum,* Sternberg. *Annals Bot.* vol. XX. p. 371.

—— (07) Fossil plants from Egypt. *Geol. Mag.* vol. IV. [v] p. 253.

—— (07²) On a collection of Permo-Carboniferous plants from the St Lucia (Somkale) coalfield, Zululand, and from the Newcastle district, Natal. *Trans. Geol. Soc. S. Africa,* vol. X. p. 65.

—— (07³) Notes on fossil plants from South Africa. *Geol. Mag.* vol. IV. p. 481.

—— (07⁴) Jurassic Plants from Caucasia and Turkestan. *Mém. Com. Géol. St Pétersbourg,* Livr. 38.

—— (07⁵) Permo-Carboniferous Plants from Kashmir. *Rec. Geol. Surv. India,* vol. XXXVI. pt. i.

—— (08) On a collection of fossil plants from South Africa. *Quart. Journ. Geol. Soc.* vol. LXIV. p. 83.

—— (09) Fossil plants from the Witteberg series of Cape Colony. *Geol. Mag.* vol. VI. p. 482.

Seward, A. C. and E. Dale. (01) On the structure and affinities of *Dipteris,* with notes on the geological history of the Dipteridinae. *Phil. Trans. R. Soc.* vol. CXCIV. p. 487.

Seward, A. C. and S. O. Ford. (03) The anatomy of *Todea,* with notes on the geological history and affinities of the Osmundaceae. *Trans. Linn. Soc.* vol. VI. pt. v.

—— (06) The Araucarieae, recent and extinct. *Phil. Trans. R. Soc.* vol. CXCVIII. p. 305.

Seward, A. C. and J. Gowan. (00) The Maidenhair Tree. (*Ginkgo biloba,* L.) *Ann. Bot.* vol. XIV. p. 109.

Seward, A. C. and A. W. Hill. (00) On the structure and affinities of a Lepidodendroid stem from the Calciferous Sandstone of Dalmeny, Scotland. *Trans. R. Soc. Edinb.* vol. XXXIX. pt. iv. p. 907.

Seward, A. C. and T. N. Leslie. (08) Permo-Carboniferous plants from Vereeniging. *Quart. Journ. Geol. Soc.* vol. LXIV. p. 109.

Seward, A. C. and A. Smith Woodward. (05) Permo-Carboniferous Plants and Vertebrates from Kashmir. *Mem. Geol. Surv. India,* vol. v. Mem. 2.

Shattock, S. G. (88) On the scars occurring on the stem of *Dammara robusta,* Moore. *Journ. Linn. Soc.* vol. xxix. p. 441.

Shove, Rosamund. (00) On the structure of the stem of *Angiopteris erecta. Annals Bot.* vol. xiv.

Smith, G. O. and D. White. (05) The geology of the Perry basin in South Eastern Maine. *U. S. Geol. Surv.* No. 35.

Sollas, Igerna B. J. (01) Fossils in the Oxford Museum. On the structure and affinities of the Rhaetic plant *Naiadita. Quart. Journ. Geol. Soc.* vol. lvii. p. 307.

Solms-Laubach, H. Graf zu. (83) Zur Geschichte der *Scolecopteris,* Zenker. *Nachr. K. Ges. Wiss. Univ. Göttingen,* p. 26.

—— (92) Ueber die in den Kalksteinen des Kulm von Glätzisch-Falkenberg in Schlesien erhaltenen Structurbietenden Pflanzenreste. *Bot. Zeit.* p. 49.

—— (94) Ueber *Stigmariopsis,* Grand'Eury. *Palaeont. Abh. (Dames and Kayser)* [N. F.] Bd. ii. *Jena.*

—— (96) Ueber die seinerzeit von Unger beschriebenen Strukturbietenden Pflanzenreste des Unterculm von Saalfeld in Thüringen. *Abh. K. Preuss. Geol. Landes.* Heft xxiii.

—— (99) Ueber das Genus *Pleuromeia. Bot. Zeit.* p. 227.

—— (99²) Beiträge zur Geologie und Palaeontologie von Südamerika. *Neues Jahrb. Min.,* Beilageband xii. p. 593.

—— (02) *Isoetes lacustris,* seine Verzweigung und sein Vorkommen in den Seen des Schwarzwaldes und der Vogesen. *Bot. Zeit.* p. 179.

—— (04) Ueber die Schicksale der als *Psaronius brasiliensis* beschriebenen Fossilreste unserer Museen. *Festsch. P. Ascherson's Siebzigstem Geburtstage. Berlin.*

—— (06) Die Bedeutung der Palaeophytologie für die systematische Botanik. *Mitt. Philo-math. Ges. Elsass-Loth.* Bd. iii. p. 353.

Spieker, T. (53) Zur *Sigillaria Sternbergi* Münster, des bunten Sandsteins zu Bernburg. *Zeits. Gesammt. Naturw.* Bd. ii. *Halle.*

Sprengel, A. (28) Commentatio de Psarolithis. *Halle.*

Spruce, R. (08) Notes of a botanist on the Amazon and Andes (Edited by A. R. Wallace). *London.*

Staub, M. (87) Die Aquitanische Flora des Zsilthales im comitate Hunyad. *Mitt. Jahrb. K. Ungar. Geol. Anst.* Bd. vii. Heft vi.

Stenzel, C. G. (54) Ueber die Staarsteine. *Nova Acta Leop. Carol.* Bd. xxiv.

—— (86) *Rhizodendron Oppoliense,* Göpp. *Jahresber. Schles. Ges. Vaterl. Cultur.* Ergänzungsheft lxiii.

—— (89) Die Gattung *Tubicaulis. Bibl. Bot.* Heft xii.

—— (97) Verkieselte Farne von Kamenz in Sachsen. *Mitt. K. Mineralog. Geol. und prähistorisch. Mus. Dresden.* Heft xiii.

—— (06) Die Psaronien, Beobachtungen und Betrachtungen. *Beit. Paläont. Geol. Öst.-Ung.* Bd. xix.

Sterzel, J. T.   (78) Ueber *Palaeojulus dyadicus* Geinitz und *Scoleco-pteris elegans* Zenker. *Zeitsch. Deutsch. Geol. Ges.* p. 417.

——   (80) Ueber *Scolecopteris elegans* Zenker und andere fossile Reste. *Zeitschr. Deutsch. Geol. Ges.*

——   (86) Die Flora des Rothliegenden im Plauenschen Grunde. *Abh. K. Sächs. Ges. Wiss.* Bd. xix.

——   (86^2) Neue Beitrag zur Kenntniss von *Dicksonites Pluckeneti* Brongn. sp. *Zeitschr. Deutsch. Geol. Ges.* p. 773.

——   (96) Gruppe verkieselten Araucariten Stämme. *Ber. Naturwiss. Ges. Chemnitz*, 1896–99.

——   (96^2) See Weber, O. and J. T. Sterzel.

Stiehler, A. W.   (58) Beiträge zur Kenntniss der Vorweltlichen Flora des Kreidegebirges im Harze. *Palaeont.* Bd. v.

——   (59) Zu *Pleuromeia*. *Zeit. Gesammt. Nat. Halle*, Bd. iii. p. 190.

Stokey, A. G.   (07) The roots of *Lycopodium pithyoides*. *Bot. Gaz.* vol. xliv. p. 57.

——   (09) The anatomy of *Isoetes*. *Bot. Gaz.* vol. xlvii. p. 311.

Stopes, Marie C.   (06) A new fern from the Coal-Measures: *Tubicaulis Sutcliffii*, spec. nov. *Mem. Proc. Manch. Lit. Phil. Soc.* vol. l.

Strasburger, E.   (73) Einige Bemerkungen über Lycopodiaceen. *Bot. Zeit.* p. 81.

——   (74) Ueber *Scolecopteris elegans*, Zenk. *Jen. Zeitsch. Naturwiss.* vol. viii. p. 88.

Strzelecki, Count.   (45) Physical description of New South Wales &c. *London.*

Stur, D.   (81) Die Silur-Flora der Étage H-h in Böhmen. *Sitzb. Akad. Wiss. Wien*, 1 Abth. Bd. lxxxiv. p. 330.

——   (84) Zur Morphologie und Systematik der Culm- und Carbonfarne. *Sitzb. Akad. Wiss. Wien*, Bd. lxxxviii.

——   (85) Die Carbon-Flora der Schatzlarer Schichten. *Abh. K. K. Geol. Reichs.* Bd. xi. Abth. i.

Sykes, M. Gladys.   (08) The anatomy and morphology of *Tmesipteris*. *Annals Bot.* vol. xxii. p. 63.

——   (08^2) Note on an abnormality found in *Psilotum triquetrum*. *Ibid.* vol. xxii. p. 525.

——   (08^3) Notes on the morphology of the Sporangium-bearing organs of the Lycopodiaceae. *New Phyt.* vol. vii. p. 41.

——   (09) Note on the Sporophyll of *Lycopodium inundatum*. *Ibid.* vol. viii. p. 143.

Szajnocha, L.   (88) Ueber fossile Pflanzenreste aus Cacheuta in den Argentinischen Republik. *Sitzb. K. Akad. Wiss. Wien*, Bd. xcvii. Abth. i. p. 219.

——   (91) Ueber einige Carbone Pflanzenreste aus der Argentinischen Republik. *Sitzb. K. Akad. Wiss. Wien*, Bd. c. Abth. i. p. 203.

Tansley, A. G.   (08) Lectures on the evolution of the filicinean vascular system. (Reprinted from the *New Phytologist.*) *Cambridge.*

—— and **Edith Chick.**   (01) Notes on the conducting tissue-system in Bryophyta. *Annals Bot.* vol. xv. p. 1.

Tansley, A. G. and **F. E. Fritsch.** (05) The flora of the Ceylon littoral. *New Phyt.* vol. IV. p. 1.

—— and **R. B. J. Lulham.** (05) A study of the vascular system of *Matonia pectinata. Annals Bot.* vol. XIX. p. 475.

Thiselton-Dyer, Sir W. T. (05) *Cycas Micholitzii,* Dyer. *Gard. Chron.* Aug. 19, p. 142.

Thoday, D. (06) On a suggestion of heterospory in *Sphenophyllum Dawsoni. New Phyt.* vol. V. p. 91.

Thomas, A. P. W. (02) The affinity of *Tmesipteris* with the Sphenophyllales. *Proc. R. Soc.* vol. LXIX. p. 343.

Thomas, Ethel N. (05) Some points in the anatomy of *Acrostichum aureum. New Phyt.* vol. IV. p. 175.

Thomas, H. H. (08) See Arber, E. A. N. and H. H. Thomas.

Thompson, D'Arcy W. (80) Notes on *Ulodendron* and *Halonia. Edinb. Geol. Soc.*

Trautschold, H. (60) See Auerbach, J. and H. Trautschold.

Treub, M. (84—90) Études sur les Lycopodiacées. *Ann. Jard. Bot. Buitenzorg,* vol. IV. V. VII. VIII.

Underwood, L. E. (00) See Lloyd, E. and L. M. Underwood.

—— (07) A preliminary review of the North American Gleicheniaceae. *Bull. Torrey Bot. Club,* vol. XXXIV. p. 243.

Unger, F. and **R. Richter.** (56) Beitrag zur Paläontologie des Thuringer Waldes. *Denksch. Wien. Akad.* Bd. XI.

Velenovský, J. (85) Die Gymnospermen der böhmischen Kreideformation. *Prag.*

—— (88) Die Farne der böhmischen Kreideformation. *Prag.*

Vines, S. (88) On the systematic position of *Isoetes,* L. *Annals Bot.* vol. II. pp. 117, 223.

Wanklyn, A. (69) Description of some new species of fossil ferns from the Bournemouth leaf-beds. *Ann. Mag. Nat. Hist.* vol. III. p. 10.

Ward, L. F. (99) The Cretaceous Formation of the Black Hills as indicated by the fossil plants. *U. S. Geol. Surv.,* 19*th Ann. Rep.* pl. II.

—— (00) Status of the Mesozoic Floras of the United States, I. *U. S. Geol. Surv.,* 20*th Ann. Rep.*

—— (04) Palaeozoic seed-plants. *Science,* Aug. 26, p. 279.

—— (05) Status of the Mesozoic Floras of the United States, II. *U. S. Geol. Surv. Monographs,* vol. XLVIII.

Watson, D. M. S. (06) On a "fern" synangium from the Lower Coal-Measures of Shore, Lancashire. *Journ. R. Micr. Soc.* p. 1.

—— (07) On a confusion of two species (*Lepidodendron Harcourtii,* Witham, and *L. Hickii,* sp. nov.) under *Lepidodendron Harcourtii,* With. in Williamson's XIX. Memoir, with a description of *L. Hickii* sp. nov. *Mem. Proc. Manch. Lit. Phil. Soc.* vol. LI.

—— (08) On the Ulodendron Scar. *Ibid.* vol. LII.

—— (08²) The cone of *Bothrodendron mundum. Ibid.* vol. LII.

Watson, D. M. S. (09) On *Mesostrobus*, a new genus of Lycopodiaceous cones from the Lower Coal-Measures &c. *Annals Bot.* vol. XXIII. p. 379.

Weber, O. and **J. T. Sterzel.** (96) Beiträge zur Kenntniss der Medulloseae. *Ber. Naturwiss. Ges. Chemnitz*, 1893–96.

Weiss, C. E. (69) Fossile Flora der jüngsten Steinkohlen-formation und des Rothliegenden im Saar-Rhein-Gebiete. *Bonn*, 1869–72.

—— (70) Studien über Odontopteriden. *Zeitsch. Deutsch. Geol. Ges.*

—— (79) Bemerkungen zur Fructification von *Nöggerathia*. *Zeitsch. Deutsch. Geol. Ges.*

—— (84) Zur Flora der ältesten Schichten des Harzes. *Jahrb. K. Preuss. Geol. Landes. Berlin.*

—— (86) Ueber eine Buntsandstein *Sigillaria* und deren nächste Verwandte. *Ibid.* 1885.

—— (88) Ueber neue Funde von Sigillarien in der Wettiner Steinkohlengrube. *Zeitsch. Deutsch. Geol. Ges.*

—— (89) Beobachtungen an Sigillarien von Wettin und Umgegend. *Neues Jahrb.* Bd. XLI. p. 376.

—— and **J. Sterzel.** (93) Die Sigillarien der Preussischen Steinkohlen und Rothliegenden Gebiete. *K. Preuss. Geol. Landes.* [N.F.], Heft 2.

Weiss, F. E. (02) On *Xenophyton radiculosum* (Hick), and on a Stigmarian rootlet probably related to *Lepidophloios fuliginosus* (Will.). *Mem. Proc. Manch. Lit. Phil. Soc.* vol. XLVI. pt. 3.

—— (03) A biseriate Halonial branch of *Lepidophloios fuliginosus*. *Trans. Linn. Soc.* vol. VII. pt. 4.

—— (04) A probable parasite of Stigmarian rootlets. *New Phyt.* vol. III. p. 63.

—— (06) On the tyloses of *Rachiopteris corrugata*. *New Phyt.* vol. V. p. 82.

—— (07) The Parichnos in Lepidodendraceae. *Mem. Proc. Manch. Lit. Phil. Soc.* vol. LI. pt. ii.

—— (08) A Stigmaria with centripetal wood. *Annals Bot.* vol. XXII. p. 221.

Weiss, F. E. and **J. Lomax.** (05) The stem and branches of *Lepidodendron selaginoides*. *Mem. Proc. Manch. Lit. Phil. Soc.* vol. XLIX.

White, D. (93) A new Taeniopteroid Fern and its allies. *Bull. Geol. Soc. America*, vol. IV. p. 119.

—— (95) The Pottsville series along New River, West Virginia. *Bull. Geol. Soc. America*, vol. VI. p. 305.

—— (98) *Omphalophloios*, a new Lepidodendroid type. *Ibid.* vol. IX. p. 329.

—— (99) Fossil flora of the Lower Coal-Measures of Missouri. *U. S. Geol Surv. Mon.* XXXVII.

—— (02) Description of a fossil alga from the Chemung of New York. *Rep. New York State Palaeontologist*, 1901.

—— (04) The seeds of *Aneimites*. *Smithsonian Miscell. Coll.* vol. XLVII. pt. iii. p. 322.

White, D. (05) See Smith, G. O. and D. White.

—— (05²) Fossil plants of the group Cycadofilices. *Smiths. Misc. Coll.* vol. XLVIII. pt. iii.

—— (07) Permo-Carboniferous changes in South America. *Journ. Geol.* vol. XV. p. 615.

—— (07²) A remarkable fossil tree trunk from the Middle Devonic of New York. *New York State Mus. Bull.* 107. *Albany.*

—— (08) Fossil flora of the Coal-Measures of Brazil. *Rio de Janeiro.*

White, I. C. (80) See Fontaine, W. M. and I. C. White.

Wickes, W. H. (00) A new Rhaetic section at Bristol. *Proc. Geol. Assoc.* vol. XVI. p. 421.

Wigglesworth, Grace. (02) Notes on the rhizome of *Matonia pectinata.* *New Phyt.* vol. I. p. 157.

Wild, G. and J. Lomax. (00) A new Cardiocarpon-bearing strobilus. *Annals Bot.* vol. XIV. p. 160.

Williamson, W. C. (72) On the organization of the fossil plants of the Coal-Measures. III. Lycopodiaceae. *Phil. Trans. R. Soc.* vol. CLXII. p. 283.

—— (76) *Ibid.* pt. vii. *Phil. Trans. R. Soc.* vol. CLXVI. p. 1.

—— (77) *Ibid.* pt. viii. *Phil. Trans. R. Soc.* vol. CLXVII. p. 213.

—— (83) Presidential address. *Brit. Assoc.*

—— (87) Note on *Lepidodendron Harcourtii* and *L. fuliginosum.* *Proc. R. Soc.* vol. XLII. p. 6.

—— (92) *Sigillaria* and *Stigmaria.* *Nat. Science,* p. 214.

—— (93) On the organization &c. pt. xix. *Phil. Trans. R. Soc.* vol. CLXXXIV. p. 1.

—— (93²) General morphological and histological index to the author's collective memoirs on the fossil plants of the Coal-Measures, pt. ii. *Mem. Proc. Manch. Lit. Phil. Soc.* [7] vol. VII.

—— (95) On the light thrown upon the question of growth and development of the Carboniferous arborescent Lepidodendra by a study of the details of their organisation. *Ibid.* vol. IX. p. 31.

—— (96) Reminiscences of a Yorkshire Naturalist. (Edited by Mrs Crawford Williamson.) *London.*

Woodward, A. Smith. (05) See Seward, A. C. and A. Smith Woodward.

Worsdell, W. C. (95) On transfusion-tissue; its origin and function in the leaves of Gymnospermous plants. *Trans. Linn. Soc.* vol. v. p. 301.

Wünsch, E. A. (67) Discovery of erect stems of fossil trees in trappean ash in Arran. *Trans. Geol. Soc. Glasgow,* vol. II. p. 97.

Yabe, H. (05) Mesozoic plants from Korea. *Journ. Coll. Sci. Imp. Univ. Japan,* vol. XX.

Yapp, R. H. (02) Two Malayan 'Myrmecophilous' ferns, *Polypodium* (*Lecanopteris*) *carnosum* (Blume), and *P. sinuosum.* *Annals Bot.* vol. XVI. p. 185.

—— (08) Sketches of vegetation at home and abroad. IV. Wicken Fen. *New Phyt.* vol. VII. p. 61.

Yokoyama, M. (89) Jurassic plants from Kaga, Hida, and Echizen. *Journ. Coll. Sci. Imp. Univ. Japan*, vol. III.

—— (06) Mesozoic plants from China. *Ibid.* vol. XXI.

Zalessky, M. (04) Végétaux fossiles du Terrain Carbonifère du Bassin du Donetz. *Mém. Com. Géol. St Pétersbourg.* Livr. XIII.

—— (07) Sur la présence de *Mixoneura neuropteroides*, Göpp. avec *Neuropteris Scheuchzeri*, Hoffmann, et *N. rarinervis*, Bunbury &c. *Bull. Com. Géol. St Pétersbourg*, tome XXVI.

—— (08) ,Végétaux fossiles du Terrain Carbonifère du bassin du Donetz. II. Étude sur la structure anatomique d'un Lepidostrobus. *Mém. Com. Géol.*, Livr. XLVI.

Zeiller, R. (79) Note sur quelques fossiles du terrain permien de la Corrèze. *Bull. Soc. Géol. France*, tome VIII. p. 196.

—— (79²) Note sur le genre *Mariopteris*. *Bull. Soc. Géol. France* [3], tome VII. p. 92.

—— (83) Fructifications de Fougères houillères. *Ann. Sci. nat.* [6], vol. XVI.

—— (84) Cônes de fructification des Sigillaires. *Ibid.* vol. XIX. p. 256.

—— (85) Sur les affinités du genre *Laccopteris*. *Bull. Soc. Bot. France*, tome XXXII. p. 21.

—— (86) Présentation d'une brochure de M. Kidston sur les *Ulodendron* et observations sur les genus *Ulodendron* et *Bothrodendron*. *Bull. Soc. Géol. France* [3], tome XIV. p. 168.

—— (89) Sur les variations de formes du *Sigillaria Brardi*, Brongn. *Ibid.* [3], tome XVII. p. 603.

—— (90) Bassin Houiller et Permien d'Autun et d'Épinac. *Études des Gîtes Min. France.*

—— (94) Notes sur la flore des Couches Permiennes de Trienbach (Alsace). *Bull. Soc. Geol. France* [3], tome XXII. p. 163.

—— (95) Note sur la flore fossile des Gisements houillers de Rio Grande do Sul. *Bull. Soc. Géol. France* [3], tome XXIII. p. 601.

—— (97) Observations sur quelques fougères des Dépôts houillers d'Asie Mineure. *Bull. Soc. Bot. France* [3], tome XLIV. p. 195.

—— (97²) The reference of the genus *Vertebraria*. (Translation from the *Compt. Rend.* tome CXXII. p. 744.) *Rec. Geol. Surv. India*, vol. XXX. pt. i.

—— (97³) Les Provinces botaniques de la fin des temps primaires. *Rev. Gén. Sci.* (*Jan.* 15).

—— (97⁴) Revue des travaux de paléontologie végétale. *Rev. Gén. Bot.* tome IX.

—— (98) Sur un Lepidodendron silicifié du Brésil. *Compt. Rend.* (*July* 25).

—— (98²) Sur la découverte, par M. Amalitzky, de Glossopteris dans le Permien supérieur de Russie. *Bull. Soc. Bot. France*, tome XLV. p. 392.

—— (98³) Contribution à l'étude de la flore ptéridologique des schistes permiens de Lodève. *Bull. Mus. de Marseille*, tome I. Fasc. II. p. 9.

Zeiller, R.   (99)   Étude sur la flore fossile du Bassin houiller d'Héraclée. *Mém. Soc. Géol. France* (*Paléont.*), *Mém.* 21.

—— (00)   Sur une Sélaginellée du terrain houiller de Blanzy.   *Compt. Rend.* vol. CXXV. p. 1077.

—— (00²)   Éléments de Paléobotanique.   *Paris.*

—— (02)   Observations sur quelques plantes fossiles des Lower Gondwanas.   *Mem. Geol. Surv. India* [New Series], vol. II.

—— (03)   Flore fossile des Gîtes de Charbon du Tonkin.   *Études des Gîtes Min. France.   Paris.*

—— (03²)   Revue des travaux de Paléontologie végétale.   *Rev. Gén. Bot.* vol. XV.

—— (05)   Une nouvelle classe de Gymnospermes : les Ptéridospermées. *Rev. Gén. Sci.* p. 718.

—— (06)   Bassin houiller et Permien de Blanzy et du Creusot (Fasc. ii).   *Études Gîtes Min. France.*

—— (09)   Observations sur le *Lepidostrobus Brownii.   Compt. Rend.* vol. CXLVIII. p. 890.

—— (09²)   Revue des travaux de Paléontologie végétale (1901–06). *Rev. Gén. Bot.*, vols. XXI, XXII.

Zenker, J. C.   (37)   *Scolecopteris elegans,* Zenk. ein neues fossiles Farrngewächs mit Fructificationen.   *Linnaea,* vol. XI. p. 509.

INDEX

The Index includes the names of Authors and plants mentioned in this volume. No references are, however, given to the following Authors, whose names occur too frequently to render special reference of use to the reader: A. Brongniart, R. Kidston, A. G. Nathorst, H. Potonié, B. Renault, D. H. Scott, H. Graf zu Solms-Laubach, D. Stur, W. C. Williamson, R. Zeiller.

CAMBRIDGE: PRINTED BY JOHN CLAY, M.A. AT THE UNIVERSITY PRESS.

Printed in the United States
By Bookmasters